LIB051005 R3/78

SIMON FRASER UNIVERSITY LIBRARY

Unles[illegible]rials

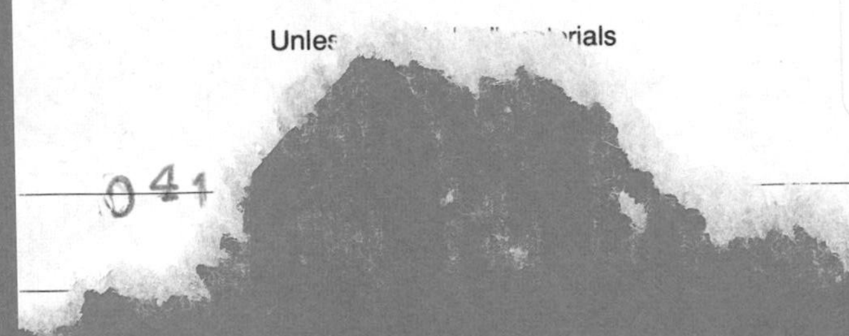

Turfgrass Insects of the United States and Canada

Turfgrass Insects of the United States and Canada

HARUO TASHIRO

Professor Emeritus, Department of Entomology
New York State Agricultural Experiment Station, Geneva
A Division of the College of Agriculture, Cornell University

Comstock Publishing Associates a division of
Cornell University Press | Ithaca and London

Copyright © 1987 by Cornell University

All rights reserved. Except for brief quotations in a review, this book, or parts thereof, must not be reproduced in any form without permission in writing from the publisher. For information, address Cornell University Press, 124 Roberts Place, Ithaca, New York 14850.

First published 1987 by Cornell University Press.

Library of Congress Cataloging-in-Publication Data

Tashiro, Haruo, 1917–
Turfgrass insects of the United States and Canada.

Bibliography: p.
Includes index.
1. Turfgrasses—Diseases and pests—United States. 2. Turfgrasses—Diseases and pests—Canada. 3. Insect pests—United States. 4. Insect pests—Canada. I. Title.
SB608.T87T37 1987 635.9′642 86-6368
ISBN 0-8014-1814-3

Printed in the United States of America.
Color plates printed in Japan.

The paper in this book is acid-free and meets the guidelines for permanence and durability of the Committee on Production Guidelines for Book Longevity of the Council on Library Resources.

Dedicated to my wife, Hatsue, for her patient
and understanding help with this book, when
other pursuits are normally expected during retirement.

Contents

Preface

This book attempts to fill a long-standing need for a comprehensive text-reference on turfgrass insects that will bring together under one cover a discussion of practically all insects and other arthropods that are destructive to turfgrass in the United States, including Hawaii but excluding Alaska, and in southern Canada bordering the United States. Turfgrass insects of the latter region are included because climates near the border are similar.

I have provided substantial technical detail to make the book useful to the professional entomologist who might seek background information on biology and behavior as a basis for further study and research. Measurements in most cases are given in SI (metric) units, followed by their English equivalents in parentheses. Conversion to the metric system in the United States is fully expected to take place, however slowly, at least for the more commonly used measurements. I hope that the technical nature of the text does not dissuade turfgrass managers with limited or no entomological training from using this book to help them understand some aspects of the turfgrass insects they encounter. They will find many distribution maps and life history charts as well as other illustrations that are self-explanatory.

The full-page color plates, which depict some phase of practically all turfgrass insects found in the United States and southern Canada, should be helpful to both professional entomologists and laymen, as they aid in the identification of many turfgrass insects and promote understanding of the insects' habits.

Chapter 1, an introduction, discusses the turfgrasses that are most important agronomically. Chapter 2 provides fundamental information about insects and related arthropods. In Chapters 3 through 21, the orders and families of pest arthropods are covered in the same sequence used in most introductory textbooks in entomology; this sequence affords a logical framework for the treatment of the entire insect fauna. Chapters 3 through 20, and the appropriate sections within those chapters, discuss the taxonomy, importance, history and distribution, host plants, stages, life history and habits, and finally natural enemies of each insect or group of closely related species. Treatment of these topics in the same sequence throughout the book facilitates comparison of the habits of different species or groups. Chapter 22 considers vertebrate pests of turfgrass, since their presence and destruction of turf in most cases is directly related to the presence of insects. Finally, Chapters 23 through 26 provide a general overview of all the interrelationships between insects and the turfgrass ecosystem.

Control recommendations for insects were purposely omitted because these change frequently as new insecticides are approved and as older insecticides become ineffective for various reasons. Chapter 26, a discussion of insect control principles and strategies, should provide sufficient background to permit effective management of turfgrass insects.

Many persons and organizations have contributed to the realization of this book. The inclusion of the full-page color plates was made possible by private funds that helped subsidize their cost. Generous financial support for them, which I wish to acknowledge with gratitude, has been granted by the following organizations: Ciba-Geigy Corporation, Greensboro, North Carolina; Mobay Chemical Corporation, Kansas City, Missouri; New York State Turfgrass Association, Massapequa Park, New York; O. M. Scott & Sons, Marysville, Ohio; The Lawn Institute, Pleasant Hill, Tennessee; and Union Carbide Company, Inc., Research Triangle Park, North Carolina.

The manuscript review process was conducted with help from entomologists and other biologists who have been or are presently involved in turfgrass research or extension activities. I owe much to the following reviewers for their time and effort: entomologists Sami Ahmad and Herbert T. Streu, Rutgers—The State University of New Jersey, New Brunswick; William A. Allen, Virginia Polytechnic Institute and State University, Blacksburg; Arthur L. Antonelli, Western Washington Research and Extension Center, Puyallup; James R. Baker, North Carolina State University, Raleigh; Paul B. Baker, Paul J. Chapman, Timothy J. Dennehy, Charles J. Eckenrode, and Michael G. Villani, New York State Agricultural Experiment Station, Cornell University, Geneva (NYSAES); Robert L. Crocker, Texas A & M Agricultural Research and Extension Center, Dallas; Paul R. Heller, Pennsylvania State University, University Park; John L. Hellman, University of Maryland, College Park; Milton E. Kageyama, O. M. Scott & Sons, Marysville, Ohio; James A. Kamm, U.S. Department of Agriculture at the Oregon State University, Corvallis; M. Keith Kennedy, S. C. Johnson & Sons, Inc., Racine, Wisconsin; S. Dean Kindler, U.S. Department of Agriculture at the University of Nebraska, Lincoln; Thyril L. Ladd and Michael G. Klein, U.S. Department of Agriculture at the Ohio Agricultural Research and Development Center, Wooster; Kenneth O. Lawrence, ChemLawn Services Corporation, Boynton Beach, Florida; Wallace C. Mitchell and Charles L. Murdoch, University of Hawaii, Honolulu; Harry D. Niemczyk, Ohio Agricultural Research and Development Center, Wooster; Daniel A. Potter, University of Kentucky, Lexington; Roger H. Ratcliffe, U.S. Department of Agriculture, Beltsville, Maryland; Donald L. Schuder, Purdue University, West Lafayette, Indiana; Mark K. Sears, University of Guelph, Guelph, Ontario, Canada; David J. Shetlar, ChemLawn Corporation, Columbus, Ohio; Patricia J. Vittum, University of Massachusetts Suburban Experiment Station, Waltham; Joseph E. Weaver, West Virginia University, Morgantown; and Turf Management Specialist A. Martin Petrovic, Cornell University, Ithaca, New York.

Many entomologists, including some who were also reviewers, lent color transparencies of turfgrass insects for color plates, especially in the case of the southern and western insects; the captions indicate the source of borrowed transparencies. I am most grateful to the following individuals, who have contributed to the plates: William A. Allen; Arthur L. Antonelli; Leland R. Brown, University of California, Riverside; Scott R. Cameron, Texas Forest Service, Lufkin; Patricia P. Cobb and Costas Kouskolekas, Auburn University, Auburn, Alabama; Sharon J. Collman, Cooperative Extension Service, Seattle, Washington; Robert L. Crocker; M. Keith Kennedy; Michael G. Klein; Charles L. Murdoch; Harry D. Niemczyk; Asher K. Ota, Hawaii Sugar Planters' Association, Aiea; Daniel A. Potter; Roy W. Rings, Ohio Agricultural Research and Development Center, Wooster; Mark K. Sears; David J. Shetlar; Herbert T. Streu and Louis M. Vasvary, Rutgers—The State University of New Jersey; and Michael G. Villani (for slides he prepared before joining the Department of Entomology, NYSAES). Roughly half of the color slides attributed to the NYSAES were photographed by Gertrude Catlin (now retired) and Joseph Ogrodnick, and the remaining slides were photographed by myself except for single slides by John Andaloro, Shiu-Ling Chung, and James Larner, all former employees of the NYSAES.

Various individuals supplied live subjects for many photographic transparencies. Some specimens were available from routine rearing colonies, while for others, concerted efforts were made to locate infestations so that live specimens could be collected and forwarded. These contributions have significantly improved the scope of the color illustrations. In this connection I must thank James R. Baker; William C. Buell, veterinarian, Geneva, New York; Frank Consolie, NYSAES; Kenneth O. Lawrence; Wayne C. Mixson, O. M. Scott & Sons, Opapka, Florida; Charles L. Murdoch; Roger H. Ratcliffe; David J. Shetlar; Constance Strang, Cooperative Extension Service, Plainview, New York; Clyde Sorensen, North Carolina State University, Raleigh; and John Zukowski, Eisenhower Park, East Meadow, New York.

Important turfgrass insects for which neither color transparencies nor live specimens were available for photography included mainly pyralid and noctuid moths and the more common beetles of the genus *Phyllophaga*. Illustration of these was possible thanks to Paul J. Chapman and Siegfred E. Lienk, NYSAES, who collected and prepared the noctuid moths as museum specimens. Thanks are also due James K. Liebherr, Curator of the Cornell University Collection at Ithaca, who supplied museum specimens of pyralid moths, *Phyllophaga* adults, and a variety of secondary insect pests. These contributions are indicated in the source notes in the captions.

Mary Van Buren, Librarian of the NYSAES, has most helpfully assisted in searching the literature, obtaining publications through interlibrary loan, and tracking down publications and dissertations, often with the vaguest of clues. I am immensely grateful for her time and efforts in adding to the fund of information available to me.

The photographic and illustrative services for the Departments of Entomology and Plant Pathology, NYSAES, are provided by Joseph Ogrodnick, Bernadine Aldwinckle, and Rose McMillen-Sticht. Gertrude Catlin also provided this service until her retirement. In addition to supplying color slides, they photographed many of the black-and-white illustrations. Bernadine Aldwinckle also did much of the intermediate work involved in the preparation of the color plates. Rose McMillen-Sticht's artistry and illustrative skills are evident in the many drawings she prepared in their entirety or adapted from earlier publications. The many significant contributions made by these station staff members have greatly increased the value of the book.

Two members of the secretarial staff of the Department of Entomology, NYSAES, made essential contributions to this book. I owe a debt of gratitude to Janice Allen for typing almost the entire manuscript, entering it on the word processor, and preparing the many revisions. Preparation and completion of the book also entailed much correspondence, particularly to obtain permission to use copyrighted material. I am grateful to Donna Price for doing most of this work.

To the Department of Entomology, NYSAES, I express my most sincere gratitude for allowing me unlimited access to the services of all the staff members whose special skills I sought and for permission to use the fruits of their labor—the many color slides, black-and-white photographs, and drawings, as well as the secretarial and stenographic output so efficiently and willingly rendered. I also thank the department for the use of its facilities and supplies.

Finally, I am grateful to Hélène Maddux and Robb Reavill of Cornell University Press, who were most patient and helpful during the preparation and completion of this book.

HARUO TASHIRO

Geneva, New York

Turfgrass Insects of the United States and Canada

1

Turfgrass in the Modern Environment

Of the more than 7,500 species of plants in the grass family Graminae, not more than 40 species serve mankind as major turfgrasses in various ornamental, recreational, and functional uses throughout the world. By *turfgrass* I mean uniform stand of grass or a mixture of grasses maintained at a relatively low height, usually less than 10 cm (4 in.), and serving any of the above mentioned purposes. By *lawn* I mean any area covered with fine-textured grass and kept closely mowed. In accordance with common usage, I have called turfgrass maintained around a residential property a lawn. Other synonymous or near-synonymous terms are *sod, sward,* and *green. Sod* refers to plugs, squares, or strips of turfgrass used for vegetative planting. A sward is a grassy surface composed of one or more species. *Green* refers to a smooth, grassy area maintained at the lowest cut for golf, lawn bowling, or other sports (Beard 1973, Hanson et al. 1969, Smiley 1983).

Many turfgrass species are used as pasture, field, and forage crops associated with the livestock industry. In this usage particular grasses often have the same pest problems that are associated with them as turfgrasses. Discussions in this treatise do not, however, cover grasses grown for agricultural use.

Grass Structure

All turfgrasses have much in common with respect to basic structure and variations. A typical grass plant (Figure 1) has a crown located at or near the soil surface where most of the meristem activity occurs. A fibrous adventitious root system originating from the crown permeates throughout the surface soil. The stem, also arising from the crown, is enclosed in a leaf composed of the leaf sheath and unfolding leaf that is called the *blade* as well. If it is allowed to grow naturally, without mowing, the stem usually terminates in a seedhead. Some mowed turf may also produce seedheads below the cutting height. Lateral growth and maturation of a grass plant consists of tillers (secondary shoots) that emerge directly from the original crown and/or stolons (surface runners) or rhizomes (long underground

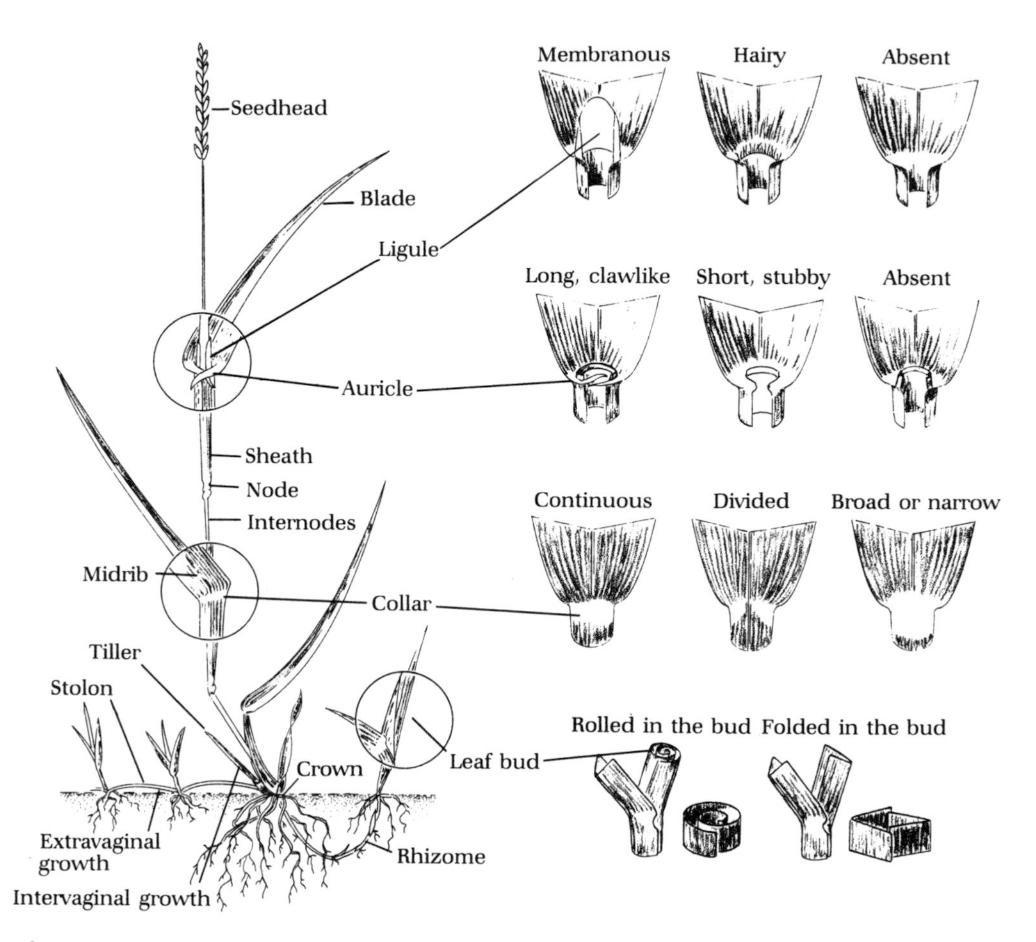

Figure 1. Schematic diagram of a grass plant. (Adapted from O. M. Scott & Sons 1979, p. 46, courtesy of the O. M. Scott & Sons Company.)

runners) that produce secondary and tertiary crowns at the internodes. The tissues in the crown are the most vital portion of the turfgrass. A plant can recover from loss of roots or death of leaves and stems but not from the death of its crown.

Seemingly minute differences can be found at the junction of the sheath and blade and are of utmost importance in the vegetative identification of grass species (Figure 1). These include various differences found in structures called the *ligules,* the *auricles,* and *collars.* The vernation of leaf buds (rolled or folded) is also an important taxonomic character (Beard 1973, O. M. Scott & Sons 1979).

Turfgrass Climatic Adaptations

Turfgrasses grown in the United States are designated as either cool-season or warm-season grasses, depending on their climatic adaptations. Cool-season grasses grow best at temperatures between 15.5° and 24.0°C (60°–75°F) and become at least partially dormant above this range. Warm-season grasses grow best at temperatures between 26.6° and 35.0°C (80°–95°F) and are usually dormant below 10°C (50°F). Warm-season grasses are best adapted to the climate in roughly the southern third of the country, designated by a line that proceeds east of the 100th meridian, then westward through the southern half of each state bordering Mexico to the Pacific Ocean (Figure 2).

A relatively narrow belt 161–483 km (100–300 mi) wide extending along the line of cool-season and warm-season adaptation zones constitutes a transitional climatic zone. In this region either group of turfgrasses may survive and may do well under favorable weather conditions, but under adverse conditions, it may not survive at all. Each of these two zones is divided according to precipitation into the humid and arid-semiarid zones. The dividing line lies near the 100th meridian. Thus the four distinct turfgrass adaptation zones in the United States (Figure 2) are governed by temperature and precipitation. Two distinct coastal influences occur. Along the Atlantic coast the warm, humid zone extends north as far as Delaware. Along the Pacific coast a narrow, cool, humid zone extends from the Canadian border to southern California. The four zones and their dominant features as given by Beard (1973, 1975, 1984) are listed below.

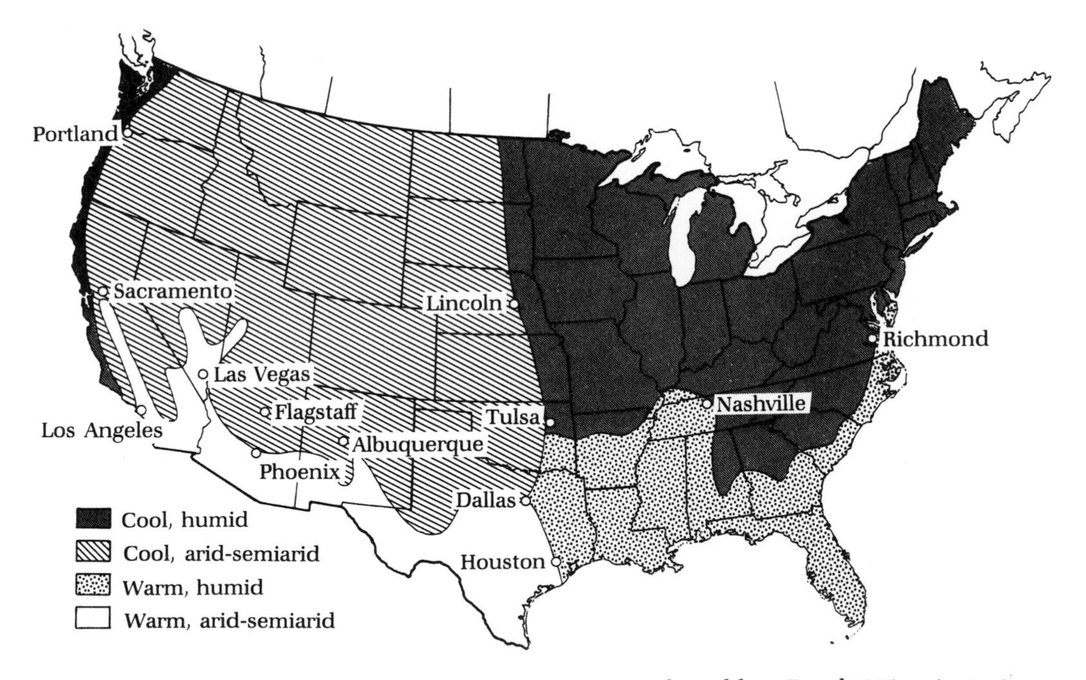

Figure 2. Turfgrass adaptation zones in the United States. (Adapted from Beard 1975, p. 4, courtesy of Beard Books, College Station, Tex.)

Cool, Humid Zone

The cool, humid zone is characterized by cold winters and mild to hot summers in the east and by mild temperatures along the Pacific coast. Rainfall ranges from 51 cm to 114 cm (20–45 in.) throughout the zone in the east and from 51 cm to >250 cm (20–>98 in.) along the Pacific coast. Kentucky bluegrass is the dominant species, followed by bentgrasses, tall and fine fescues, perennial ryegrass, and annual ryegrass. Zoysiagrass and bermudagrass are sometimes planted in the transition belt.

Cool, Arid-Semiarid Zone

Two contrasting regions consist of the vast central plains area to the east and the mountains to the west. Both are characterized by cold winters and hot summers. Rainfall ranges from less than 25 cm (10 in.) in the intermountain plateaus to about 64 cm (25 in.) along the eastern edge of the zone. In the drier areas, irrigation is required to maintain turf of good quality. With irrigation, Kentucky bluegrass and creeping and colonial bentgrasses are dominant through most of the region, with bermudagrass and zoysiagrass used sparingly in the transition belt in Kansas, Oklahoma, and Texas. Where there is no irrigation, buffalograss, *Buchloe dactyloides* [(Nutt.) Engelm.], and wheatgrass, *Agropyron smithii* Rydb., are used in the transition belt because both have good to excellent drought resistance.

Warm, Humid Zone

The climate is characterized by mild temperatures in the northern portion and subtropical areas along the Gulf of Mexico and in Florida. Rainfall ranges from 102 cm (40 in.) along the Atlantic seaboard, to 178 cm (70 in.) along the eastern Gulf Coast, and is as low as 64 cm (25 in.) in Texas and Oklahoma. Bermudagrass is grown throughout the zone. Zoysiagrass is planted mostly in the north, while bahiagrass, centipedegrass, and St. Augustinegrass are used primarily in the southern third of the region.

Hawaii belongs to this zone. The turfgrasses grown there are nearly the same as those of southern Florida except for bahiagrass.

Warm, Arid-Semiarid Zone

This zone is a belt occupying the southern half of each state from Texas into Southern California, with northward projections into Nevada and central California. Precipitation ranges from less than 13 cm (5 in.) to about 50 cm (20 in.); summers are generally dry. Bermudagrass is widely used with and without irrigation. Zoysiagrass is used to a limited extent. St. Augustinegrass, centipedegrass, and dichondra are grown in southern California. Irrigation is generally required for lawns of good quality.

Major Turfgrasses in the United States

Cool-Season Grasses

The most commonly used cool-season grasses include some 11 species in four genera, as listed below (also see Plate 1). All except the ryegrasses are native to Europe or Eurasia. The ryegrasses are native to North Africa and Asia Minor (Beard 1973, Hanson et al. 1969).

Bluegrasses	*Poa* spp.
Kentucky	*P. pratensis* L.
Annual	*P. annua* L.
Rough	*P. trivialis* L.
Ryegrasses	*Lolium* spp.
Perennial	*L. perenne* L.
Italian (annual)	*L. multiflorum* Lam.
Fescues	*Festuca* spp.
Red (creeping)	*F. rubra* L.
Chewings	*F. rubra commutata* Gand.
Tall	*F. arundinacea* Schreb.
Bentgrasses	*Agrotis* spp.
Creeping	*A. palustris* (Huds)
Colonial	*A. tenuis* Sibth
Velvet	*A. canina* L.

All of the bentgrasses and bluegrasses listed (except some forms of annual bluegrass) plus red fescue have extravaginal growth and produce rhizomes or stolons, so that they have a creeping, spreading habit of growth. Some ecotypes of annual bluegrass, the ryegrasses, and the remaining fescues have intravaginal growth, resulting in a tufted or bunch type of growth (Beard 1973). Growth habits have an important bearing on the turf's recovery from a mild attack by insects. Patches of dead grass in recovery fill in much more rapidly with grasses of extravaginal growth because of their creeping, spreading habits.

The bluegrasses are the most widely distributed turfgrasses in the cool, humid region primarily because of the dominance of Kentucky bluegrass. The most distinguishing vegetative character of the genus *Poa* is the boat-shaped leaf tip. Most bluegrasses are adapted to moist, fertile soils of pH 6.0–7.0. Except for annual bluegrass, *Poa* spp. show their best quality at a cutting height of 2.5–5.0 cm (1–2 in.).

Kentucky bluegrass is the most widely used general-purpose turf and lawn grass. No species has received as much attention in the development of new cultivars for specific characteristics as this species has, and there are no fewer than 40 proven superior cultivars, with more being developed all the time.

Annual bluegrass is generally considered a weed in turfgrass culture, but many golf courses in the Northeast maintain *Poa annua reptans* as a perennial grass. It is light green to greenish yellow, with shorter, broader, softer leaves than Kentucky bluegrass. It is well adapted to moist, shaded environments and grows best in fertile soils of pH 6.5–7.5. A cutting height of 2.5 cm (1.0 in.) or less makes it a most aggressive, competitive turfgrass. Its prolific seedhead production even at a cutting height of 0.6 cm (0.24 in.) or less makes it an undesirable grass on putting greens, which it often invades.

Rough bluegrass produces turf of relatively poor quality and does best in shaded areas that are wet and poorly drained. It is included in many lawn and turf seed mixes for such areas. Rough bluegrass often invades Kentucky bluegrass much as creeping bentgrass does, producing a light green patch.

Ryegrasses establish themselves at the most rapid rate of any commonly used cool-season turfgrasses and are often used to produce a quick stand. *Perennial ryegrass* is used widely for most general purposes, including home lawns and shoulders alongside roads. It wears well, so that it is an important turfgrass for athletic fields. Many improved cultivars are compatible with Kentucky bluegrass in color, fineness of leaf, and shoot growth and often make excellent polystands. Perennial ryegrass is commonly used in the fall and winter for winter overseeding where warm-season turfgrasses are going into dormancy.

Italian ryegrass, often called *annual ryegrass,* has limited usage in general-purpose seed mixtures for obtaining a quick stand and is being replaced by perennial ryegrass.

The fescues, adapted to the cool, humid regions, tolerate droughty, infertile acid soils with a pH of 5.5–6.5. Their establishment rate lies between those of Kentucky bluegrass and ryegrass. *Red fescue* is one of the three most widely used turfgrass species in the cool, humid region. Seed mixtures with Kentucky bluegrass are used to supplement the latter in the shade and in droughty, sandy areas.

Chewings fescue is also used in seed mixtures with Kentucky bluegrass for lawns and for general-purpose turf. In appearance it resembles red fescue and has similar performance characteristics. *Tall fescue* has a bunch-type growth habit, with leaf blades coarser than those of any other cool-season turfgrass. It makes a turf of poor quality if it is mixed with other grasses, but it is well adapted to the transition belt between the cool, humid zones and the warm, humid zones and is used mainly in this area.

Bentgrasses are the cool-season grasses that most tolerate the continuous, close mowing heights characteristic of golf greens, lawn bowling greens, and tennis courts, where they are most frequently found. Several improved cultivars are available. *Creeping bentgrass* is one of the most vigorous cool-season grasses, with rapid stolon growth. It forms the best quality of turf at a cutting height of 1.8 cm (0.7 in.) or less and is the outstanding grass available for putting and bowling greens maintained at 0.5–0.8 cm (0.2–0.3 in.).

Colonial bentgrass, one of two most widely used *Agrostis* species, is in very limited usage today. It requires a high maintenance level for the best turf and does best at a cutting height of 0.8–2.0 cm (0.3–0.8 in.). *Velvet bentgrass* is one of the finest textured grasses, with almost needlelike leaves. It is used primarily for putting and bowling greens and does best in New England. Turf of the highest quality is produced with frequent mowing at a height of 0.5–1.0 cm (0.2–0.4 in.).

Warm-Season Grasses

The most commonly used warm-season grasses include nine species in five genera, listed below (see Plate 2). In origin they are somewhat more diverse than the cool-season turfgrasses; the bermudagrasses come from Africa, zoysiagrass and centipedegrass originate in East Asia, St. Augustinegrass is native to the West Indies, and bahiagrass hails from South America (Beard 1973, Hanson et al. 1969).

Bermudagrasses	*Cynodon* spp.
Common	*C. dactylon* (L.) Pers
African	*C. transvaalensis* Burtt-Davey
Magennis	*C. magennisii* (Hurcombe)
Zoysiagrass	*Zoysia* spp.
Japanese lawn grass	*Z. japonica* Steud.
Manilagrass lawn grass	*Z. matrella* (L.) Merr.
Korean velvet lawn grass	*Z. tenufolia* Willd. ex Trin.
Centipedegrass, Chinese lawn grass	*Eremochloa ophuiroides* (Munro.) Hack.
St. Augustinegrass	*Stenotaphrum secundatum* (Walt.) Kutze
Bahiagrass	*Paspalum notatum* Flugge

All of the above-mentioned warm-season grasses have extravaginal growth, with rhizomes and/or stolons producing a creeping, spreading habit.

Bermudagrass, also known as *couchgrass,* is the most important and widely adapted of the warm-season turfgrasses. The improved turf-type bermudagrasses form a vigorous and aggressive turf of high shoot density with fine-textured leaves. Bermudagrass tolerates low temperatures poorly; discoloration occurs at soil temperatures below 10°C (50°F). It does tolerate a wide range of soil acidity (pH 5.5–7.5) and can withstand flooding and high levels of soil salt. Bermudagrass is used in the warm, humid and warm, semiarid regions in lawns and in other general-purpose turf areas. Medium to high maintenance is required, with a cutting height of 1.3–2.5 cm (0.5–1.0 in.).

Common bermudagrass, as far as I know, is the only turf type that is propagated by seed. All others are sterile and must be vegetatively propagated. 'Common' is relatively coarse in texture compared with the improved cultivars of *C. dactylon* or the hybrids of *C. dactylon* × *C. magennisii* or *C. dactylon* × *C. transvaalensis*. *African bermudagrass* has the finest texture and highest shoot density of any bermudagrass. *Magennis bermudagrass* is a hybrid between *C. dactylon* and *C. transvaalensis*, resembling the latter, and does not produce viable seed. In Florida more than 80% of all the bermudagrass turf was devoted to golf courses in 1974 (Cromroy and Short 1981).

Zoysiagrasses are adapted to the warm, humid and transitional regions. Three species used as turf are dense and low growing, with tough and stiff stems and leaves. All go dormant at 10°–13°C (50°–55°F). They perform well in full sun on well-drained fertile soils of pH 6–7. They require medium intensity of culture with a cutting height of 1.3–2.5 cm (0.5–1.0 in.).

Centipedegrass is a medium, coarse-textured, slow-growing species that spreads by thick, leafy stolons. It is propagated vegetatively or by seed and prefers soil pH of 4.5–5.5. It is used on lawns and in other turfgrass areas where traffic is light and a relatively low intensity of culture is desired.

St. Augustinegrass is a versatile, sod-forming warm-season grass of very coarse leaf texture. Propagation is vegetative, with sprigs, plugs, or sod. It is the least hardy of the warm-season turfgrasses at low temperatures but has outstanding shade tolerance. This grass is used primarily in the warmer portions of the warm, humid region for lawns where a fine-textured turf is not required. A cutting height of 4–6 cm (1.5–2.5 in.) is preferred. St. Augustinegrass thatches readily. More than 90% of the entire St. Augustinegrass turf acreage in Florida was devoted to home lawns in 1974 (Cromroy and Short 1981).

Bahiagrass forms a very coarse-textured erect turf. Propagation is by seed. It is adapted to the warmer areas of the warm, humid region and is used for turf of relatively low quality that requires a low intensity of culture. About 90% of the entire bahiagrass turf acreage in Florida was devoted to home lawns in 1974 (Cromroy and Short 1981).

Summary of Grass Characters

The several minute but distinct characters found at the junction of the stem and leaf blades—ligule, auricles, and collar—plus the vernation of the leaf bud and growth characteristics are useful diagnostic characters (Figure 1). Their differences are summarized for the 10 most common cool- and warm-season grasses (Table 1). The appearance of these structures in live green plants should help identify these grasses (Plates 3 and 4).

Drought Dormancy and Its Relationship to Insect Damage

Turfgrasses can survive extended periods of drought by becoming dormant. Cessation of shoot growth is accompanied by death of leaves and stems, and the

Table 1. Characteristics of 10 major turfgrasses

	Vernation	Ligule and length	Collar	Auricle	Leaf blade and width	Growth and spread
Cool-season grasses						
Kentucky bluegrass	Folded	Membranous, short, blunt, 0.2–1.0 mm	Medium broad, divided	Absent	V-shaped to flat, boat-shaped tip, 2–4 mm	Rhizomes
Annual bluegrass	Folded	Membranous, thin, white, acute tip, 1–3 mm	Conspicuous, divided	Absent	V-shaped light green, boat-shaped tip, 2–3 mm	Bunch and stolons
Perennial ryegrass	Folded	Membranous, blunt, 0.5–1.5 mm	Distinct, divided, smooth	Small, claw-like soft	Flat dull above, glossy below, 2–5 mm	Bunch, noncreeping
Red fescue	Folded	Membranous, blunt, 0.2–0.5 mm	Narrow, indistinct, smooth	Absent	Narrow, folded to involute, 0.5–1.5 mm	Rhizomes
Creeping bentgrass	Rolled	Membranous, acute to oblong, 1–2 mm	Narrow to medium broad, continuous	Absent	Narrow, flat, acuminate tip, 2–3 mm	Stolons
Warm season grasses						
Bermudagrass	Folded	Fringe of white hairs, 1–3 mm	Continuous, smooth, ciliate	Absent	Flat, stiff, acute tip, 1.5–3.0 mm	Stolons and rhizomes
Zoysiagrass	Rolled	Fringe of hairs, 0.2 mm	Continuous, broad, ciliate at margins	Absent	Flat, stiff, 2–4 mm	Stolons and rhizomes
Centipedegrass	Folded	Membranous, short, ciliate, 0.5 mm	Continuous, broad, ciliate at margins	Absent	Flat, blunt, ciliate, 3–5 mm	Stolons
St. Augustinegrass	Folded	Fringe hairs inconspicuous, 0.3 mm	Continuous, broad, smooth	Absent	Flat, petioled, smooth, 4–10 mm	Stout stolons
Bahiagrass	Rolled or folded	Membranous, blunt, entire, 1 mm	Continuous, broad to divided	Absent	Flat to folded, ciliate margin at base, 4–8 mm	Flattened rhizomes and stolons

Source: adapted from Beard 1973, 1982.

turfgrass appears dead (Plate 5). Buds in the crowns, stolons, and rhizomes of dormant grasses survive and initiate new growth when soil moisture is replenished. Kentucky bluegrass, for example, can break dormancy rapidly, with recovery visible in 3–5 days (Plate 5; Beard 1973).

Summer drought-induced dormancy is often associated with turfgrass insect damage that is nearly impossible to detect except under very close scrutiny. See Chapters 5, 7, and 26 for further discussion of this phenomenon.

Dichondra Lawns

In addition to grasses, at least one broadleaf plant is used as a lawn (Plate 5). Dichondra, *Dichondra* spp., a member of the morning-glory family, is a native of the southern coastal plains of North America. It occurs in pure stand in lawns in southern California and finds limited usage in central California and north to the San Francisco Bay and Sacramento areas. Mowing height determines some of its characteristics. When it is close mowed at 2.0 cm (0.75 in.), a small-leaved wear-resistant stand develops. Cutting at 3.8–5.0 cm (1.5–2.0 in.) produces a mixed stand of large- and small-leaved dichondra. Without mowing, large-leaved dichondra dominates as a prostrate ground cover of about 7.6 cm (3 in.) in height (Gibeault et al. 1977; V. A. Gibeault, Cooperative Extension, University of California, Riverside, personal communication, 1985).

Economic Impact of Turfgrass Culture

Unlike most agricultural crops, whose production costs and sales values can be readily estimated, turfgrass has a value that is difficult to measure, primarily because most turfgrass acreage is not grown for sale. One way of measuring the value of the turfgrass industry, however, is by comparing annual maintenance costs. The entire United States in 1983 had an estimated 6–8+ million ha (15–20 million acres) of turf, with an annual maintenance expenditure of about $15 billion.

In New York State, the annual maintenance expenditure was estimated to be near $600 million in 1977 to maintain 1.2 million acres of turf. A 1982 update has increased the estimated expenditure of the New York state turfgrass industry (production and service) to a $1 billion a year. These figures may be compared with those for the dairy industry, New York State's largest agricultural industry; astonishingly, dairy products in 1981 amounted to $1.5 billion in sales, or 56% of the total cash farm receipts (Chapman and Kohut 1985, Gruttadaurio et al. 1978, Roberts 1983b, Smiley 1983).

Available data from several other states show equally startling maintenance expenditures when the year of study is taken into consideration (Table 2). No esti-

Table 2. Estimates of total annual expenditures for turfgrass maintenance for several states since 1975

State	Year of estimate	Total expenditures ($ millions)
California	1977	690
California	1983	1,080
New York	1977	595
New York	1982	1,050
Florida	1976	539
Tennessee	1983	100
Rhode Island	1983	43

Sources: California: Roberts 1984; New York and Tennessee: Roberts 1983b; Florida: Roberts 1983a; and Rhode Island: Duff 1984.

mates for Florida in the 1980s were available, but comparisons with California and New York for the late 1970s indicate that the projected estimates for Florida in the early 1980s could be near $1 billion annually, in line with the estimates for California and New York.

In these and other similar studies, the dominance of expenditures for residential lawns is obvious (see Table 3). This dominance varies relatively little, the lowest figure being 64% and the highest 75%, and compares favorably with the U.S. average of 69%.

The studies cited and older ones plainly show that turfgrass culture in its entirety as an industry contributes significantly to the economy of the country. Protection of the existing turfgrass plantings from various pests, including insects and other closely related organisms, is thus an important concern.

Table 3. Analysis of annual turfgrass expenditures in several areas for 1956–1965, 1977, 1982, and 1983 (percent)

	Percentage of total expenditures for year of survey				
Turf uses	Calif., 1977	New York, 1977	Rhode Island 1982	Tenn., 1983	United States, 1956–1965
Residential properties	68	64	75	66	69
Public lands	5	10	6	10	11
Golf courses	11	7	17	6	6
Institutional lands	1	6	1	8	8
Miscellaneous	15	13	1	10	6
Total	100	100	100	100	100

Sources: California: Roberts 1984; New York: Gruttadario et al. 1978; Rhode Island: Duff 1984; and Tennessee: Roberts 1983b.

2

Insects and Near Relatives

Phylum Arthropoda

Insects and mites belong to a larger category of related animals of the phylum Arthropoda. It includes: insects; arachnids (mites, ticks, spiders, and scorpions); chilopods, or centipedes; diplopods, or millipedes; and crustaceans (crabs, lobsters, shrimps, and barnacles). Except for the crustaceans, all arthropods are fairly similar in size, occupy relatively the same niche in the environment, and are fairly similar in general appearance. Arthropod bodies and legs are jointed; the word *arthropod* is derived from two Greek words meaning *jointed feet.* The exterior of the body (the integument) is covered by a horny layer known as chitin, which functions as the framework of the body and is the *exoskeleton.*

Of all arthropods, mites and spiders are the most closely related to insects in form and function and in their status as pests that damage turfgrass. Compared with insects they are relatively minor turfgrass pests. In the following general discussion I draw extensively on four sources: Borror et al. (1981), Jeppson et al. (1975), Krantz (1978), and Matheson (1955). These are excellent sources of more detailed information on insects and mites in general.

Form and Function of Insects and Mites

External Characteristics of Adults

Insects are distinguished from other arthropods by a body that has three distinct divisions—the head, thorax, and abdomen (Figure 3). The insect head typically possesses mouthparts, simple eyes (ocelli) and/or compound eyes, and a pair of antennae that have a sensory function. The three-jointed thorax includes the *pro-*, *meso-*, and *metathorax;* each has a pair of legs. The most primitive insects and some of the more specialized are wingless, but the vast majority have two pairs of wings, the front pair being attached to the mesothorax and the hind pair to the metathorax. In insects with a single pair, the wings are attached to the mesothorax. The flies appear to have a single pair attached to the mesothorax, but the methathorax has a pair of minute stubs known as *halteres* that aid in balance and are actually minute wings. The abdomen is typically 11-segmented, with the terminal 3 segments modified for reproductive functions.

The body of a typical mite is separated into an anterior gnathosoma and a posterior idiosoma, which is further divided into the propodosoma and the hysterosoma (Figure 3). The gnathosoma resembles the head of a typical insect only in that the mouthparts are attached to it. Chelicerae and palps are the main external feeding organs that rasp and pierce the epidermis of host plants, causing damage. The function of the idiosoma parallels that of the abdomen, the thorax, and portions of the head of insects. Division of the propodosoma and hysterosoma may or may not be evident, but the anterior two pairs of legs are attached to the former and the posterior two pairs of legs are attached to the latter. Mites of the superfamily Eriophyoidea (with only one family, Eriophyidae) have only two pairs of legs (Figure 3). These mites are basically wormlike and minute, varying from 0.1 mm to 0.3 mm in length and essentially invisible to the unaided eye. This family contains at least one turf pest, the bermudagrass mite.

Internal Structure and Function

The central nervous system of insects comprises a brain located in the head and a ventrally located double nerve cord that extends the entire length of the body (Figure 4). A series of ganglia, nerve centers composed of cell masses, unite the nerve cord at various regions along the nerve cord. Nerve cords radiate out from each ganglion to other parts of the body.

The digestive system is a relatively simple tube (Figure 4), extending from the mouth to the terminally located anus, or is a highly convoluted, complex system (Figure 4). Whether simple or complex, the digestive tract has three sections: the foregut, midgut, and hindgut, also called the *stomodaeum, ventriculus,* and the *proctodaeum,* respectively. Excretory organs called *Malpighian tubules* arise at the anterior end of the proctodaeum and usually contain many long, stringlike structures that permeate the body cavity.

The circulatory system consists of the dorsal vessel (Figure 4), which runs the full length of the body, and all the space between the various organs in the body cavity. The heart is merely a chambered dilation of the dorsal vessel. Blood, which may be clear or any of various shades of yellow or green, enters the dorsal vessel at the posterior end of the arthropod and is pumped forward to the head, where it flows out into the body cavity and into all the appendages.

Insects have no lungs. Oxygen is delivered to cells throughout the body by diffusion. Oxygen enters through small openings called *spiracles,* which are located laterally on the body (Figure 3), moves through a network of small tubes called *trachea,* and finally diffuses into the tissues, eventually reaching cells. Much of the

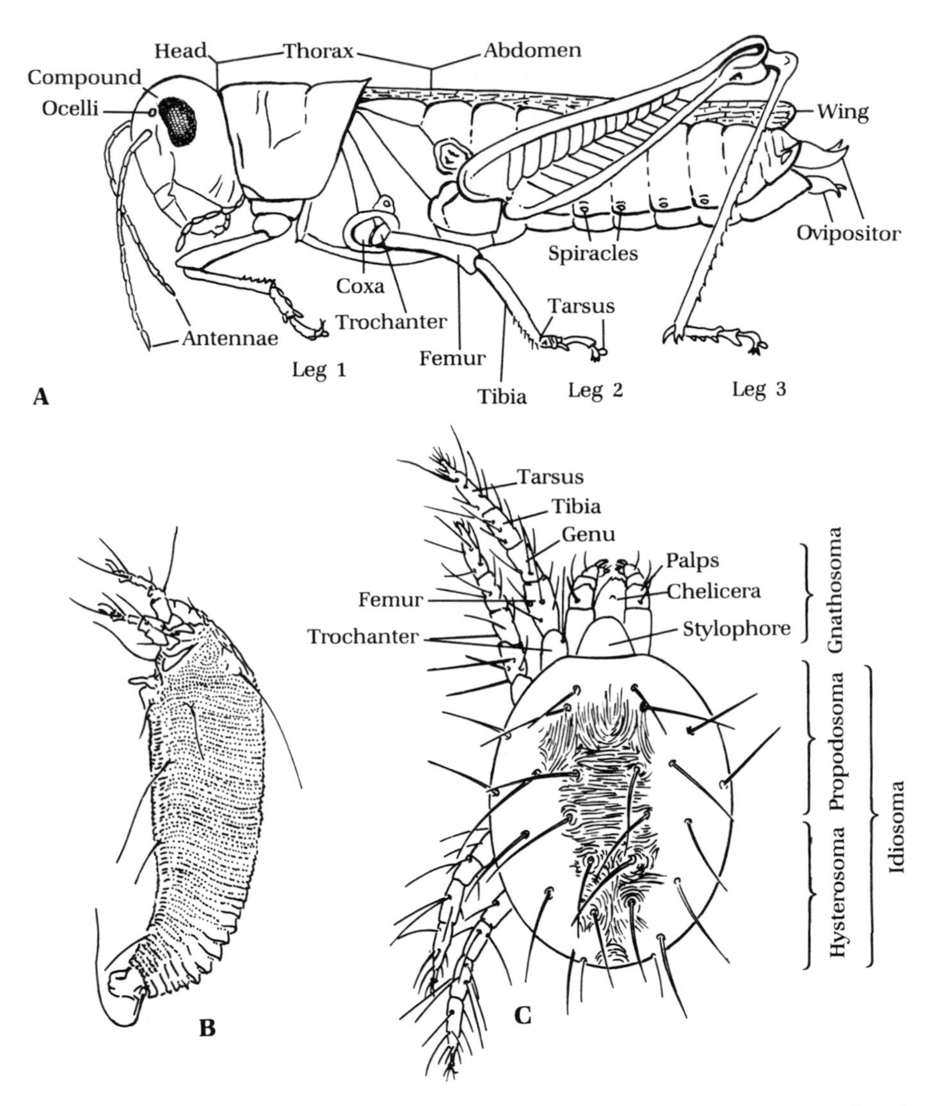

Figure 3. External anatomy of arthropods. **A.** Grasshopper (insect). **B.** Body of eriophyid mite. **C.** Body divisions of typical mite. (Part A adapted from Matheson 1951, fig. 31, courtesy of Cornell University Press; parts B and C adapted from Jeppson et al. 1975, figs. 3, 121A, courtesy of the University of California Press.)

preceding discussion of the internal structure and function in insects applies to mites as well.

Growth and Development

In insects and mites, growth and development are accomplished through a process known as *metamorphosis*. There are two major types. Those with a simple

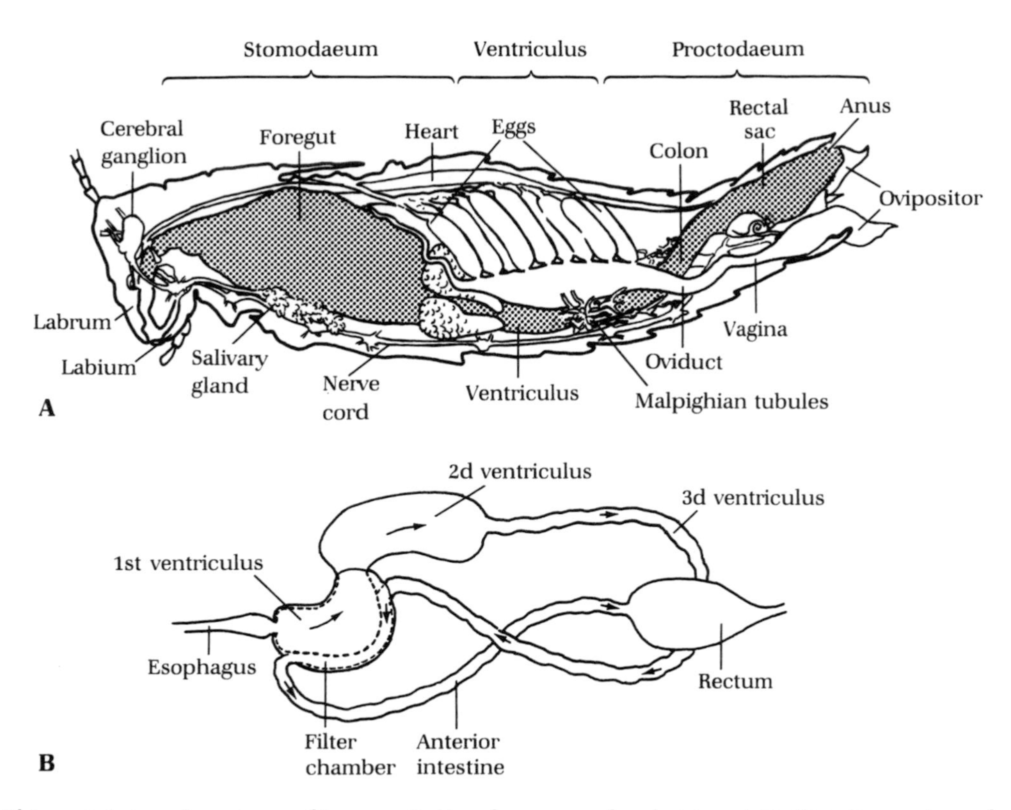

Figure 4. Internal anatomy of insects. **A.** Grasshopper, a chewing insect. **B.** Digestive system of piercing, sucking insect (Homoptera). (Part A adapted from Matheson 1951, fig. 51, courtesy of Cornell University Press; part B adapted from Snodgrass 1935, fig. 209A.)

metamorphosis develop through three stages, involving, respectively, the egg, the immature stage, called a *nymph,* and the adult (Figure 5). In simple metamorphosis the immatures look like small, wingless adults, and relatively simple reorganization occurs between the nymphal stage and the adult. Nymphs and adults occupy a similar niche. Arthropods with a complete metamorphosis develop through four stages—the egg, the immature form, called the *larva,* the resting stage, called the *pupa,* and the adult (Figure 5). In complete metamorphosis the immatures differ radically from the adults, and the latter have a niche very different from that of the larva. Growth of insects from egg to adult occurs in discrete stages, or steps. After hatching from the egg, insects will molt only twice or as many as eight or nine times, depending on the species. Molting for growth occurs in the nymphal or larval stages. Between each molt, an insect's stage of growth is designated *first instar, second instar,* and so forth.

Mites develop through a slightly different metamorphosis. The egg hatches into a six-legged larval stage. Upon molting, the larval stage becomes an eight-legged

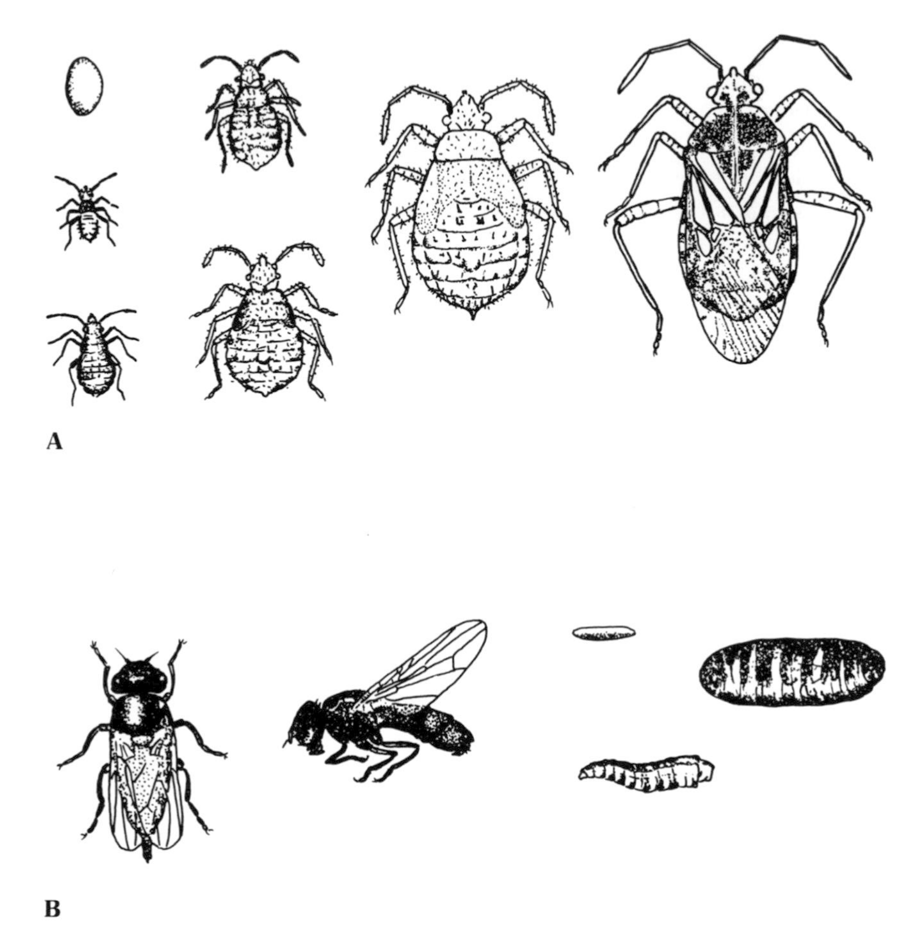

Figure 5. Insect metamorphosis. **A.** Simple metamorphosis, illustrated by grassbug. **B.** Complete metamorphosis, illustrated by sugarbeet root maggot. (Part A from Borror et al. 1981, fig. 58, courtesy of the Entomological Society of America; part B from Borror et al. 1981, fig. 59, courtesy of G. F. Knowlton, Utah State University Agricultural Experiment Station.)

nymph that resembles an adult except that it is smaller. There are generally two nymphal instars, the protonymph and the deutonymph, before the adult stage.

Insects and mites are cold-blooded; that is, their body temperature is near that of the surrounding environment, and their activity is strongly influenced by the ambient temperature.

Reproduction

Reproductive capacities of insects are highly variable and specialized. Egg production of insects may vary between species, with as few as several dozen eggs per female produced during the life of a female and as many as several thousand. The biology of egg production varies greatly, depending on the insect. Copulation and transfer of sperm are required in most insects, but in some, parthenogenetic reproduction is the rule, and no males are present in the species.

Many species lay eggs (are oviparous) on a substrate such as leaves or in the soil. Some species deposit live young that have hatched from an egg within the female (such insects are viviparous). Most insect eggs develop into a single offspring. A group of parasitic wasps, however, produces multiple offsprings from a single egg (a phenomenon known as polyembryony) by one of the most specialized and highly efficient reproductive systems known.

Generations

Insects that have one generation each year are said to be *univoltine,* while those with two or more generations each year are said to be *multivotine.* In discussions of generations of multivoltine insects, certain ambiguities arise. First, the terms *generation* and *brood* have been rather loosely and interchangeably used. Schurr and Rings (1964) have attempted to clarify this ambiguity and others, and I have adopted their naming and numbering system. The stage that overwinters is designated as the overwintering stage, whether it involves adult, egg, larva, or pupa; for example, we speak of *overwintering larvae.* With the arrival of spring, the overwintering stage and any succeeding stage in that generation (ending with adult) are called the *spring generation.* A new generation is considered to start when eggs are laid. These offspring, as eggs and in all succeeding stages through the adult, are called *first-generation* eggs, larvae, and so forth. First-generation adults lay eggs, thereby starting the second generation. Generations are thus numbered consecutively through the rest of the season. When winter returns, the hibernating insect again consists of the overwintering stage, whether its members belong to the first generation, the second, or any later one. These authors consider the term *broods* applicable only to different generations of insects that have life cycles longer than a year and yet have annual adult emergences. May or June beetles and periodic cicadas are examples of such organisms.

Types of Mouthparts and Turf-Feeding Damage

Since insects and mites damage turf principally in their quest for food, a brief review of the principal types of mouthparts will provide an understanding of the types of damage caused.

Chewing Insects

Biting-chewing mouthparts are more common among insect pests of turf than other types are (Figure 6). The main mouthparts of chewing insects consist of (1) the

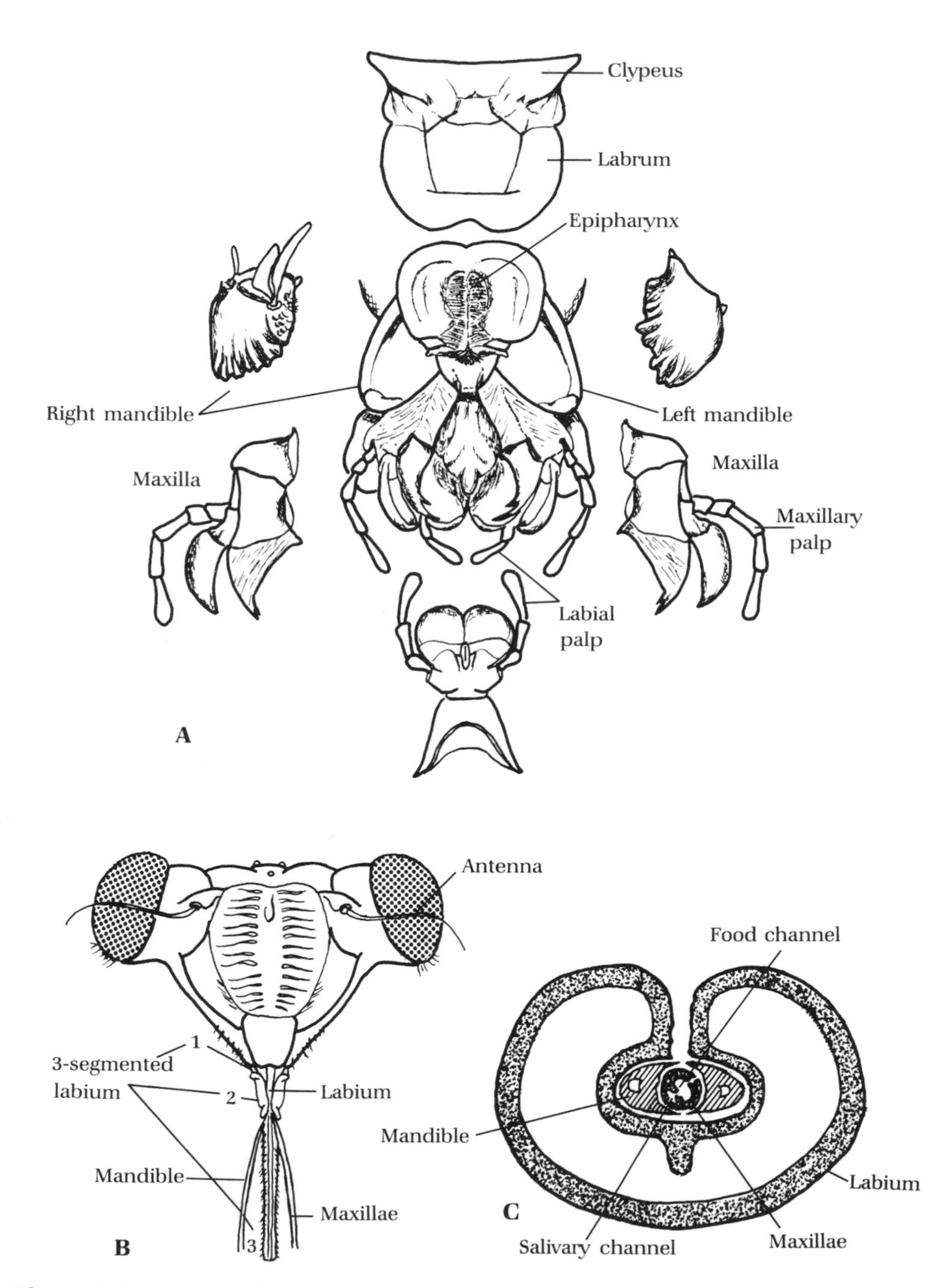

Figure 6. Mouthparts of insects. **A.** Chewing insect (grasshopper). **B.** Piercing-sucking insect (cicada). **C.** Cross section of cicada mouthparts. (Adapted from Matheson 1951, part A: fig. 37, parts B and C: fig. 64, courtesy of Cornell University Press.)

upper lip (labrum), (2) a pair of opposed laterally moving jaws (mandibles), (3) a second pair of opposed laterally moving jaws (maxillae), and (4) a lower lip (labium).

The labrum acts as an organ for manipulating and moving food into the jaws. Mandibles lie immediately behind the labrum. They are triangular, strongly sclerotized appendages that gradually taper outward to a cutting edge. Maxillae, more complex in structure, lie directly behind the mandibles and act as a second pair of jaws to manipulate and break up the food. The labium lies at the back of the mouthparts and forms a typical lower lip. Collectively, these mouthparts form the preoral cavity.

Turfgrass damage by insects with biting-chewing mouthparts exhibits physical removal of plant tissues, such as stripping away of the epidermis of leaves, notching of leaves and stems, complete severing of plant parts, hollowing out of stems and crowns, or pruning of roots.

Most of the chewing insect pests of turfgrass are the immature forms (larvae and nymphs), but a few adults are also involved.

Sucking Insects

The second most common method by which insects feed involves the piercing and sucking of plant tissue (Figure 6). Piercing, sucking mouthparts are thought to represent a highly modified form of the more primitive chewing mouthparts. The labium forms a beak that surrounds the needlelike mandibles and maxillae. The maxillae unite along their margins to form a tube with two channels, the food channel and the salivary channel. Mandibles that lie outside the maxillae act primarily as the cutting and piercing organ. Once the mandibles and maxillae have been inserted deep into the plant tissues, feeding commences. Salivary secretions pumped into the plant help the insect suck up the plant sap and the cell contents.

Plants injured by this method of feeding generally remain completely intact. The entire plant starts to deteriorate because of the loss of plant sap or in response to the injection of toxic salivary secretions. Chinch bugs and greenbugs are two insects that damage plants by both methods. Early symptoms may be yellowing, wilting, blasting of leaves, and necrosis, followed eventually by browning and death. Both adults and nymphs of piercing, sucking insects are involved in turf injury.

Insects that feed by piercing and sucking serve as vectors of plant diseases, primarily those caused by viruses. Fortunately, insect-caused virus transmission in turfgrasses has not been a serious problem to date.

Other Types of Mouthparts and Feeding

Other, less common types of mouthparts are found in flies and mites. Fly maggots have mouth hooks that move vertically rather than horizontally and are used primarily to break and tear plant tissues. The liquified plant tissues and small particles

Table 4. Orders, habits, and representative groups of turfgrass-infesting arthropods

Classes and orders	Primary and secondary turf pests: Type of feeding	Turf-damaging stage	Groups
Class Arachnida			
Order Acari	Rasping, sucking	Adults, larvae, nymphs	Mites
Class Insecta (Hexapoda)			
Order Orthoptera	Chewing	Adults, nymphs	Mole crickets, crickets, grasshoppers
Order Hemiptera (Heteroptera)	Sucking	Adults, nymphs	Chinch bugs, plant bugs
Order Hemiptera (Homoptera)	Sucking	Adults, nymphs	Aphids, leafhoppers, mealybugs, scales, spittlebug
Order Lepidoptera	Chewing	Larvae	Webworms, cutworms, armyworms, skippers, butterflies
Order Coleoptera	Chewing	Larvae	Scarabaeid grubs, weevils, wireworms, flea beetles
Order Diptera	Tearing	Larvae	Cranefly, frit fly
Order Hymenoptera	Chewing	Adult	Ants, bees, wasps

are then sucked in. Mites have still another type of mouthpart, a rasping, sucking type. Their feeding process ruptures the epidermal cells, giving injured tissues a silvery to gray appearance.

Orders of Turfgrass-Damaging Insects and Mites

Relatively few orders of insects and mites have species destructive to turfgrass. In number of destructive species present, the orders Lepidoptera and Coleoptera are by far the most important. The orders Diptera and Hymenoptera are less important. Table 4 condenses information relating to orders of turf-infesting arthropods.

3

Acarine Pests

Bermudagrass Mite

Taxonomy

The bermudagrass mite, or BGM, *Eriophes cynodoniensis* Sayed, family Eriophyidae, was previously named *Aceria neocynodonis* Keifer. It is also called the *bermudagrass stunt mite* (Keifer et al. 1982, Reinert 1982a).

Importance

This mite is a serious pest of bermudagrass on golf courses. Since about 87% of all the bermudagrass grown as turf in Florida is on golf courses representing more than 20,000 ha (49,420 acres), the BGM constitutes a serious threat to the welfare of this sport. This mite is found over the entire state and is active the entire year in areas roughly south of Palm Beach. During the 1970s, just one Florida course spent $25,000 a year to control this pest. Treatments are not always dependable, and repeat applications may be necessary (Cromroy and Short 1981, Johnson 1975, Reinert and Cromroy 1981, Tuttle and Butler 1961).

History and Distribution

The BGM is widely distributed in New Zealand and North Africa and has been collected in Australia. In the United States it was first discovered infesting a bermudagrass lawn in Phoenix, Arizona, in 1959 and was found later the same year also in Tucson and Yuma, where it was causing extensive bermudagrass damage. It was first reported in Florida in 1962 and has since been collected in California, Nevada, New Mexico, Texas, Alabama, and Georgia (Butler 1963, Keifer et al. 1982, Reinert 1982a, Reinert et al. 1978).

Host Plants and Damage

The BGM feeds only on bermudagrass. In Florida, all commonly used varieties are highly susceptible. Florida golf course infestations were most abundant where close mowing was not practical, for example, at the edge of bunkers, on the lips of sand traps, and around trees. In Arizona, lawns under flood irrigation were injured less than lawns irrigated with sprinklers (Butler 1963).

Damage is caused by mites that feed under the leaf sheaths. The first signs to appear are a slight yellowing of the leaf tips and a twisting of the leaves, with the margins rolling upward and inward. Shortening of the internodes produces a thick rosette (Plate 6). When rosettes are numerous, clumps that resemble cabbage heads develop, and the grass no longer appears to have internodes. Finally all leaves die back to the point of their insertion on the stem. The distorted growth is believed to be caused by a toxin that is injected into the developing grass buds. Death of leaves, stems, and stolons soon follows (Butler 1963, Johnson 1975, Keifer et al. 1982).

In the West, damage is first noticed in the spring when lawns fail to begin normal growth in spite of irrigation and fertilization. Well-fertilized lawns are more attractive to mites than starved grass. In Florida, mite injury is more pronounced during dry weather, when grass is under stress. Severe injury causes large areas of turf to become thinned or killed (Cromroy and Short 1981, Tuttle and Butler 1961).

Description of Stages

Adult

Mites are 165 μm to 210 μm long, wormlike in shape, creamy white to yellowish, and barely visible with a 10× to 20× lens. Under higher magnification, two pairs of legs are apparent (Figure 7, Plate 6). Males can be distinguished by the presence of five-rayed feather claws, while those of females are six rayed (Baker 1982, Johnson 1975, Keifer et al. 1982, Niemczyk 1981).

Egg and Nymph

Eggs are about 70 μm long and transparent to opaque white (Figure 7). Nymphs are about two-thirds the size of adults, are also wormlike, and have four pairs of legs (Baker 1982).

Life History and Habits

Life Cycle

Mites are active primarily during late spring and summer. They require only 5–10 days to complete their development from eggs to adults. After hatching, they pass through two nymphal instars and molt, becoming egg-laying adults in 7–10 days. All life stages are found together under leaf sheaths (Plate 6). Optimum temperatures

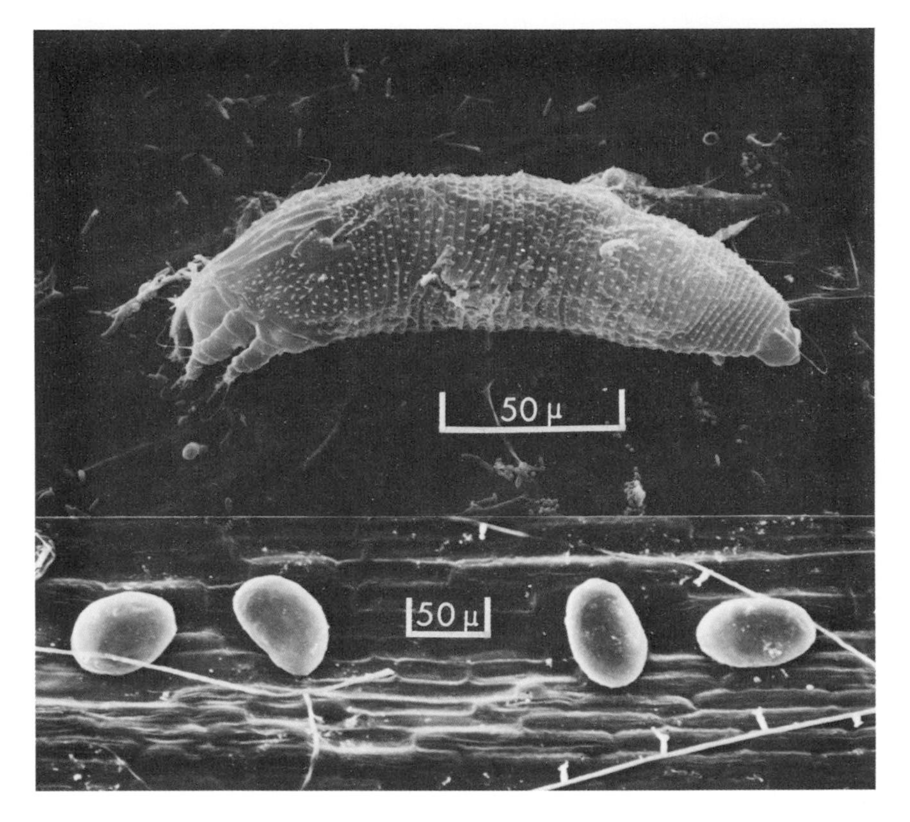

Figure 7. Bermudagrass mite adult and eggs. (Photo by T. M. Bourett, Plant Pathology, New York State Agricultural Experiment Station, Geneva [NYSAES].)

for growth and reproduction are 26.5°–44.0°C (80°–112°F). Mites are spread through normal cultural operations, such as mowing that spreads clippings, and may hitch-hike on other insects present in the bermudagrass. Wind is one major means of dispersal; movement by water is another (Baker 1982, Johnson 1975, Reinert 1982).

Adult Activity

Adults infest bermudagrass by settling under leaf sheaths and depositing eggs. A few mites or more than 100 can congregate under a single leaf sheath. Large numbers of mites are found in the rosettes (Baker 1982, Johnson 1975, Keifer et al. 1982).

Miscellaneous Features

Evaluations of many cultivars and accessions of bermudagrass for resistance to the BGM have been made in Arizona and Florida. 'Tifway' and FB-141 showed moderate resistance to the mite, with nearly half the plants tested remaining free of infestations. 'Midiron' and 'Tifdwarf' have shown high degrees of resistance to the mite. FB-119 showed no infestations during 6 years of field observations. This common type of bermudagrass is being developed for release by the Florida Agricultural Experiment Station (Reinert 1982a).

Two other cultivars, 'Tifgreen (328)' and 'Tifway (419)', were highly resistant to the BGM. More than 200 golf course greens were examined, from Jacksonville to Miami, Florida, over a 2-year period. None of the greens planted to 'Tifgreen (328)' cultivar showed any evidence of mites, or damage. When bermudagrass turf was cut at a height of 3.8–5.0 cm (1.5–2.0 in.) the 'Common' cultivar was severely infested, while 'Tifway (419)' was completely free of an infestation. Bermudagrass cultivars and accessions showing BGM resistance have parentage of *Cynodon transvaalensis,* which confers resistance (Johnson 1975, Reinert 1982a, Reinert et al. 1978).

Natural Enemies

A predacious mite, *Neocunoxoides andrei* (Baker and Hoffman), attacks the BGM and is widely distributed in Florida. It belongs to the suborder Prostigmata, family Cunaxidae. In Arizona the mite *Stenotarsonemus spirifex* (Marchal) of the suborder Trombidiformes, family Tarsonemidae, has been most frequently associated with reduced BGM populations (Butler 1963, Johnson 1975).

Winter Grain Mite

Taxonomy

The winter grain mite, WGM, *Penthaleus major* (Dugés) is a member of the family Eupodidae (Penthaleidae), which consists of seven genera. *Penthaleus* is easily separated from the other genera by the presence of a dorsal anus. It was first described under *Notophallus dorsalis* n. sp. by Banks (1904). Other common names ascribed are *red-legged earth mite, pea mite,* and *blue oat mite.* This mite occurs during the winter and causes more damage to small grain than to any other crop. It is not blue. Its habits make *winter grain mite* an appropriate name (Chada 1956, Streu 1981).

Importance

The WGM is a common and important pest of small grains and certain cool-season vegetables throughout the world. It affects grass seed production in Oregon and was discovered relatively recently to cause winter injury to cool-season grasses in the midwestern and eastern United States. Its widespread distribution in the United States (Figure 8) makes it more than an incidental turfgrass pest (Krantz 1957, Streu 1981, author's personal observations).

There are many indications that the use of the insecticide carbaryl against other

Figure 8. States in which the winter grain mite has been observed. (Data from Chada 1956 and Streu and Gingrich 1972; drawn by R. McMillen-Sticht, NYSAES.)

turfgrass insects increases the severity of WGM buildup. Reduction of predatory mites is the probable cause (Streu and Gingrich 1972).

History and Distribution

The WGM is widely distributed throughout both north and south temperate zones of the world. In the United States it was reported mainly west of the Mississippi River. Although the WGM was first described from specimens collected in Washington, D.C., it was rarely recorded east of the Mississippi River. It was not known to infest turfgrasses in the Northeast prior to 1968. Pitfall trap catches on a red fescue Kentucky bluegrass utility-type turf established the presence of the WGM in New Jersey for the first time during October and November and again during February 1968–1970. Of all the arthropods collected, the WGM accounted for more than 95% of the total specimens trapped during peak population periods when feeding damage was apparent (Streu and Gingrich 1972).

During March and April 1978–1980, several incidents involving significant WGM injury to turfgrasses in the Midwest and Northeast were noted. Reports included damage to bentgrass golf course fairways in Pennsylvania; bentgrass fairways and greens in Cincinnati, Ohio; Kentucky bluegrass and fine fescue home lawns south of

Cleveland, Ohio; and Kentucky bluegrass parkways on Long Island, New York. From its widely scattered locations, reported from 19 states (Figure 8), it is logical to assume that the WGM is present in practically every mainland state (Streu and Niemczyk 1982, author's personal observations).

Host Plants and Damage

Kentucky bluegrass, bentgrass, chewings red fescue, and perennial ryegrass are damaged. In Oregon seed production, perennial ryegrass is more susceptible to damage than Kentucky bluegrass and bentgrass (Kamm and Capizzi 1977, Streu and Niemczyk 1982).

The host plants recorded, exclusive of turfgrasses, include the small grains barley (*Hordeum vulgare*), oats (*Avena sativa*), rye (*Secale cereale*), and wheat (*Triticum sativum*); the legumes alfalfa, several clovers, lupine, and peas; the vegetables lettuce and potatoes; and weeds (Chada 1956).

The rasping of the leaf surface with mouthparts called *chelicerae* and the sucking up of the plant sap produce a silvered, scorched appearance caused by loss of chlorophyll (Plate 6). The most severe grass damage appears from mid-December through mid-March, the period of highest mite population (Figure 9). Heavy damage to turf resembles typical winter freezing damage due to desiccation, a resemblance that no doubt contributed to the failure to identify this problem until relatively recently. Feeding by the WGM does not cause the yellowing that is characteristic of feeding by tetranychid mites. The highest WGM populations are found in turfgrass that was treated with carbaryl during the previous summer (Figure 9; Niemczyk 1978, Streu 1981).

A laboratory host preference study that used petri dishes as containers for isolating the WGM and plants and holding them at 10°C (50°F) and 8:16 L:D period showed that mites survived the longest on Kentucky bluegrass, living for about 7 weeks and depositing eggs at the base of the plants and in the root system. Perennial ryegrass and chewings red fescue supported mites for about 4 weeks. On bentgrass, the least suitable host, the mites died in 1–2 days (Streu and Gingrich 1972).

Description of Stages

Adult

Males have not been observed, and their occurrence has not been definitely established. Females average 1 mm in length. They have a dark brown to black body that is tinged with green. Mouthparts and legs are red to reddish brown. Adults have four pairs of six-segmented legs. The first and fourth pairs are longer than the middle two. A pair of silvery eyes are located just behind the second pair of legs. The entire body is sparsely covered with very small, white setae, and each leg is covered with more pronounced setae. The most unique feature is a dorsal anus, present in

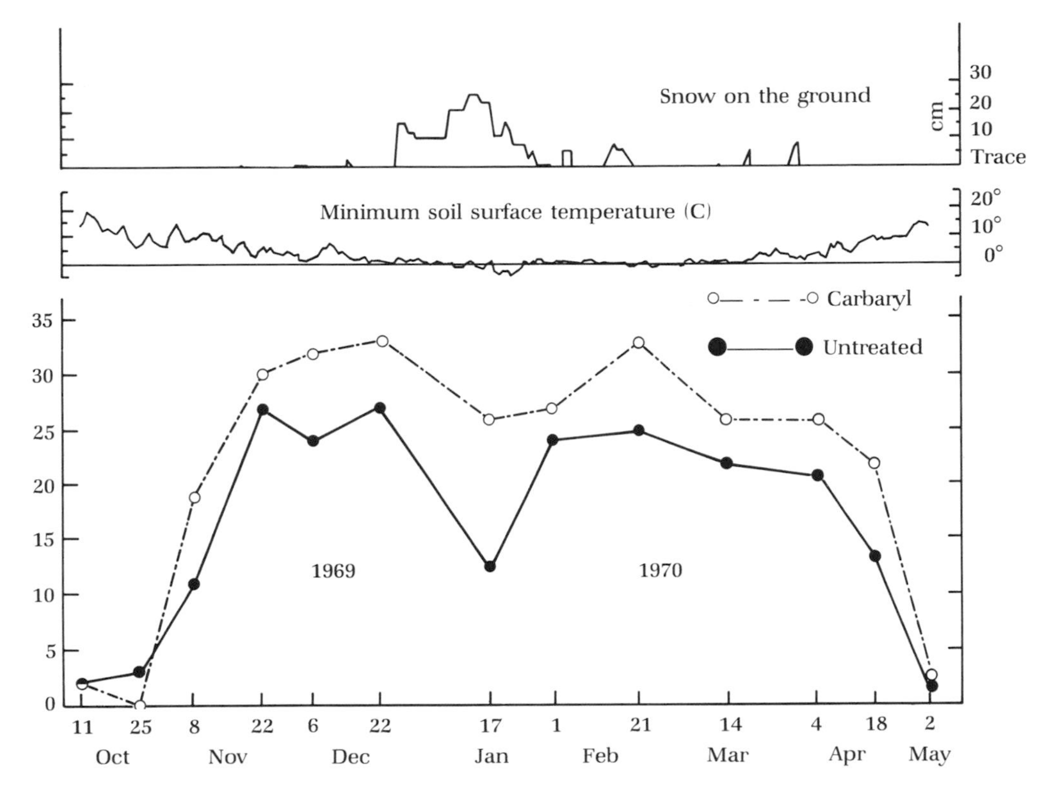

Figure 9. Seasonal abundance of the winter grain mite recorded in turf with and without summer carbaryl treatment in New Jersey, 1969–1970. (Adapted from Streu and Gingrich 1972, fig. 1, courtesy of the Entomological Society of America.)

no other mite associated with cool-season grasses. It is surrounded by a conspicuous reddish orange spot (Plate 6). A droplet of fluid that is clear to light yellow is frequently seen exuding from the anus. The genital opening is ventral (Chada 1956, Streu 1981).

Egg

Freshly laid eggs are plump and kidney shaped. They average 0.25 mm in length and 0.14 mm in diameter. They are orange to reddish brown and are glued singly to the base of grass plants, to the root system, or to the thatch (Plate 6). The brightly colored smooth surface becomes opaque and wrinkled within 1–2 days (Chada 1956, Streu 1981).

Larva

The sole larval instar is readily recognizable by its three pairs of legs. Larvae are 0.18 mm in length, 0.11 mm in width, and sparsely covered with white setae. They are reddish orange when first hatched but after a day become light brown, with legs and mouthparts yellowish orange. Just before molting the body becomes darker brown and tinged with green (Chada 1956).

Nymph

First-instar nymphs resemble full-grown larvae except that they have four pairs of legs. The body is dark brown, with a light brown spot surrounding the dorsal anus. Legs and mouthparts are yellowish orange. The three nymphal instars have bodies that change from dark brown to black, with legs and mouthparts that change to reddish orange as they complete their nymphal stage (Chada 1956, Streu 1981).

Life History and Habits

Seasonal Cycle

The WGM has an unusual biology in that it feeds on grasses only during the cold winter months and oversummers as eggs. Two generations a year are reported in north central Texas and in New Jersey; the second-generation mites deposit eggs that aestivate (Chada 1956, Streu 1981).

In Texas, mites are normally present from about November 1 to April 15. Maximum populations occur about February 1 for the first generation and about April 1 for the second. In New Jersey, eggs hatch in early October, and females are present by November. Populations increase rapidly during November and early December, decline slightly in January between generations, and then peak at several thousand mites per 0.1 m^2 (1 ft^2) by late February or early March. Mites decline in April and disappear by May after having deposited aestivating eggs (Chada 1956, Streu 1981).

Daily Activity

Most feeding by adults (and presumably also by larvae and nymphs) takes place during the night. As soon as the sun rises, mites move to shaded areas or descend into the base of the plants. During the heat of the day they are found on the moist soil surface under foliage. Mites migrate upward as the sun declines. After sunset the entire plant becomes covered with feeders. Mites may also feed during dark, cloudy days or under snow cover (Chada 1956).

Oviposition

Females lay an average of 1–2 eggs a day for a total of 30 to 65 eggs during nearly a 40-day oviposition period. A secretion cements eggs to sheath leaves, to the stems, or on or in the soil. For the second generation, egg laying begins in March. By May all the mites are dead except for the aestivating eggs. The WGM is not seen again until the following October (Chada 1956, Niemczyk 1978).

Environmental Effects

Cold rather than warmth favors the development of the WGM. Individuals are not harmed by short periods of sleet, ice, or frozen ground. Oviposition is heaviest between 10.0° and 15.5°C (50°–60°F), while optimum temperatures for hatching are 7°–13°C (45°–55°F). Adult activities are greatest at 4°–24°C (40°–75°F). When temperatures go beyond these optimums, mites stop feeding and descend to the ground or burrow into the soil. Hot, dry days force mites to penetrate 10–13 cm (4–5 in.) into the soil to seek moisture and relief from heat (Chada 1956).

Dispersion

The mite is probably spread by transportation of aestivating eggs on vegetal debris or soil. Eggs may also be windborne (Chada 1956).

Natural Enemies

Many predatory mites have been implicated. Larvae of *Chrysopa californica* (Cog.) and the predatory mite *Balaustium* sp. feed on several stages of mites (Chada 1956, Wildermuth 1916).

4

Orthopteran Pests: Family Gryllotalpidae

Southern and Tawny Mole Crickets

Taxonomy

The southern mole cricket, SMK, *Scapteriscus acletus* Rehn and Nebard, and the tawny mole cricket, TMK, *S. vicinus* Scudder, are members of the order Othoptera, family Gryllotalpidae, subfamily Gryllotalpinae (see Plate 7). Two genera are listed in this subfamily. The genus *Scapteriscus* can be distinguished by the front tibia with two dactyls; the genus *Gryllotalpa* has front tibia with four dactyls. Before 1984, the TMK was called the *changa mole cricket,* but this name is now reserved for *S. didactylus* (Latreille), another introduced species. The common name *changa* originated in Puerto Rico, where the face of the TMK is thought to resemble that of a pet monkey called *chango.* The TMK has also been called the Puerto Rican and West Indian mole cricket. Both the SMK and the TMK are sufficiently similar in appearance, habits, and destructive nature to be discussed together (Blatchley 1920, Nickle and Castner 1984, Reinert 1983b, Van Zwaluwenburg 1918).

Another species introduced into Brunswick, Georgia, in about 1904 and into at least four Florida coastal areas (Tampa, Key West, Miami, and Fort Myers) is the short-winged mole cricket, *Scapteriscus abbreviatus* Scudder (Plate 8). It has also been implicated as a pest of turfgrasses and pasturegrasses and other agricultural crops. This species is easily recognized by its short tegmina, which cover only about one-third of the abdomen. The hind wings are vestigial. The pronotum is more elongate than in other species of *Scapteriscus,* and it has a distinctly mottled color pattern (Nickle and Castner 1984, Walker and Nickle 1981).

Importance

The SMK and the TMK are the most destructive insect pests of bahiagrass and bermudagrass turf and bahiagrass pastures in Florida. Florida has millions of acres of these two favorite host grasses, with sandy soils that promote the development and spread of mole crickets. Golf courses are among the areas hardest hit. The two species of mole crickets were responsible for >$100 million damage to Florida turfgrass during 1976–1978. About 4.9 million ha (12 million acres) of pasturegrass exist in Florida alone. Statewide, about 30% of the bahiagrass pastures have been seriously damaged; in some locations virtually 100% of the bahiagrass, including that in lawns, was damaged severely (Koehler and Short 1976, Reinert and Short 1981, Short and Reinert 1982).

Infestations are sporadic, unpredictable, and largely undetected until tunneling activity is visible. Economically significant damage appears when nymphs are too old to be controlled easily. In southern Alabama the mole crickets have become the major insect problem on bermudagrass since about 1976, with the TMK considered the more serious (Cobb 1982, Reinert and Short 1981, Short and Koehler 1979, Short and Reinert 1982).

History and Distribution

Introduction and Spread

Neither of the two species is native to the United States. The common belief that the TMK was introduced into the United States from Puerto Rico has been refuted; the species in the United States have calling songs that vastly differ in pulse rate from those of species in Puerto Rico. The TMK's place of origin is not known for certain but is thought to be somewhere in South America. Uruguay, Argentina, and Brazil seem particularly likely original sources of the U.S. population. This assumption is strengthened by the fact that the bahiagrass cultivars attacked by the TMK in Florida originated in these areas of South America (Walker and Nickle 1981).

The TMK was introduced into Brunswick, Georgia, in about 1899, presumably in ship ballast, from a source other than Puerto Rico. By 1960 the original colony had spread into southernmost South Carolina, across southern Georgia and all of Florida, and into southeastern Alabama (Figure 10). No additional known spread has been evident since about 1960 (Walker and Nickle 1981).

The SMK was introduced at Brunswick, Georgia, in about 1904, also presumably in ship ballast. The subsequent spread has been slow. Individuals were present in Jacksonville, Florida, in 1920, in southern Florida in the 1940s, and in western Florida in about 1955. During the 1960s the SMK spread northward into North Carolina and westward into eastern Louisiana, with a large disjunct population in western Louisiana and eastern Texas (Figure 10). It was discovered at College Station, Texas, in 1980 and was collected on a golf green at Longview in northeastern Texas in 1982. Morphological variations in the SMK from various areas support the hypothesis that a number of introductions occurred (Crocker and Beard 1982, Walker and Nickle 1981).

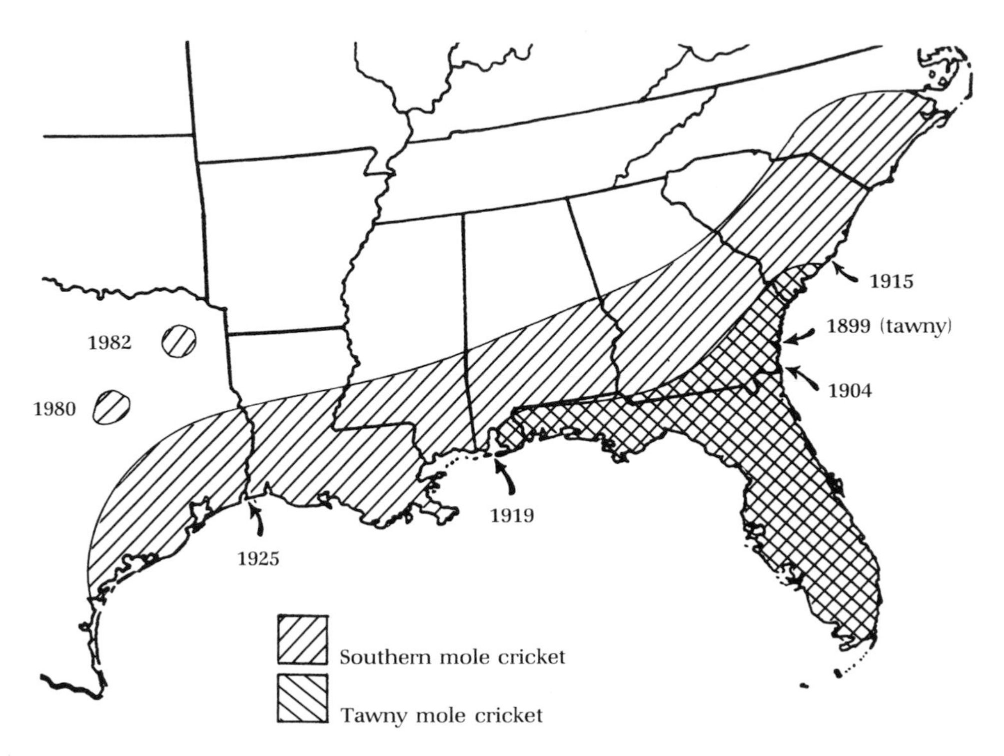

Figure 10. Introduction and spread of the southern and tawny mole crickets in the United States. Years and arrows indicate apparently independent introductions. (Adapted from Walker and Nickle 1981, figs. 1, 2, courtesy of the Entomological Society of America.)

Host Plants and Damage

All of the warm-season turfgrasses are attacked by mole crickets, although bahiagrass and bermudagrass are the two damaged the most severely. The texture of the grass is related to the degree of damage. In a study of turfgrass varieties with potential resistance, the coarser selections of St. Augustinegrass, bermudagrass, and bahiagrass sustained the least damage, while the greatest reduction in growth was exhibited by 'Tifway' and 'Tifgreen', two fine-textured bermudagrasses (Short and Reinert 1982).

Mole crickets are never found in heavy soil. They occur only in soils that range from light sand to loam; the soil must be compressible enough to allow tunneling without the need to remove loosened material (Van Zwaluwenburg 1918).

The most severe damage occurs during late summer and early fall, when the nymphs are approaching maturity and are actively foraging for food. Overwintering adults often cause severe damage in March and April. Burrowing in the upper soil levels mechanically dislodges roots and leaves mounds of soil on the surface, causing the soil to dry out excessively (Plate 7). Mole crickets also feed directly on the root system and seriously weaken the turf. Growth habits and cultural practices influence damage. Bahiagrass with its open growth allows greater dessication of the disturbed root system. Golf course bermudagrass, being closely cut, reduces the root system and thereby makes it more susceptible to uprooting and dessication. In Alabama, the presence of mole crickets on turf has been noted for many years, but turf-damaging populations started in about 1975. The severely damaged bermudagrass turf appears plowed. On thick sod the tunneling activities produce a fluffy turf. On Jekyll Island off Georgia, a population >1 cricket per 0.1 m^2 (per ft^2) had the capacity of destroying a tee overnight (Cobb 1982, Duff 1982, Koehler and Short 1976, Reinert 1983b, Reinert and Short 1981).

Description of Stages

Adult

Mole cricket bodies, well adapted for burrowing, have strong shovellike forelegs and a greatly enlarged, heavily chitinized prothorax for shaping and packing the soil in tunnels. The forewings overlap and are shorter than the abdomen (Plate 7). Wings of the short-winged mole crickets cover only about a third of the abdomen; individuals cannot fly. Males have a dark spot on the forewings that is made by the coalescence of wing veins, which also form the stridulating organ (Plate 7). The short-winged species cannot be sexed by wing vein patterns. The sexes are about equal in ratio in the TMK and are presumed to be similar in the SMK (Nickle and Castner 1984, Short and Reinert 1982, Van Zwaluwenburg 1918).

Adults of both the SMK and the TMK are similar in appearance. Both average about 3.2 cm in length and about 1 cm in width (Reinert and Short 1981, Short and Reinert 1982).

Separation of Species. The TMK is slightly larger and more robust than the SMK and has a broader thorax. The TMK is light creamy brown, while the SMK is reddish to dark brown, with four distinct lighter spots on the thorax (Plate 7; Short and Reinert 1982).

The most reliable diagnostic character for separation of the two species relates to the tibial dactyls on the foreleg (Plate 7). In the TMK the two dactyls are separated by a V-shaped space narrower than their width. In the SMK the space is wider and U-shaped (Short and Reinert 1982).

The two species can also be separated by differences in the maxillary laciniae (Figure 11). The lacinia bears a secondary toothlike process in the TMK, which is absent in the SMK. This difference is noted in the first instar, in the middle instar, and in adults (Matheny and Kepner 1980).

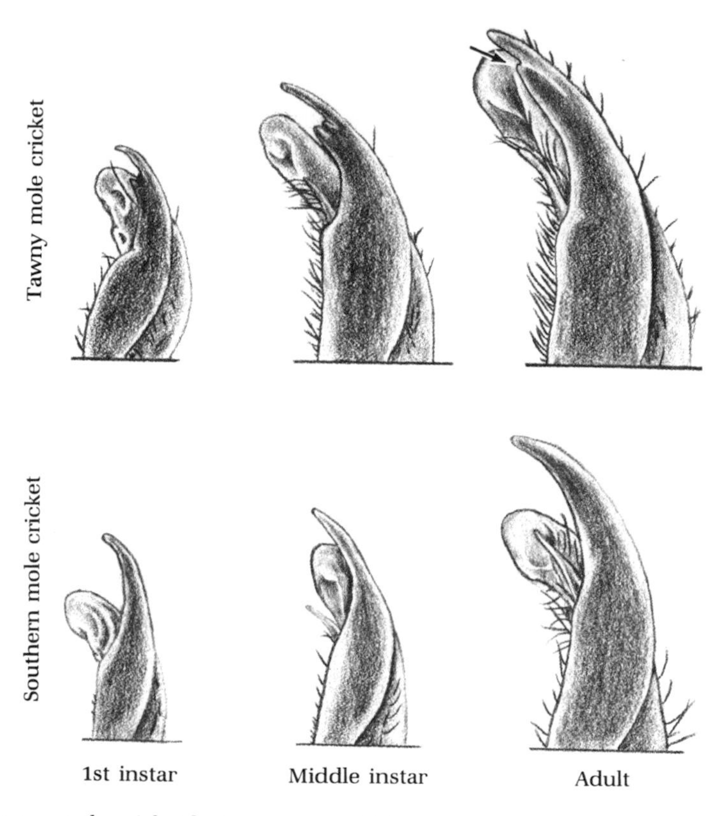

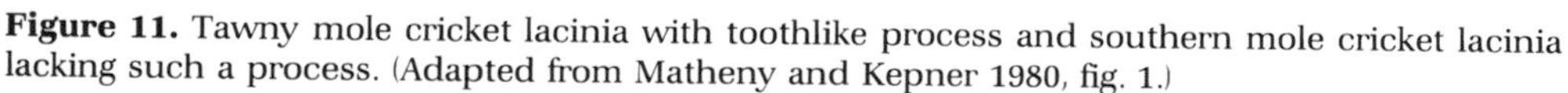
Figure 11. Tawny mole cricket lacinia with toothlike process and southern mole cricket lacinia lacking such a process. (Adapted from Matheny and Kepner 1980, fig. 1.)

Egg

Freshly deposited eggs of the TMK are gray and change to a dirty, yellowish white. They are oblong-oval, with a shiny, unsculptured surface. As eggs develop, they change to milky white or light brown. At maturity the reddish brown appendages become visible through the chorion (eggshell). Young eggs are about 3.0 mm × 1.7 mm and increase about 25% in size to about 3.9 mm × 2.8 mm in width before hatching (Hayslip 1943, Van Zwaluwenburg 1918). No description of the TMK eggs appears in the literature.

Nymph

First instars of the TMK are about 6 mm long. Budlike wing pads are first noticeable in second instars, increase in size with each succeeding molt, and become plainly visible in the fifth instar. The number of nymphal instars is not certain, but six to eight are recognizable (Short and Reinert 1982, Van Zwaluwenburg 1918).

Seasonal History and Habits

Seasonal Cycle

As determined in Florida, the SMK and the TMK have generally similar life cycles and require about a year to complete a generation (Figure 12). Both overwinter as adults and nymphs in north and central Florida. Oviposition usually begins in the latter part of March, and 75% of the eggs are deposited during May to mid-June. Eggs deposited in May and June require about 20 days to hatch. Peak hatching occurs during the first half of June in northern Florida and continues through August in southern Florida. Some oviposition occurs throughout the year, as evidenced by the presence of a few first instars every month of the year (Short and Reinert 1982).

Adult Activity

Flight. Both SMK and TMK adults have a major spring flight and a minor fall flight. While large numbers of both species often fly on the same night, flights of the two species are seasonally separated during part of the spring. The TMK flies from March to May, and the SMK flies from April to July. Large flights of both species on the same night are not uncommon. Fall flights normally occur on warm evenings from October into December (Ulagaraj 1975).

Heavy flights normally occur after heavy rains during warm weather, starting soon after sunset and continuing for about 1 hour. The insects are strongly attracted to fluorescent, incandescent, and mercury-vapor lights (Ulagaraj 1975, Short and Reinert 1982).

Adult Sound Production and Trapping. Males of the SMK and the TMK produce calling songs after sunset for 1.0–1.5 hr in specially constructed subterranean chambers. Tegminal stridulation produces the sound. Adults also produce sound by tapping the soil with their forelegs. The specially constructed bulbous earthen chambers are 2.5 × 1.0 × 2.0 cm (1.0 × 0.4 × 0.8 in.), some 3–5 cm (1.2–2.0 in.) below the soil surface. The function of male calling songs (as with all crickets and katydids) is to attract sexually responsive females (Ulagaraj 1976).

The natural calling songs produced by males of each species were tape recorded in the field (natural environment) and in the laboratory (synthetic environment). Taped songs were broadcast from 0.5 hr after sunset to the end of the flight period, about 1 hr. It was found that crickets flying toward lights 100 m away would alter direction to fly toward the source of the broadcast sound. Playbacks of taped natural songs and synthetic songs showed that both species were attracted to their own songs. Many more females than males were trapped, and more than 60% of the trapped females bore sperm in their spermatheca (Ulagaraj and Walker 1973).

Marking and release of captured adults showed that at least 2% of the adults fly

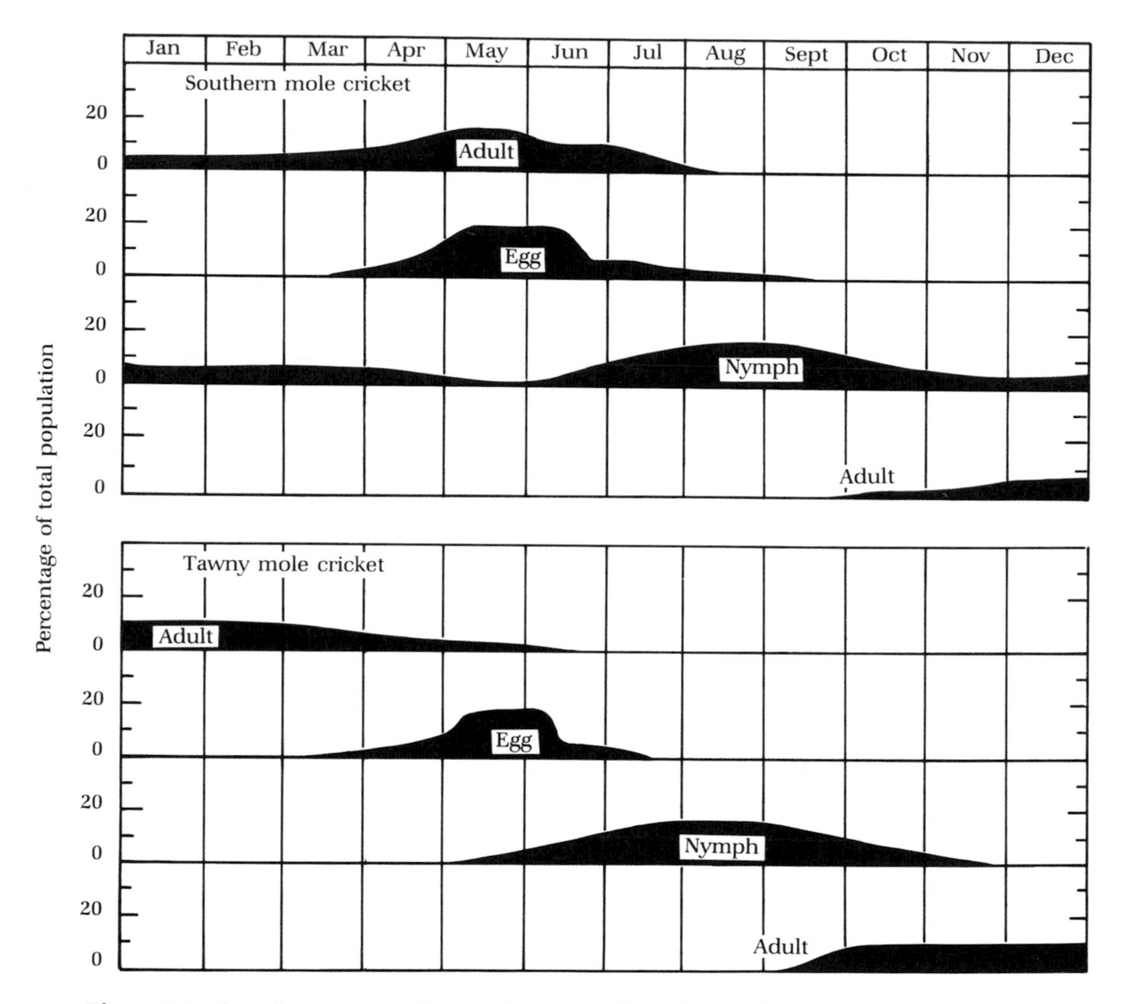

Figure 12. Life cycles of the southern and tawny mole crickets. (Adapted from Hayslip 1943, fig. 1.)

more than once. Some were recaptured twice, indicating at least three flights as much as 6.5 weeks after the first flight. Some were recaptured at least 0.7 km (0.4 mi) away from their first capture (Ulagaraj 1975).

Mating and Oviposition. Mating flights occur during spring. By mid-June, the majority of the crickets have mated. After mating, the females enter the soil for oviposition. Eggs are deposited in oval chambers about 3.8 cm long × 2.5 cm high and 2.5 cm wide (1.5 × 1.0 × 1.0 in.) off the main galleries. The entrances to the chambers are concealed by a packing of loose earth after eggs are deposited. Most chambers are in the upper 15 cm (6 in.) of warm, moist soil, but cool temperatures and dry soil force females to construct chambers to a depth of about 30 cm (12 in.).

Each female excavates three to five chambers and deposits about 35 eggs in each. The range for both species is 10–59 eggs (Hayslip 1943, Short 1973, Short and Reinert 1982, Van Zwaluwenburg 1918).

Nymphal Activity

Immediately upon hatching, the young nymphs search and fight for food in the egg cell, eating eggs, empty chorions, and weaker living nymphs. Many young nymphs perish through cannibalism. Soon the young nymphs escape from their egg cells and burrow to the soil surface to begin feeding on roots, organic matter, other insects, and other small organisms. Nymphs develop rapidly throughout the summer, and the first adults appear in September (Hayslip 1943, Reinert and Short 1981, Short 1973).

Most crickets, remaining as nymphs in November when cold weather arrives, overwinter as nymphs and become adults the following spring. During the two winters of 1970–1972, about 75% of the SMK and 15% of the TMK overwintered as nymphs in central Florida. Eggs deposited late during the oviposition periods tend to produce overwintering nymphs, while those deposited early become adults in the fall (Hayslip 1943).

Feeding Activity

All nymphs as well as adults come to the surface at night to feed. Tunneling of >3–6 m (10–20 ft) a night has been observed. Most of the feeding occurs during warm nights following rain or irrigation. Crickets return to permanent burrows during the day and may remain there for long periods during dry periods or cold weather. Each individual has its own burrow, which may extend to a depth of 36 cm (14 in.; Hayslip 1943, Reinert and Short 1981, Short and Reinert 1982).

The TMK does more root feeding; its gut contains mostly plant food material, while that of the SMK has primarily remains of insects (Plate 8; Reinert and Short 1981, Ulagaraj 1975).

Miscellaneous Features

Sampling Techniques

The linear pitfall trap, a useful general sampling and collection device designed by Kenneth Lawrence of ChemLawn's Boynton Beach, Florida, research and development center, is used mainly for collecting mole crickets for research needs. The trap (Plate 8) works equally well in sod or bare ground to capture mole crickets of all stages as well as other insects that crawl over the soil surface. This trap is described in detail in Chapter 25 (Lawrence 1982).

Routine monitoring of mole cricket flights is accomplished using a timer-operated sound synthesizer for each species. A standard trapping station consists of one SMK trap and one TMK trap, each with a wading pool 1.5 m in diameter that has been partially filled with water (Plate 8). Captured crickets remain floating on the surface of the water and are routinely removed the next morning (Walker 1982).

Threshold Populations

While some researchers believe that there is no reliable method of determining definite population levels and damage thresholds, others have applied irritant solutions to the turfgrass for determining damage thresholds. These solutions are a mixture of 30 ml (1 oz) of liquid soap in 7.6 l (2 gal) of water applied to 0.4 m^2 (4 ft^2) or a mixture of 15 ml (0.5 oz) synergized pyrethrins (1.2% pyrethrins + 9.6% piperonyl butoxide) to 7.6 l (2 gal) of water, also applied to 0.4 m^2 (4 ft^2) (Reinert and Short 1981, Short and Koehler 1979).

If irritant solutions cause about seven mole crickets to surface within 3 min, an insecticidal treatment is recommended. Late June through July is considered optimum period for insecticidal control in central Florida. Eggs are all hatched, nymphs are small and close to the surface, and extensive damage has not yet occurred. For best results, night temperatures should be above 15.6°C (60°F), and the soil should remain moist (Reinert and Short 1981, Short 1973, Short and Koehler 1979, Short and Reinert 1982).

Natural Enemies

Microorganisms

Two fungi have been observed infecting mole crickets. *Metarrhizium anisopliae* (Metch.) infections produce a carcass covered with white hyphae that are later covered with light green spores (Plate 8). The fungus *Sorosporella uvella* (Kass.), which produces a distinct brick red carcass, has been observed to cause some mortality in mole crickets (Hayslip 1943, Short and Reinert 1982).

Invertebrate Parasites and Predators

The most promising of parasites is the sphecid *Larra bicolor* (*americana*), found in abundance in Brazil, Venezuela, Dutch Guiana, and Puerto Rico, where it is an effective parasite. There have been recent attempts to colonize it in Florida (Short and Reinert 1982, Wolcott 1941).

The adult wasp enters mole cricket tunnels and chases the cricket to the surface, where it is captured, is stung at the base of the prothoracic legs, and is temporarily paralyzed. The wasp then deposits an egg ventrally behind the prothoracic legs. After recovery the mole cricket burrows back into the soil. Upon hatching, the parasite larva develops externally on the host. About 2 weeks are required for the parasite to complete its development as it simultaneously devours the entire internal contents of the host. Pupation of the parasite occurs in the mole cricket gallery (Anonymous 1983).

Fire ants, ground beetles, *Labidura* earwigs, and *Lycosa* spiders are predators of mole crickets. Mole crickets are very cannibalistic. The young nymphs devour each other and unhatched eggs as well (Short and Reinert 1982).

Vertebrate Predators

Raccoons, skunks, red foxes, armadillos, and toads feed on mole crickets. Of the birds, grackles, cow egrets, and white ibis will search for crickets on turf (Short and Reinert 1982).

5

Hemipteran Pests: Suborder Heteroptera

Chinch Bugs

Taxonomy

Chinch bugs infesting turfgrasses are members of the family Lygaeidae, subfamily Blissinae. The family Lygaeidae was originally known as the chinch bug family, with about 1,400 known species. Since only three chinch bugs are of any real concern to our welfare, the common name was not considered appropriate for the family. The common name for this family that is currently accepted is *lygaeid bugs*. The subfamily Blissinae, previously considered to have only two genera, now has three recognized genera. Only the genus *Blissus* contains serious turfgrass pests. A discussion of the turfgrass-infesting chinch bugs would not be complete without a discussion of the *leucopterus* complex of Leonard (Blatchley 1926; Leonard 1966, 1968; Werner 1982).

The *leucopterus* Complex

Five species or subspecies make up the complex, including two incidental species of *Blissus* that inhabit the Atlantic coastal dunes. Nothing more need be said of these. The two turfgrass-infesting chinch bugs, the hairy chinch bug (HCB), *Blissus leucopterus hirtus* Montandon, and the southern chinch bug (SCB), *B. insularis* Barber, plus the chinch bug *B. leucopterus leucopterus* (Say), complete the complex. The last named is the most serious of insect pests of small grain and corn in the heartland of the United States and throughout the central Atlantic region. All three chinch bugs occur in the eastern half of the United States, with some overlapping of geographic ranges between *hirtus* and *leucopterus* and between *insularis* and *leucopterus* (Figure 13). Much has been published on the subspecies *leucopterus*, and much in the literature on it has been used to understand the two turfgrass-infesting species (Leonard 1966).

It is possible to distinguish between *hirtus*, *leucopterus*, and *insularis* only when they are treated as populations. Morphological similarities and large individual variations make it very difficult to differentiate between individual specimens or a short series (Leonard 1966).

Differences. A comparison of mean width and length of adults of *leucopterus, hirtus,* and *insularis* from various geographic regions shows that *hirtus* is generally more robust than *insularis* but is similar in length or somewhat shorter (Leonard 1966).

The setae of *hirtus* are golden yellow, while the setae in *leucopterus* are silver or light straw yellow. The abdomen of *hirtus* is darker than that of *leucopterus* (Leonard 1966).

The much higher incidence of brachypters (short-winged adults) in *hirtus* and *insularis* than in *leucopterus* is the most characteristic difference. Macroptery (a long-winged form) in *leucopterus* is apparently associated with agricultural ecosystems where migration is often necessary. Such migration is rarely necessary in the turf habitat of *hirtus* and *insularis* (Leonard 1966).

Speciation. In isolation, both males and females of *hirtus* successfully cross with *leucopterus,* supporting the concept of two subspecies. Copulation occurs between *insularis* and *leucopterus,* resulting in production of fertile eggs, but no nymphal development occurs. The genetic enviability of *insularis* when crossed with other members of the *leucopterus* complex elevates the former to a distinct species (Leonard 1966).

Hairy Chinch Bug

Importance

The hairy chinch bug, HCB, is a major pest of turfgrasses in Connecticut, New Jersey, New York, and Ohio. The period July–August is the single most critical time for turf damage. The semidormancy of the turf due to drought often masks HCB damage. Damage may not become apparent until turfgrass previously infested with the HCB fails to respond to late summer rains (Maxwell and McLeod 1936, Niemczyk 1982, Polivka 1963, Schread 1970a, Streu and Vasvary 1966). In New York it was considered to be a serious pest only in the southeastern portion of the state, but beginning in about 1970, it also became a serious home lawn pest in much of upstate and western New York (author's personal observations).

History and Distribution

Unusually heavy infestations of the HCB occurred in lawns and golf courses on Long Island during 1932–1935. It is found in all the northern states in and east of Minnesota and throughout all the New England and mid-Atlantic states south into Virginia (Figure 13). In Canada, the HCB occurs in all of the provinces that border the

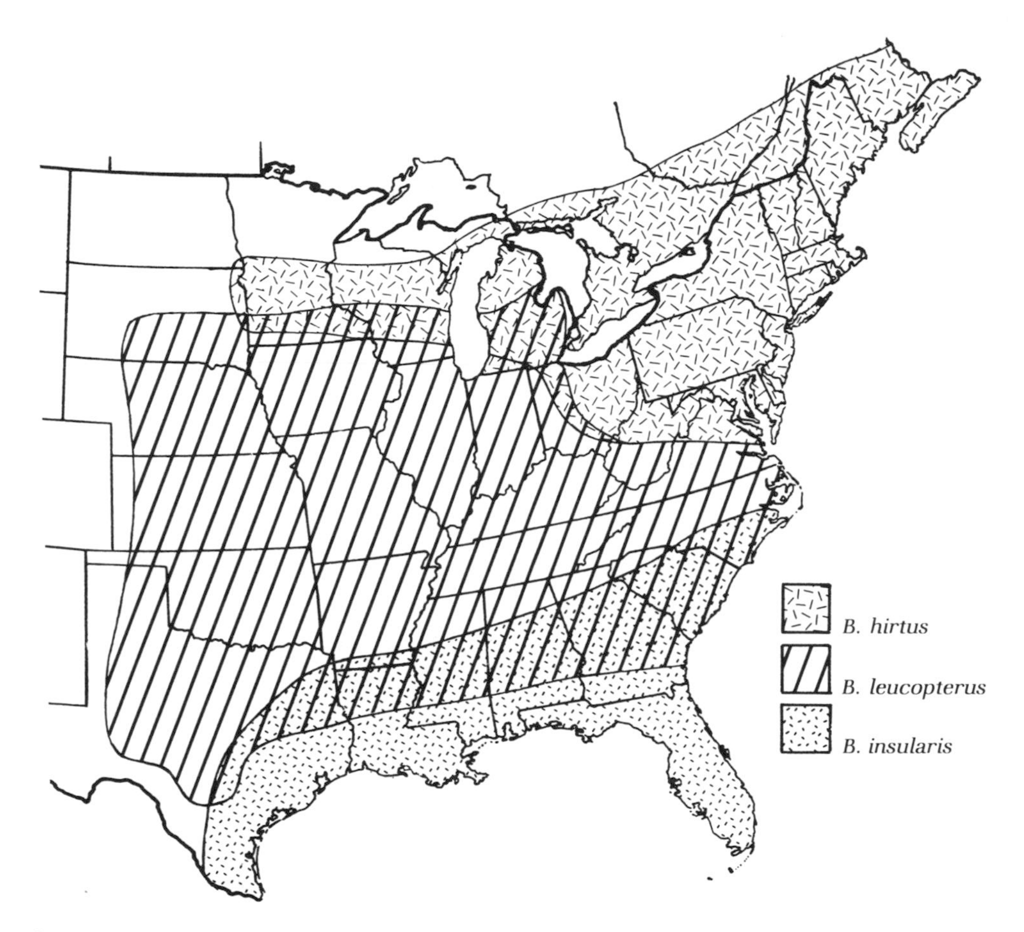

Figure 13. Distribution of species and subspecies of the *leucopterus* complex in the United States and Canada. (Adapted from Leonard 1966, fig. 2, courtesy of the Connecticut Agricultural Experiment Station, New Haven.)

United States in and east of Ontario. In Ontario the first HCB damage to lawns was reported in 1971. Damage has since occurred in many locations (Leonard 1966, Mailloux and Streu 1981, Maxwell and McLeod 1936, Sears 1978).

Host Plants and Damage

The HCB feeds on most of the cool-season turfgrasses, including red fescue, perennial ryegrass, bentgrass, and Kentucky bluegrass, as well as on zoysiagrass, which is considered to be a warm-season grass. Creeping bentgrass maintained at lawn height is most susceptible to injury because of the thick mat of stolons, but at putting-green height, with the stolons eliminated, it is relatively immune from damage (Baker et al. 1981, Maxwell and McLeod 1936).

Chinch bugs have piercing-sucking mouthparts and suck the sap from the crown and stems of grasses. They tend to aggregate. Their feeding results in localized turfgrass injury—as yellow grass that soon turns brown. Coalescence of localized injured areas can cause large patches of dead or dying grass. Most susceptible to damage are lawns on sandy locations in direct sunlight, where moisture deficiency leaves the grass less tolerant to injury (Maxwell and McLeod 1936, Niemczyk 1980a).

Description of Stages

Much of the description of chinch bugs comes from accounts of *B. leucopterus leucopterus* and is considered to describe the HCB reliably. When chinch bugs are crushed, they emit an odor resmbling that of stink bugs.

Adult. HCB adults are black with shiny white wings. They are slightly more than 1.0 mm in width and from slightly more than 3.0 mm to nearly 3.6 mm in length (Plate 9). The four segmented antennae are mostly black but have reddish proximal segments. The legs vary from red to yellowish red. The most conspicuous feature of adults is the pattern of the forewings, which are folded over the body. Near the middle of the costal margin of each hemelytron (front wing), there is a black spot with a somewhat Y-shaped black line extending diagonally toward the head. The black spot and line separate the front wings into a large median posterior white area and two smaller laterally located white areas (French 1964, Leonard 1968). Chinch bugs can occur as macropterous or brachypterous adults (Plate 9). In large HCB populations, brachyptery dominated by about 64% (Leonard 1966).

The male chinch bug is slightly smaller and less robust than the female. The most conspicuous differences are evident when individuals are viewed ventrally (Plate 9). The abdomen of the male is rounded in cross section, while that of the female forms an inverted V produced by a distinct median ridge that is part of the ovipositor (Luginbill 1922).

In general there were more females than males in macropters, but there was an excess of males in brachypters in a 1974 New Jersey field population. The largest difference in macropters occurred in spring adults, which showed a male-to-female ratio of 1:1.5. In brachypters the largest difference occurred in summer adults, which had a male-to-female ratio of 1.4:1. The sex ratio of the overall population was 1.1:1 male to female (Mailloux and Streu 1981).

Egg. Eggs are elongate, ovate, slightly reniform, rounded at the anterior end, and truncated at the other (Plate 9). The truncated end has three to six, usually four, micropyles 0.1 mm in length. Eggs average 0.31 mm by 0.86 mm in width and length. Freshly deposited eggs are whitish and turn yellow in a few days, becoming deep red several days before hatching. Just before hatching, the embryo is visible through the smooth, shiny, somewhat iridescent chorion (Chambliss 1895, Chobam and Gupta 1972, Luginbill 1922).

Nymph. Chinch bugs have five nymphal instars (Plate 10). They grow from a width and length of 0.23 mm by 0.90 mm in first instars to 0.96 mm by 2.97 mm in fifth instars (Luginbill 1922).

Morphological differences distinguish the nymphal instars. In the first the head width is greater than or subequal to the thoracic width. In the second the head is narrower than the thoracic width. In the third, mesothoracic wing pads appear. In the fourth the wing pads extend over the abdomen no farther than the posterior area of the first abdominal segment, which is white. In the fifth and final nymphal instar, the wing pads become prominent and conspicuous and extend at least onto the second abdominal segment, which is white, and sometimes onto the third abdominal segment, which is brown. There is also a distinct color variation as nymphs grow. The first and second instars are bright red, with a distinct white band

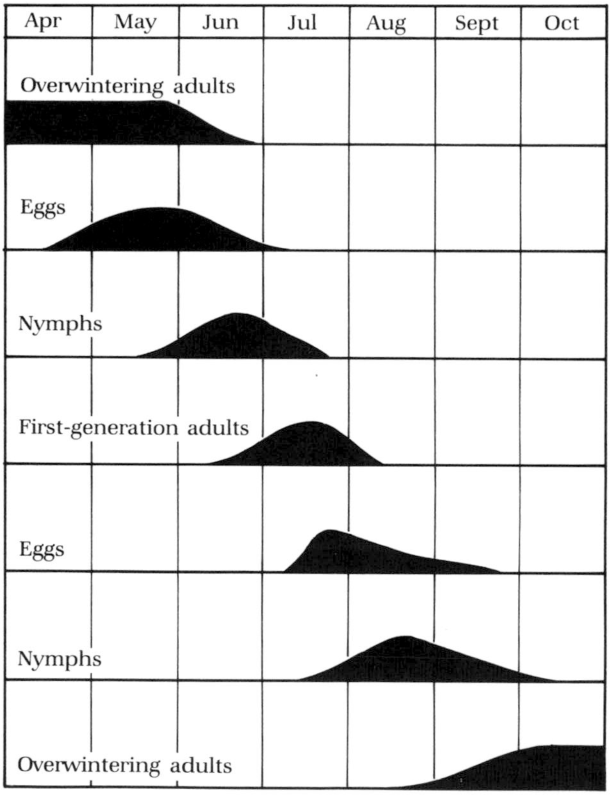

Figure 14. Seasonal history of the hairy chinch bug that infested turfgrass on Long Island, New York, during 1934–1935. (Adapted from Maxwell and McLeod 1936, fig. 2, courtesy of the Entomological Society of America.)

Table 5. Hairy chinch bug, mean numbers per 0.1 m^2 (per ft^2) in turf in Wooster, Ohio, in 1978

Month	Chinch bug stages[a]					
	Adult	1st	2d	3d and 4th	5th	Total nymphs
April	3	0	0	0	0	0
May	2	0	0	0	0	0
June	4	63	30	2	<1	95
July	9	23	31	27	30	111
August	40	14	23	11	19	81
September	19	7	11	10	17	45
Ocotber	26	<1	<1	<1	2	3
November	4	0	0	<1	<1	1

[a]Based on 12 samples of turf 10.8 cm in diameter × 7.6 cm deep (4.25 in. in diameter × 3.00 in. deep, using a standard golf cup cutter), taken on a weekly basis.

Source: Compiled and adapted from Niemczyk 1982.

on the anterior two abdominal segments. The red changes to orange in the third instar, then orange brown in the fourth instar, then blackish in the fifth instar (Luginbill 1922, Mailloux and Streu 1981, Niemczyk 1981).

Seasonal History and Habits

Seasonal Cycle. The HCB has two generations a year in southern New England, in the mid-Atlantic states, including New Jersey and Long Island, and westward through Ohio (Figure 14). It has only a single generation in upstate New York around Rochester and in southern Ontario, Canada. Eggs from overwintering females were present during May and June and from summer females from mid-July through August and into September on Long Island during 1934–1935. In New Jersey, overwintering females lay eggs from the 3d week in April to the end of May, and summer females do so from the 3d week in July to the end of August, precisely the same period as that recorded on Long Island (Liu and McEwen 1979, Mailloux and Streu 1981, Maxwell and McLeod 1936, Niemczyk 1981).

The predominance of nymphs during the second half of June and the second half of August in Ohio (Table 5) corresponds to the pattern seen on Long Island. Second-generation nymphs complete their development by early October, becoming overwintering adults that seek hibernation quarters (Mailloux and Streu 1981, Maxwell and McLeod 1936).

Adult Activity. Adults seek hibernation quarters in late summer and fall. Common sites include infested turf with sufficient undamaged grass to supply shelter and food before hibernation. Thatch or tall grass near the edge of lawns and putting greens is sought. Other overwintering sites include plant debris and space around the foundations of houses, under shingles and clapboards (Leonard 1966, Maxwell and McLeod 1936).

Overwintering adults become active and leave their hibernation sites when a threshold temperature of 7°C (45°F) is reached. This threshold temperature was evident in New Jersey, where there are two generations a year, and in southern Ontario, Canada, where there is one. The HCB migrates primarily by crawling, but the presence of specimens on clothes on a clothesline indicates at least a limited flight capability. Individuals immediately start to feed and copulate. The HCB mates repeatedly at 5- to 8-day intervals. Copulation lasts 12 hr or more, with the female alone being active, dragging the male behind her. Females start to lay eggs after a preoviposition period of about 2 weeks (Leonard 1966, Liu and McEwen 1979, Luginbill 1922, Mailloux and Streu 1981).

Eggs are deposited in leaf sheaths and in the ground on roots of host plants. Estimates of egg production vary considerably, ranging from an average of 1.5 eggs per day, for a total of 54 eggs, to 20 eggs per day for 2–3 weeks. During observation in New Jersey in spring 1974, overwintering females averaged 15.6 eggs and summer adults 6.9 eggs per female (Kennedy 1981, Mailloux and Streu 1981).

During 1974, a peak population of 196 eggs per 0.1 m^2 (1 ft^2) was found during the period from May 6 to June 1, when two common legumes associated with turf were in early bloom. These were white clover, *Trifolium repens* L., and bird's-foot trefoil, *Lotus corniculatus* L. Field-collected females are reported to lay as many as 170 eggs, with a female longevity of about 100 days (Baker et al. 1981, Mailloux and Streu 1981).

Nymphal Activity. During early spring, eggs may not hatch for a month or more, but during midsummer, eggs hatch in as short a time as 7–10 days. Immediately upon hatching, young nymphs pierce the turfgrass stems and leaves and feed on the sap of plants. Development through the five nymphal instars requires about 4–6 weeks during the summer. Nymphs may seek hibernation quarters but will die unless they transform to adults (Kennedy 1981, Leonard 1966).

Miscellaneous Features

Degree-Day Developmental Relationship. Degree-day accumulations and relationship to HCB development have yielded useful information for management of HCB infestations. In Ontario, Canada, there is one generation per season. When 7°C (45°F) was used as the base temperature for development, third-instar nymphs peaked at 750–900 degree-day accumulations (in degrees Celsius) in the thatch. This occurred during July 4–8, 1977. Sampling for treatment/nontreatment decisions was started at 897 degree-day accumulations and terminated at 950 degree days. The peak third-instar period was chosen for the decision because at that time most eggs have hatched but no adults have developed to disperse out of the area. Populations of third-instar nymphs per 225 cm^2 (36 $in.^2$) of turf were determined. Fewer than 20 nymphs per sample required no treatment, 20–30 nymphs per sample required a decision to treat or not, and >30 nymphs required treatment. When no damage is yet evident, a single insecticidal treatment during the third-instar period provides maximum control (Liu and McEwen 1979).

By comparison, in New Jersey, where two generations occur per season, an air temperature of 14.6°C (58.3°F) was used as the base temperature for egg development. Degree-day accumulations (in degrees Celsius) for development of the two generations of the HCB in relation to calendar periods were determined. The egg hatch of the first generation was completed at 115 degree-days during early June, and the peak adult presence occurred at about 630 degree-days in late July. The second-generation egg hatch was completed near 850 degree-days after mid-August, and peak adult presence occurred at above 1,159 degree-days during mid-October (Mailloux and Streu 1981). These determinations would be very useful in estimating the develpoment of various stages of the HCB in turf without the need for laborious and continuous field sampling.

Mortality Factors. Studies made in New Jersey during 1974 and 1975 revealed that a high incidence of mortality occurs in HCB eggs—59% and 48% in the spring and summer generations, respectively. Six mortality factors with regard to the overall population were identified as follows: (1) infection by *Beauveria bassiana* (Bals.) Vuillemin, (2) parasitism by *Eumicrosoma benefica* Gahan, (3) predation by *Amara* sp., (4) desiccation, (5) failure to hatch, and (6) wet conditions at eclosion (Mailloux and Streu 1981). I discuss the first three more thoroughly in connection with natural enemies.

The winter mortality of adults can be significant; it was 68% and 28% in 1974 and 1975, respectively. Winter mortality is considered to relate to the presence of moisture in the overwintering sites. Lack of snow cover increases mortality. Generally speaking, the survival of chinch bugs is promoted by high humidity at low temperatures and by low humidity at high temperatures (Guthrie and Decker 1954, Mailloux and Streu 1981).

Laboratory Rearing. HCB rearing has been accomplished on a year-round basis on sections of young corn (*Zea mays*) treated with 2% sodium hypochlorite. Egg survival was greatest following treatment with the same sterilant. The rearing environment had a constant temperature of 26°C (79°F), with 40%–75% RH and a 16-hr photoperiod. The mean preoviposition period was about 11 days, with about 80% of the females ovipositing between day 5 and day 14. Under these conditions, the nymphal periods averaged 12.3, 5.4, 5.2, 4.9, and 7.1 days from the first through the fifth instars, respectively. Diapause in the HCB can be broken by 10–14 days of continuous exposure to a temperature of 29.5°C (85°F) in the absence of light (Baker et al. 1981, Leonard 1966).

Population Management. The commonly accepted period for insecticidal control of the HCB is during the summer, when chinch bugs are actively feeding. Tests conducted during 1978 and 1979 in Ohio support the theory that the HCB can be controlled by an early spring application of an insecticide to eliminate overwintering adults. The life history chart (Figure 15) illustrates the periods when the two generations of adults and nymphs are in their greatest numbers and shows why elimination of a small population of adults in early spring may destroy the potential for severe damage (Niemczyk 1982).

Sampling Techniques. Several sampling techniques have been employed to determine populations in turfgrass. The most common involves a metal flotation cylinder that is open at both ends, is driven into the soil, and is filled with water. Nymphal and adult chinch bugs present float to the surface within 10 min and can be counted (Streu and Vasvary 1966).

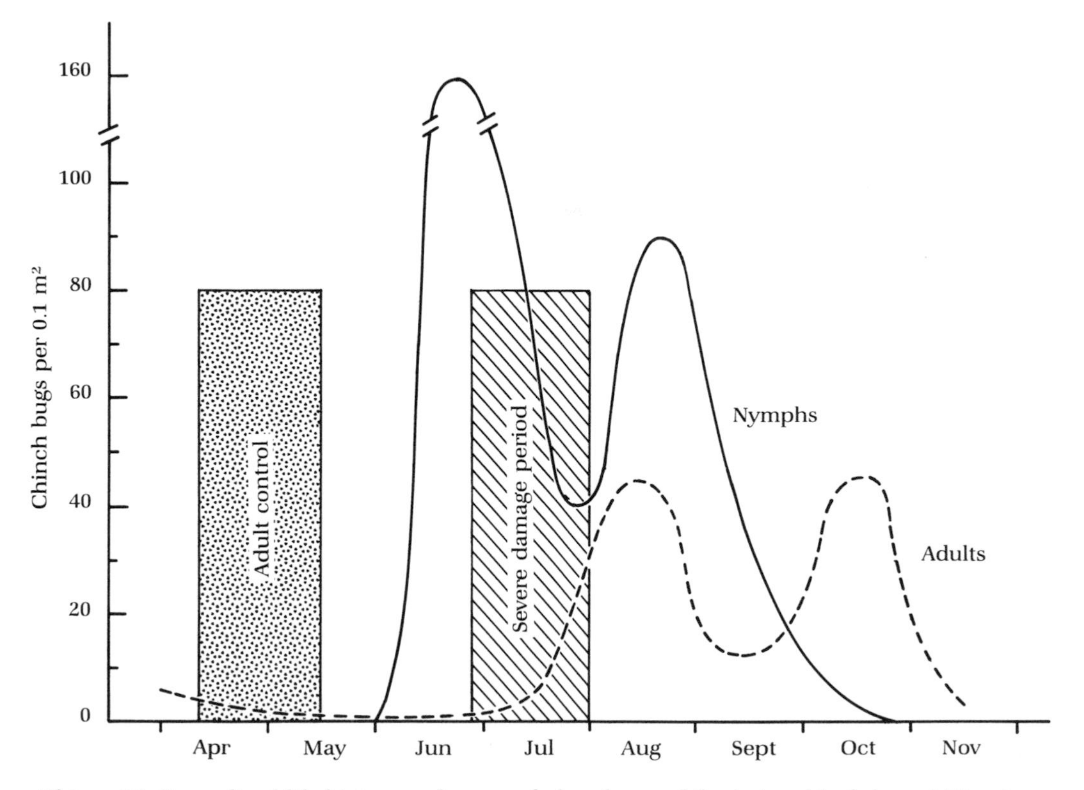

Figure 15. Generalized life history and seasonal abundance of the hairy chinch bug at Wooster, Ohio, showing the time of severe damage and the period when adult control programs may be most effective. (Adapted from Niemczyk 1982, fig. 3, courtesy of the ChemLawn Services Corporation.)

A modification of this technique, using water pressure to remove thatch, made it possible to collect eggs as well. Centrifugation separated the chinch bug stages from thatch. A 90%–100% recovery of planted eggs demonstrated the efficiency of this field-sampling technique (Mailloux and Streu 1979). A turfgrass area can also be flooded and then covered with a piece of white cloth. If chinch bugs are present, they will crawl to the underside of the cloth (Schread 1970a). In a third sampling technique, soil plugs 10.8 cm in diameter and 7.6 cm deep (4.25 in. × 3.00 in.), obtained with a standard golf cup cutter, are placed grass side down in a Berlese funnel fitted with a 25-watt lamp. During a 16-hr period, chinch bugs will be forced downward by the heat and may be collected in a jar of 70% ethyl alcohol (Niemcyzk 1982).

Plant Resistance. Laboratory and field evaluations for selection of HCB-resistant strains indicated preferential response of several cool-season grasses. In Kentucky bluegrass, 'Baron' and 'Newport' were more tolerant to feeding injury, as compared with 'Adelphi', which was the most susceptible. In perennial ryegrasses, 'Score', 'Pennfine', and 'Manhattan' had significantly lower infestations than 12 other cultivars in field plots during 2 years of observation (1979–1980). In fine-leaf fescues, 'Jamestown' had the heaviest HCB populations of 11 cultivars in field plots (Ratcliffe 1982).

Natural Enemies

Microorganisms. Warm, moist weather is necessary before the entomophagus fungus *Beauveria bassiana* becomes active, killing adults and nymphs (Plate 10). Infection can occur at any time from May to October. During 1973 and 1974, maximum incidence of infection ranged from 80% to 90%, with minimum incidence of about 20%. Early stages of infection are not externally detectable, but when the atmosphere is very moist the insects become covered with white mycelium that later sporulates on the surface of the dead insect (Mailloux and Streu 1981).

Insect Parasite. *Eumicrosoma benefica* Gahan (Hymenoptera: Scelionidae) is a recently discovered egg parasite of the HCB in New Jersey. Adults were present throughout the year. Parasitized HCB eggs can easily be distinguished by the parasite pupa, which is visible through the host chorion (Mailloux and Streu 1981).

Insect Predators. Some eight species of arthropods in the turfgrass fauna were found to feed on early stages of the HCB when they were taken to the laboratory (Plate 10). A predatory mite and the lygaeid *Geocoris bullatus* (Say) are important predators. In Connecticut, *G. uliginosus* (Say) was found in association with *G. bullatus.* Adults of the latter fed on bentgrass, bluegrass, and ryegrass but did little damage. *Amara* sp., the most common adult carabid of the HCB habitat, was the only egg predator detected (Dunbar 1971, Mailloux and Streu 1981).

Southern Chinch Bug

Importance

The southern chinch bug, SCB, was formerly known as the lawn chinch bug. Its present common name was adopted when it was designated a distinct species (Stringfellow 1969). The SCB is the most injurious pest of St. Augustinegrass turf in Florida. More than 368,700 ha (911,000 acres) of St. Augustinegrass grow in Florida, or about 37% of the total turfgrass area. The money spent on the SCB exceeds that for all other insect pests in Florida with the possible exception of the citrus rust mite. More than $25 million go annually to control this pest in Florida lawns, with as many as six insecticidal applications to a given lawn (Kerr 1966, McGregor 1976, Reinert 1978, Stringfellow 1969, Strobel 1971).

In Louisiana, the SCB is also considered the most injurious pest of St. Augustinegrass, which has been adopted as a lawn grass throughout the state. In Texas, the SCB is the most injurious turfgrass pest in the entire St. Augustinegrass area (Hamman 1969, McGregor 1976, Oliver and Komblas 1981).

Insecticidal Resistance. The SCB's continuous development of resistance to insecticides greatly increases the cost of control and makes necessary the constant evaluation of new compounds. With the need for as many as six applications of insecticide to manage 7–10 generations of SCB infestations each year, the species' deveopment of resistance to the chemicals is not surprising. Resistance developed to chlordane and DDT after 8–12 years, to parathion after 7 years, to diazinon after about 20 years, and to chlorpyrifos (Dursban) after 11 years. Propoxur, a carbamate, has been used for more than 15 years without an indication of resistance (Reinert 1982b).

In spite of resistance, chlorpyrifos is still the standard chemical widely used by the lawn spray industry in Florida. SCB populations vary, showing no resistance in some localities and as much as 3.19×10^8-fold resistance in others, as measured by LC_{50} values. The greatest problem areas are on the east coast of Florida, around Fort Lauderdale (Reinert and Portier 1983).

History and Distribution

The SCB is distributed wherever suitable habitats exist, from southern North Carolina southward throughout most of South Carolina, Georgia, and all of Florida. Westward it is present in most of Alabama and Mississippi, in all of Louisiana, and over the southeastern third of Texas (Figure 13). In Florida, Fort Lauderdale is considered the center of SCB activity. In Texas, damaging populations have been present since the early 1950s, and populations considerably increased during the early 1970s. The St. Augustinegrass–growing areas of Texas extend south from San Antonio to the lower Rio Grande Valley and eastward to Louisiana (Hamman 1969, Stringfellow 1969).

Host Plants and Damage

St. Augustinegrass is by far the foremost host plant of the SCB. It also feeds to some extent on centipedegrass, zoysiagrass, bahiagrass, and bermudagrass but only where St. Augustinegrass may be growing in close association. Because little damage occurs on these grasses, the SCB is a serious threat only to St. Augustinegrass (French 1964, Kerr 1966).

SCB damages the grass by sucking the sap from the nodes and basal parts of the plant, causing the grass to become dwarfed, to yellow, and eventually to die. Both nymphs and adults are destructive to the grass. Damage occurs any time from May to November but is most evident during dry conditions, when even 25–30 insects per 0.1 m^2 (1 ft^2) can cause severe damage. Many of the nymphs remain hidden for as much as 1.5 weeks, feeding where the grass blades come together at the nodes (French 1964, Kerr 1966, Oliver and Komblas 1981, Stringfellow 1969).

Since the SCB aggregates in scattered patches rather than being evenly distributed, small spots of damaged grass initially become noticeable. The aggregated colonies do not move to new areas until the infested patch has completely been killed, so that there is ample time to apply controls before large patches of grass are dead (Kerr 1966).

Influence of the Turfgrass Environment. The condition of the host plant has a marked effect on injury. A heavily fertilized turfgrass with lush growth suffers the greatest injury from chinch bugs. Populations develop more rapidly and cause injury more quickly on heavily fertilized grass than on grass that receives moderate amounts of nitrogen (Kerr 1966).

Weather and the thickness of the thatch appear important in affecting chinch bug development. In areas of Florida where frost rarely occurs, the grass grows continuously, so that there is a thick, spongy thatch. It is commonly 10–15 cm (4–6 in.) deep and may be as much as 30 cm (12 in.) deep. Such thatch provides an ideal habitat for chinch bugs (Reinert and Kerr 1973).

Moisture has a marked paradoxical effect on the chinch bug. While an abundance of moisture makes the grass lush and susceptible to greater damage from chinch bugs, the moisture also has an effect in suppressing chinch bug populations (Kerr 1966).

Description of Stages

The description of adults, eggs, and nymphs of the SCB is identical to that of the hairy chinch bug and is taken, for the most part, from that of the chinch bug, *B. leucopterus leucopterus*.

Brachypterous adults dominate, but macropters are also found. Macropters can fly, but there is no spring dispersal flight like that of *B. leucopterus leucopterus* (Reinert and Kerr 1973).

Seasonal History and Habits

Seasonal Cycle. In southern Florida, six to seven generations occur each year. During the winter, adults compose 80%–90% of the population, followed by nymphs, with eggs also present in smaller numbers. In northern Florida and Louisiana, there may be three and four generations, with only adults present during the winter. Adults sometimes hibernate (Reinert and Kerr 1973).

The first large surge of first-instar nymphs occurs during February in southern Florida and in late March to early April in northern Florida. In sunny open areas of lawns, aggregations of 500–1,000 per 0.1 m^2 (1 ft^2) are common, and as many as 2,300 per 0.1 m^2 have been found (Reinert and Kerr 1973).

During warm weather, a generation passes from the egg stage to the egg-laying stage in 5–8 weeks in southern Florida and in 7.5–8.0 weeks in northern Florida. In the laboratory at constant temperatures, a generation is completed in 6 weeks at 28°C (83°F) and in 17 weeks at 21°C (70°F; Reinert and Kerr 1973).

In Louisiana, where four generations per year are considered normal, first-generation eggs are deposited in early April, second-generation eggs in early June, third-generation eggs in August, and fourth-generation eggs during August into September (Oliver and Komblas 1981).

Adult Activity. Migration flights do occur but are of minor importance. The SCB migrates short distances mainly by walking. Literally streams of bugs can be seen moving from heavily infested areas, where they may walk several hundred feet in a half hour. They are numerous in grass on sandy soil but not on muck. Adults tend to be aggregated more than evenly dispersed. They are vertically distributed through the turf thatch and into the upper, largely organic layer of soil and in cracks in the soil. When populations are large, and on hot days, adults may be seen running over St. Augustinegrass blades, but they are not feeding or resting on the blades (Crocker and Simpson 1981, Kerr 1966).

Members of the genus *Blissus* have a definite courtship behavior. Males and females approach each other and make first contact with their antennae. Once they have paired, the bugs characteristically face in opposite directions, with the female the more active, walking about and sometimes feeding. Copulation, which occurs first in spring after warm weather has prompted activity, lasts as much as 2 hours (Leonard 1966).

Females begin depositing eggs 7–10 days after mating. Eggs are inserted into the crevices of grass nodes and at the junction of blades and stems, either singly or in groups of two or three (Kerr 1966). Adults live for about 70 days, the females depositing a few eggs a day over several weeks for a total of 100–300 eggs before they die. Since chinch bugs mated repeatedly in a cage, they are assumed to do so also in the field. The sex ratio is considered to be about 1:1 (Kerr 1966).

Nymphal Activity. Eggs hatch in about 2 weeks under normal conditions, but development and hatching varies with temperature. In Louisiana, where there are four generations from April into September, incubation periods average about 29 days for the first generation and 14–15 days for the remaining three because of warmer weather (Oliver and Komblas 1981).

After hatching, first-stage nymphs feed largely on tender basal growth and nodes of runners. Thick, spongy thatch that is 10–15 cm (4–6 in.) thick is an ideal habitat for SCB development. Nymphs require about 30 days to transform through five instars and reach maturity with a total life cycle of 7–8 weeks in Louisiana. First instars found during the winter indicate oviposition during warm winter days (Oliver and Komblas 1981, Reinert and Kerr 1973).

Miscellaneous Features

Threshold Populations. Efficiency in applications of insecticide has been increased by monitoring populations in order to identify an economic threshold. Treatment only when populations exceeded 22–28 bugs per 0.1 m^2 (1 ft^2) reduced effective pesticide applications by 90% (Reinert 1982c).

Plant Resistance. Work by Reinert et al. (1980) indicates that the most practical method of managing the SCB results from the development of insect-resistant cultivars of St. Augustinegrass. In laboratory feeding tests, 'Floratam' and two accessions, FA-108 and TX-33, produced 66%–80% adult mortality, compared with only 11% mortality feeding on 'Florida Common'. Also, significantly fewer eggs were produced by feeding on the first three grasses than on 'Florida Common'.

Natural Enemies

Reinert (1978) made a comprehensive study of the natural enemies of the SCB during 1971 to 1977. Parasites and predators associated with or observed feeding on various stages of the SCB in the field were collected, were taken to the laboratory for observation, and were confined with life stages of the SCB on stolons of St. Augustinegrass. Those found to be natural enemies of the SCB are listed in Table 6.

Microorganisms. The fungus *Beauveria bassiana,* pathogenic on all life stages of the SCB, produced epizootics only when high populations and high moisture levels were present (Reinert 1978).

Insect Parasites. The only parasite of the SCB observed in Florida is an egg parasite, *Eumicrosoma benefica* (Reinert 1972). The wasp, on finding an egg, examines the entire egg, first hurriedly, then carefully, by tapping it with her antennae. If it is found suitable, she thrusts her ovipositor into the host egg and deposits her own inside. The developing parasite usually consumes the entire content of the host egg before pupation. Unlike a normal SCB egg, which turns reddish when it is about 3 days old, a parasitized egg remains tan until the parasite pupates, then becomes blackish, with the parasite pupa clearly visible. This parasite has been observed throughout the year in southern Florida. An average population of about 35 wasps

Table 6. Natural enemy complex of the southern chinch bug and the life stages attacked by each organism

Organism	Eggs	Nymphs					Adults
		1	2	3	4	5	
Microorganism							
Moniliales							
Moniliaceae							
Beauveria bassiana (Bals.) Vuillemin	X	X	X	X	X	X	X
Parasite							
Hymenoptera							
Scelionidae							
Eumicrosoma benefica Gahan	X						
Predator							
Hemiptera							
Lygaeidae							
Geocoris uliginosus (Say)	X	X	X	X	X	X	X
Geocoris bullatus (Say)	X	X	X	X	X	X	X
Nabidae							
Pagasa pallipes Stal.		X	X	X	X	X	X
Anthocoridae							
Xylocoris vicarius (Reuter)	X	X	X	X	X		
Lasiochilus pallidulus Reuter	X	X	X	X	X		
Reduviidae							
Sinea sp.		X	X	X	X	X	X
Dermaptera							
Labiduridae							
Labidura riparia Pallas		X	X	X	X	X	X
Hymenoptera							
Formicidae							
Primarily *Solenopsis geminata* (F.)		X	X	X	X	X	X
Araneida							
Lycsidae							
Lycosa sp.		X	X	X	X	X	X

Source: From Reinert 1978, table 1, courtesy of the Entomological Society of America.

was associated with about 90 SCB per 0.1 m^2 (1 ft^2) of St. Augustinegrass (Reinert 1972).

Insect Predators. The most numerous and most frequently encountered predator in turf in Florida is *Geocoris uliginosus,* one of two big-eyed bugs. In size, shape, and color, it superficially resembles its host, for which it is often mistaken. Big-eyed bugs are more robust, with large eyes that protrude at the sides of the head, which is the widest part of the body. They move more rapidly among the grass blades and stolons than chinch bugs. Densities as high as 17 per 0.1 m^2 (1 ft^2) may be present. In the laboratory each predator fed on an average of 9.6 nymphs per 5-day period. A less frequently observed predator of SCB, *G. bullatus,* did not exceed 2 per 0.1 m^2 (1 ft^2) and was found in thinner grass with frequent bare areas. Both species of *Geocoris* fed on all stages of the SCB, including adults, but only when they were freshly molted (Reinert 1978).

A dermapteran, *Labidura riparia,* was one of the most active predators of the SCB. Although only one or two were present per 0.1 m^2 (1 ft^2), they range over a wide area in search of prey. In the laboratory an adult consumed 50 adult chinch bugs in 24 hr (Reinert 1978).

In a 1971 observation in Fort Lauderdale, Florida, where an area damaged by the SCB was not spreading at the typically rapid rate during the summer, the three most numerous biotic agents present were the egg parasite *E. benefica* and predators *G. uliginosus, Xylocoris vicarius,* and *Lasiochilus pallidulus.* The latter two are predators on eggs and on early-instar nymphs. The combined parasite-predator complex is considered to have been responsible for the relatively low density of the SCB during the summer and to have caused the population to collapse during August (Reinert 1982b).

In light of many observations of this type, parasites and predators are now considered to prevent rapid population increases where insecticides are not applied. Conversely, lawns receiving repeated insecticidal applications continue to have chinch bug problems. In yards where SCB populations were monitored and were treated only when economic thresholds of slightly more than 20 bugs per 0.1 m^2 (1 ft^2) had been exceeded, pesticide applications were reduced by 90% (Reinert 1978, 1982c).

6

Hemipteran Pests: Suborder Homoptera

Greenbug

Taxonomy

Greenbug, GB, *Schizaphis graminum* (Rondani), family Aphididae, was placed in the genus *Toxoptera,* but since the mid-1960s it has belonged to the genus *Schizaphis.* Its common name was the spring grain aphis, or greenbug (Webster and Phillips 1912), but it is currently called only *greenbug.*

Much of the description and bionomics of the GB necessarily comes from literature that treats this insect as a small-grain pest. Nevertheless, this information is largely pertinent to the GB as a turfgrass pest.

Importance

Kentucky bluegrass has long been known as a host of GB in the United States and Canada, and occasionally this aphid has caused minor damage to lawns and pastures of this grass species. Not until 1970 was GB reported in epidemic levels on Kentucky bluegrass lawns in many midwestern states. Since then, certain lawns, particularly in Ohio, have been devastated for at least 4 successive years (Garman 1926, Niemczyk 1980b, Webster and Phillips 1912).

The importance of the GB as a serious pest of turfgrass has been accentuated by its development of resistance to some of the common insecticides used on turfgrasses (chlorpyrifos, diazinon, and malathion). A level of resistance to chlorpyrifos as great as ninefold has been recorded (Niemczyk and Moser 1982, Potter 1982a).

Few insects have had such a long history of devastation to American agriculture. Since the turn of the century, it has caused an estimated annual loss of at least 50 million bushels of oats and wheat in the three states of Kansas, Oklahoma, and Texas. Because of its general distribution and its prolificacy, the GB causes a loss of 1% to possibly 3% of the annual wheat crop of the entire world (Metcalf et al. 1962).

History and Distribution

The GB is widely distributed in North and South America and in Europe, Africa, and Asia. It was first recorded not as a pest of small grain but, because of its abundance, as a nuisance to humans in Italy in 1847 (Street et al. 1978, Webster and Phillips 1912).

The first serious outbreak of the GB in the United States occurred in Virginia in 1882, when it was a pest on oats. As early as 1907, GB damage to bluegrass lawns was observed in Washington, D.C. By 1912 the insect was known to occur throughout the wheat-growing areas of the United States and southern Canada (Figure 16). The New England states are about the only area of the country where it is nearly absent (Hunter 1909, Webster and Phillips 1912).

During the 1970s, GB damage to Kentucky bluegrass was reported from Illinois, Indiana, Kansas, Missouri, Ohio, and Wisconsin. In 1981, Kentucky bluegrass and fine fescue lawns were damaged in the additional states of Iowa and Minnesota (Figure 16; Niemczyk 1980c, Niemczyk and Power 1982, Street et al. 1978).

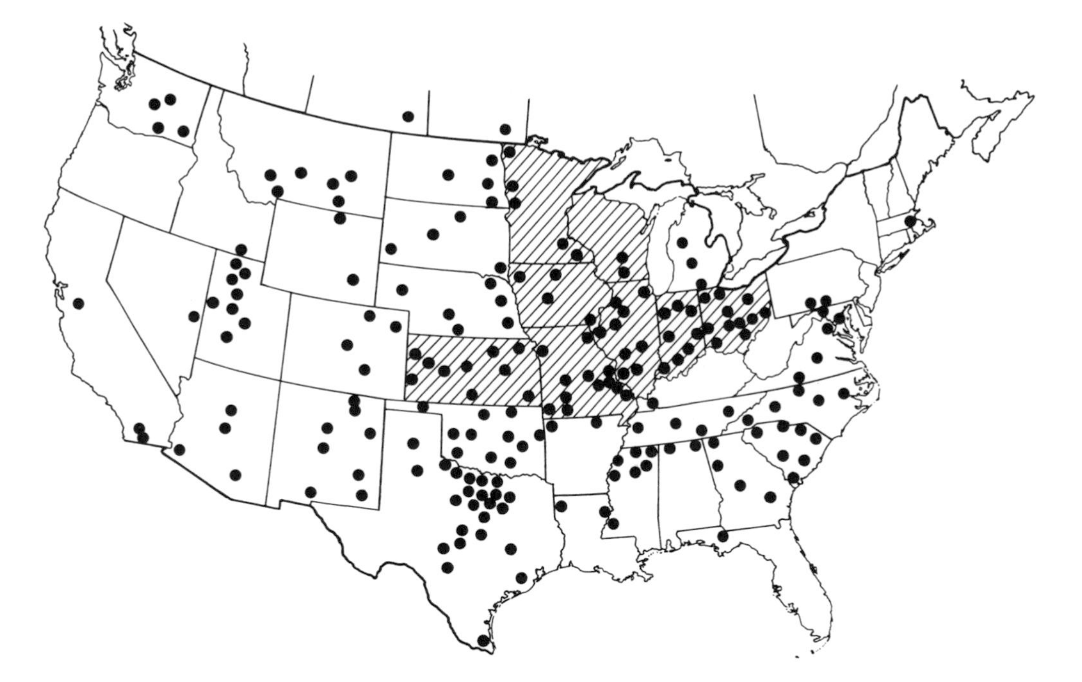

Figure 16. Known distribution of the greenbug to 1912 and states (striped) in which turfgrass damage has occurred since 1970. (Data from Street et al. 1978 and Webster and Phillips 1912, fig. 2; drawn by R. McMillen-Sticht, NYSAES.)

Host Plants and Damage

The GB has a wide range of host plants in the family Graminae. At least 60 species of grasses are listed as hosts. It feeds only rarely on plants outside this family. Turfgrasses that serve as hosts include Kentucky bluegrass, Canada bluegrass, annual bluegrass, fescues, and perennial ryegrass. Reproduction has been observed on Kentucky bluegrass, chewings fescue, and tall fescue. Small grains serving as the dominant host plants include wheat; oats; rye; rice, *Oryza sativa;* and sorghum, *Sorghum vulgare* (Potter 1982a, Webster and Phillips 1912).

Damage is caused when the aphid pierces grass blades with its needlelike mouthparts and feeds on the phloem tissue (Plate 11). Feeding alone weakens the plants, but the GB also injects its toxic salivary secretions, causing yellow spots with necrotic centers, followed by death of the tissues surrounding the feeding site, which turns burnt orange. The translocation of the salivary toxin weakens the entire plant, including the root system, as the enzymes break down the plant cells (Niemczyk and Moser 1982).

On home lawns, damage usually begins in shaded areas under trees, taking the form of circular to irregular brown patches as much as 4.6 m (15 ft) in diameter (Plate 11). A narrow band of yellow to burnt orange grass exists just beyond the areas most densely settled by the aphid in live grass at the edge of the damaged turf. From a distance, damaged turf typically has a burnt orange cast. Injury can also begin in the open. In Kentucky, injury begins to appear in May and early June and continues until fall. In 1981 the most severe damage occurred in mid-November. Some 10–50 aphids may line the midrib on the upper surface of a single grass blade. Populations of 4,000 aphids per 0.1 m^2 (1 ft^2) are not uncommon (Niemczyk and Moser 1982, Potter 1982). The GB does not infest trees.

Biotypes

At least four biotypes of GB are recognized on small grain, and good evidence indicates that other biotypes occur on turfgrasses. On small grain, biotypes A and B are recognized by their feeding habits. Biotype A makes intercellular penetration and feeds invariably in phloem tissue. Biotype B penetrates intra- and intercellularly and preferentially feeds in the mesophyll parenchyma of leaves. It causes greater damage than biotype A. Biotype C attacks sorghum. A strain of the GB occurring in Bushland in southeastern Texas, found to be other than biotypes A, B, or C, was designated as biotype E (Porter et al. 1982, Saxena and Chada 1971). Plant resistance studies conducted in recent years in widely scattered locations have in fact demonstrated the probability of new biotypes (Niemczyk and Moser 1982).

The rather suddenly appearing damaging populations of the GB on Kentucky bluegrass lawns suggest the development of a new biotype adapted to this grass. Not only is there evidence that such an adaptation is occurring, but within Kentucky bluegrass–infesting GB alone, at least two biotypes, separable by visual differences, are currently recognized. The normal green dorso-median stripe is present (except in the youngest nymphs) in the Kentucky bluegrass–infesting Maryland biotype (Plate 11). The stripe is absent in the Kentucky bluegrass–infesting Ohio biotype (Plate 11; R. H. Ratcliffe, U.S. Department of Agriculture, Beltsville, Md., personal communication, 1985).

Description of Stages

Adult

GB adults are soft bodied, pear shaped, 1.5–2.5 mm long, and pale yellow to light green, usually with a dark green dorsal median stripe. The tips of the legs, the tips of the cornicles, and the antennae are black (Plate 11).

Adult Forms. At least three distinct forms of females occur: (1) a winged viviparous female, (2) a wingless viviparous female, and (3) a wingless oviparous female (Figure 17). Males occur only as winged insects and can be distinguished from the winged female by the differences in the terminal abdominal segments (Figure 17). The male is also smaller. The wingless oviparous female is the largest of the adults. It can be distinguished from the wingless viviparous female by the evidence of eggs in the abdomen and by the slightly wider hind tibia (Webster and Phillips 1912).

Overall body length for the winged male is 1.3 mm (wingspan 4.5 mm); for the winged female is 1.5–2.0 mm (wingspan 5–7 mm); for the wingless viviparous female is 1.0–1.8 mm; and for the wingless oviparous female is 2.0–2.5 mm (Hunter 1909, Potter 1982a, Webster and Phillips 1912).

Egg

Freshly deposited eggs, cemented to the food plants, are pale yellow and change within a few hours to faint green, with a circular area of darker green at one pole due to the *ovarian yolk.* After a day this region turns dark green, and during the second day the entire egg turns darker green. By the end of the 3d day, the egg has become jet black. Eggs are broadly elliptical, slightly reniform, 0.70–0.78 mm long, and 0.33–0.45 mm wide (Webster and Phillips 1912).

Nymph

There are four nymphal instars, and all resemble the wingless adults in general shape and color. Upon hatching they are green with legs that have black tips.

Seasonal History and Habits

Seasonal Cycle

In seasonal cycle the GB in the southern states differs considerably from that in the North. These two regions can be separated by the 35° latitude, a line that

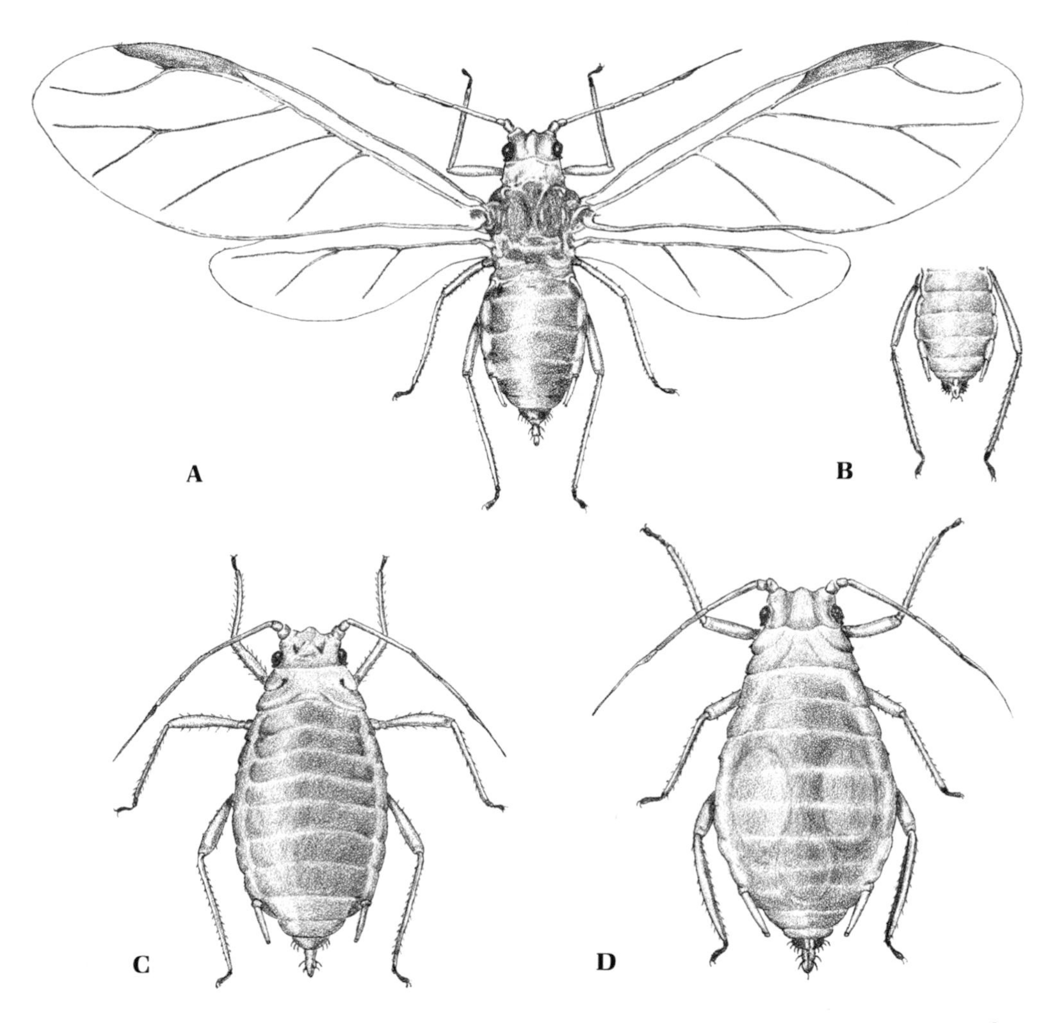

Figure 17. Greenbug forms. **A.** Winged viviparous female. **B.** Abdomen of winged male. **C.** Wingless viviparous female. **D.** Wingless oviparous female showing egg development. (From Webster and Phillips 1912, figs. 1, 3, 4, 5; redrawn by R. McMillen-Sticht, NYSAES.)

extends along the southern edge of North Carolina and Tennessee through the midsection of Arkansas and Oklahoma and the Texas panhandle.

South of Latitude 35°. The life cycle is very simple. Active nymphs and wingless viviparous females reproduce during winter warm spells (Figure 18). Reproduction continues during the warmer months. All nymphs become wingless viviparous females in 7–18 days and begin producing 30–40 living young during an average life span of about 35 days. The hot, dry summer periods are the least favorable, with a food shortage as the grains ripen (Walton 1921, Webster and Phillips 1912).

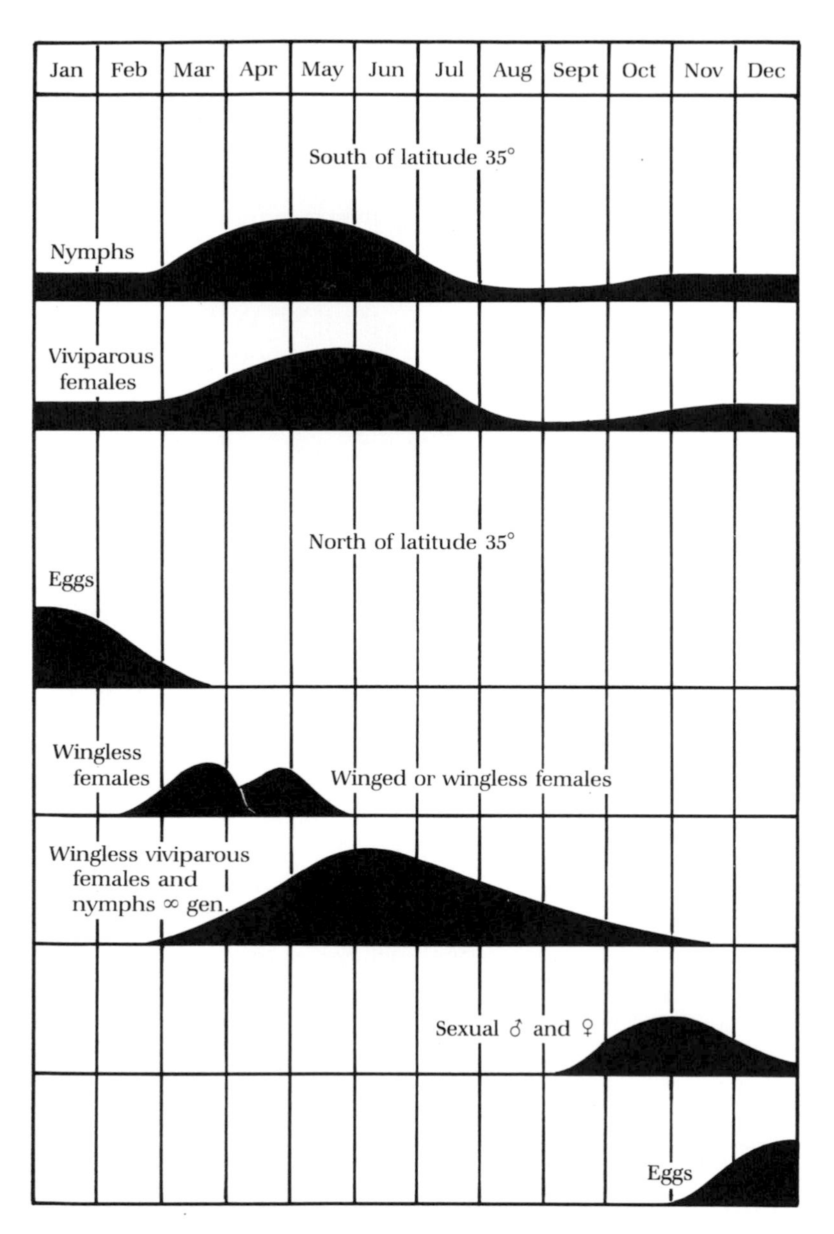

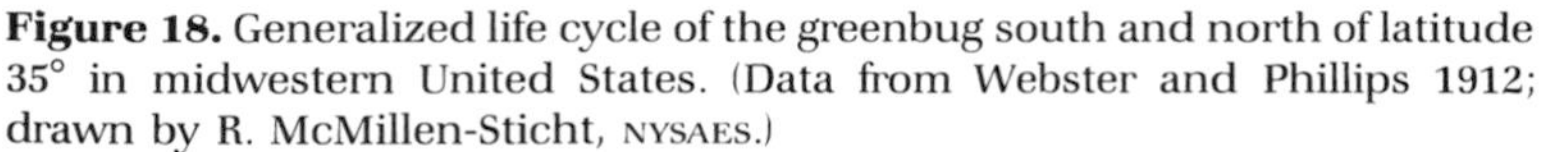

Figure 18. Generalized life cycle of the greenbug south and north of latitude 35° in midwestern United States. (Data from Webster and Phillips 1912; drawn by R. McMillen-Sticht, NYSAES.)

North of Latitude 35°. The life cycle of the GB is much more complex in the northern states (Figure 18). It overwinters primarily as eggs deposited on leaves of host plants. Sexual males and females may also overwinter under the protection of vegetation. Eggs hatch during late winter or early spring, and in 7–18 days nymphs become wingless females, producing living young that may become winged or wingless females. Both types of females give birth to living young that become reproductive in as little as 6–7 days or as long as 7–18 days. A single female may produce from 1–8 young per day for 14–21 days, giving birth to 50–60 young per female. Parthenogenetic reproduction (reproduction without fertilization) of wingless females continues in all seasons, with as many as 20 generations a year in the latitude of central Indiana. The most rapid reproduction occurs at relatively low temperatures in the spring (Potter 1982a, Walton 1921, Webster and Phillips 1912).

During late September and October, winged males and females appear and mate, and females produce eggs that overwinter. From 5 to 14 generations appear each season. All except the last generation are composed entirely of females (Walton 1921, Webster and Phillips 1912).

Infestations of northern lawns were first considered to result from annual migrations from the South. This hypothesis, however, did not explain reinfestations of the same lawns year after year. Examination of previously infested Kentucky bluegrass lawns in Ohio in November 1981 revealed many GB eggs glued to grass blades, debris, and tree leaves. Overwintering eggs in turfgrass plugs collected in March 1982 and held at 24°C (75°F) hatched and thereby confirmed that the GB can overwinter as eggs in home lawns in the northern states (Niemcyzk and Power 1982).

Adult Activity

Overwintering in Northern States. A small percentage of the GB population can successfully overwinter in northern states when hibernation occurs in tall, rank Kentucky bluegrass in waste areas. The insect prefers to hibernate in dead or dying leaf blades that are buried under several inches of matted leaf cover. Adults crawl out near the tips of the leaves, where they begin to fold to deposit eggs. The temperature under such vegetative cover aided by a snow cover may be −12° to −11°C (10° to 12°F) higher than the ambient air temperature (Webster and Phillips 1912).

Influence of Temperature. GB development takes place at temperatures between 7° and 33°C (45°–91°F). The most rapid development takes place at 30°C (86°F). Maximum reproduction occurs at about 22°C (72°F; Wadley 1931).

Development of Sexual Forms. Development of sexual forms, winged males, and oviparous females depends on and varies with a combination of temperature and day length, the latter being the dominant factor. When days are less than 12 hr long and temperatures average less than 22°C (72°F), the sexual forms appear and continue to develop as long as days remain short (Wadley 1931).

Oviparous females deposit an average of only 5.4 eggs per female, far less than the prolific viviparous females. Their longevity depends on weather conditions and on the presence or absence of males. When the weather is favorable and males are present, females live 60–70 days, but in the absence of males, females may live nearly 90 days (Webster and Phillips 1912).

Development of Winged Forms. Differences between winged and wingless forms seem less marked than differences between parthenogenetic and sexual forms. Winged forms appear most frequently when the parent aphids received poor nutrition and when the temperature averages about 15°C (59°F), for example during spring or fall. In the absence of this temperature, limited nutrition of parents produces winged forms and can occur any time during the season. Winged females seem inherently migratory. They appear more restless than the wingless females and leave plants even when nutrition is adequate. Flights seem to require strong effort during calm periods but become easy when there is wind; the wind directly assists flight. The shaking of plants helps the aphids become airborne. Migration from wintering areas to the north appears to occur in stages involving successive generations; no one generation traverses the entire distance (Wadley 1931).

Miscellaneous Features

Plant Resistance

In Kentucky greenhouse trials, tall fescue, chewings fescue, and three genetically diverse Kentucky bluegrass cultivars—'Kenblue', 'Vantage', and 'Adelphi'—all supported heavy GB populations and suffered severe feeding damage. Perennial ryegrass, creeping bentgrass, zoysiagrass, and bermudagrass suffered no visible feeding damage; virtually no GB survived on these grasses (Jackson et al. 1981).

In Maryland, seedlings of 48 Kentucky bluegrass cultivars screened for resistance to the Maryland biotype of the GB showed that all cultivars were highly susceptible. Individual plants in 36 of these cultivars, however, showed varying degrees of antibiosis or tolerance ranging from 0.1% to 7.1% of individual seedlings. Cultivars with 1% or greater frequency of resistant plants included, in descending order of frequency, 'Piedmont', 'A-34', 'Troy', 'Rugby', 'Adelphi', 'Kenblue', and 'Sydsport'. There was also some indication of GB biotypes on turfgrass. Progeny from Ohio GB generally caused more severe feeding injury on susceptible plants than did that of the Maryland GB. In addition, the latter could be reared on Kentucky bluegrass and barley, *Hordeum vulgare* L., but the Ohio GB did not establish well on barley (Ratcliffe and Murray 1983).

In Nebraska, GB from the sorghum-feeding biotypes C and E, the two most currently prevalent biotypes in the United States, were selected for screening Canada bluegrass and 23 cultivars of Kentucky bluegrass to feeding injury. Response to both biotypes was similar. The highest levels of resistance were shown by Canada bluegrass, followed by Kentucky bluegrass cultivars 'South Dakota Common', 'Nebraska Common', 'Vantage', and 'Sydsport' (Kindler et al. 1983).

Natural Enemies

The literature contains no indication that the GB is affected by any entomogenous microorganisms. The GB are hosts, however, to many parasites and predators.

Insect Parasites

The most important parasite is *Aphidius testaceipes* Cress, a small black hymenopteran parasite that prefers the wingless GB (Figure 19). A single female may parasitize more than 300 GB but averages roughly 100. Two other species of the same genus and at least three species of the hymenopteran genus *Aphelinus* are recorded as parasites (Webster and Phillips 1912).

The worst outbreaks of the GB occur when mild winters are followed by cool, late springs. Reproduction begins slowly at slightly above 4°C (39°F) and increases

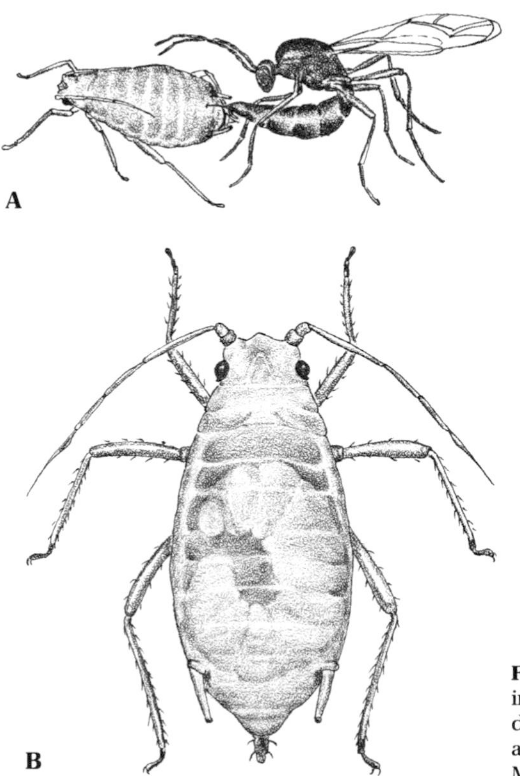

Figure 19. *Aphidius testaceipes.* **A.** Ovipositing in the body of the greenbug. **B.** Parasite larva developing within the greenbug. (From Webster and Phillips 1912, figs. 8, 9; redrawn by R. McMillen-Sticht, NYSAES.)

rapidly at 13°–18°C (55°–64°F). During such conditions temperatures are too low for the most effective parasite *Aphidius testaceipes* to develop (Webster and Phillips 1912).

Insect Predators

Chief among the predatory insects is the coccinellid *Hippodamia convergens* Guerin-Meneville, with both adults and larvae feeding on the aphid. Other coccinellids include *Coccinella 9-notata* Hbst. and *Megilla maculata* (DeG.). Larvae of a syrphid fly of the genus *Syrphus*, a lacewing, *Chrysopa plorabunda* Fitch, and a cecidomyiid fly, *Aphidoletes* spp., are natural enemies (Webster and Phillips 1912).

Vertebrate Predators

Birds listed as effective predators include the goldfinch, *Carduelis tristis*, and at least four species of sparrows (several genera; Webster and Phillips 1912).

Rhodesgrass Mealybug

Taxonomy

The rhodesgrass mealybug, RMB, *Antonina graminis* (Maskell), family Pseudococcidae, was known until very recently as the rhodesgrass scale. Most of the information presented comes from Chada and Wood (1960). I cite additional work below only when it supplements that of these authors.

Importance

Actual damage by the RMB is difficult to assess because drought and close mowing as well as the scale contribute to damage. This is the most widely distributed and damaging of the grass mealybugs or scales in Florida. The RMB is established in a few locations in southern California, but the severity of its damage has not been fully determined (Bowen 1980, Kelsheimer and Kerr 1957).

History and Distribution

The RMB occurs in all tropical and subtropical regions of the world. It was first identified in the United States in 1942 infesting rhodesgrass, *Chloris sayana* Kunth, in southern Texas. The RMB is currently known to occur in all the states bordering the Gulf of Mexico, Mexico, and Hawaii (Figure 20). Records of infestations in Georgia and Maryland appear doubtful. No winged forms occur, and dispersion takes place when crawlers are carried by air currents, when sod is moved, and when crawlers hitchhike on animals.

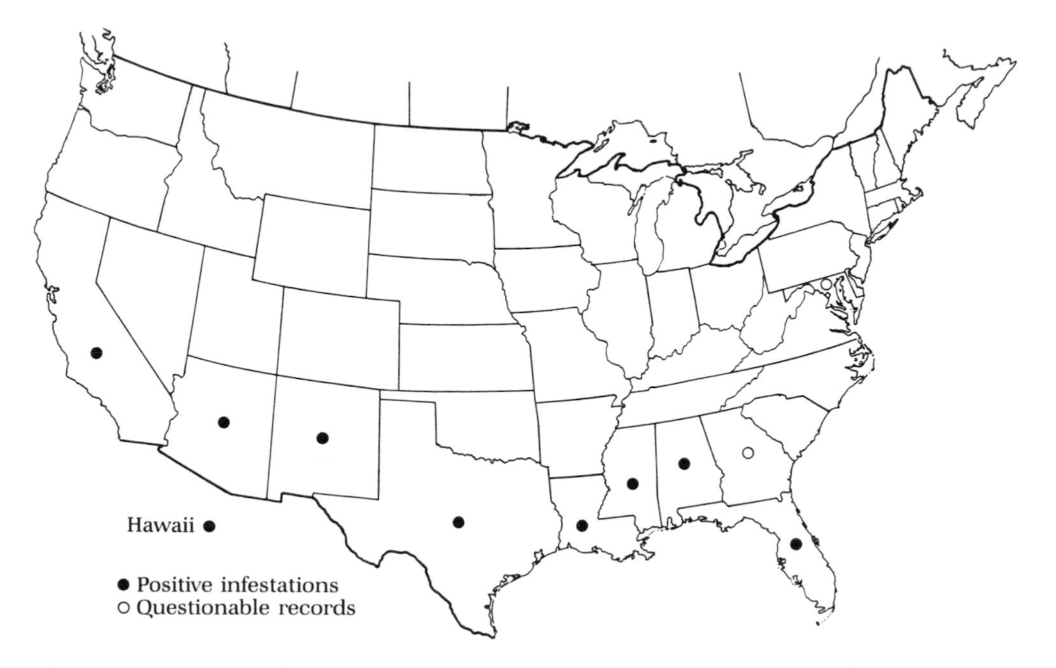

Figure 20. Nine states known to have rhodesgrass mealybug infestations and two states with questionable records of infestation. (Data from Chada and Wood 1960; drawn by R. McMillen-Sticht, NYSAES.)

Host Plants and Damage

Rhodesgrass, a coarse-textured pasturegrass widely used in the Deep South, is the most preferred host. Preferred turfgrasses include bermudagrass and St. Augustinegrass, followed by tall fescue and centipedegrass. Several weed grasses associated with turf are also infested. Lawns and golf course turf mowed at a height of 4 cm (1.5 in.) are less prone to injury than grass that has been cut shorter. Irrigation and fertilization help prevent damage.

Destruction of grass is not dramatic, but infested plants slowly lose vitality. The loss becomes apparent during periods of drought as the insect feeds on the plant sap and causes cells to collapse. The base of infested plants, including the crown, nodes, and leaf axils, appears covered with tufts of cotton. These tufts result from a waxlike secretion produced by the mealybug that encloses the actual insect. Eventually, after the plant cells have collapsed, the plant appears to be suffering from drought and fades to a dull, lifeless brown. Injury is most severe during extended hot, dry periods (Converse 1982).

Description of Stages

Adult

The most obvious sign of infested grass is the white, cottony masses that the insect secretes. Adults found inside these tufts are dark, purplish brown, with saclike, broadly oval to subcircular bodies about 3 mm by 1.5 mm. The caudal extremity is strongly chitinized. Appendages include minute, two-segmented antennae, long, styletlike mouthparts, and a white, waxy, tubular filament protruding from the anal end of the insect itself and protruding outside the white cottony mass (Plate 11). This is an excretory organ, and some filaments may exude droplets of excrement. The body is enclosed in a felted waxy sac that turns yellow with age. Openings in the anterior and posterior ends expose the actual insect.

Egg

Eggs are observable only when the female is dissected. They are cream colored and oblong and measure about 500 μm by 180 μm.

Larva and Nymph

Larvae, or crawlers, the only motile form, are born alive (Plate 11). They are oblong-oval and cream colored, the median area being tinged with purple. Appendages include the long, styletlike mouthparts, six-segmented antennae, three pairs of legs, and two long caudal appendages. The second and third instars are nymphs and have saclike bodies bearing little resemblance to larvae (Plate 11). They become sessile and lose legs and appendages in the first molt. The body becomes enclosed in a waxy sac, so that they resemble adults except for their smaller size.

Life History and Habits

Life Cycle

No males have been seen. The parthenogenetic, ovoviviparous females reproduce over about 50 days. Emerging first-instar larvae move upward onto leaf sheaths, settle beneath a leaf sheath at a node, insert their mouthparts, and become sessile. A life cycle takes 60–70 days, and there are about five generations annually. The winter generation in southern Texas requires about 3.5–4.0 months for completion. Spring, summer, and fall generations require about 2 months each. There is no winter diapause.

Scales thrive under moisture conditions that promote good plant growth. Temperature is most influential in affecting development. Optimum temperatures are 29°–32°C (84°–90°F), but the insect does well between 38° and 0°C (100°–32°F). Freezing temperatures cause high mortality; −2.2°C (28°F) for 24 hr is fatal to all stages.

Natural Enemies

The only recorded parasite is *Anagyrus antoninae* Timb. (Encrytidae: Hymenoptera), established in Hawaii. Its importation into Texas did not appear successful (Dean and Schuster 1958).

Ground Pearls

Taxonomy

Ground pearls, *Margarodes* spp., suborder Homoptera, family Margarodidae, are subterranean scale insects infesting roots of turfgrasses in the southeastern United States. There are several species; one present in Alabama is *M. meriodionalis* Morr. Their name is derived from the pearllike appearance of the cysts that enclose the immature stages. Most of the information comes from Kouskolekas and Self (1974); I cite other works below only when they supplement that of these authors.

Importance, History, and Distribution

Ground pearls infest roots of turfgrasses and cause widespread, extensive damage to home lawns and golf courses in the southeastern United States. The insect has the potential for causing serious turf damage in the southwestern states (Baker 1982).

Host Plants and Damage

The ground pearls attack the roots of bermudagrass, St. Augustinegrass, zoysiagrass, and centipedegrass. Nymphs extract plant sap from the roots. The damage appears as irregular patches. During summer dry spells, the grass yellows, browns, and usually dies by fall (Plate 12). Cysts are present in larger numbers at the interface between damaged and healthy grass (Baker 1982).

Description of Stages

Adult and Egg

Females are wingless, pinkish scale insects with well-developed forelegs and claws and are about 1.6 mm long (Plate 12). Males, considered rare, are gnatlike and vary from 1 mm to 8 mm in length (Plate 12). Clusters of pinkish white eggs are enclosed in a white waxy sac (Plate 12; Baker 1982).

Nymph

Nymphs, called ground pearls, have a hard, globular, yellowish purple shell (Plate 12). Most cysts are 0.5–2.0 mm in diameter. The sucking mouthparts extend through the wall of the cyst and are inserted into grass roots (Plate 12; Baker 1982).

Life History and Habits

Life Cycle

The life cycle of ground pearls is not completely understood. They overwinter in the cysts. Females mature in late May, emerge from the cysts, and, after a short period of mobility, secrete a waxy filament that covers the bodies completely. They remain about 5.0–7.5 cm (2–3 in.) deep in the soil and deposit eggs within the waxy coat. Oviposition begins in June and continues into July, with the egg hatch extending into August. Young crawlers start feeding on grass roots and develop the globular appearance. There is usually one generation each year, but under unfavorable conditions two or three years may be required to complete a life cycle (Baker 1982).

Adult and Nymphal Activities

Ground pearls have parthenogenetic reproduction. Each female deposits about 100 eggs within their waxy secretion. After hatching, young crawlers start feeding on the grass roots and form their pearllike cysts. Ground pearls are found as deep as 25 cm (10 in.) in the soil, a depth that eliminates the possibility of developing any practical means of control (Niemczyk 1981).

7

Lepidopteran Pests: Family Pyralidae

Temperate-Region Sod Webworms

Taxonomy

The name *sod webworm* refers to a large number of grass-infesting moths and larvae of the family Pyralidae, subfamily Crambinae. Adults are often called *lawn moths* because of their habitat or *snout moths* because of the prominent labial palpi that extend in front of the head. Most of the turfgrass-infesting species were placed in the genus *Crambus*, which is distributed practically worldwide. About 100 species are recognized in North America (Bohart 1947).

The generic designation of these insects is in a state of flux. Original members of the genus *Crambus* are now divided into more than a dozen genera. Species reported to infest turfgrass in the United States are listed in Table 7. The binomials accepted and used here are those occurring in the *Check List of Lepidoptera of America North of Mexico* (Hodges et al. 1983). Since many of the turfgrass-infesting species of sod webworms have not been assigned common names, the scientific names will be used for sake of uniformity. The common names, if known, appear only in Table 7 and in my next paragraph.

The six most important of the sod webworms in the temperate regions of the United States include: the bluegrass webworm, *Parapediasia teterrella* (Zinck.); the striped sod webworm, *Fissicrambus mutabilis* (Clem.); the silver-striped webworm, *Crambus praefectellus* (Zinck.); the larger sod webworm, *Pediasia trisecta* (Wlk.); the corn root webworm, *Crambus caliginosellus* Clem.; and the subterranean webworm, also known as the cranberry girdler, *Chrysoteuchia topiaria* (Zell.) (Plate 13; Ainslie 1923a, 1923b, 1927, 1930; Kennedy 1980).

Burrowing sod webworms, *Acrolophus* spp. (Acrolophidae), occasionally damage turfgrass. These are considered secondary pests; I discuss two species briefly in Chapter 20.

Importance

Sod webworms restrict their feeding with rare exceptions to plants of the family Graminae, and turfgrasses serve as ideal host plants. The most serious problems

Table 7. Sod webworms of the subfamily Crambinae infesting turfgrass in the United States

Binomial	Common name	Major reference
Agriphila ruricolella (Zell.)	—	Robinson and Tolley 1982
A. vulgivagella (Clem.)	Vagabond crambus	Felt 1894, Werner 1982
Chrysoteuchia topiaria (Zell.)	Cranberry girdler, subterranean webworm	Kamm 1973, Kennedy 1980
Crambus agitatillus Clem.	—	Robinson and Tolley 1982
C. caliginosellus Clem.	Corn root webworm	Dominick 1964, Kennedy 1980
C. laqueatellus Clem.	Paneled crambus	Matheny 1971, Felt 1894
C. leachellus (Zinck.)	Leach's crambus	Felt 1894, Robinson and Tolley 1982
C. luteolellus (Clem.)	Yellow crambus	Felt 1894, Robinson and Tolley 1982
C. perlellus (Scop.)	—	Robinson and Tolley 1982
C. praefectellus (Zinck.)	Silver-striped webworm	Ainslie 1923a
C. sperryellus Klots	Silver-barred lawn moth	Bohart 1947
C. tutillus McD.	—	Kamm 1971
Fissicrambus haytiellus (Zinck.)	—	Wylie 1944
F. mutabilis (Clem.)	Striped sod webworm	Ainslie 1923b
Microcrambus elegans (Clem.)	Pretty crambus	Robinson and Tolley 1982, Tolley 1983
Parapediasia decorella (Zinck.)	—	Robinson and Tolley 1982
P. teterrella (Zinck.)	Bluegrass webworm	Ainslie 1930, Robinson and Tolley 1982
Pediasia trisecta (Wlk.)	Larger sod webworm	Ainslie 1927, Robinson and Tolley 1982
Surattha identella Kearfott	Buffalograss webworm	Sorensen and Thompson 1971
Tehama bonifatella (Hulst)	Western lawn moth	Bohart 1947
Urola nivalis (Drury)	—	Robinson and Tolley 1982

Note: Dashes indicate that no common name has been assigned.

with sod webworms have occurred in the midwestern, eastern, and southeastern portions of the United States, with Illinois, Indiana, Ohio, Kentucky, and Tennessee routinely having serious webworm problems (Ainslie 1930, Heinrichs 1973, Matheny 1971).

In the Pacific Northwest, webworm problems occurring in commercial turfgrass seed production receive greater attention than problems occurring on home lawns. In addition to being turfgrass pests, some species are recognized as being of even greater importance to agriculture, as they feed on field and forage crops. The most diverse destructive sod webworm in feeding habits is *Chrysoteuchia topiaria*, which destroys the roots of turfgrasses, cranberries, and coniferous seedling plants (Kamm 1971, 1973; Kamm et al. 1983; Kennedy 1980).

Increased problems with sod webworms on turfgrasses occurred several years after the chlorinated hydrocarbon insecticides—chlordane and dieldrin—came into general use. The destruction of natural enemies was considered to be the primary cause of severe webworm outbreaks. In addition, development of almost complete resistance to the chlorinated hydrocarbons complicated the webworm problem in the 1960s (Heinrichs and Southards 1970, Schuder 1964).

History and Distribution

A wide-scale drought occurred during 1928–1934. After 1929, sod webworms, coupled with the drought, became recognized as a serious lawn and golf course turf pest. An outbreak of unprecedented magnitude occurred in many states during 1931 (Bohart 1947).

Unlike many of our most destructive introduced pests, the webworms are all native to America. The single most important webworm species infesting turfgrass is *Parapediasia teterrella.* Its geographic range covers an area south and west of a line from Massachusetts and Connecticut through southeastern New York, westward into eastern Colorado, and southward to the southern tip of Texas (Figure 21). It is more abundant in the bluegrass region of Kentucky and Tennessee than in any other part of its range (Ainslie 1922, 1930).

One of the most voracious feeders is *Pediasia trisecta,* injurious to lawns, pastures, meadows, and cornfields. This is a major species widely distributed from coast to coast in the northern United States and across southern Canada (Figure 21). It is most abundant in the region encompassing Ohio and Iowa. Southern extensions include North Carolina, Texas, and New Mexico (Ainslie 1927).

Fissicrambus mutabilis is one of the most common species and is widespread over much of the United States east of the Rocky Mountains (Figure 21). It is most abundant in a triangular area encompassing Illinois, Tennessee, and Pennsylvania. A prominent species, *Crambus praefectellus,* is distributed widely over the eastern half of the United States in practically every state east of the Mississippi River, in at least seven states west of the river, and in two provinces in Canada (Figure 21; Ainslie 1923a, 1923b).

Another important webworm of turfgrass that is widespread across the northern portion of the United States is *Crambus caliginosellus.* It is troublesome in the north central and northeastern states and southern Ontario in Canada. To the south it extends into Virginia and North Carolina (Dominick 1964, Kennedy 1980).

A pest of turfgrasses in Michigan, Ohio, and western Pennsylvania, *Chrysoteuchia topiaria* is an important species that occurs over much of northern United States. It is most important in the Pacific Northwest, where greatest attention is paid to it because it infests turfgrasses grown for seed production. It burrows into the crown of grasses and also eats roots. It is also a serious pest of cranberries because it girdles roots (McDonough and Kamm 1979).

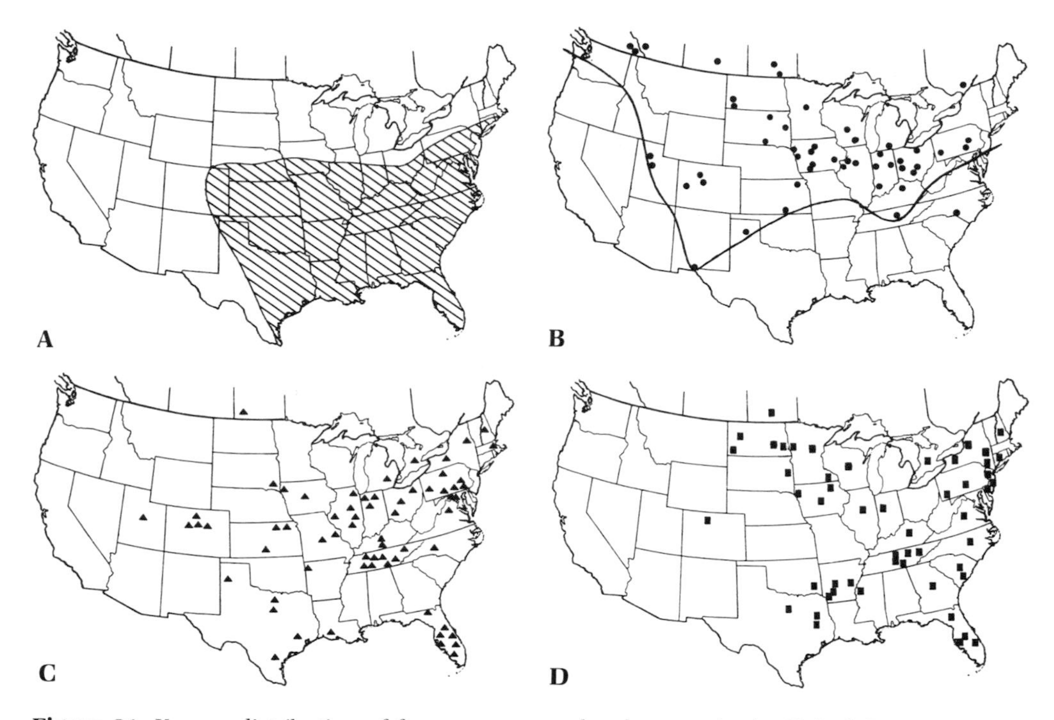

Figure 21. Known distribution of four common sod webworms in the United States. **A.** *Parapediasia teterrella.* **B.** *Pediasia trisecta.* **C.** *Fissicrambus mutabilis.* **D.** *Crambus praefectellus.* (From Ainslie 1930, fig. 1; 1927, fig. 2A; 1923b, fig. 1; and 1923a, fig. 1; redrawn by R. McMillen-Sticht, NYSAES.)

The two most serious turfgrass pests of California are *Tehama bonifatella* (Hulst) and *Crambus sperryellus* Klots. The former is confined to a coastal area in California that is 82 km (50 mi) wide. It occurs in the entire Rocky Mountain area, in the Great Basin (occupying the western half of Utah and most of Nevada, with extensions into California, Oregon, and Idaho), and from Oregon to British Columbia. *Crambus sperryellus* occupies the coastal, inland, and desert valleys of California. In coastal southern California these two species are the most common lawn insect pests and favor bluegrass and bentgrass lawns where larval populations may exceed 22 per 0.1 m^2 (per 20 ft^2) (Bohart 1940, 1947; Jefferson and Eades 1952; Jefferson et al. 1964).

At least 14 species of sod webworms were associated with turfgrass in Virginia as indicated by black-light trap catches during May to October 1981. Traps were located in Blacksburg in the western part of the state and Virginia Beach on the Atlantic coast. The three most abundant species were, in descending order, *Parapediasia teterrella, M. elegans* (Clem.), and *Pediasia trisecta* (Robinson and Tolley 1982).

Host Plants and Damage

Host Plants

With rare exceptions, sod webworms feed primarily on plants of the family Gramineae. Turfgrasses most commonly recorded as primary host plants include Kentucky bluegrass, perennial ryegrass, fine fescue, and bentgrass. Records of damage to warm-season grasses are relatively few. *Fissicrambus haytiellus* (Zinck.) damages bermudagrass and zoysiagrass in Florida. In addition to lawns and other turfgrasses, most webworms feed on corn; wheat; rye; oats; timothy, *Phleum* L.; pastures; and meadows, with greatest destruction to areas of permanent sod. Webworm damage becomes most pronounced at times of drought (Ainslie 1930, Bohart 1940, Wylie 1944).

Webworms feeding on entirely different types of host plants include *Crambus caliginosellus,* which feeds on turfgrasses but also on corn and tobacco seedlings. Its larvae concentrate most in the root zone of the narrow-leaf plantain, *Plantago lanceolata L.,* where more than 30 larvae have been found around a single plant (Dominick 1964, Kennedy 1980). *Chrysoteuchia topiaria* feeds in the crown and roots of grasses in fields of commercial seed production and girdles the roots of cranberry plants. Seedlings of Douglas fir, *Pseudotsuga menziesii* (Mirb.) Franco, and several true firs, *Abies* species, that are grown in nurseries are girdled by these larvae (Kamm et al. 1983, Kennedy 1980, McDonough and Kamm 1979).

Damage

First-instar sod webworm larvae feed only on the surface layers of leaves and stems (Plate 14). The first evidence of feeding damage to a normally growing lawn may be small patches of leaves that are yellow to brown during the summer. The patches grow daily in size. Holes pecked by birds may become evident (Plate 14). Since larvae are nocturnal, they are found during the day only in their burrows in the thatch or in surface soil along with accumulations of green fecal pellets (Plate 14). Under heavy and widespread infestations, the browning localized patches coalesce to produce large irregular areas of dead or dying grass in sunny locations, with shaded areas and weedy patches other than grasses remaining green (Plate 14). When populations of at least 100 larvae per yd^2 (equivalent to 12 per 0.1 m^2 or 11 per ft^2) can be flushed out in 10 min with a pyrethrum or detergent drench, there is a serious infestation (Bohart 1947).

Presence of a large number of moths flying over the lawn at dusk does not necessarily mean that a heavy larval infestation will occur, nor does the presence of flocks of birds feeding on the lawn during the day signify a heavy infestation (Bohart 1947). These phenomena, however, should cause the presence of webworm larvae to be suspected and should not be ignored.

The presence of webworms under a drought condition constitutes the most serious situation for potentially serious turfgrass damage. Not only can the dormancy of the grass restrict the manifestation of early feeding symptoms, but all too often the dead turf does not become evident until fall rains revitalize the normal turf. Normally sod webworm feeding in itself does not kill the crown. A normal lawn can often be revived if the infestation is controlled and if fertilizer and water are added.

Description of Stages

Adults

Moths of sod webworms are distinguished from all other moths by their appearance. When they are at rest, their very long labial palpi extend, snoutlike, in front of their heads (Plate 15). Moths also fold their wings partially around their abdomen, so that they appear very slender and are often very difficult to see after they have come to rest on a grass stem or leaf (Ainslee 1922).

In general coloration they may be whitish or light gray to tan, but on closer observation their forewings often show designs of silver, gold, yellow, brown, and black. These colors occur as longitudinal stripes on a whitish or dull gray background. The hindwings are usually white or grayish. Both pairs of wings have delicate fringes on the outer margin. Moths of most species are about 12 mm in length, with wing expanses of 20–25 mm. Table 8 gives the dimensions of three webworms of common size together with that of *Pediasia trisecta,* one of the largest of webworms, with a wing expanse of 21–35 mm. The larger size is also evident for larvae as well (Ainslie 1923a, 1923b, 1927; Felt 1894).

Eggs

Sod webworm eggs are completely dry and nonadhesive. The many species have eggs that are very similar in shape, size, sculpturing, and coloration (Plate 15). Felt (1894) and Matheny and Heinrichs (1972) measured eggs of a total of 20 species and examined them to eclosion. Depending on species, most eggs are oval to elliptical oval in form, with varying numbers (from 12 to 30) longitudinal ridges on the surface from pole to pole. Length and width varied from 0.598 mm by 0.503 mm for species

Table 8. Size of the more common sod webworms in the United States (mm)

Species	Adult wingspan	Eggs, width × length	Larval head cap. width		Larval total length	
			First instar	Final instar	First instar	Final instar
Parapediasia teterrella	18–21	0.31 × 0.51	0.21	1.23	1–2	9–13
Fissicrambus mutabilis	18–24	0.31 × 0.48	0.19	1.73	1.8	18
Crambus praefectellus	18–25	0.31 × 0.51	0.23	1.40	?	?
Pediasia trisecta	21–35	0.34 × 0.52	0.21	2.21	1–2	24–28

Source: Ainslie 1923a, 1923b, 1927, 1930.

with the largest eggs to 0.390 mm by 0.300 mm for species with the smallest eggs. The eggs of most species are white to creamy white when they are first laid and turn to various shades of bright orange to bright red, or becoming ocherous at eclosion (Plate 15).

Scanning electron micrographs were taken of the chorion of eggs from 15 species of sod webworms in Tennessee. Sufficient differences were present in the longitudinal carinae, polar areas, and various other features to permit the development of a key to Crambinae eggs (Matheny and Heinrichs 1972).

Larvae

In general appearance, sod webworm larvae vary in color, from a greenish hue (reflecting food content) to beige, brown, or gray. Most have characteristic dark circular spots scattered over the entire body (Plate 15). At least two common species lack the darker circular spots, *Crambus caliginosellus* (Plate 15) and *Chrysoteuchia topiaria* (Kennedy 1980).

Head capsules in some species are black during the first two or three instars. In most species, mature larvae have light brown head capsules, with varying shapes of black sculpturing (Plate 15). Table 8 shows that head capsule widths of four common species increase from 0.19–0.23 mm in first instars to 1.23–2.21 mm in ultimate instars. Total larval lengths for the final instar vary from 9 mm to 13 mm in the smallest and from 24 mm to 28 mm in the largest webworm (Ainslie 1923a, 1923b, 1927, 1930).

Larval instars vary from 6 to as many as 10, with 8 being most common; *Parapediasia teterrella* has 7 over a period of about 50 days, and *Pediasia trisecta* has 8 (Robinson and Tolley 1982).

Pupae and Cocoons

The cocoons are made from soil particles, plant debris, and fecal pellets and may be a part of the larval tunnel or may be constructed separately. In size and shape the cocoon resembles a peanut meat. It is firm, smoothly lined with soft gray silk, and outwardly covered with soil and grass. Its depth depends on the looseness of soil, and it may be mistaken for a lump of earth. The pupae developing within the cocoon are pale yellow when first formed and darken to mahogany brown. A day or two before their emergence, the color patterns of the forewing become plainly visible through the integument. Pupae of *Parapediasia teterrella* are about 8–10 mm long and about 2.5 mm wide (Ainslie 1923a, 1923b, 1927, 1930).

Seasonal History and Habits

Seasonal Cycle

Sod webworms overwinter as larvae in their hibernacula in the thatch or soil. Most species do so in their penultimate or ultimate instar, but others may overwinter in earlier instars. Larvae of *Pediasia trisecta* are known to overwinter in the second to the fifth instar. With the advent of warmer weather, larvae resume feeding or, if in the ultimate instar, will pupate. Moths emerge during late spring or early summer, depending on species. Moths of *Crambus praefectellus* are the first to appear in the spring in the eastern United States (Ainslie 1923a, 1927).

Some species such as *Chrysoteuchia topiaria* and *Crambus caliginosellus* are univoltine wherever they occur. Why *Pediasia trisecta* is univoltine in the Pacific Northwest and bi- or trivoltine in the eastern states is not known (Figure 22). It is reported to have two generations a year in the Midwest, two or three in Virginia, and three in New Jersey (Ainslie 1927, Kamm 1970, Mailloux and Streu 1982, Robinson and Tolley 1982).

Parapediasia teterrella has two generations in Virginia and three in Tennessee, a slightly more southern latitude (Figure 22). Figure 23 presents a detailed life history of *Parapediasia teterrella* as it often occurs in Tennessee. The overwintering larvae pupate in April and May, and moths make their first appearance in early May, steadily increase throughout the month, and then decrease through June. First-generation eggs are laid during June. Larvae are present during June and July, with pupation in July. Adults are present in July–August, depositing second-generation eggs during August. Larval and pupal development occurs throughout August and September. With favorable weather conditions, moths of the second generation lay eggs into early October for overwintering larvae (Ainslie 1930).

In the Los Angeles area, *Crambus sperryellus* has three generations a year and *T. bonifatella* has four, completing a life cycle within 6 weeks in warm weather (Bohart 1940).

Both species were reared from eggs to adults at 24°C (75°F). *Crambus sperryellus* required 6, 38, and 11 days to complete the egg, larval, and pupal stages, respectively, for a total of 55 days. *Tehama bonifatella* required 4.5, 24.0, and 8.0 days to complete the three stages, for a total of 36 days, showing that it is a much more rapidly developing species (Bohart 1947).

Adult Activity

Sod webworm moths are inactive during the day. They rest on turfgrass, on associated weeds, or on the leaves and stems of trees or shrubs (Plate 15). If disturbed they will take flight, but flight is weak. Fluttering moths soon come to rest. In Virginia, daily sampling for moths during August–October 1982 revealed that *M. elegans, Parapediasia teterrella,* and *Agriphila ruricolella* (Zell.) preferred shrubs over turfgrass for daylight resting sites, with no preference for shrub species (Tolley 1983).

Moths of *Parapediasia teterrella* prefer broad-leaved plants to grass for resting places. They often alight on the upper surface of leaves and immediately move underneath and rest, with the head pressed close to the leaf surface and the wings and abdomen elevated about 25°. This habit together with the white scales on top of the head helps to identify the species (Ainslie 1930).

During the day, if one walks across a turfgrass area, webworm moths will fly for a

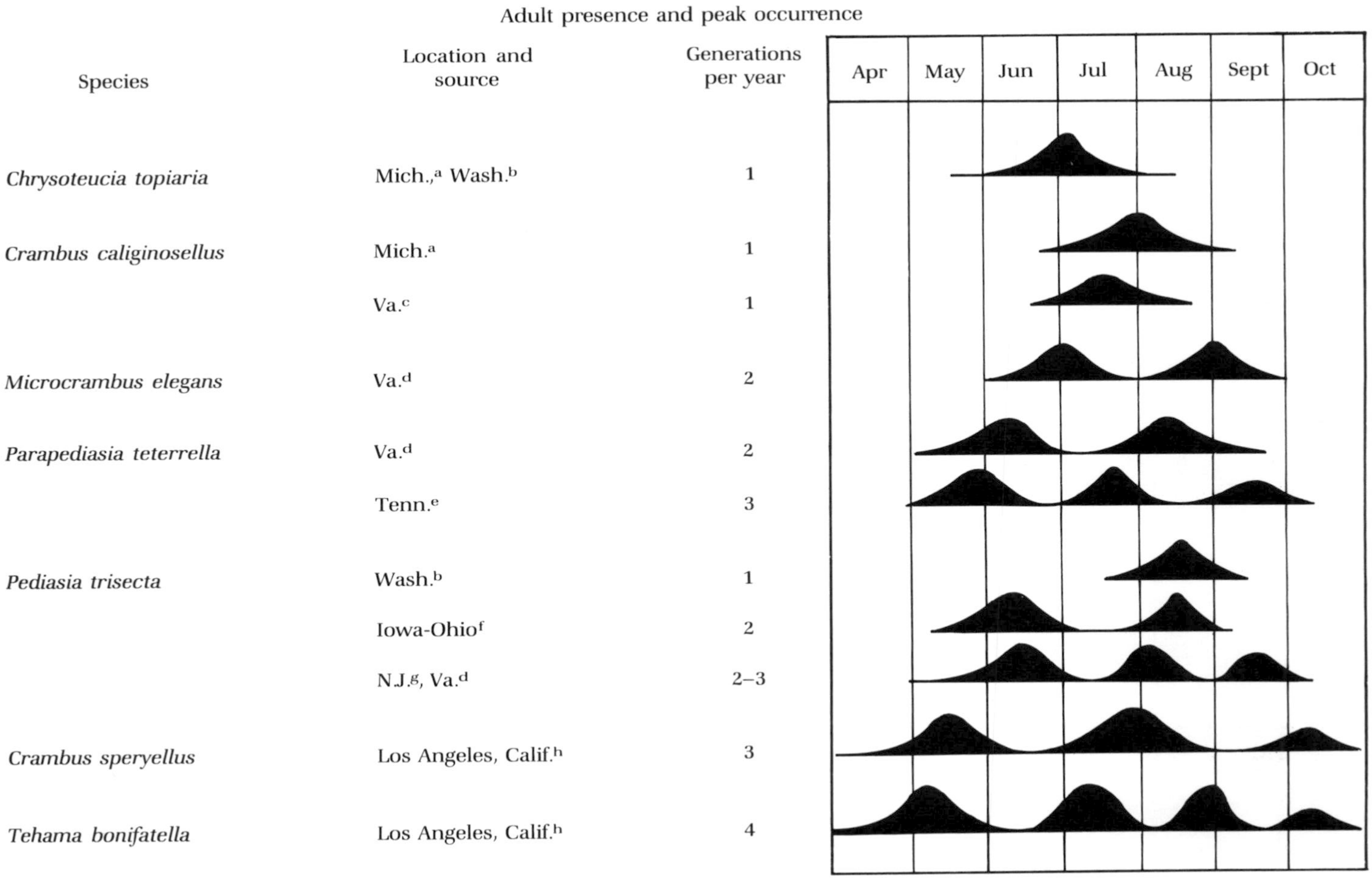

Species	Location and source	Generations per year	Adult presence and peak occurrence: Apr	May	Jun	Jul	Aug	Sept	Oct
Chrysoteucia topiaria	Mich.,[a] Wash.[b]	1							
Crambus caliginosellus	Mich.[a]	1							
	Va.[c]	1							
Microcrambus elegans	Va.[d]	2							
Parapediasia teterrella	Va.[d]	2							
	Tenn.[e]	3							
Pediasia trisecta	Wash.[b]	1							
	Iowa-Ohio[f]	2							
	N.J.[g], Va.[d]	2–3							
Crambus speryellus	Los Angeles, Calif.[h]	3							
Tehama bonifatella	Los Angeles, Calif.[h]	4							

Figure 22. Seasonal presence (approximate) of common sod webworm adults at various locations in the United States. (Data from [a]Kennedy 1980, [b]Crawford and Harwood 1964, [c]Dominick 1964, [d]Robinson and Tolley 1982, [e]Ainslie 1930, [f]Ainslie 1927, [g]Mailloux and Streu 1982, and [h]Bohart 1947; drawn by R. McMillen-Sticht, NYSAES.)

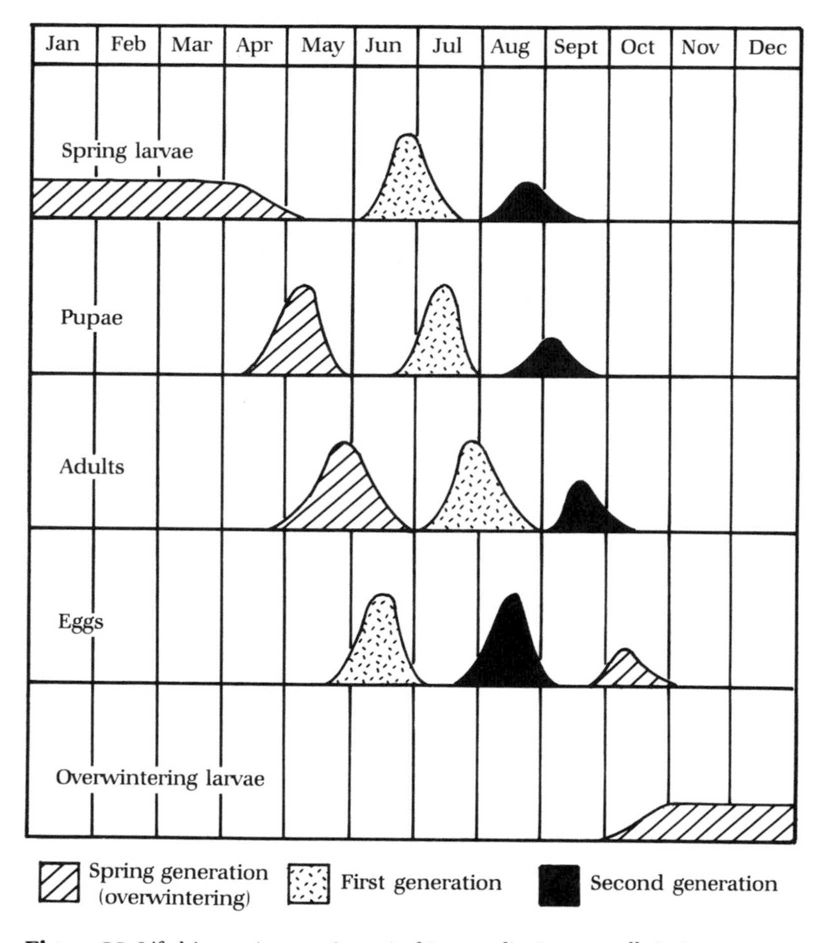

Figure 23. Life history (approximate) of *Parapediasia teterrella* in Tennessee. (Data from Ainslie 1930 and Robinson and Tolley 1982, fig. 3; drawn by R. McMillen-Sticht, NYSAES.)

short distance (few meters) in a zigzag pattern before again alighting on the turf. Toward evening the moths are more easily flushed from their resting place and fly freely.

Nightly Abundance. Webworm moths are most active during the early hours of the night, as revealed in light trap catches. Under favorable weather conditions, females of *Parapediasia teterrella, Pediasia trisecta,* and *F. mutabalis* are most abundant from about 30 min after dark to 1.5–2.0 hr after sunset, while males reach maximum flights from 11:00 PM to 1:00 AM (Ainslie 1930, Banerjee 1967a, Banerjee and Decker 1966).

Mating. Females are considered to mate only once. Females of some species mate during their 1st night. Others, such as *Parapediasia teterrella, Pediasia trisecta,* and *M. elegans,* mate 2 or 3 nights following emergence. Males may mate more than once but usually not on the same night (Banerjee 1969b, Robinson and Tolley 1982).

Mating habits of different species of webworm moths appear to be similar. Males do not arrive in numbers until the evening is well advanced. In the case of *P. teterrella,* males appear at about 9:00 PM, and by midnight the two sexes may be present in roughly equal numbers. Mated pairs are tail to tail and generally show little or no movement, but occasionally the female drags the smaller male about. Observations have indicated that the union may last as short a time as 12 min or may last more than 2 hr, with an average of nearly an hour. Many of the males die within 12–24 hr after mating (Ainslie 1930).

Sex Pheromones. The presence of species-specific sex pheromones was demonstrated in a turfgrass area in Illinois when virgin females of *Parapediasia teterrella* and *Pediasia trisecta* of known age were placed in traps and were exposed to natural flights of moths. Virgin females were most attractive during the first 3 or 4 days after emergence (Banerjee 1969a).

The sex pheromone of *Chrysoteuchia topiaria* has been identified as (Z)-11-hexadecenal (Z)-9-hexadecenal. When it was used to monitor the presence of males by placing a trap just above the canopy of the grass, it was found to be six times as attractive as a single live virgin female. One trap per 1–2 ha (2.5–5.0 acres) gave a good indication of the population density. Since the larva enters the crown of grasses and its presence is undetectable until damage has occurred, the pheromone is a valuable monitoring tool to determine population densities (Kamm and McDonough 1979, 1980).

A combination of (Z)-11-hexadecenal and (Z)-13-octadecenal is a sex attractant for *T. bonifatella.* Traps baited with the two-component lure caught eight times more males than traps baited with females (McDonough et al. 1982).

Oviposition. Nightfall is the primary stimulus for oviposition. Females of most species drop their dry, nonadhesive eggs indiscriminately as they fly over the turfgrass. Females of *Parapediasia teterrella* start ovipositing during their flights at dusk and continue for 2–3 hr as they fly low across the lawn, scattering their eggs over the grass. They may also drop eggs while at rest. The area below fluttering moths is usually thickly strewn with eggs. Unlike other webworms, *Chrysoteuchia topiaria* females deposit their eggs while resting in the grass (Ainslie 1930, Crawford 1968, Kennedy 1980).

Oviposition by *Pediasia trisecta* begins on the 2d night after emergence, at the approach of darkness when the light intensity approaches 0 ft-c. About 60% of the eggs are laid the 1st hr and about 90% during the 2d hr (Banerjee and Decker 1966).

Fecundity. Fecundity varies greatly among species and is related somewhat to the size of the moths. *Crambus laqueatellus* Clem., one of the larger, lays more than 480 eggs per female. *Microcrambus elegens* (Clem.), a small moth, lays eggs of the same size as those of the former (0.28 × 0.42 mm) but averaged only 32 per female. Females of *Parapediasia teterrella* normally deposit 200–250 eggs with oviposition of as many as 60 eggs per day until death. Females of *Crambus caliginosellus* deposited an average of slightly over 200 eggs per female in Michigan. In New Jersey, a mated female of *Pediasia trisecta,* common in Kentucky bluegrass and red fescue lawns, lived about 8 days and laid an average of 470 eggs per female (Ainslie 1930, Kennedy 1980, Mailloux and Streu 1982, Matheny 1971).

Adult Longevity. Normally in nature the bluegrass webworm is considered to live 7–10 days. The females live slightly longer than the males. Caged moths had a maximum longevity of 28 days when they were given only water or honey in water. Captive moths without moisture live only a third as long (Ainslie 1930).

Egg Activity

Eggs of various species generally hatch in 6–10 days in the field. Eggs of *Parapediasia teterrella* hatched in 5–7 days during summer but required 41 days at 10°C (50°F). Eggs of *Pediasia trisecta* did not hatch below 10°C and required 29 days at 14°C (57°F) and 5.4 days at 32°C (90°F). Eggs of *Crambus caliginosellus* hatched in 5–9 days during the summer (Ainslie 1930, Banerjee 1969a, Dominick 1964, Heinrichs 1973).

Widely distributed species may show adaptations to local climatic conditions. Eggs of *Parapediasia teterrella* from moths collected at Lexington, Kentucky, and Albuquerque, New Mexico, both had >90% hatch at >60% RH. At RH below 60%, egg hatch from Lexington moths was decidedly less than hatch from Albuquerque moths at any and all RH percentages, with no hatch of eggs from either location at 0% RH (Morrison et al. 1972).

Larval Activity

As soon as they have hatched, first instars of many species find suitable hiding places and begin to conceal themselves by webbing together particles of debris. First instars feed only on the surface tissues of tender leaves (Plate 14).

Habitat and Feeding. As they grow, third- and fourth-instar larvae prepare burrows by webbing together plant debris, soil particles, and their own excrement and lining the interior with their webbing (Plate 15). Larvae of *Crambus caliginosellus* seldom construct a complete burrow of webbing as the others do. It often constructs only one side, using the stem of the plant on which it is feeding as the other side. Burrows open not to the surface but downward, where much of the feeding takes place. During the day, larvae remain concealed in their burrows but during the night wander out to feed. They notch leaves and during the later instars may cut off entire leaves and draw them into their silk-lined tubes. The tube of *Parapediasia teterrella* measures 10 × 30 mm (0.4 × 1.2 in.). Larvae of most species feed above the crown of the plant and above ground, but at least two well-known species feed below ground. These are *Crambus caliginosellus* and *Chrysoteuchia topiaria*, the former infesting the crown and roots of grasses and the latter infesting cranberry and fir seedling roots as well as those of grasses (Ainslie 1922, Kennedy 1980).

Young larvae of *T. bonifatella* and *Crambus sperryellus* curl into a compact ball when they are disturbed but after they are half grown try to escape when disturbed. They crawl backward as rapidly as they do forward and thrash wildly when touched. With the aid of a flashlight, larval feeding can be observed between 8 PM and 10 PM; leaves are being notched, and blades are being cut off and drawn into burrows. Larvae feed on the greener portions of the crown but do not molest roots. Fifth instars consume twice their weight in bluegrass daily (Bohart 1947).

Larval Instars. Larval instars vary among species, but 7 or 8 are common. Larvae of *Parapediasia teterrella* normally have 8 instars but may have 6–10. Those of *F. mutabalis* normally have 7, those of *Crambus praefectellus* have 6, and those of *Pediasia trisecta* have 7 or 8 (Ainslie 1923a, 1923b, 1927, 1930).

Diapause. The sod webworm overwinters as larvae in their final instar in some species and as larvae in earlier instars in others. The photoperiod was the primary factor in the induction and termination of diapause in larvae of *Pediasia trisecta*, but both the photoperiod and temperature were involved for *Crambus leachellus* (Zinck.). Both species diapaused in the second-fifth instar. Exposure lasting 25–35 days to short days 10L:14D at 15° (59°F) and 24°C (75°F) induced 100% diapause in *Pediasia trisecta* (Kamm 1970).

In a univoltine species, *Crambus tutillus* McD., diapause was induced by exposing seventh instars to short days, 10L:14D, which initiated molt to the eighth instar, which diapaused. Short days and warm temperature (18° and 24°C; 64° and 75°F) killed larvae. Those exposed to long days and warm temperature terminated diapause, completed the ninth instar, pupated, and became adults. In early instars, growth rate accelerated as day length decreased (Kamm 1971).

Pupal Activity

Upon completion of the larval stage, *Parapediasia teterrella* abandons its feeding burrows. It then constructs cells in the soil lined loosely with gray silk and forms a cocoon. Pupation lasts from 5 days to 15 days (Ainslie 1930).

Temperature influenced pupation of *Pediasia trisecta*. Larvae failed to pupate at 18°C (65°F) but pupated after 59 days at 21°C (70°F) and after 29 days at 32°C (90°F) (Banerjee 1969b).

Miscellaneous Features

Plant Resistance

'Scaldis' hard fescue and 'Dawson' red fescue have shown some tolerance to sod webworms, as have the Kentucky bluegrasses 'Windsor' and 'Park'. Crossing resistant × resistant, resistant × susceptible, and susceptible × susceptible cultivars yielded highly resistant, moderately resistant, and susceptible strains of grass, respectively, and demonstrated the heritable nature of resistance (Sargent 1982).

In New Jersey, resistance of perennial ryegrasses to sod webworm was found to be associated with the presence of a *Lolium* endophyte fungus. Cultivars 'GT-11', 'Pennant', and 'All-Star' have high levels of endophyte and were found to be highly resistant. The fungus is transmitted primarily by seed and vegetative propagation (Funk and Hurley 1984).

Threshold Populations

Since the presence of adults is not a positive indication of damaging larval populations to come, threshold populations of larvae should be determined before a decision is made to apply an insecticidal treatment. There are wide differences of opinion as to what constitutes a threshold population. Reduced to the common denominator of numbers per 0.1 m^2 or per ft^2, thresholds are 0.22 or more (2+ per yd^2) in Wisconsin, 1.7 or more (15+ per yd^2) in California, and 12–16 (108–144 per yd^2) as a general countrywide recommendation (Bowen 1980, Mahr and Kachadoorian 1984, Vance and App 1971). Obviously these threshold estimates are much too divergent to be of much practical value per se. When a decision is made, all other contributing factors must be considered, including turf vigor, temperature, and moisture regime, to name the more important features.

Natural Enemies

Microorganisms

Larvae and pupae of *Pediasia trisecta* in Illinois were found heavily infested with microsporidia *Nosema* spp. and *Thelohania* spp. These organisms also infected *Parapediasia teterrella* (Banerjee 1968).

The fungus *Beauveria bassiana* was present in all field collections but infected only 3.1% of *Chrysoteuchia topiaria* larvae in Oregon (Kamm 1973).

Insect Parasites

Larvae and pupae of *Pediasia trisecta* in Illinois were parasitized by braconids *Macrocentrus crambivorus* Vier and *Orgilus detectiformis* Vier and by the tachinid *Stomatomyia floridensis* (Tris). Another tachinid, *Lydina polidoides* (Townsend), is reported to parasitize 8.3% of the mature larvae of *Chrysoteuchia topiaria* in Oregon. In California, webworms in lawns were parasitized by two braconids, *Orgilus* spp.

and *Apanteles* spp., and by a tachinid, *Aplomyia confusionis* Sllrs. (Banerjee 1967b, Bohart 1947, Kamm 1973).

Insect Predators

The robber fly *Erax aestuans* L. has been observed capturing webworm moths. Vespid wasps, native earwigs, carabid beetles, and rove beetles are considered important predators of webworms in California. In Kentucky predators consumed or carried off as many as 75% of sod webworm eggs (*Crambus* and *Pediasia* spp.) within 48 hr of exposure. Four species of ants, including *Pheidole tysoni* Forel as the major predator, and a mite, *Macrocheles* spp., foraged on the eggs in turfgrass (Banerjee 1968, Bohart 1947, Cockfield and Potter 1984).

Vertebrate Predators

Brewer's blackbird is considered to be an important predator of sod webworms in California. In Oregon, overwintering larvae of *Chrysoteuchia topiaria* as prepupae in their hibernacula were prey to starlings, killdear, sandpipers, and blackbirds, which reduced the population in excess of 80% and, combined with parasites and diseases, reduced the overwintering population about 91% (Bohart 1947, Kamm 1973).

On lawns, probing holes made by birds may be a clue to webworm presence. These holes are not always a positive indication, but such clues should certainly be pursued.

Tropical-Region Sod Webworms

Grass Webworm

Taxonomy and Importance

The grass webworm, GWW, *Herpetogramma licarsisalis* (Walker), is a member of the order Lepidoptera, family Pyralidae, subfamily Pyrustinae. The insect was placed previously under the genera *Psara* and *Pachyzancla*. Since its discovery in Hawaii, it has been considered the most serious turfgrass pest in the state. At least 90% of all turfgrass insects in bermudagrass have been GWW larvae (Swezey 1946, Tashiro 1976a).

History and Distribution

The GWW is widely distributed in southeast Asia, in the adjoining islands, and in Australia. It was first found in Hawaii in 1967 on Oahu, where it was damaging pasturegrass. In 1968 it was collected from four other islands, Hawaii, Kauai, Maui, and Molokai (Davis 1969). Hawaii is its only known location in the United States.

Host Plants and Damage

All of the important turfgrasses in Hawaii are attacked, including bermudagrass, centipedegrass, and St. Augustinegrass. Kikuyugrass, *Pennisetum cladestinum*

Hochst ex Chiov, a relatively minor turfgrass but a major pasturegrass, is the most heavily infested, with larval counts as high as 55 per 0.1 m^2 (50/ft^2.; Davis 1969).

GWW feeds on leaves, stems, and crowns of turfgrass. During early feeding, the turf appears ragged but is still green. With continued feeding and the passage of time, large brown patches develop (Plate 16).

In a species and varietal feeding test, there was no evidence of oviposition preference. Common bermudagrass and its cultivar 'Tifway' were the least injured. Feeding injury developed faster on fine-textured grasses than on those with coarse texture (Murdoch and Tashiro 1976, Tashiro 1977).

Description of Stages

Adult. Moths are uniformly fawn to light brown, with a mean wingspan of 23.9 mm. The only apparent pattern on the front wing is a faint zigzag line of slightly darker scales paralleling the apical margin and near it. Males have a more slender abdomen than females, with seven visible segments, as compared with six visible segments, ending in a recessed opening for females. Unlike sod webworm moths of the subfamily Crambinae, which wrap their wings around the body when at rest, the GWW holds its wings horizontally, so that the moth has a triangular outline (Plate 16; Tashiro 1976a).

Egg. The flat, elliptical eggs, measuring 0.64 mm by 0.91 mm, are laid singly or in masses generally glued to the upper surface of leaves along the midrib or on nonliving objects where there is free moisture (Plate 16). In masses, the eggs are laid so that they overlap each other like shingles. The chorion has fine reticulations. When it is fresh it is creamy white. It becomes light yellow on the 2d day, a light orange embryo shows on the 3d day. It becomes dark orange, with the black head capsule visible, on the 4th day, the day before hatching (Tashiro 1976a).

Larva. GWW larvae have five instars, with head capsule widths of 0.25 mm in the first to 1.8 mm in the fifth instar. Total body length grows from 2.32 mm in the first to 20 mm in the fifth instar. First-instar larvae have black head capsules; all others have brown (Plate 16). The prothoracic shield is lighter brown than the head. Mature larvae are brown to greenish, depending on the food consumed, and many have a rose tint over part or most of the body. Each body segment has a conspicuous ring of dark brown spots (Plate 17; Tashiro 1976a).

Pupa. Pupation occurs in a hibernaculumlike case (Plate 17). It cannot be considered a true hibernaculum, since the GWW does not diapause. The pupa is creamy white when first formed and turns light brown, then dark brown (Plate 17; Tashiro 1976a). The pupae can be sexed in the way described by Butt and Cantu (1962).

Life History and Habits

Life Cycle. A generation is completed in about 32 days at a mean temperature of 24.5°C (76°F) and 60%–80% RH. Females have a 3-day to 6-day preoviposition period, eggs hatch in about 5 days, larval development takes about 14 days, and the pupal period lasts about 7 days. During an average life span of 13 days, each female lays about 250 eggs. The maximum observed was 556 eggs. There is no diapause, and the rate of development depends mainly on temperature and food supply. The GWW is strictly nocturnal; adult eclosion, mating, oviposition, hatching, feeding, molting, and pupation occur at night (Tashiro 1976a, 1977).

Adult Activity. During daylight, moths remain at rest, mainly in tall grass, unless they are disturbed. They will then fly for a short distance before returning to the grass. Literally hundreds of GWW moths can be flushed out of tall grass simply by walking through it (author's personal observations).

Moths require some nutrients for maximum longevity and fecundity (Table 9). In screen cages with no water, females lived <4 days and laid no eggs, while females supplied with a 10% honey or sucrose solution as food lived >12 days and laid an average of >340 eggs (Tashiro 1977).

Larval Activity. First and second instars feed on the upper surface of the leaf, leaving the lower epidermis intact (Plate 16). From the third instar onward, larvae notch leaves, start to eat entire leaves, and spin copious amounts of webbing. Mature larvae in preparation for pupation become slightly shortened and web together bits of grass, feces, and other debris to form cocoons in which to pupate (Tashiro 1976a).

During the day, larvae are seldom, if ever, seen in the field but can be found secreted in the thatch. They can, however, be flushed to the surface within minutes with a pyrethrin or detergent solution in copious amounts of water (Tashiro et al. 1983).

A 10% injury to hybrid bermudagrass turf in Hawaii is considered the level at and

Table 9. The effect of adult diet on the fecundity and longevity of female grass webworm moths

Liquid provided moths	Eggs deposited per female on days following emergence and pairing (average)						Mean total eggs per female
	3–4	5–6	7–8	9–10	11–12	13–18	
None	0[a]	0	0	0	0	0	0[c]
Water only	41	0[a]	0	0	0	0	41[c]
Honey sol.[b]	97	77	54	43	3	0[a]	274[d]
Sucrose sol.[b]	87	124	109	71	16	14[a]	421[d]

Note: Means followed by the same letter were not significantly different at the 5% level, according to Duncan's multiple-range test.

[a]Death of last female.

[b]10% solutions + 0.1% methyl paraben.

Source: After Tashiro 1977.

beyond which insecticidal treatments would be desirable (Mitchell and Murdoch 1974).

Miscellaneous Features

Rearing. The GWW was an easy insect to rear when 60%–80% RH was maintained. Glassware and rearing equipment were washed with hot water and detergent. No further sanitary measures were taken, and there was no loss of colony due to pathogens. Field-collected moths placed in a screen cage and supplied with honey or sugar solution oviposited on the grass blades and on any moist surface. Colonies were maintained on potted 'Sunturf' bermudagrass and on kikuyugrass, but the succulence of the latter made it the better host plant. Any warm-season grass with soft blades was apparently suitable for rearing. Grass that had been heavily damaged suddenly began to recover usually because larvae had completed their feeding in preparation for pupation. The rapidity of development was governed by temperature, with time between hatching and adulthood about being 22 days at a constant 24.5°C (76°F) and 16 days at 31°C (88°F).

Natural Enemies

Microorganisms. No reports note the presence of any microorganism infecting the GWW.

Insect Parasites and Predators. The GWW is susceptible to several parasites, some of which were introduced to Hawaii for other lepidopterous pests. *Trichogramma semifumatum* (Perkins), accidentally introduced, has parasitized as many as 96% of GWW eggs. Larval parasites include the tachinid *Eucelatoria armigera* (Coq.) and a braconid, *Meteorus laphygmae* Vier. The pupae are parasitized by an ichneumonid. An ant, *Pheidole* sp., destroys GWW eggs (Davis 1969).

Vertebrate Predators. The cattle egret, *Bubulcus ibis,* which forages in infested grass, consumes larvae (Davis 1969). Other avian predators include the common mynah bird, *Acridothermes tristis tristis* (L.), the red-crested or Brazilian cardinal, *Paroaria coronata,* and the Pacific golden plover, *Pluvialis dominica fulva.* Head capsules of the GWW have been found in scats of the giant toad, *Bufo marinus* (L.) (Mitchell and Murdoch, personal communications, 1985; author's personal observations).

Tropical Sod Webworm

Taxonomy

The tropical sod webworm, TSW, *Herpetogramma phaeopteralis* Guenée, is also a member of the order Lepidoptera, family Pyralidae, subfamily Pyrustinae. As in the case of the grass webworm, the TSW was previously listed under the genera *Pachyzancla* and *Psara* (Hodges et al. 1983, Kerr 1955).

Importance

The TSW, like the southern chinch bug, is considered to be one of the two most destructive turfgrass pests in Florida. The larval attack severely damages all major southern turfgrass species. Late April to December is the period of greatest potential damage. Bermudagrass grown under high-maintenance programs has the highest populations and suffers the most. In mixed larval populations the TSW is generally the dominant species. Populations of 11–22 larvae per 0.1 m^2 (10–20 per ft^2) occur frequently, and counts of 85–100 per 0.1 m^2 (79–93 per ft^2) are sometimes present (Kerr 1955, Reinert 1974, 1983a).

History and Distribution

The TSW has a wide tropical distribution throughout the southeastern United States and the Caribbean archipelago. In the United States it has caused severe turf damage in Louisiana since 1933 and in Georgia since 1953. The severe turf damage caused by a mixed population of lepidopteran pests in Florida in 1953 was due primarily to the TSW (Kerr 1955).

Host Plants and Damage

The turfgrasses damaged by the TSW include bermudagrass, centipedegrass, St. Augustinegrass, zoysiagrass, and bahiagrass, with the first three most widely grown as fine turf. Damage is caused when the larvae eat leaves, giving the turf a notched, ragged appearance. The notches along the edge of grass blades are a sure sign that these larvae are present. Continued feeding gives the turf an extremely close-cropped appearance. Damage appears as patches that become yellowish, then brown. Turf adjacent to flower beds and shrubs usually shows the first signs of damage, since adults rest in such foliage and moths lay more eggs in nearby turf (Kerr 1955).

Damage may be seen in southern Florida in the spring. By late summer, webworms are active throughout the state and may continue to cause injury into November. Much of the damage blamed on armyworms is done by the TSW (Kelsheimer and Kerr 1957).

Description of Stages

Adult. The dingy brown moths have a wingspread of about 20 mm. Like a near relative, the grass webworm, TSW moths can be distinguished from members of the Crambinae by the fact that they do not roll their wings about their bodies when they are at rest (Plate 17). As in the grass webworm, the males have a slimmer abdomen and six visible segments, while the females have five visible segments, with the terminal segment ending in a large, fusiform opening (Kerr 1955).

Egg. TSW eggs share several characteristics with the grass webworm. Eggs appear in clusters of 6–15, but some are deposited singly on grass blades or on moist surfaces. Eggs in a group partially overlap each other, like shingles. Each egg is flat, rounded, about 0.70–0.73 mm in widest diameter, and about 0.1 mm high. Fresh eggs are whitish and become brownish red just before hatching. The head of the larva is plainly visible (Plate 17; Kerr 1955).

Larva. The body is dingy cream but appears mostly green when larvae are feeding. The head is dark yellowish brown. There are seven or eight larval instars that grow from just over 1 mm long at hatch to about 19 mm long at maturity (Kelsheimer and Kerr 1957, Kerr 1955).

Pupa. Pupae lie freely or partially buried in duff or are enclosed in a shapeless bag that the mature larva spins, using bits of grass and soil particles. At maturity pupae are reddish brown, about 8.5–9.5 mm long and about 2.1–2.9 mm wide (Kerr 1955).

Life History and Habits

Life Cycle. Populations of the TSW are present throughout the year in southern Florida, with the highest numbers present in late summer and fall. In Gainesville (in north central Florida), peak adult emergence occurs in October and November. It appears that pupae and adults cannot survive the winters here. South of Gainesville there appears to be no single overwintering stage; all stages are present throughout the year. At 25.5°C (78°F), the TSW develops from egg to adult in about 6 weeks, spending 6–10 days as an egg, about 25 days in the seven larval instars, and about 7 days as a pupa during the summer (Kerr 1955, Reinert 1973).

Adult Activity. During the day most moths rest in shrubbery around lawns and when disturbed will make weak short flights before settling down again. At dusk, moth flights are stronger and longer. Moths lay their eggs in the grass largely at dusk (Kelsheimer and Kerr 1957, Kerr 1955).

Moths require liquid food for survival and for oviposition. No eggs were laid until moths were fed a sucrose and yeast solution; females lived about 14 days. Females that received only water lived only 7 days (Kerr 1955).

Larval Activity. Larvae feed only at night. Newly hatched larvae feed along the midrib lengthwise, and consume only surface cells. After larvae are 10–12 days old, they devour all portions of leaves. Typically, larvae with seven instars require about 25 days for completion of larval life at 25.5°C (78°F). Those with eight instars require 45–50 days for completion at 23°C (73°F). Larvae typically rest in a tightly curled position. Mature larvae spin shapeless silken bags with bits of grass and soil (Kerr 1955).

Miscellaneous Features

Plant Resistance. There is some evidence that certain bermudagrass strains have resistance to TSW oviposition and feeding. When seven clonal selections were exposed for oviposition, fewer adults emerged from two plant introduction selections of South African origin. Least foliage damage occurred to one of these selections plus common bermudagrass and to the hybrid FB-119 (Reinert and Busey 1983).

Natural Enemies

The only parasite mentioned is an ichneumonid wasp, *Horogenes* sp. (Kerr 1955).

8

Lepidopteran Pests: Family Noctuidae

Cutworms and Armyworms

Taxonomy

The popular names *cutworm* and *armyworm* describe larval habits. In many situations cutworms cut off young plants at ground level or a few centimeters above, do no other feeding on the plant, and proceed to the next plant to do the same. This wasteful habit contributes greatly to the highly destructive nature of these insects.

Larvae of some species under high population pressures move armylike across a field, devouring all tender plants in their path. Species with this habit are called armyworms.

Of the many armyworms and cutworms that feed on grasses, relatively few are reported as pests of the same grasses grown under turfgrass cultural conditions in the United States. The few that are listed as turfgrass pests belong to the order Lepidoptera, the family Noctuidae (Phalaenidae), and three subfamilies (Crumb 1956, Hodges et al. 1983).

Subfamily Noctuinae
- *Agriotis ipsilon* (Hufuagel), black cutworm, BCW
- *Peridroma saucia* (Hubner), variegated cutworm, VCW

Subfamily Hederinae
- *Pseudaletia unipuncta* (Haworth), armyworm, AW
- *Nephelodes minians* Guenée, bronzed cutworm, BZCW

Subfamily Amphipyrinae
- *Spodoptera frugiperda* (J. E. Smith), fall armyworm, FAW
- *Spodoptera mauritia* (Boisduval), lawn armyworm, LAW

There have been some name changes, which I mention to avoid later confusions. The BCW was called the *greasy cutworm* in North America, and in Great Britain its approved common name is the *dark sword grass (moth)*. The VCW has also been called the *common cutworm* and the *alfalfa cutworm*, and the moth has been called the *unarmed rustic* in the United States. In England the adult is named the *pearly underwing moth. Peridroma margaritosa* (Haworth) is a binomial synonym of VCW often used in past and in relatively recent literature. The AW was listed under the genus *Cirphis,* and the FAW has been listed under the genus *Laphygma* (Okamura 1959; Rings et al. 1974, 1976).

Importance

All six species feed on grasses as a primary host. In turfgrass culture, the BCW, VCW, AW, and BZCW are considered as relatively minor pests that have an ever-present potential for serious outbreaks on turfgrass. The FAW is a principal turfgrass pest on the mainland only in the Southeast. This species and the four above-mentioned species have a wide host range as major pests of agricultural crops over vast areas of the United States. The LAW, however, was the most serious turfgrass pest in Hawaii. It does not occur on the mainland. Because it is restricted to bermudagrass and to Hawaii, the LAW is discussed separately in the second section of this chapter (Oliver 1982b, Tanada and Beardsley 1958).

History and Distribution

The five species under discussion here have varied but wide distribution over large areas of the United States. One of the most cosmopolitan of species, the BCW is distributed throughout much of the Americas, Europe, Asia, and Africa. Nevertheless, much remains unknown about this important pest. Both adults and larvae are commonly encountered in the field, but there is a poor correlation between light trap catches and the true abundance of larvae at any given location. It is most troublesome in temperate regions east of the Rocky Mountains where rainfall is ample. In California it is distributed throughout the state (Chapman and Lienk 1981, Okamura 1959, Troester et al. 1982).

Another widely distributed species, the VCW, is found throughout North and South America and over much of Eurasia. Adult trap catches in New York correlated well with its pest status as larvae. This species is generally found throughout California (Chapman and Lienk 1981, Okamura 1959, Rings et al. 1976).

The AW extends south of the United States into Mexico and northwestern South America. It is found throughout the United States east of the Rocky Mountains and rather sparingly in the Pacific coast states and into British Columbia (Crumb 1956, Okamura 1959, Walkden 1950).

The BZCW is found generally in North America east of the Rocky Mountains between latitude 35° and 50°. Its greatest abundance occurs from Missouri to Maine and from Minnesota to West Virginia, with Colorado, Kansas, Missouri, Tennessee, and Virginia its approximate southern range (Chapman and Lienk 1981, Rings et al. 1974, Walkden 1950).

The FAW is the most susceptible to cold of the five species. During mild winters it is capable of surviving in the United States only in the southernmost regions of Florida and Texas. It is a permanent resident in South and Central America and in the West Indies. Every season it manages to spread from these areas throughout the United States east of the Rocky Mountains and westward into southern California, Arizona, and New Mexico (Baker 1982, Luginbill 1928, Vickery 1929).

Host Plants and Damage

Cutworms and armyworms feed on grasses as one of their primary host plants, but most species are equally serious pests of other agricultural crops, feeding principally on foliage at the soil surface or on any portion above ground. In some species larvae become subterranean, feeding in the crown and underground fleshy structures.

The BCW, an economically important pest, feeds primarily on grasses. On California lawns it feeds on all common grasses, on dichondra, and on white clover. It is a major pest of corn in the United States. On vegetables it can be a serious pest of roots and tubers (Harris et al. 1962, Rings 1977, Troester et al. 1982).

On lawns, the VCW feeds on bentgrass and white clover in California, but since it has a wide host range, it probably feeds on most grasses, whether it is used in lawns or elsewhere. It attacks many field and forage crops, vegetables, and ornamental plants. At times it becomes a climbing cutworm, feeding on buds and leaves of fruit trees (Okamura 1959, Rings 1977, Walton 1929).

Larvae of the AW feed almost exclusively on grasses, including all lawn grasses and also white clover in California, and are especially abundant in damp situations. The AW is a more serious pest of corn and small grains and is known to cut off the seedheads of wheat. When grass and small grains are not available, it will eat vegetables and ornamental plants. Periodically such large numbers of larvae develop that rather than starve they will move armylike from decimated fields to new fields (Baker 1982, Chapman and Lienk 1981, Okamura 1959, Oliver and Chapin 1981, Rings 1977).

The BZCW is fond of bluegrass and frequently appears on turf and pastures throughout its region. On bluegrass it may be found deep within the crown. It also feeds on clover, small grains, and corn. Often the BZCW becomes a climbing cutworm to eat the buds and leaves of fruit trees (Chapman and Lienk 1981, Rings 1977, Walkden 1950).

FAW larvae prefer grasses and feed on bermudagrass, fescue, ryegrass, and bluegrass, consuming all plant parts that are above the ground. The tips of the grass blades may appear transparent where the plant cells have been eaten, leaving only a thin membrane. The FAW is one of the major turfgrass pests in the southern United States. Home lawns and golf course fairways of bermudagrass have been damaged in Louisiana, southern Texas, and California, but the FAW also feeds on bentgrass, bluegrass, and white clover. In addition, it is a serious pest of small grains and corn. Moths attracted to light may become so numerous around buildings that they soil windows and white walls, so that the surfaces require scrubbing or even repainting (Baker 1982, Okamura 1959, Oliver 1982b, Oliver and Chapin 1981).

Description of Stages

Adults

Moths in this group have forewings with a rather common background color of dull brown, gray, and black, with no outstanding prominent identifying marks. Figure 24, adapted from Forbes (1954) and Rings (1977), showing a schematic wing pattern of noctuid moths, should assist in identifying the various markings on forewings of the species being discussed. The orbicular and reniform are rather indistinct but distinguishable in these species and form the single most reliable identifying mark. The descriptions of Chapman and Lienk (1981) appear below in abbreviated form (also see Plate 18).

BCW. Moths are dark gray to black, some having brown also. Of the five species, this is the only one to show distinct sexual differences in the antennae; the male has

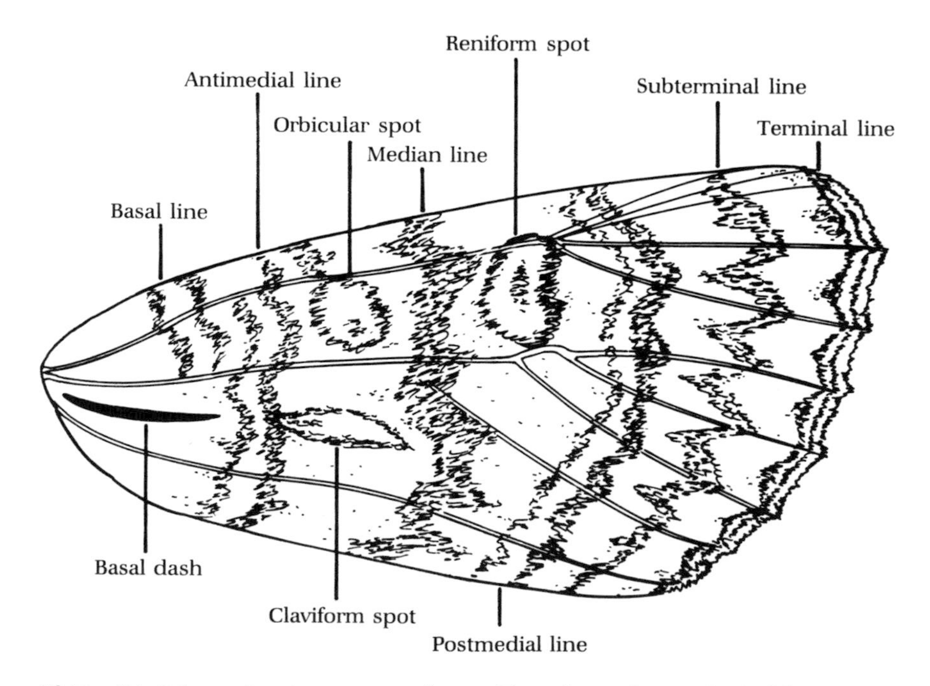

Figure 24. Schematic wing pattern of noctuid moths to show principal lines, spots, and dashes. (Adapted from Rings 1977, fig. 1, courtesy of the Ohio Agricultural Research and Development Center.)

pectinate antennae and the female has filiform. On the forewings a dagger-shaped marking appears at the outer edge of the black-lined reniform.

VCW. The ground color is reddish brown except for the blackish costal margin. Although they appear obscure to the unaided eye, under some magnifications the reniform, orbicular, and the claviform spots appear well formed and are outlined incompletely with black and some white.

AW. Forewings are a fairly uniform light reddish brown. The most distinctive diagnostic character is the small conspicuous white diamond-shaped dot in the center of the forewings in the lower angle of the reniform.

BZCW. The ground color is somewhat variable but is usually rose, purplish gray, or brown. A wide, rich, darker brown band crosses the center of the wing. Large diffuse patches make up the orbicular and reniform. This moth is found only from early September to early October.

FAW. Moths of this species show distinct sexual differences on the forewings similar to those of the lawn armyworm discussed in the next section, with the wings of the male being more vividly marked. Males have a prominent, pale diagonal marking over the orbicular and extending to the costa. Forewings of the female are dull gray brown with small oblique, oblong inconspicuous orbiculars.

The hindwings of all five species have a similar appearance, with a whitish to brownish background color and brownish visible veins. The wingspan of all five species falls within a rather close range and is of little value in separating these species. The FAW, the smallest, has a wingspan of 25–40 mm, while BZCW, the largest, has a wingspan of 34–50 mm (Rings 1977).

Eggs

Eggs of the BCW, VCW, and FAW are circular, oblate, slightly wider than tall, and fairly similar in size. They are 0.47–0.58 mm broad by 0.39–0.50 mm high for the three species; the eggs of the FAW are the smallest. When freshly laid, the eggs of the BCW and the VCW are white, while they are greenish to greenish white in the AW and the FAW. As they age, eggs darken to tan, gray, and dark brown to blackish before hatching.

Eggs are deposited generally in masses by the BCW, VCW, AW, and FAW. The egg masses of the BCW and the VCW remain naked, but those of the LAW are covered with the abdominal hairs of the female. Females of all three of the last-named species deposit eggs indiscriminately on their host plants, on twigs, on fence posts, on buildings, and on other objects (Baker 1982, Crumb 1929, Luginbill 1928, Oliver 1982b).

Larvae

Larvae of the five species have background colors of essentially dull gray or brown to nearly black but interspersed with minor amounts of brighter colors, mainly in the stripes. The dullest of the larvae are those of the BCW, which are dark gray to nearly black; others are brighter, with shades of light pink, yellow, and green. All have a dorso-median stripe that may be continuous or broken. All except the BCW have conspicuous subspiracular stripes of various widths and shades. All have black spiracles except the FAW, which has pale spiracles surrounded by a whitish ring. When full grown the larvae are fairly large, ranging in length from about 30 mm to 45 mm and in maximum width from 4.5 mm to 9.0 mm (Rings and Musick 1976).

The general color of each species and the color and location of the markings, especially of the stripes, are apparent (Plate 19). Specific information on coloration, markings, and measurements for the five species that was given by Rings and Musick (1976) appears below.

BCW. The general color above the spiracles is nearly uniform, ranging from light gray to nearly black. The lower half of body is lighter gray but not strikingly so. A pale and indistinct middorsal line occurs. The spiracles are black. Body length is 30 mm to 45 mm, and the width is 7 mm.

VCW. Individuals range in ground color from dark brown to gray. The middorsal line is broken, leaving from four to seven whitish to yellowish dashes. Both a narrow orange-brown spiracular stripe and a black W-shaped mark on the dorsum of the eight abdominal segments are usually present. Spiracles are black. The body length is about 40 mm; the width is 6 mm.

AW. Its yellowish or gray ground color is more or less tinged with pink. The dorsum of the larva is greenish brown to black, with a narrow, broken, light median stripe. A dark stripe includes the black spiracles in its lower edge. The subspiracular stripe is pale orange, mottled, and edged with white. The body tapers posteriorly and is about 35 mm in length and 5 mm wide in the middle.

BZCW. The general color is light to dark brown to blackish above and paler below, usually with a distinct bronzy sheen. Its middorsal stripe is yellow, broad, and sharply defined. The subspiracular stripe is broad and pale. All spiracles are black. The body length is 35 mm to 45 mm and the width is 9 mm at the middle.

FAW. The general color ranges from pinkish to yellowish, greenish, and dull gray to almost black. There is a faint, narrow, pale middorsal stripe. A broad, sharply defined, yellowish or whitish subspiracular stripe is mottled with reddish brown. The body length is about 30 mm and the width about 4.5 mm, with all abdominal segments about equal in width.

Life History and Habits

Seasonal Cycles

Four of the five species discussed in this section overwinter only in the southernmost regions of North America, and the annual infestations in the temperate regions result from annual northward migrations of moths. The annual migrant species include the BCW, VCW, AW, and FAW. Only the BZCW is indigenous to areas where it occurs in the United States. It overwinters as eggs (Chapman and Lienk 1981, Walkden 1950).

Black-light trap catches determined the annual appearance of moths in the Midwest, namely in Ohio, and in the Northeast, namely in New York. Figure 25 illustrates the seasonal presence of adults. In Ohio, three species whose adults are present from April into November, the AW, BCW, and VCW, have three peaks of adult abundance, indicating three generations each season. Moths of the same three species are present in New York for a shorter duration because of their later arrival. These three species have only two generations a season in New York (Chapman and Lienk 1981, Rings 1977).

The FAW, also dependent on annual northward migration of moths, has a single generation rather late in the season in both Ohio and New York (Chapman and Lienk 1951, Rings 1977). The only indigenous species, the BZCW, has moths appearing mainly in September and peaking at the same time in both states (Chapman and Lienk 1981, Rings 1977).

The number of generations of each species at other locations depends primarily on latitude. The AW has up to four generations in Louisiana and overwinters primarily as pupae, with the largest populations of late-instar larvae appearing from late March into May (Oliver and Chapin 1981).

The BCW in Louisiana has five to six overlapping generations each year, with adults found every month. Larvae are present from March into December and overwinter primarily as pupae. Four broods are reported in northern Tennessee and three in the central Great Plains and in Missouri (Crumb 1956, Oliver and Chapin 1981, Satterthwait 1933, Walkden 1950).

Adults of the VCW are also active throughout the year in Louisiana and have four to five generations annually. The VCW overwinters primarily as pupae. Four generations occur in Tennessee, and three with a partial fourth occur in Kansas. Pupal survival is low even during mild southern winters, but the VCW's enormous reproductive capacity ensures high annual populations (Crumb 1929, Oliver and Chapin 1981, Walkden 1950).

The AW has at least five generations each year in North Carolina. Outbreaks appear and disappear suddenly (Baker 1982).

The species that is most susceptible to the cold is the FAW. The ability to survive mild winters in Louisiana is so uncertain that more southern tropical areas are the main source of annual migrants. As few as 4 broods occur each season in Louisiana,

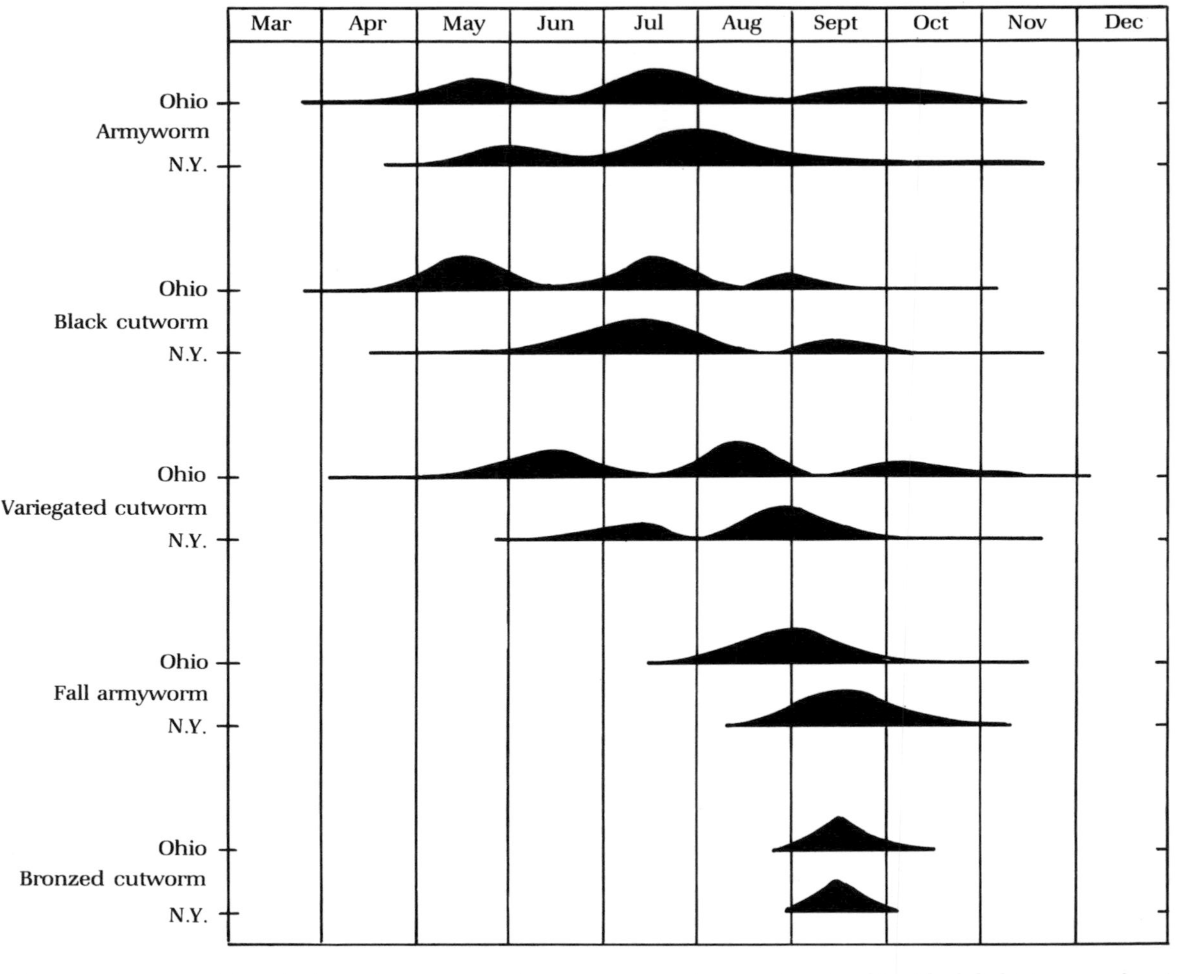

Figure 25. Seasonal distribution of cutworm and armyworm moths as determined by black-light trap catches in Ohio and New York. (Data from Chapman and Lienk 1981 and Rings 1977; drawn by R. McMillen-Sticht, NYSAES.)

while 9–11 generations occur in Brownsville, Texas, where all stages are present during the entire year. There is only 1 generation in Kansas, and larvae are most numerous during late August and September (Oliver and Chapin 1981, Vickery 1929, Walkden 1950).

Moths of the BZCW, the only indigenous species, are present from mid-September to early October in Kansas, oviposit shortly thereafter, and overwinter as eggs. Full-grown larvae are present by late April and pupate about mid-August (Walkden 1950).

The period required to complete a full generation varies considerably among the multibrooded species. The AW requires approximately 60 days, the BCW about 40–65 days, and the FAW only 23–28 days to complete a generation. The VCW requires 58–69 days in spring, 57–78 days in summer, and 101–152 days in the fall (Oliver and Chapin 1981, Vickery 1929, Walkden 1950). Stages in the life cycle of the BCW are illustrated in Plate 20.

Adult Activity

Adults of the five species have only a few habits in common. All are attracted to light, especially black light. Adults of the BCW have an oviposition preference for curled dock, *Rumex crispus* L., and yellow rocket mustard, *Barbarea vulgaris* R. and Br. The basis for the attraction appears to be the low, dense growth forms of these plants. The petioles and the lower surface of leaves receive the most eggs. During August and September in western New York, dirty brown to black moths that are flushed out when the lawn is being mowed and fly only a step or two before alighting are most commonly BCW moths (Busching and Turpin 1976, Chapman and Lienk 1981, author's personal observations).

In Kansas, adult VCW moths are taken every month except during December–February. It is difficult to identify generations, because all stages are found the rest of the year. Of all the cutworms and armyworms observed, the VCW has the greatest fecundity of all. Each female lays 1,185–2,696 eggs, with a mean of 2,111. AW moths oviposit at night in clusters with as many as 130 eggs between the sheath and the blade of grass. Each female can lay as many as 2,000 eggs (Baker 1982, Walkden 1950).

Moths of the FAW are noticed mostly at night, as they are attracted to lights. Egg masses are glued to the undersurfaces and overhang of facial boards as well as to the leaves of trees. Each female lays about 1,000 eggs in masses of as many as 400 eggs two to four layers thick and covered with abdominal hairs. Oviposition begins shortly after dark and lasts until about midnight. Females live about 5 days (Luginbill 1928, Oliver 1982b, Vickery 1929, Walkden 1950).

Larval Activity

Generally speaking, cutworms, and armyworms are nocturnal in habit. The BCW larvae in early instars feed only on foliage, but later instars develop a subterranean habit, cutting off plants at night and pulling them into a burrow in the ground before feeding. At times the BCW adopts the climbing habit to eat the seeds of bushes and fruit trees. In the fall, after aeration of golf greens, larvae often use these holes about 1 × 6 cm (0.4 × 2.5 in.) as a hiding place during the day and migrate out at night to feed on the grass around these holes. The quantity of food eaten by first–third instars is small but increases abruptly between the fifth and sixth instars. Larvae bite off entire plants only to discard them, which greatly increases the insect's destruction. Larvae that are crowded are highly cannibalistic. Pupal cells are formed in the soil by rolling movements (Archer and Musick 1976, Crumb 1929, Harris et al. 1962, Satterthwait 1933).

When large numbers are present, the VCW assumes armyworm habits. Larvae are not strictly nocturnal and frequently feed during cloudy days. The AW invades grass in wet areas and may become most abundant after flooding has occurred. Plants that have lodged to make a dense canopy are often infested with the highest numbers. This species is a true armyworm; the larvae migrate en masse from a decimated area to enter an area of abundant food supply (Oliver and Chapin 1981, Walkden 1950).

First-instar FAW larvae spin silken threads as soon as they have hatched. The threads assist in wind dispersal or allow the insects to lower themselves to the turf below. First instars feed only on the undersurface of leaf blades, leaving behind the upper, clear epidermal layer. Larvae feed any time of the day but are most active early in the morning or late in the evening. Later-instar larvae eat the entire leaf area, giving the turf a ragged appearance. The larvae prefer lush, dense grass. They leave cut leaf particles strewn about, and when abundant, the FAW assumes the true armyworm's habit, moving en masse from a decimated area to a fresh source of food (Baker 1982, Oliver 1982b).

Miscellaneous Features

The BCW has been reared by collecting moths from light traps to start cultures. The adult diet consisted of mainly water, beer, and honey. The larval diet was mainly water and pinto beans plus the usual minor ingredients, which were blended, cooked, and placed in containers. At 21° ± 1°C (70° ± 2°F), 16:8-hr L:D cycle, moths began ovipositing 4 days after emergence (Harris et al. 1962, Reese et al. 1972).

Natural Enemies

All five species have hymenopteran parasites, among which *Apanteles* spp. were common. The five species were also parasitized by tachinid flies of many genera (Walkden 1950).

Lawn Armyworm

Taxonomy

The lawn armyworm, LAW, *Spodoptera mauritia* (Boisduval), belongs to the order Lepidoptera, family Noctuidae, subfamily Amphipyrinae. Most of the information about it comes from Tanada and Beardsley (1958). I cite other work only when it supplements that by these authors.

Importance

Within 1 year of its discovery in Hawaii, the LAW became the most serious pest of bermudagrass lawns. It continued to be the most severe lawn and turfgrass pest during the 1960s, but in more recent years its populations have stabilized, apparently because of numerous parasites and predators. The LAW is now considered one of four major turfgrass pests in Hawaii (LaPlante 1966a, author's personal observations).

History and Distribution

The LAW is a native of the oriental, Indo-Australian, and Pacific regions. It apparently arrived on the island of Oahu in Hawaii well before 1953, when it was first correctly recorded. It is now present on all the Hawaiian Islands but is not known to occur anywhere else in the United States (Fletcher 1956, Pemberton 1955, Tanada 1955).

Host Plants and Damage

In Hawaii, the LAW has inflicted most of its damage on bermudagrass lawns. It will also feed on sedges, sugarcane seedlings, zoysiagrass, and several grassy weeds. Severe damage to lawns is characterized by a sharply defined front of undamaged turf and a more or less completely denuded area (Plate 21). With heavy populations of actively feeding larvae, the front may move about 30 cm (1 ft) each night. The insects leave no leaves or stems behind them in their path (LaPlante 1966a).

Description of Stages

Adult

Males of the LAW are more vividly marked than females and have a conspicuous white diagonal mark in the anterior median area of the forewing between the whitish to buff orbicular spot and the roughly reniform dark spot (Plate 21). In the female the dark reniform spot on the forewing is well defined. Hindwings are pale except for a darker costal margin and outer margins in both sexes. The dorsum of the thorax is covered with elongate grayish to reddish brown scales. The front legs, clothed with brushes of hairlike scales, are much more strongly developed in the males than in the females. Wingspan is 30–37 mm in males and 34–40 mm in females.

Egg

Egg masses (Plate 21), cemented to leaves of trees, on buildings or other objects, are elongate-oval in outline, with five or more irregular layers of eggs. Young females cover their egg masses with long, light brown hairs from their abdomen, so that individual eggs are not visible. As the female ages and her abdominal hairs are exhausted, the last egg masses have more or less naked eggs (Plate 21). There may be 600–700 eggs in a mass. Some egg masses acquire a greenish or pinkish cast (LaPlante 1966a).

Individual eggs are light tan with a pearly luster and darken to gray or dark tan before hatching. Eggs are circular and somewhat flattened and sculptured, with fine longitudinal striations. Each egg is about 0.5 mm in diameter and about 0.4 mm through the polar axis.

Larva

The larvae of the LAW have seven or eight instars in either sex. First instars are about 1.24 mm long, with a head capsule width of about 0.30 mm. Mature larvae are 35–40 mm long, with head capsule widths of about 2.8 mm in the seventh instars and 3.5 mm in the eighth instars.

First instars become greenish soon after feeding and remain predominantly green as second and third instars (Plate 21). Patterns and stripes characteristic of mature larvae develop in the fifth instar. Mature larvae are typically smooth skinned and vary considerably in color, from brown to purplish brown and even blackish (Plate 21). The head capsule and pronotal shield are dark brown. The dorso-median stripe varies in color. A pair of prominent jet black marks occur on each body segment except the prothorax and terminal-segment-located mesad of the prominent yellow dorso-lateral stripes. The spiracles are black.

Pupa

Pupae of the LAW have the same general appearance as those of other army-worms and cutworms and are reddish brown when fully hardened (Plate 21). They average 16 mm in length and 4.5 mm in width.

Life History and Habits

Life Cycle

Inasmuch as this is a tropical species, development is continuous; there is no overwintering stage. The entire life cycle from egg to adult requires about 42 days.

Moths have a preoviposition period of nearly 4 days, eggs hatch in about 3 days, the larval period lasts nearly 28 days, and the pupal period averages nearly 11 days.

Adult Activity

Moths mate within a day after eclosion and start laying eggs about 4 days later. Oviposition begins shortly after dusk and is generally completed before midnight. Eggs are deposited on the foliage of shrubs or small trees, on the lower leaves of tall trees, or on buildings. The moths rarely lay eggs on grass. Since adults are attracted to light, egg masses are often on buildings and foliage near outdoor lights. When the moths are fed sugar water, they live 9–14 days.

Larval Activity

Only first to fifth instars are seen feeding during the day. The older larvae are nocturnal and hide during the day. No cannibalism was observed.

Natural Enemies

The LAW has a whole host of natural enemies that have adapted to it in Hawaii.

Microorganism

A polyhedrosis virus was found infecting larvae a year after discovery of this insect in Hawaii, and the virus was probably introduced by the insect. Attempts to obtain reciprocal infections with a virus of the armyworm *Pseudaletia unipunata* that was present in Hawaii were unsuccessful. The microsporidia *Nosema* spp. were found in eggs (Bianchi 1957, Tanada and Beardsley 1957).

Insect Parasites

Two species of hymenopteran egg parasites attacking LAW eggs in Hawaii are *Telenomus nawai* Ashmead and *Trichogramma minutum* Riley. *Apanteles marginiventris* (Cress.), a larval parasite, appears to be one of the most important natural enemies of the LAW in Hawaii. Three species of tachinid flies were found parasitizing larvae (Laigo and Tamashiro 1966).

Insect Predators

Two species of ants have been observed attacking eggs of the LAW, and coccinellids are found feeding on eggs.

Vertebrate Predators

The giant toad and the common mynah bird have been observed feeding on LAW larvae.

9

Lepidopteran Pests: Family Hesperiidae

Fiery Skipper

Taxonomy

The fiery skipper, FS, *Hylephila phyleus* (Drury), belongs to the order Lepidoptera, family Hesperiidae, subfamily Hesperiinae (Hodges et al. 1983). Members of this family, called skippers, were named for their fast, erratic flights. Members of the subfamily are known as tawny skippers, and their larvae are chiefly grass feeders (Borror et al. 1981).

Importance

In Hawaii, the FS is considered the third or fourth most serious lepidopterous pest of turfgrass but has the potential of being the most serious pest during the warmest period of the year. It is considered one of the five most injurious lepidopterous pests attacking lawns in California. Turfgrass damage by the FS has not, to my knowledge, been reported from any other state (Okamura 1959, Tashiro and Mitchell 1985).

History and Distribution

The FS is a wide-ranging butterfly found from South America northward to Connecticut, Michigan, and Nebraska and abundant in much of the southeastern United States. It is probably a permanent resident south from the coastal Carolinas and lower Mississippi Valley and regularly migrates to more northern areas each summer. In California it is abundant from the lowlands of southern California to the San Francisco Bay area, and it is generally distributed throughout the rest of the state in residential and agricultural areas (Comstock 1927, Klots 1951, Okamura 1959, Opler and Krizek 1984).

In Hawaii, the FS was first found on the island of Oahu in 1970, and since 1973 it has been present on all the islands except Lanai (Tashiro and Mitchell 1985).

Host Plants and Damage

The larvae of the FS feed on all common lawn grasses but may prefer bermudagrass for oviposition as well as for food. Other foods include St. Augustinegrass, bentgrass, and weedy grasses, especially crabgrass. Early stages of infestation are marked by isolated round spots measuring 2.5–5.0 cm (1–2 in.) in diameter where grass blades have been removed by a single larva. Coalescence of individual spots causes large areas of lawn to die (Bohart 1947, Okamura 1959, Opler and Krizek 1984).

In Hawaii, larvae were taken from bermudagrass during a survey that focused primarily on other turfgrass pests. Adults at rest were observed most frequently on the bermudagrass fairways of golf courses on the island of Oahu and on bermudagrass lawn bowling greens in Honolulu as females landed for oviposition. Flying adults were most frequently observed as they visited the flowers of lantana, *Lantana camara* L., and the flowers of other plants to feed on nectar (Tashiro and Mitchell 1985).

Description of Stages

Adult

The FS adults are predominantly orange, yellow, and brown butterflies with a wingspread of about 25 mm (Plate 22). The males are slightly smaller than the females but can be distinguished more readily by coloration. The males are predominantly bright orange yellow above and pale yellow, with submarginal dark spots on the underside of both forewings and hindwings. The females are predominantly dark brown on the upper surface, with coloration similar to that of males on the underside but much overlaid with olivaceous dusting (Bohart 1947, Klots 1951).

Egg

The hemispherical eggs (Plate 22), glued singly to grass blades, turn from white to powder blue to greenish blue in 1–2 days. Before hatching, they become nearly white again, and the black head becomes plainly visible. Eggs range from 0.70 mm to 0.75 mm in breadth and from 0.50 mm to 0.55 mm in height (Bohart 1947, Tashiro and Mitchell 1985).

Larva

The body of the first-instar larva (Plate 22) is pale greenish yellow, with a granular surface appearance that becomes more pronounced in the second-instar larva. The most distinctive feature of a skipper larva is the strongly constricted neck and the presence of a narrow but prominent pronotal shield in all five instars (Plate 22). Both head and shield are coal black in all five instars. The black head of a mature larva is finely pitted, with minute setae, and the front bears several reddish brown markings, two elongate spots near the base and two parallel lines, one on each side of the epicranial suture (Okamura 1959, Tashiro and Mitchell 1985).

In later instars, the body becomes yellow brown to gray brown, with an indistinct to distinct median longitudinal stripe. Even fainter lateral stripes may be present. Many short secondary setae cover the entire body, and the granulated appearance is maintained. The prepupa differs little from the mature larva except that it becomes rigidly straight (Bohart 1947, Okamura 1959, Tashiro and Mitchell 1985).

In size, FS larvae grow in head capsule widths from a mean of 0.44 mm in first instars to 1.75 mm in fifth instars. Mean body lengths increase from 2.5 mm in first instars to 24–25 mm in fifth instars (Tashiro and Mitchell 1985).

Pupa

Pupae of the FS are 15–18 mm long and have abdominal and head areas covered with sparse, bristly hairs (Plate 22). Young pupae have a head and thoracic area colored light green and an abdomen of light tan. As they mature, they turn an overall brownish shade, with an olivaceous abdomen and considerable black in the head. The color of the forewings becomes apparent before eclosion. The pupae are found either free in the grass-root zone or partially enclosed in loosely webbed debris in the grass near the soil surface. Males and females can be separated according to the methods of Butt and Contu (1962; also see Plate 22, Bohart 1947, Tashiro and Mitchell 1985).

Life History and Habits

Seasonal Cycle

In areas where the FS is a permanent resident, it appears to have three to five generations per year. It may be seen throughout the year in Florida but is not seen during January and February in Mississippi. In the northern states, the FS is not seen until May and is not common until late August. In California, adults were most numerous in late August and September, especially around low flowers such as red clover, *Trifolium pratense* (Bohart 1947, Opler and Krizek 1984). In Hawaii during 2 years, 1975 and 1982, I saw them more abundantly during May–July than earlier in the year.

When it was reared on bermudagrass at 24°C (75°F), the insect required an average of 48 days to pass from egg to adult. When reared at 27.5°–29.0°C (81°–84°F), it required an average of only 23 days to complete the same development (Bohart 1947, Tashiro and Mitchell 1985).

Adult Activity

Adults of the FS are most often seen as rapidly flying butterflies frequenting the flowers of lantana, honeysuckle, alfalfa, clover, and other plants to feed on nectar. Their most rapid flights occur when males are pursuing females but they are much too rapid in flight for easy netting, even when they are flying more slowly to seek flowers (Tashiro and Mitchell 1985).

Males perch all day close to the ground, waiting for females. The flight of a passing female elicits male response. Males mate primarily with virgin females, and mating pairs have been seen in late afternoon. The union lasts about 40 min, and if she is disturbed, the female takes flight, carrying the male. During the heat of the day, females alight on the turfgrass to deposit an egg before flying a short distance to repeat the process. Eggs are generally cemented singly to the underside of grass blades, but smaller numbers, at least in captivity, were placed on stems and on the upper surface of blades. When ovipositing, females can be caught if they are approached slowly and a net is dropped over them (Shapiro 1975, Tashiro and Mitchell 1985).

Larval Activity

Newly hatched larvae notch the edge of leaves (Plate 22) and in later instars eat entire leaves. Starting in the third instar, the larvae actively spin profuse amounts of strong webbing. Larvae are seldom seen even when they are abundant because they remain concealed in the thatch area in lightly woven silken shelters. During two separate 6-month periods in Hawaii, January–July, I did not see a single larva in undisturbed turf except those forced to the surface by survey techniques (Tashiro and Mitchell 1985).

Pupal Activity

Pupation occurs in loosely woven shelters (Plate 22), or pupae are free near the soil surface if debris is not readily available (Bohart 1947, Tashiro and Mitchell 1985).

Miscellaneous Features

Greenhouse Rearing

Field-collected females, all presumed to have mated, laid no eggs under artificial light in the laboratory. They readily laid viable eggs when held in screen cages, measuring 24 × 24 × 24 cm (a cube with a 9.4 in. side), in the greenhouse under natural light at diurnal temperatures of 26.5°–35.0°C (80°–95°F). Oviposition began about 10:00 AM, continued throughout the afternoon, and conformed with limited observations made in the field. Potted FB-137 bermudagrass was placed in each cage as an oviposition medium, and blooming potted lantana was used as a source of nectar.

Adults from laboratory-reared larvae (under artificial illumination) were paired and were placed in screen cages of two sizes and held in the greenhouse. Potted FB-137 bermudagrass was placed in each cage as an oviposition medium and potted lantana in bloom as a source of nectar. The pairs held in a cage of 0.4 m^3 (14 ft^3) oviposited readily, producing 83% viable eggs, and compared favorably with field-collected females, which deposited 99% viable eggs. Pairs held in a cage of 0.014 m^3 (0.5 ft^3) produced only a few eggs of only 7% viability (Table 10). This difference

Table 10. Oviposition and the viability of eggs of the fiery skipper from field-collected and laboratory-reared butterflies

Days from oviposition to hatching	Source adults, cage size, and total eggs		
	Field,	Laboratory	
	24 × 24 × 24 cm, 586 eggs	61 × 99 × 66 cm, 149 eggs	24 × 24 × 24 cm, 212 eggs
2	0	1	0
3	271	50	2
4	181	71	7
5	114	0	5
6	15	1	0
Total hatched	581	123	14
Viable eggs (%)	99	83	7

Source: Adapted from Tashiro and Mitchell 1985.

seemingly indicates that greater flight activities are needed for normal reproductive development (Tashiro and Mitchell 1985).

Laboratory-reared females began ovipositing on the 3d day following eclosion and peaked between day 5 and day 9. Their egg production varied from 9 eggs to 71 eggs per female, while field-collected females laid 14–168 eggs per female. The maximum life of laboratory-reared females was 11 days (Tashiro and Mitchell 1985).

Development of eggs was relatively rapid. When they were held at 26.6°C (80°F) and a saturated atmosphere, they began to hatch in 2 days, with more than 80% hatching on the 3d and 4th days; by the 6th day hatching was completed (author's personal observations).

Natural Enemies

Very little is known about natural enemies of the FS. Only two hymenopteran insect parasites are recorded—a braconid, *Apanteles* sp., and an ichneumonid, *Amblyteles* sp. (Bohart 1947).

10

Coleopteran Pests: Family Scarabaeidae

Overview

The larvae of turfgrass-infesting species of the family Scarabaeidae form a grub complex whose members are similar in general appearance, in habits, and in the turfgrass damage they cause. Because of their similarities, a general account will serve for an understanding of the group as a whole.

Taxonomy and Nomenclature

No fewer than 10 species belonging to five subfamilies are pests of turfgrass in the United States. The larvae of this family are known also as *grubs,* a term more specifically applied to the larvae of Coleoptera and Hymenoptera. This is the only stage that is destructive of turfgrasses as well as of the underground portions of many other plants. According to Ritcher (1966) *white grubs* is the common name applied in many countries to larvae of the family Scarabaeidae and in particular to those important to agriculture. Ritcher subsequently states, however, that larvae belonging to the subfamily Melolonthinae are known universally as white grubs. In current usage, many individuals engaged in turfgrass research and management refer to the masked chafers of the genus *Cyclocephala,* subfamily Dynastinae, as *annual white grubs.* This confusion regarding *white grubs* poses no real conflict as long as we recognize that the general term is loosely used.

Some authors have divided the family into two groups, calling them lamellicorn scavengers and lamellicorn leaf chafers. This designation reflects the most notable morphological character, the lamellate antennae (described later) of the family, and the varied food habits of both adults and larvae.

Importance

Grubs of the Scarabaeidae are the most serious turfgrass pests we have in the Northeast and are considered a major pest in the Midwest. Their strictly subterranean habits make them the most difficult of turfgrass insects to control. Of necessity, applications of chemicals must be made to the surface of established turf; gravity and precipitation (including irrigation) are needed to move chemicals to the target zone. These requirements greatly restrict the availability of effective insecticides.

In addition to the destructive nature of grubs, adults of some species are voracious leaf feeders on the trees and shrubs used in the turfgrass environment. Both the adults and the grubs of some turfgrass pests are serious pests on plants outside the turfgrass environment, making certain species such as the Japanese beetle and June beetles general agricultural pests (Fleming 1972, Ritcher 1940).

History and Distribution

Some of our most serious northeastern turfgrass pests of the family Scarabaeidae are introduced species, the most notable being the Japanese beetle, followed (in order of importance) by the European chafer, the oriental beetle, and the Asiatic garden beetle. All of these were accidentally introduced into the United States from the Orient or Europe, as their various names imply, and were discovered along the eastern seaboard between 1916 and 1940. With the exception of the Japanese beetle, which is a serious pest in the Midwest, Southeast, and the Northeast, these beetles are currently problems predominantly in the Northeast.

Some of the native Scarabaeidae are turfgrass pests as serious as the introduced species. In general, the native species cover a much wider geographic distribution. The most notable of these are the northern and/or southern masked chafers that are present as destructive pests from the Northeast through the Midwest and extending south and west into Florida, Texas, and California. The black turfgrass ataenius is currently an important turfgrass pest from Colorado to the Atlantic Coast in the northern half of the country. The most widespread species are the group known commonly as May or June beetles in the genus *Phyllophaga*. There are 152 valid species that occur in the United States and Canada. They are distributed throughout the entire country but are most serious in the northeastern quarter of the United States, in Texas, and in the Canadian provinces of Ontario and Quebec. Not all are serious pests (Luginbill and Painter 1953, Ritcher 1940, 1966).

Host Plants and Damage

Adult Feeding

While all turfgrass-infesting scarabaeid grubs cause similar damage, the adults have food habits that vary according to species. Some feed on dead and decaying organic matter, including carrion and dung (for example, the black turfgrass ataenius). Others feed on foliage, flowers, and fruits (for example, the Japanese beetle). Some adults do not feed at all (for example, masked chafers).

Of the turfgrass-infesting species that damage the parts of plants that are above the ground, the Japanese beetle is by far the most notorious. It feeds on nearly 300 species of plants, damaging the foliage, flowers, and fruit. Oriental beetles feed on some of the same plants but with a comparatively more restricted diet. Many of the May or June beetle adults feed on foliage of many deciduous shade and forest trees. Some also feed on petals of flowers (for example, the oriental beetle). Scarabaeids that injure turfgrass but do not feed as adults, or feed so little that they do not damage plants, include the Asiatic garden beetle, the black turfgrass ataenius, the European chafer, and the northern and southern masked chafers (Fleming 1972, Hammond 1940, Johnson 1941, Ritcher 1940, Tashiro et al. 1969).

Larval Feeding

Turfgrass-infesting scarabaeids are injurious because they feed on the roots of all species and cultivars of the commonly used turfgrasses. Indiscriminate pruning of the roots at or just below the soil-thatch interface is the primary cause of turf damage. When severe root pruning is accompanied by drought, death of the turfgrass can be rapid and certain.

Damage: Seasonal Aspects. The season of turfgrass damage is directly related to life cycle. With species having 1-year life cycles, severe turfgrass damage occurs during late summer and early fall, after grubs have become third instars (at this stage they are most destructive because of their size), and fall rains stimulate grubs upward to the soil-thatch interface. Damage continues during the spring as overwintering grubs return to the surface and feed on the turf damaged in the fall. Feeding during the spring, prior to the maturation of third instars and prior to pupation, is of relatively short duration but vigorous, and serious turf damage may not become obvious until this feeding period. Turfgrass damaged by scarabaeid grubs during midsummer cannot be attributed to members of this group except for the black turfgrass ataenius (Fleming 1972, Tashiro et al. 1969).

Invasion by the black turfgrass ataenius causes damage to golf course fairways during June, July, and August where two generations a year are present and during late July and August where only one generation a year is present (Wegner and Niemczyk 1981, author's personal observations).

Members of the genus *Phyllophaga* have 1-, 2-, or 3-year life cycles, the last being the most common. Major turf damage occurs throughout the summer of the second year, when mature second instars and vigorously feeding young third instars are present. Minor turf damage occurs during late summer of the first year and late spring of the third year as mature third instars complete their feeding preparatory to pupation during midsummer (Hammond 1940, Ritcher 1940).

Damage: Progression. Turf-damaging populations of grubs produce symptoms varying from turf weakness to death of large patches (Plate 23). A gradual thinning and weakening of the stand may be one of the earliest symptoms as grubs feed on roots near the soil surface. Damage may progress from thinning to small patches of dead grass. When the population of grubs is large, there may be a sudden wilting of the grass caused by the severe root pruning even with adequate soil moisture. As damage continues, the surface of the soil may become very spongy, the sponginess being evident as one walks over the infested area. Birds and mammals that are predators cause further damage, as they tear the turf in search of grubs.

With high grub density, root pruning in the soil-thatch interface may be so complete that the sod can be lifted in large sections to reveal the actively feeding grubs. The exposed grubs may be only a part of the population, since many will be present just under the soil surface (Plate 23).

Description of Stages

Adults

Beetles of this large family have a number of distinguishing characters that are illustrated in Figure 26. They are mostly short and stout bodied and are usually convex with shortened elytra, the leathery or chitinous forewings serving as a covering to the hindwings. In many species, as in the European chafer, the tip of the abdomen is exposed (Figure 26). The membranous hindwings are folded under the elytra and are visible only when the beetle is in flight. The tibias of the forelegs are fossorial and are fitted with broad teeth on the outer edge. Most beetles are predominantly dull colored (for example, the oriental beetle), but some are brightly colored, with a metallic sheen (for example, the Japanese beetle).

The lamellate club (flagellum) of the antennae, consisting of the last three segments (rarely more), is the most distinguishing character and indicates the origin of the name *lamellicorn beetles* (Figure 26). Beetles readily spread these segments apart or unite them to form a compact club. In many species the antennal club of males is about twice as long as that of the female, making separation of sexes fairly simple (Figure 26). In other species, sexual differentiation of adults (exclusive of sex organs) occurs on other parts of the body, for example on the legs.

Eggs

All turfgrass-infesting scarabaeids deposit eggs that are nearly identical in shape and appearance except for size. Freshly deposited eggs are shiny, milky white, and ellipsoidal, about 1.3–1.5 times as long as they are wide. Eggs absorb water as they mature and become more spherical. The surface of the chorion may be smooth or textured. The elastic chorion stretches to accommodate the growing embryo, making the egg resilient and able to bounce without injury from hard surfaces. In fully mature eggs, the tan mandibles of the larva are visible through the translucent chorion (Figure 27).

Larvae

Range in Size. Turfgrass-infesting scarabaeid grubs in the United States vary in size, from the small, black turfgrass ataenius to the large, green June beetle grubs.

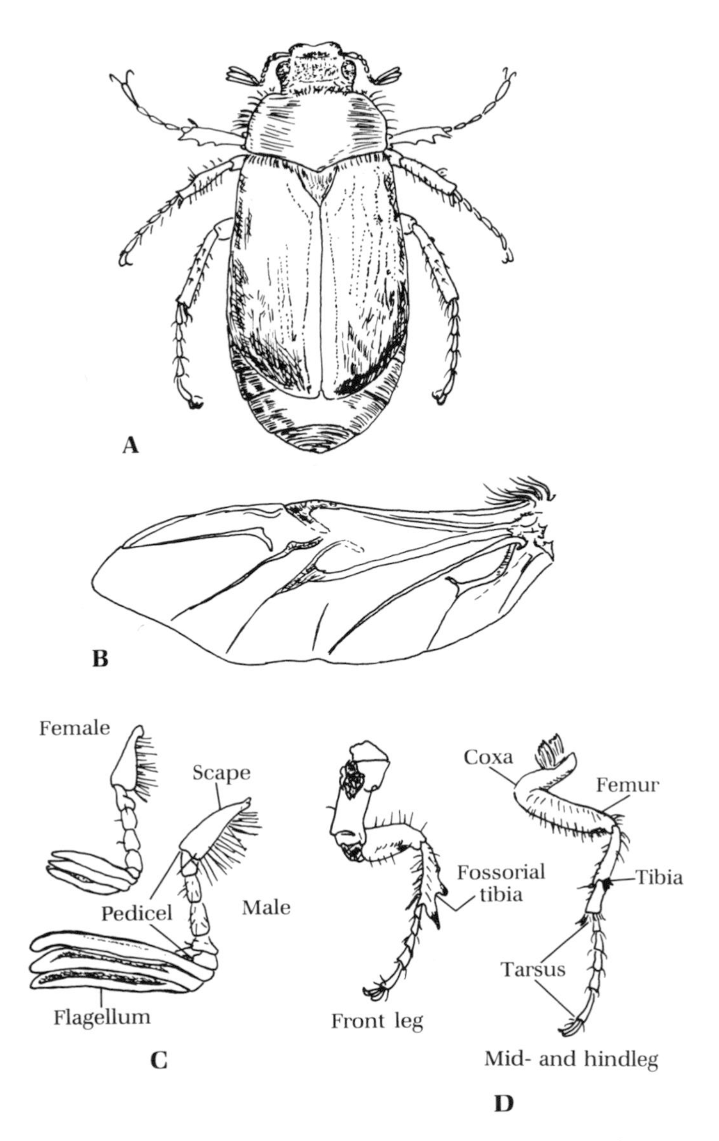

Figure 26. Adult scarabaeid characteristics, illustrated by the European chafer. **A.** Adult. **B.** Membranous hindwing folded under elytra at rest. **C.** Lamellicorn antennae. **D.** Legs of beetles. (Adapted from Butt 1944, plates 6b, 7c, 10c, 10d, 10f, courtesy of the Cornell University Agricultural Experiment Station, Ithaca.)

Mature grubs of the former are less than 10 mm in length, while the latter are nearly 50 mm in length (Figure 28). Species with medium-sized mature grubs represented by the Japanese beetle are about 20–25 mm in length and weigh about 90–270 mg (Ahmad et al. 1983).

Figure 27. Development of scarabaeid (oriental beetle) eggs from oviposition to maturity, showing parts of the first-instar grub through the translucent chorion. (Photo by B. Aldwinckle, NYSAES.)

General External Features. Newly hatched larvae are practically translucent white, but after feeding, the posterior region turns various shades of gray to brown, depending on the color of the ingested soil and the plant material in the rectal sac.

The most distinguishing feature of scarabaeid larvae is their C-shaped contour (Figure 29). The only time larvae are seen straightened to their full body length is when they have been removed from their larval medium and are seeking to burrow back into it (Figure 29).

The body consists of a brown head with strongly developed mandibles, the three thoracic segments, each having a pair of legs, and a 10-segmented abdomen. The integument is transversely wrinkled and is covered with rather long, scattered brown hairs. In some species, there is a slight constriction of the abdominal segments of the midsection (for example, in masked chafers; see Plate 24), while other species show no evidence of constriction (for example, the oriental beetle). Common to all members of this family is a pair of spiracles on the prothoracic segment and on each of the first 8 abdominal segments. Characters found on the larval head (Figure 29) and epipharynx (Figure 29) are useful in distinguishing species. To observe characters on the epipharynx, the labrum can be turned back to expose the epipharynx without injury to the living grub.

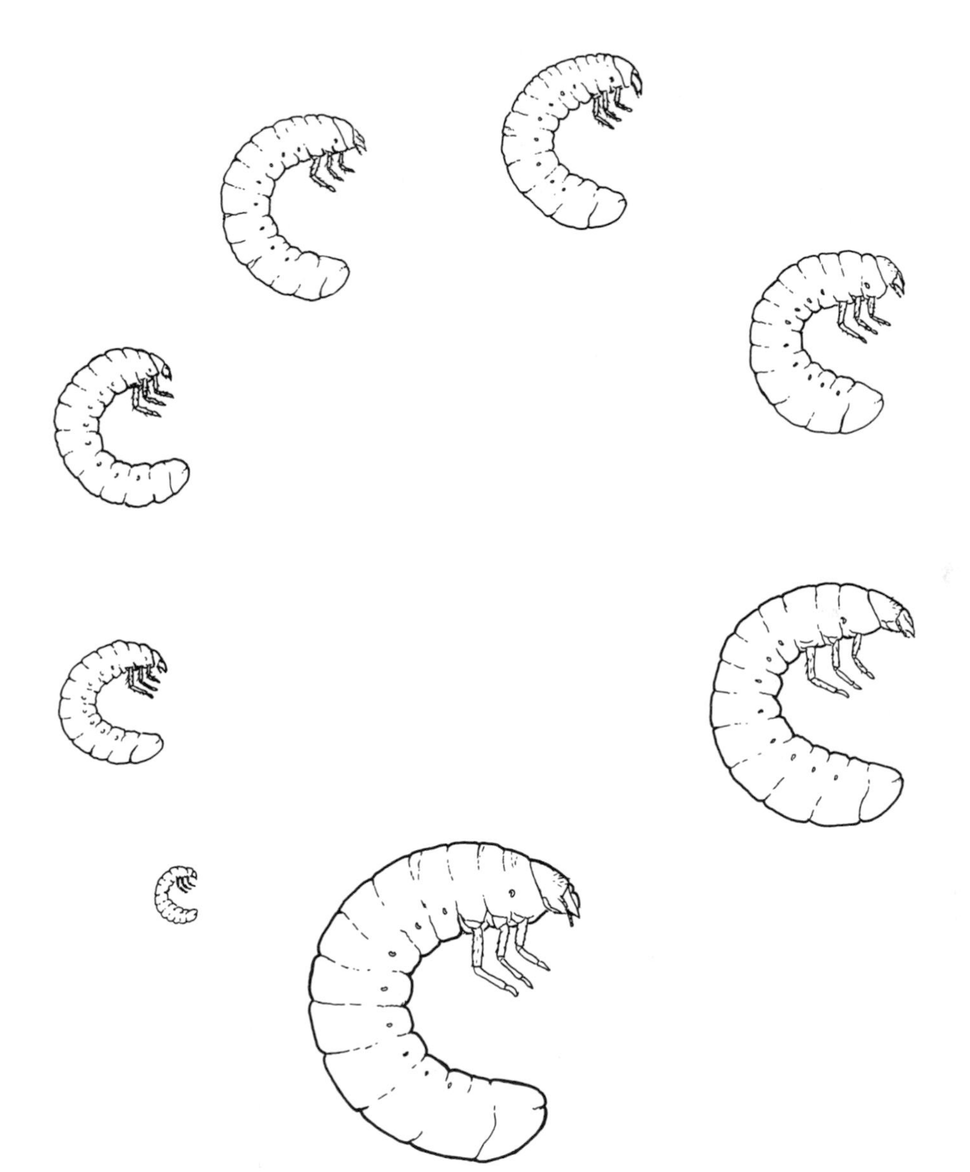

Figure 28. Relative sizes of turfgrass-infesting scarabaeid grubs. Clockwise: the black turfgrass ataenius (smallest), the Asiatic garden beetle, the Japanese beetle, the oriental beetle, the European chafer, the northern masked chafer, the May or June beetle, and the green June beetle (largest). (Drawn by H. Tashiro, NYSAES.)

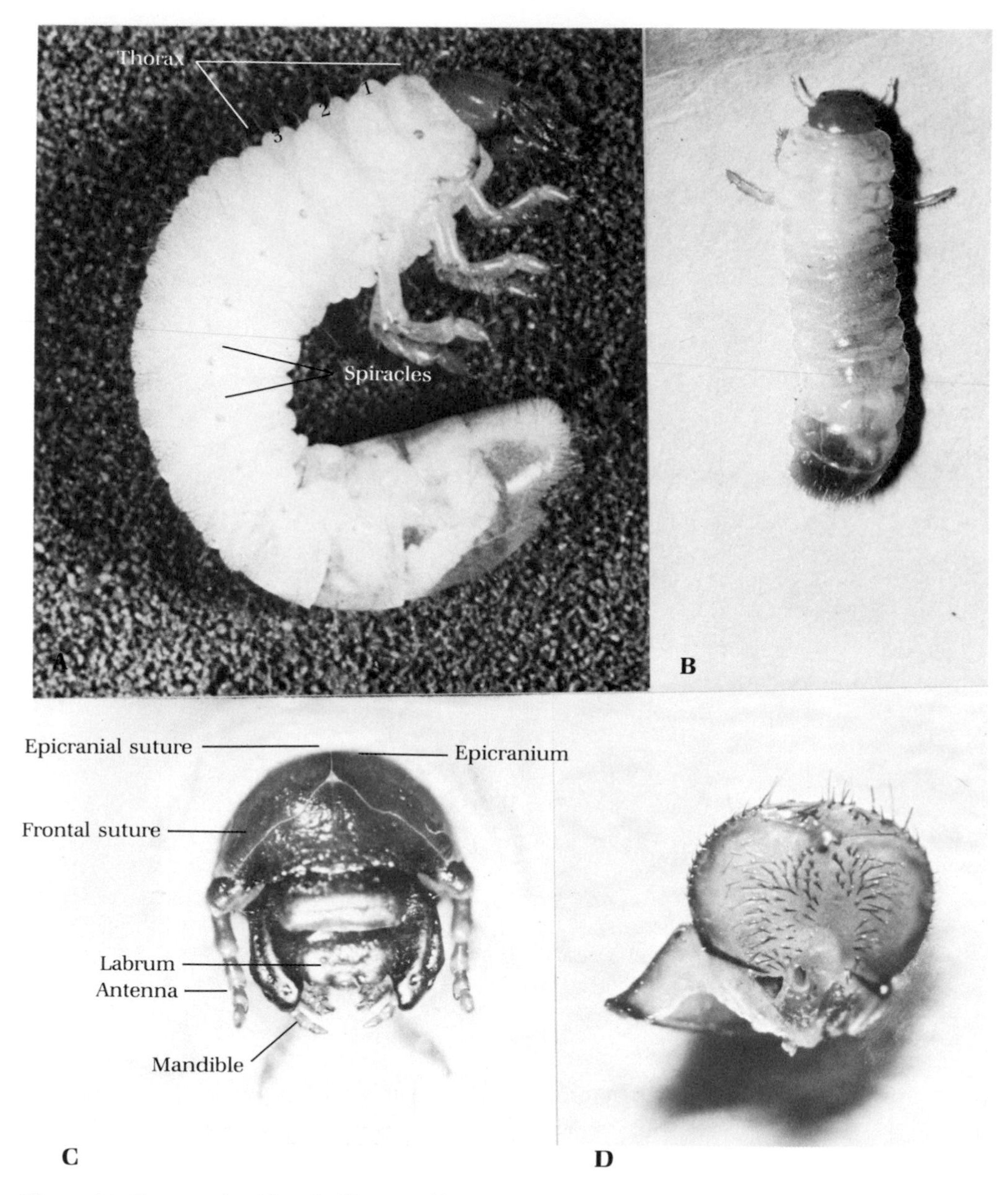

Figure 29. The scarabaeid grub, illustrated by various species. **A.** Natural position. **B.** Stretched-out unnatural position. **C.** Front view of head. **D.** The epipharynx, or inner surface of the labrum. (Photos by B. Aldwinckle, NYSAES.)

All species have three larval instars. Freshly transformed larvae of each stadium have head capsules distinctly wider than the thorax and abdomen. As growth and maturation progress, the width of the thorax and abdomen greatly expand and exceed that of the heavily chitinized head capsule, which remains constant during each stadium.

Rastral Pattern. In scarabaeid larvae, the ventral area of the last (10th) abdominal segment just anterior to the anus has definitely arranged spines, hairs, and bare spaces and is called the *raster* (Figure 30). Boving (1942) designated the median longitudinal bare area the *septula.* A longitudinal line of *pali* (spines) lying on either side of the septula is called the *palidium,* with the area lateral to each palidium and covered with scattered spines and hairs known as the *tegillum.*

The anal slit is surrounded by the upper and lower anal lobes. The anal slit may be transverse, as in the Japanese beetle; Y-shaped, with the stem generally shorter than the arms, as in May or June beetles; or essentially longitudinal, as in the Asiatic garden beetle.

The characters of the rastral patterns and anal slits are useful for identification of species, especially in the field. A hand lens 10× or stronger with good illumination is essential for identification of all three stadia of the medium-sized to larger larvae common in turfgrass.

Internal Anatomy. The alimentary tract of scarabaeid grubs is quite similar for all species and is illustrated by that of the European chafer (Figure 31). It consists of three anatomically differentiated areas—the stomodaeum (foregut), the ventriculus (midgut), and the proctodaeum (hindgut). The ventriculus occupies the largest area, extending approximately from the mesothoracic segment to the eighth abdominal segment. Caecal diverticula are present at the anterior and posterior ends of the ventriculus. Malpighian tubules originating at the junction of the anterior intestine (of the proctodaeum) and ventriculus extend anteriorly the length of the ventriculus, then posteriorly to the ventral surface of the rectal sac (Splittstoesser et al. 1973).

Prepupae

The prepupa, the terminal period of the final larval instar, is an active but nonfeeding stage of some but not all holometabolous insects (with complete metamorphosis). A prepupa occurs in all scarabaeids and occupies an earthen cell prepared by the grub (Plate 24). Some prepupae retain the C-shaped contour of the larvae (for example, the oriental beetle), while others straighten out except for a slight crook at the posterior end (for example, the European chafer; Plates 24, 33).

Pupae

Scarabaeid pupae are all similar in appearance except for size and have legs and wings that are free from the body and clearly visible (Plate 24). As pupation occurs,

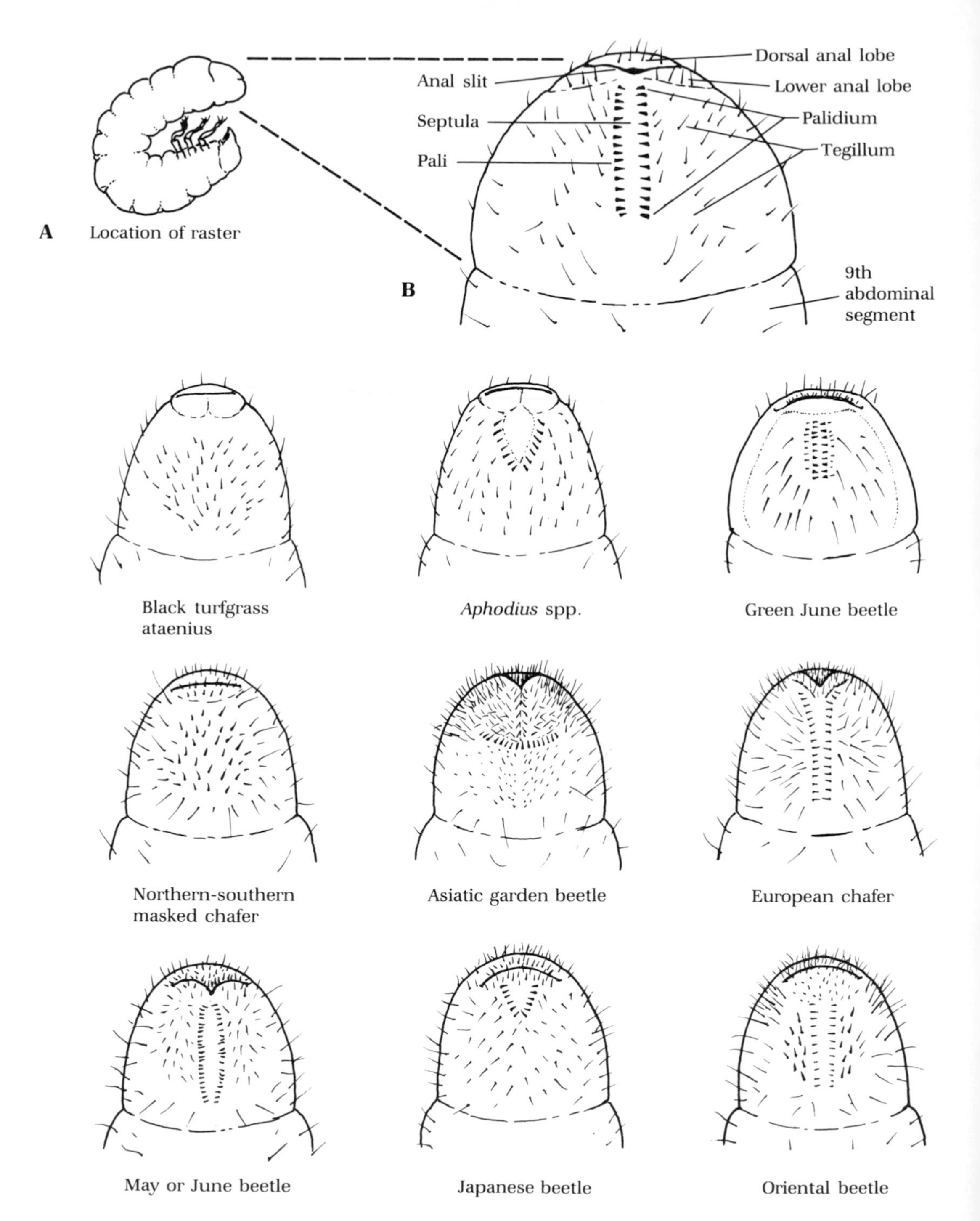

Figure 30. Rastral patterns of turfgrass-infesting scarabaeid grubs, not to scale. **A.** Position of raster. **B.** Details of raster and anal area. (Adapted in part from U.S. Department of Agriculture 1951–1980; drawn by R. McMillen-Sticht, NYSAES.)

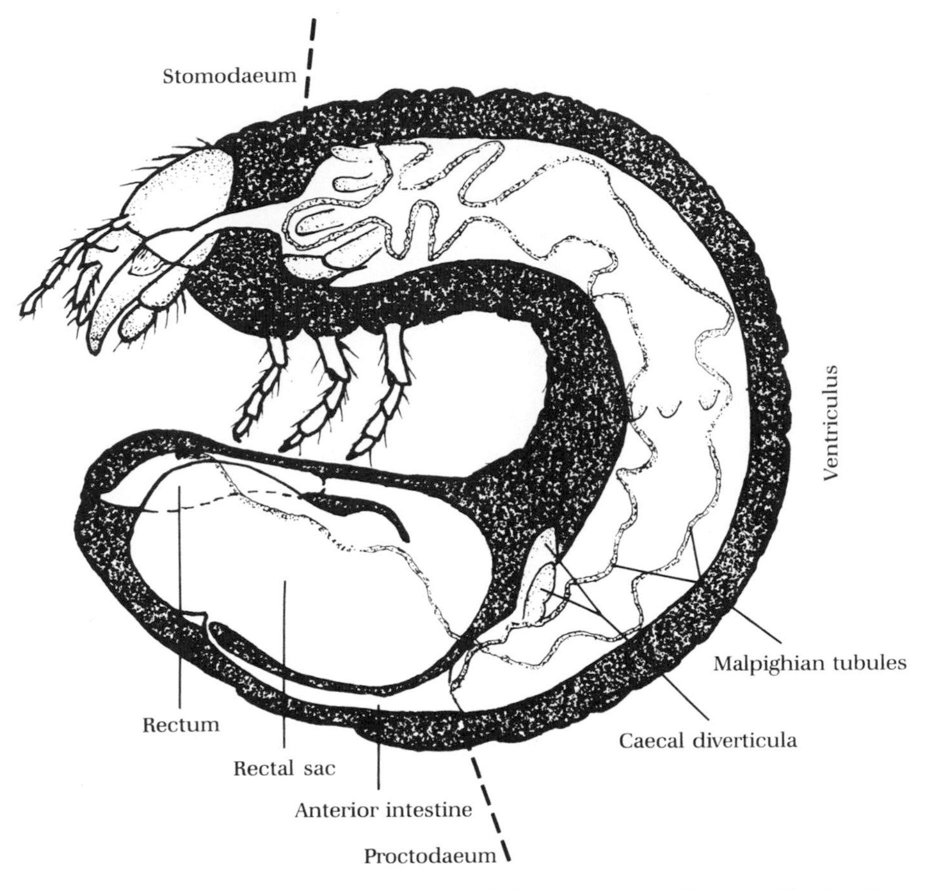

Figure 31. The internal anatomy of the scarabaeid digestive system, illustrated by the European chafer. (From Splittstoesser et al. 1973, fig. 1, courtesy of Academic Press.)

members of some subfamilies slough off the larval exuviae, which remain attached to the last abdominal segment (for example, Melolonthinae: European chafer; Plate 24). In other subfamilies, pupation occurs within the thin, meshlike exuviae, which eventually split to release the pupa (for example, the Rutelinae: Japanese beetle; Plate 24). Young pupae are creamy white and extremely tender. As they age, they gradually become the adult color just before transforming to adults.

Sexual Differentiation: Larvae and Pupae

Both males and females of scarabaeid grubs have internal sex organs called *terminal ampullae* that lie along the ventral median line of the ninth abdominal segment in males and the eighth abdominal segment in females. The ampullae are not visible externally in the females but are generally visible in males except when a layer of fat obscures these organs. When the terminal ampullae are visible, they appear as two

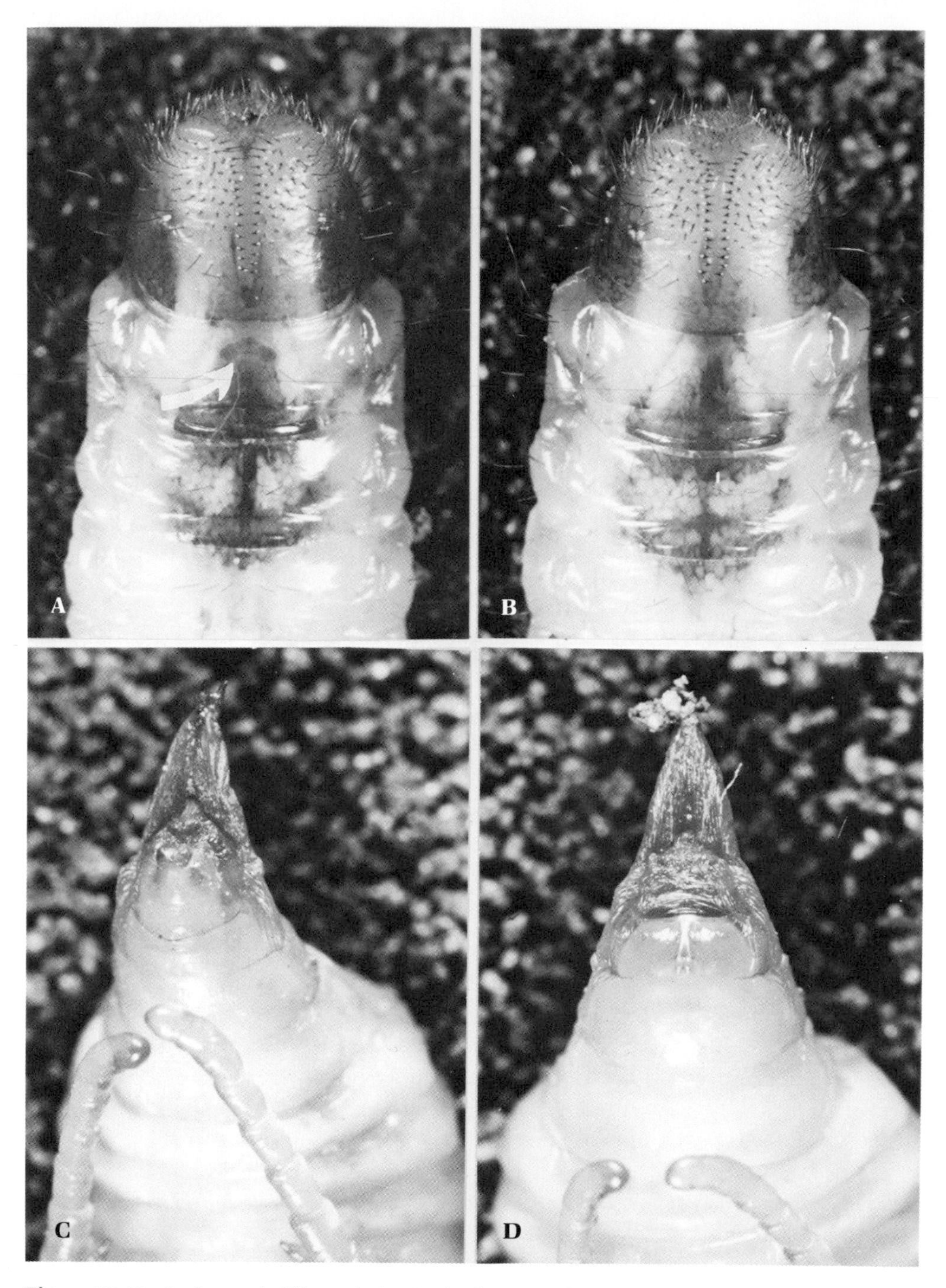

Figure 32. Typical sexual differentiation in the larvae and pupae of scarabaeids, illustrated by the European chafer. **A.** Male larva with terminal ampullae visible. **B.** Female larva with ampullae not visible. **C.** Terminal ventral view of male pupa. **D.** Terminal ventral view of female pupa. (From Tashiro et al. 1969, fig. 11, courtesy of the NYSAES.)

ovoidal structures (Figure 32) in the ninth abdominal segment, and the grub is certain to be male (Hurpin 1953). The terminal ampullae are more clearly evident in the European chafer than in several other common turfgrass species examined by the author.

The sex of pupae can be determined easily by wholly external differences in the ventral portion of the terminal abdominal segment. Male genital organs are more pronounced than those of the female (Figure 32; Tashiro et al. 1969).

Seasonal History and Habits

Seasonal Cycles

Turfgrass-infesting scarabaeids, depending on species, have two generations a year or one generation a year or require 2–4 years to complete a generation. Three-year life cycles are the most common in the last category.

The black turfgrass ataenius has two generations a year from the latitude of Connecticut southward and one generation a year north of this latitude. In either case, grubs do their major damage during the summer, and adults overwinter.

Turfgrass-infesting scarabaeids that normally have a 1-year life cycle in much of their geographic range may have a few individuals that require 2 years to complete their life cycle in the most northern latitudes of their range. Such is the case with the Japanese beetle and the European chafer. Those that complete a generation in 1 year overwinter as third instars, causing their major turf damage in the fall and again in the spring.

During the 3-year cycle, a May or June beetle spends the 1st winter as a second instar, the 2d winter as a third instar, and finally the 3d winter as an adult. Major turf damage occurs during the summer and early fall of the 2d year. Some May or June beetles complete a generation a year in their most southern range. They are suspected of requiring 2 years in their most northern range. Such is the case with *Phyllophaga crinita* Burmeister in Texas (Frankie et al. 1973).

Adult Activities

General Habits. Adults are either nocturnal or diurnal and are present above ground for short periods during late spring and summer or into early fall. Many of the nocturnal species are strongly attracted to light and are easily captured in light traps. A female sex pheromone has been demonstrated to be present in several important turfgrass species, including the Japanese beetle and the northern and southern masked chafers (Ladd 1970, Potter 1980).

Adults may feed on foliage, blossoms, and fruit or on organic matter or may not feed at all, depending on the species. I discuss feeding habits more thoroughly in chapters describing the individual species.

Oviposition. All turfgrass pests of this family deposit their eggs in moist soil singly in earthen cells or in groups of about 10 eggs in the soil or in the soil-thatch interface. The process by which they form an earthen cell and deposit an egg in it has not been previously reported in the literature. In their actions the European chafer females are considered to be essentially the same as other scarabaeids that form cells for individual eggs (Plate 25). Part of this phenomenon has been seen in at least one other species, the oriental beetle (author's personal observation). Evagination of the vaginal tract creates a balloonlike organ that compresses the moist soil and forms a cell. An egg is deposited simultaneously with the invagination of the vagina. The smooth wall of the earthen cell protects the egg against physical injury. Moisture from the surrounding soil is essential for growth and maturation of the egg.

Larval Activities

Upon hatching, young grubs of the phytophagous species begin feeding on root hairs and the fine roots of grasses and other plants, their primary source of food. They can also feed on the organic matter present in the thatch and soil but when they do they do not grow as rapidly.

Vertical movement of the grubs is governed by soil moisture and temperature. When the surface soil becomes dry, grubs migrate downward to seek moisture. After a drought, when grubs may be well below 10 cm (4 in.), a sudden precipitation or irrigation that wets the soil to the depths of the grubs will cause them to migrate to near the soil surface within 24 hr. During spring and fall when soil moisture is generally constant and high, most grubs are in the surface soil or in the soil-thatch interface.

Cold temperature in the fall forces grubs to migrate downward before the ground freezes. They will migrate as deep as necessary to avoid the frozen soil and will remain below the frost line all winter. Upward migration begins in the spring when all soil frost has disappeared and when the threshold temperature of a given species is maintained in the soil.

Larva-to-Adult Sequence

When the third instar is mature and feeding has been completed, the larva moves downward in the soil and forms an earthen cell, oscillating to compress and smooth the cell wall. Much occurs in this cell during the metamorphosis of the individual. This sequence is illustrated by the transformation of the European chafer (Plate 26).

The larva ejects its accumulated excrement and becomes a pale, flaccid prepupa that has lost its power of locomotion. It can move only by flexing and reflexing its abdomen, which further aids in smoothing the earthen cell. After a few days it becomes a pupa and, in this species, sloughs off the larval exuvia to its abdominal tip. In other species, for example in the Japanese beetle, pupation occurs within the exuvia, which later splits and frees the pupa. The pupa gradually turns from a creamy white body to one with a brown head, wings, and legs, maturing over several days. In time it becomes a callow adult with thin, lightly colored elytra and fully expanded hindwings. Within a few more days the elytra harden, the membranous hindwings are folded, and the insect becomes a mature adult.

In most species with a 1-year life cycle, the beetle digs its way to the soil surface and makes its first flight a few days after adulthood. In most *Phyllophaga* spp. with a 3-year life cycle, the adult remains in its earthen cell throughout the rest of the season and the entire winter before emerging in the spring (Ritcher 1940).

More detailed accounts of the more important scarabaeid pests of turfgrass in the United States appear in the following chapters.

11

Scarabaeid Pests: Subfamily Aphodinae

Black Turfgrass Ataenius

Taxonomy

The black turfgrass ataenius, BTA, *Ataenius spretulus* (Haldeman), order Coleoptera, family Scarabaeidae, subfamily Aphodinae, is one of 63 species recognized in the United States in the same genus. The BTA was previously named *A. cognatus* (Lec.). It is called the black fairway beetle in Canada (Cartwright 1974, Fushtey and Sears 1981, Hoffman 1935).

Importance

The BTA was considered only an incidental turfgrass pest prior to the 1970s. Since then it has become a serious pest on golf course fairways, greens, and tees over a wide area extending from Colorado eastward, in the northern half of the country, and also in the province of Ontario, Canada (Niemczyk and Dunbar 1976, Weaver and Hacker 1978).

The seriousness of the BTA's damage to turfgrass prompted two major studies on which we depend for the most detailed information that we have on this insect. The work of Weaver and Hacker (1978) was conducted in southwestern West Virginia during 1976–1977. The work of Wegner and Niemczyk (1981) was conducted during 1976–1978 in the vicinity of Cincinnati, Ohio. Both areas are within 1° latitude of each other and are similar in elevation.

Cyclodiene Resistance and Implications

Shortly after the discovery that the BTA was a serious pest, contact toxicity tests on adults gave clear evidence of resistance to chlorinated cyclodienes (aldrin, chlordane, dieldrin, heptachlor) in widely scattered populations from New Jersey, New York, Ohio, and Connecticut but not at all sites. Susceptible populations were present in Illinois and Ohio. Many courses that had experienced BTA infestations had previous histories of cyclodiene usage, but in a few courses, cyclodienes had not been used. The rather sudden appearance of the BTA as a turfgrass pest could be the result of cyclodiene resistance, but resistance alone does not explain why the BTA was not a serious pest prior to the advent of these insecticides (Niemczyk and Dunbar 1976, Niemczyk and Wegner 1982).

History and Distribution

The BTA was first described as a turfgrass pest in Minnesota during 1932, when it was found killing grass on greens and fairways. The next evidence that it was a turfgrass pest came in New York State near Rochester, Monroe County, in 1969 and in Orange County in 1970, when the insect was found damaging fairway turf. During 1973 it was discovered damaging fairways on a Cincinnati, Ohio, golf course. During the next two years, reports of fairway damage became numerous in this area. Other areas reporting damage were 12 midwestern and eastern states, Washington, D.C., and Ontario, Canada (Hoffman 1935, Kawanishi et al. 1974, Niemczyk 1976).

The BTA is not an introduced insect, like some of our most serious scarabaeid pests. It is a native insect collected from all but 7 states in the 48 contiguous states (Figure 33). By 1978 it was reported damaging golf course turfgrass in 23 states plus

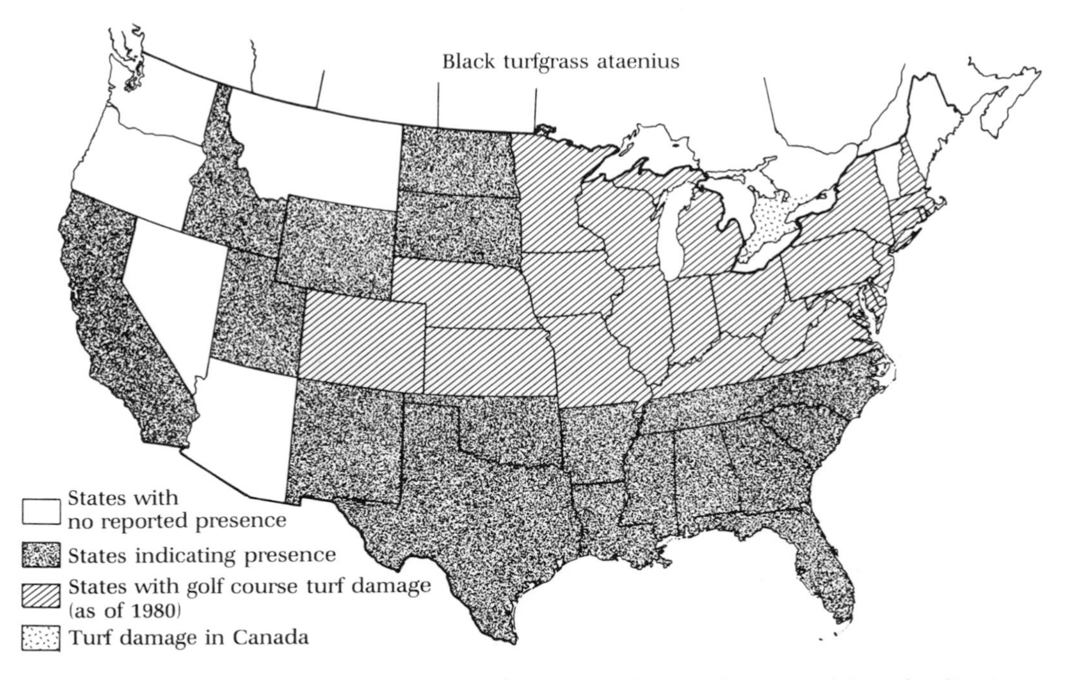

Figure 33. Distribution of the black turfgrass ataenius in the United States and Canada, showing states in which golf course turf damage occurred through 1980. (Data from Cartwright 1974 and Niemczyk and Wegner 1979, drawn by R. McMillen-Sticht, NYSAES.)

the District of Columbia and Ontario, Canada (Cartwright 1974, Niemczyk and Wegner 1979).

Host Plants and Damage

The BTA is primarily a problem on golf courses. Fairways, greens, and tees are damaged as the grubs feed on the roots of annual bluegrass, Kentucky bluegrass, and bentgrasses. The first symptoms of injury appear (in mid-June in Ohio) when turf wilts despite an abundance of moisture. Wilted areas are most visible when viewed toward the sun. Under the stress caused by summer heat and by the feeding of grubs on the roots, the turf dies in small, irregular patches that eventually coalesce to produce large dead turf areas (Plate 27). Since the roots have been devoured at the soil-thatch interface, the weakened and dead turf is easily removed to expose seemingly countless numbers of mature larvae and often prepupae, pupae, and even teneral and mature adults (Plate 27). Many additional individuals are usually present in the soil. Infestations have been found on home lawns both in Ohio and on Long Island in New York but have been a minor incidental problem (Niemczyk and Wegner 1982).

Description of Stages

Adult

The fully mature adult is a small, shiny black beetle with a mean length of 4.9 mm and a mean width of 2.2 mm. Those observed in the soil or soil-thatch interface may be mature adults or callow adults of uniform reddish brown or various combinations of reddish brown and black (Plate 27). Adults in the reddish brown phase may not be entirely callow adults, since they have been numerous in black-light trap collections in mid-July (Weaver and Hacker 1978, Wegner and Niemczyk 1981).

Egg

Newly deposited eggs grow 0.5–0.7 times in width. Mature eggs average 0.72 mm in length and 0.52 mm in width. The shiny white eggs are deposited in clusters of 11–12 within a cavity formed by the female (Plate 27; Wegner and Niemczyk 1981).

Larva

The three larval instars are typically scarabaeid (Plate 27). Head capsule widths grow from a mean of 0.5 mm in firsts, to 0.83 mm in seconds, to 1.3 mm in thirds. A full-grown third-instar grub has a mean length of 8.5 mm and closely resembles in size and appearance mature first instars of larger scarabaeids, such as the European chafer and the Japanese beetle. The third-instar BTA discovered in New York was occasionally mistaken as a mature first-instar European chafer. The two occur concurrently from July to early August in western New York. No distinct rastral pattern is present, but there are 40–45 irregularly placed hamate setae. The most distinguishing feature is the presence of two padlike structures on the tip of the abdomen just posterior to the setae and anterior to the anal slit (Plate 27; Wegner and Niemczyk 1981).

Pupa

The mean pupal length and width are 4.7 mm and 2.5 mm, respectively (Plate 27). Sexes can be differentiated in the pupae by the presence of an aedeagal protuberance in the males that is located on the ventral posterior region.

Seasonal History and Habits

Seasonal Cycle and Generations

The BTA has one or two generations a year, depending on the latitude. Two generations occur in the latitude of Connecticut and northern Ohio and southward. In the latitude of western New York (author's personal observations), Ontario, Canada, and Minnesota, the BTA is generally considered to be a one-generation insect. In southern Ohio, oviposition periods began in May and July. First-generation adults emerged between late June and early July, and second-generation adults emerged in August (Niemczyk and Dunbar 1976, Weaver and Hacker 1978, Wegner and Neimczyk 1981).

Seasonal Occurrence in Turfgrass

Wegner and Niemczyk (1981), in southern Ohio, made one of the most comprehensive studies on the presence at any stage of any turfgrass scarabaeid in the turfgrass soil proper throughout the season.

Sampling Technique. Ten cores of turfgrass and soil 10.8 cm in diameter and 7–8 cm (2.75–3.00 in.) deep were taken twice a week from each site. The numbers of all stages present were converted to numbers per 929 cm^2 (1 ft^2) by multiplying the number per sample by 10.5. (Hereafter I round off 929 cm^2 to 0.1 m^2.)

Turfgrass Populations. Figure 34 shows the mean populations of all stages present per 0.1 m^2 in turfgrass and soil at quarter-month intervals from April through October. Data from two golf courses sampled in 1977 and a third golf course sampled in 1977 and 1978 were consolidated into Figure 34. The period required for the BTA to pass through one generation was 65 days ± 5 days at soil temperatures of 25° ± 6°C (77° ± 11°F). The approximate interval between generations is evident in all stages.

The evidence of two generations a year is conclusive in Figure 34. The highest population of adults occurring in the turfgrass during July in southern Ohio correlates well with the capture of adults in black-light traps in West Virginia (Table 11

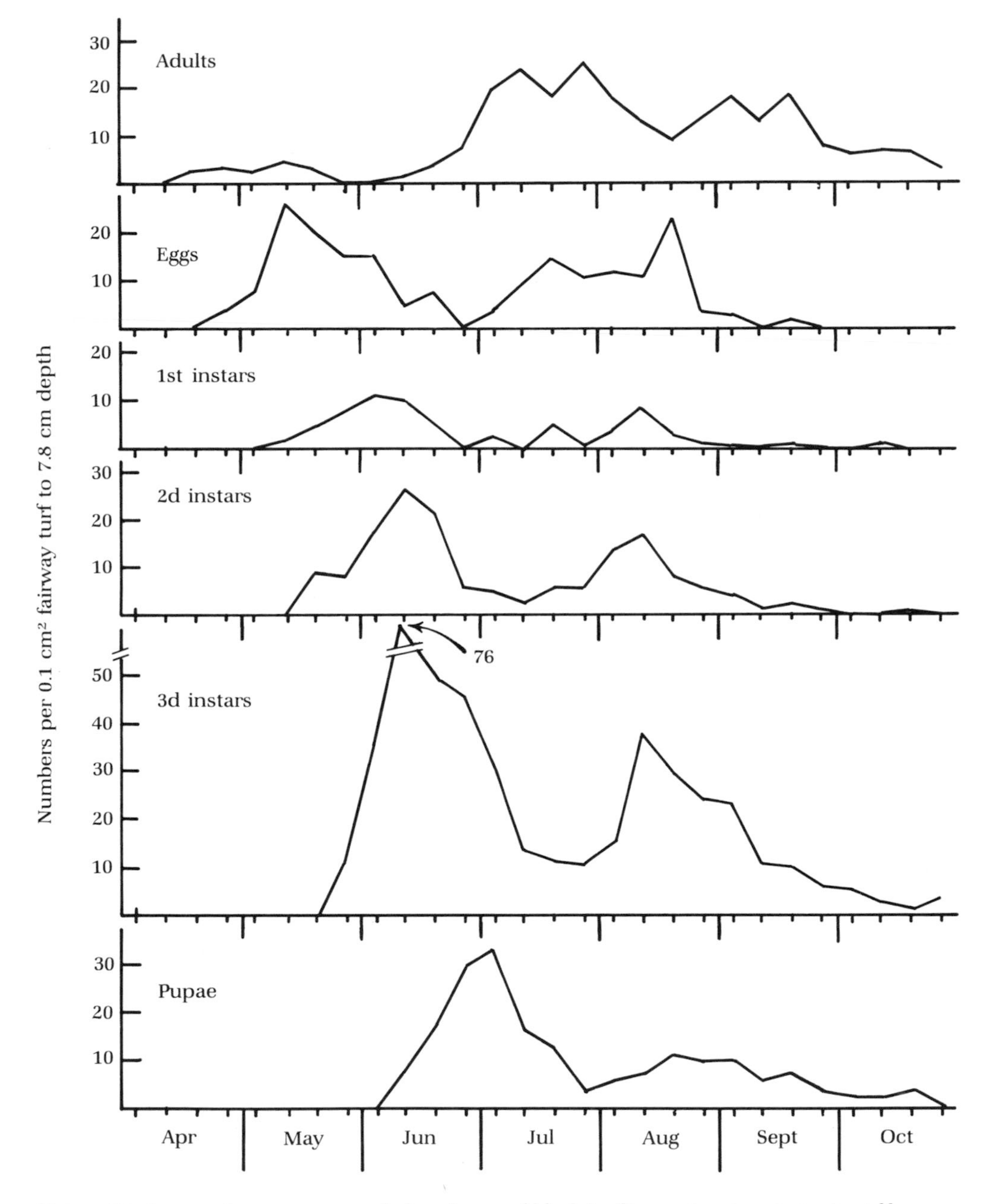

Figure 34. Seasonal occurrence and abundance of black turfgrass ataenius stages in golf course fairways in and near Cincinnati, Ohio, during 1977 and 1978. (Data from Wegner and Niemczyk 1981, figs. 8, 9; drawn by R. McMillen-Sticht, NYSAES.)

and Figure 35). First-generation eggs present peaked more sharply than second-generation eggs, which were present in numbers over a slightly longer period. Third instars are the most readily observed because of their size and position at the soil-thatch interface. These were found in greatest numbers, with the first generation peaking in June and the second generation peaking in August (Weaver and Hacker 1978, Wegner and Niemczyk 1981).

Adult Activities

Overwintering. During September and October, adults leave golf course fairways for overwintering sites at the edge of wooded areas in the roughs and perimeter of golf courses. Second-generation adults going into hibernation were 90% inseminated. Grubs that do not mature, pupate, and transform to adults by early September do not survive over the winter. Mature adults that go into hibernation have a high rate of survival, from 90% to 96%. Males and females overwinter with equal success. In West Virginia, the preferred overwintering site in wooded areas was in an accumulation of pine needle litter and in the upper 5 cm (2 in.) of loose, well-

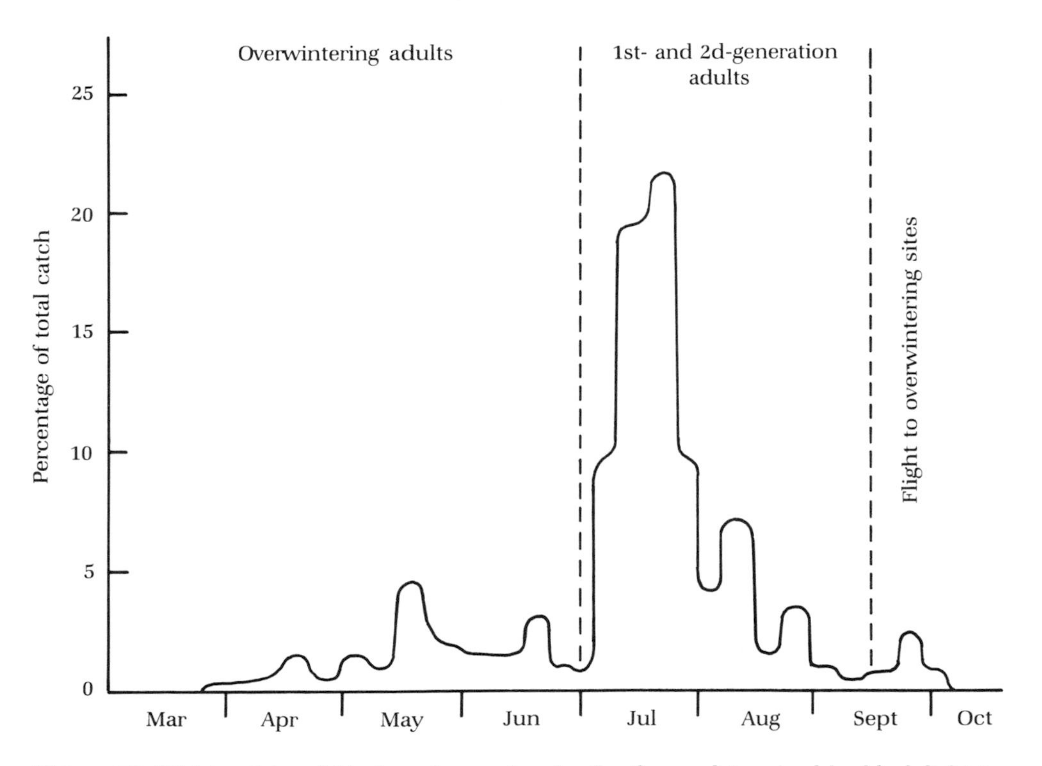

Figure 35. Flight activity of black turfgrass ataenius beetles as determined by black-light trap catches during March to October 1977 in Putnam County, southwestern West Virginia. (Adapted from Weaver and Hacker 1978, fig. 3, courtesy of the West Virginia Agricultural and Forestry Experiment Station.)

Table 11. Black turfgrass ataenius: Adult catches in two black-light traps during the 1977 season at a golf course in southwestern West Virginia

Start date of 2 week ± 1 day intervals[a]	Percentage of total seasonal catch		
	Nonoverwintering site	Overwintering site	Both traps
March 29	<1	<1	<1
April 18	2	1	2
May 2	2	3	3
May 16	7	6	7
May 29	5	2	4
June 13	6	1	4
June 27	13	5	10
July 11	42	40	41
July 25	10	19	13
August 9	7	10	8
August 22	2	5	4
September 6	<1	1	1
September 19	2	5	3
Total seasonal catch	24,244	19,163	43,407

[a]Final period September 19 through October 10, a 22-day period.
Source: Adapted from Weaver and Hacker 1978.

drained soil. As many as 77 beetles were found per 0.1 m^2 (per ft^2) in such areas (Weaver and Hacker 1978, Wegner and Niemczyk 1981).

In Ohio, extraction of adults was accomplished by obtaining samples of duff and soil 0.1 m^2 to a depth of 5 cm (2 in.), submerging them, and placing all floating debris after drainage into Berlese funnels. No immature stages were present. The greatest adult density found was 264 per 0.1 m^2. Approximately 65% of the beetles were females, and 90% had been inseminated. In Minnesota, beetles were found awaiting winter in the upper 6-in. moist layer of waste piles of milorganite and grass cuttings (Hoffman 1935, Wegner and Niemczyk 1981).

Adult Emergence and Dispersal. Adults return to the golf course fairways starting in late March and continue through April and early May. On warm evenings (4:00–6:00 PM), swarms of adults can be seen flying over the turf. Adults alight on the turf and immediately burrow into it (Niemczyk 1977a).

In West Virginia, Weaver and Hacker (1978) established two black-light traps (with 15-watt lamps) on a golf course and monitored them at weekly intervals for the entire season. One trap was located near a known overwintering site and the second was located about 1.2 km (0.75 mi) away at a nonoverwintering site. Emergence from overwintering sites began in late March and continued through mid-April. A mass flight of beetles occurred just before dusk, and adults were captured in the traps the same night. Trapped adults were removed daily except on most weekends and holidays. Table 11 gives the seasonal catch from March 29 through October 10 (modified to cover approximately 2-week periods) when adult activity declined sharply, presumably because of movement to overwintering sites. There was a sharp increase in flight during late June and early July, with peak flight activity during mid-July and tapering off into early August. These data suggest that adults disperse randomly after emerging from overwintering sites. The same trap catch information presented graphically (Figure 35) shows a sharp peak in adult captivity as a result of the simultaneous presence of both generations of adults during July (Weaver and Hacker 1978).

An increase in flight activity before and during light precipitation on warm evenings was observed in Ohio and appears to be typical of aphodiine scarabaeids. Flight studies made in Ohio showed that screen sticky traps coated with Tack Trap and traps fitted with 22- and 32-watt Circline black-light fluorescent lamps and a vacuum fan were effective devices by which to determine flights and dispersal. The largest captures of beetles occurred in April and again from July through October, with minimal captures during May and June. The largest captures ranged from 80 to more than 800 per trap-day (Wegner and Niemczyk 1981).

Oviposition and Phenological Relations. No information is available on mating habits. In addition to the high degree of insemination found in overwintering females, most females found throughout the season had been inseminated. Females examined during the months May–October were 100%, 67%, 69%, 96%, 91%, and 94% inseminated, respectively (Wegner and Niemczyk 1981).

A degree-day system of predicting BTA activity and development was based on a flight activity threshold of 13°C (55°F). The seasonal life history of the BTA was also correlated with plant phenology (Plate 28). First-generation eggs begin to appear after 100–150 degree-days (in degrees Celsius) and coincided with the full bloom of the Vanhoutte spirea, *Spiraea vanhouttei* (Briot) Zabel, the horse chestnut, *Aesculus hippocastanum* L., and the earliest bloom of black locust, *Robinia pseudoacacia* L. Normally these plants flower during the first half of May in southern Ohio and early June in New York. Second-generation eggs appeared after 650–710 degree-days (in degrees Celsius) when the rose of sharon, *Hibiscus syriacus* L., began to bloom, as it normally would during the last half of July in southern Ohio. Generation periods of 60–70 days were observed in the field where soil temperatures approached or exceeded 30°C (86°F; Wegner and Niemczyk 1981).

These phenological relationships were very useful for timing of insecticidal applications to kill adults before extensive oviposition during May in Ohio and provided an alternative for grub control of the same generation later in the season (Figure 36). Since these are very common plants growing over wide geographic areas, the relationships should be helpful in timing insecticidal applications for adult control (Niemczyk 1981, Niemczyk and Wegner 1979).

Larval Activities

Because eggs are deposited in clusters of 11–12 within a cavity formed by the females in the lower 5 mm of thatch and the upper 6 mm of soil among grass roots,

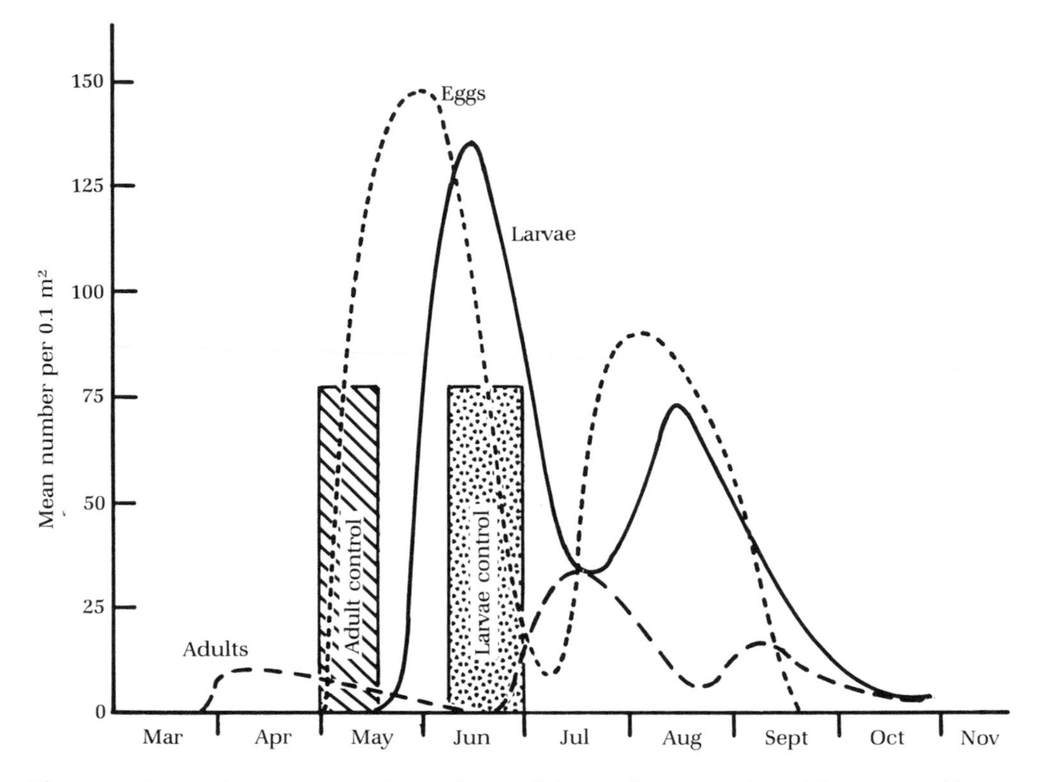

Figure 36. Seasonal occurrence and abundance of black turfgrass ataenius adults, eggs, and larvae, showing timing of treatments for adult and larval control in Ohio. (Adapted from Niemczyk and Wegner 1979, fig. 2, courtesy of H. D. Niemczyk.)

it was common to find first instars clustered together, with as many as 12 per 1–8 cm^3 (0.06–0.50 $in.^3$). Second instars were further dispersed, with 12 per 8–100 cm^3 (0.5–6.1 $in.^3$). Third instars were dispersed even further, with 12 per 100–700 cm^3 (6–43 $in.^3$), and were most common at or near the soil-thatch interface (Wegner and Niemczyk 1981).

Because of their minute sizes, first and second instars are not considered to be very detrimental to turfgrass. The third instar, found most abundantly during June, July, and August, is the most destructive stage. In the latitude of southern Ohio and West Virginia, the second-generation thirds are most prevalent during August (Figure 34), while in the latitude of Minneapolis and western New York, the first-generation thirds are prevalent during July and August (Hoffman 1935, Wegner and Niemczyk 1981, author's personal observations).

Upon cessation of feeding, larvae migrate 1–8 cm (0.4–3.0 in.) deep into the soil and excavate cells in which to remain as prepupae and pupae and finally as callow adults.

Miscellaneous Features

Threshold Populations

An economic threshold for grubs has not been determined, but levels as low as 30 per 0.1 m^2 (per ft^2) have been suggested. In Connecticut larval populations of 100–150 per 0.1 m^2 caused extensive damage. In Ohio populations of 250–300 per 0.1 m^2 were frequently associated with heavy damage. A high of 578 grubs per 0.1 m^2 is recorded (Niemczyk and Dunbar 1976, Weaver and Hacker 1978).

In personal observations on a fairway in western New York, turf was all dead to extensively damaged with populations of 200–500 grubs per 0.1 m^2 and showed no damage at roughly 100 per 0.1 m^2. During late July, insects at the soil-thatch interface represented roughly 20% to 40% of the total population, with the balance of larvae, prepupae, pupae, and adults in the soil to a depth of 10 cm (4 in.).

Natural Enemies

Microorganisms

Milky Disease. The only natural enemy encountered in BTA is milky disease. In a population of third instars, prepupae, and pupae ranging from 29 to 144 per 0.1 m^2 (per ft^2) on a golf course fairway near Rochester, New York, in 1969, about 70% of the larvae showed milky disease symptoms, and after a few days at room temperature, the balance of the grubs also became milky. Symptoms were identical to those of Japanese beetle grubs infected with *Bacillus popilliae* Dutky or *B. lentimorbus* Dutky (Plate 27). Spores of the bacillus causing milky disease in BTA resemble those of *B. lentimorbus* more than those of *B. popilliae*. The bacterium is, however, entirely different and causes no infection in grubs of the Japanese beetle, the European chafer, and the northern masked chafer, which are infected by *B. popilliae* and *B. lentimorbus*. At least three morphologically different spore forms of the bacillus infect BTA (Figure 37; Kawanishi et al. 1974, Splittstoesser and Tashiro 1977).

A wide divergence in the incidence of milky disease in BTA occurs between Ohio and New York. Incidence of milky disease in the Cincinnati, Ohio, area increased from <1% in May and June to 25%–29% in October 1977 (Wegner and Niemczyk 1981).

In New York we have consistently found a high incidence of disease during July and August wherever a population was encountered. After a few early-developing individuals in a given population had pupated, virtually all remaining grubs were infected, showing gross symptoms, or were in early stages of infection. Gross symptoms developed soon after incubation in spore-free soil. In fact we have not been able to perform reliable milky disease tests in the laboratory because sufficient numbers of healthy third instar grubs have not been available. This situation has been prevalent in populations examined on Long Island, in Westchester County just north of New York City, in the Hudson Valley, and in western New York.

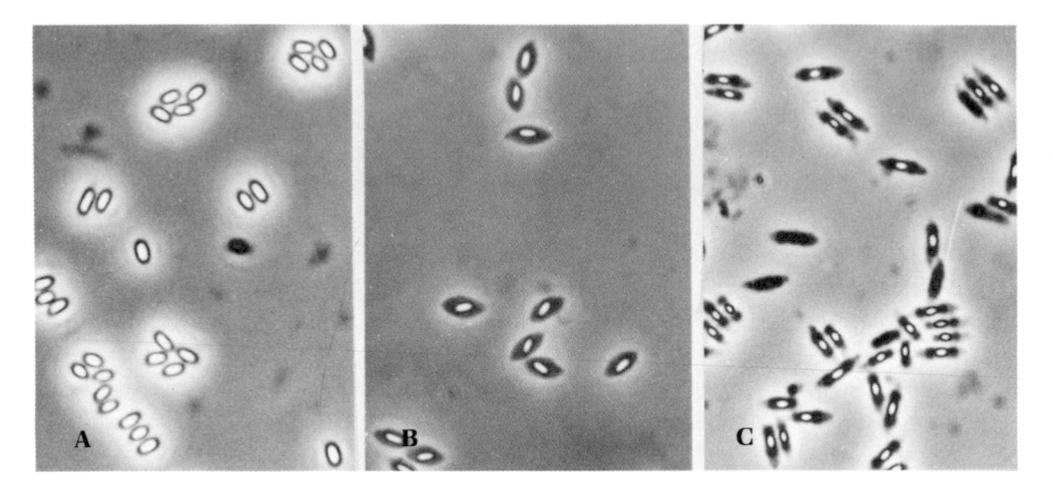

Figure 37. Morphological differences in spores of bacillus or bacilli causing a milky disease in black turfgrass ataenius grubs. **A.** Spores resembling *Bacillus lentimorbus* most closely. **B.** Spore form most commonly found. **C.** Spore form least commonly found. (Photos by G. Catlin, NYSAES.)

Vertebrate Predators

The most apparent predators have been starlings, *Sturnus vulgaris* L., and barn swallows, *Hirundo rustica* L. On Long Island, where the BTA has been more abundant than anywhere else in New York State, crows, *Corvus* spp., have been the most common predator and do considerable damage as they disrupt the turf in their search for BTA grubs (Niemczyk and Wegner 1982).

Aphodius spp.

Taxonomy

Aphodius granarius (L.) and *A. paradalis* Le Conte, order Coleoptera, family Scarabaeidae, subfamily Aphodinae, are occasionally pests of turfgrass associated with the black turfgrass ataenius. More than 100 species of the genus *Aphodius* are known to occur in America north of Mexico. The adults of most species are dung feeders, and larvae feed on dung, organic matter, and live roots (Blatchley 1910).

Importance

Grubs of *Aphodius* have often been mistaken for the black turfgrass ataenius. Like the latter, this species feeds on turfgrass roots and is capable of inflicting severe damage (Niemczyk and Power, personal communication, 1985).

History and Distribution

During 1976 and 1977, *A. granarius* grubs were found for the first time damaging golf course fairways in Toronto, Ontario, as the dominant species in association with the black turfgrass ataenius. In one mixed population, *A. granarius* made up >97% of the grub population, while the black turfgrass ataenius made up <3% (Sears 1979).

During 1978, *A. granarius* grubs were discovered in golf course fairways in Colorado and in Michigan, where the infestations were first thought to be the black turfgrass ataenius. During 1981, the same *Aphodius* grubs were detected in golf course fairways in Ohio for the first time.

Aphodius granarius is an introduced European species that is now widespread in the United States and Canada. It is found in every locality from which *Aphodius* spp. have been received for identification. To date, it has been reported from 25 widely scattered states and the District of Columbia and from two provinces in Canada (Figure 38; Niemczyk and Power 1978, Woodruff 1973).

Aphodius pardalis is a West Coast species found in the turf of lawns, golf courses, and bowling greens, where grubs cause damage by feeding on roots and rhizomes (Ritcher 1966).

Figure 38. Reported presence of *Aphodius granarius* in the United States and Canada. (Data from Woodruff 1973; drawn by R. McMillen-Sticht, NYSAES.)

Host Plants and Damage

Turfgrass-damaging *Aphodius* grubs feed on the same turfgrasses as the black turfgrass ataenius and cause damage that is essentially the same as that caused by the latter. *Aphodius* grubs are often mistaken as grubs of the latter.

Description of Stages

Adults

Aphodius adults closely resemble those of *Ataenius* but are usually broader, stouter, and variegated, with black and dull red or yellow. *Aphodius* may be distinguished from *Ataenius* by its obtuse outer apical angle of the hind tibia. When *Aphodius* adults are mounted, the hindlegs should be stretched out for easy examination (Blatchley 1910).

Aphodius granarius beetles are oblong, subcylindrical, piceous and shining, with reddish brown legs and paler antennae. They are 3–5 mm long, as compared with the BTA, which is 4.0–5.6 mm long (Wegner and Niemczyk 1981, Woodruff 1973).

Both *A. granarius* and *A. pardalis* possess transverse carinae on the tibia of the meso- and metathoracic legs (Figure 39), while *A. spretulus* does not (Woodruff 1973).

Eggs

Eggs of *A. granarius* are smooth, opaque, and oval to almost spherical. They average 0.80 mm × 0.56 mm in length and width (Wilson 1932).

Larvae

Aphodius and *Ataenius* larvae are very similar in size but can be separated by differences in the raster. Grubs of both *A. granarius* and *A. pardalis* have definite V-shaped palidia, while those of *Ataenius* have teges only, with no palidia (Figure 39; Jerath 1960, Ritcher 1966).

Larvae of the two turf-infesting *Aphodius* species show slight differences. The maximum head capsule width of *A. granarius* is 1.42–1.68 mm, while that of *A. pardalis* is 1.29–1.42 mm. The raster of *A. granarius* has a V-shaped palidium of 14–23 caudomesally directed spinelike pali (Plate 27). The V-shaped palidium of *A. pardalis* has 6–10 caudomesally directed spinelike setae. Both species have many teges scattered on each side of the palidium (Jerath 1960).

Seasonal History and Habits

Little is known about the life history of *A. granarius* and *A. pardalis*. In Ontario, Canada, *A. granarius* is thought to have two generations a year, with adults present in May and again in August–September. In Ohio, adults were first collected from mid-April to early June. Adults were again collected in mid-November, providing evidence of a second generation. In New Jersey, *A. granarius* is reported to have a

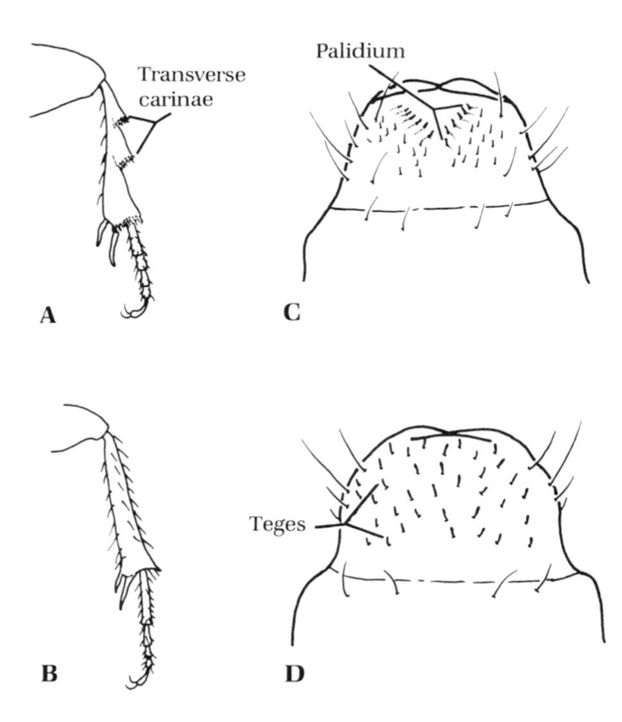

Figure 39. Adult and larval differences between *Aphodius* and *Ataenius* spp. **A.** Hindleg of *A. granarius* with transverse carina. **B.** Absence of carina in *A. spretulus*. **C.** Raster of *A. pardalis* with V-shaped palidia. **D.** Absence of palidia in *A. spretulus*. (Parts A and B drawn by R. McMillen-Sticht, NYSAES; parts C and D adapted from Ritcher 1966, figs. 66, 70, courtesy of Oregon State University Press.)

single generation that overwinters as adults (Niemczyk and Power, personal communication, 1985; Sears 1979; Wilson 1932).

Larvae of *A. granarius* were present in Ontario, Canada, in June, peaking in early July, and declining by the end of July. In Ohio, peak larval populations occurred during the 1st week in June. They gradually diminished through June and the 1st week in July, indicating pupation (Niemczyk and Power, personal communication 1985; Sears 1979).

Adult Activity

Adults of *A. granarius* are one of the most commonly observed of the genus in the Northeast. The highly polyphagous adults feed on all kinds of debris, decaying vegetation, compost, carrion, and so forth in addition to dung. For oviposition, adults prefer dung that has dried and has formed a hard crust on its surface. The eggs are laid just beneath the hard crust (Wilson 1932). Larval infestations in turfgrass indicate that oviposition apparently occurs in turf as readily as in dung.

The little that is known of the biology of *A. granarius* indicates that oviposition apparently occurs 2–3 weeks earlier than in the black turfgrass ataenius in Ohio (Niemczyk and Power, personal communication, 1985).

Aphodius spp. adults as a group are among the most numerous scavengers and occur in great numbers during the first warm days of spring, seeking the fresh dung of horses and cows. The beetles burrow into dung almost as soon as it drops from animals (Blatchley 1910).

12

Scarabaeid Pests: Subfamily Cetoniinae

Green June Beetle

Taxonomy

The green June beetle, GJB, *Cotinis nitida* L., is a member of the order Coleoptera, family Scarabaeidae, subfamily Cetoniinae (Ritcher 1966). Adults of this subfamily have the epimeron of the mesothorax visible from above (Plate 29; Chittenden and Fink 1922). This beetle is also called the *fig eater* because it is fond of ripe figs and other thin-skinned fruits.

Importance

The GJB is currently a turfgrass pest of nominal importance. Turf damage by this insect is primarily mechanical rather than being caused by feeding; the large grubs disrupt the soil surface by burrowing in and out of the soil and producing mounds. Since the beetles are attracted to dead and decaying organic matter, heavily manured fields are attractive for oviposition. Manure and decaying organic matter are also the primary food of grubs. If the turfgrass is again fertilized with manure rather than with commercial fertilizers, the GJB could become a serious pest. Adults are destructive to ripening fruits, especially those with a thin skin.

The most comprehensive literature on the biology of the GJB was published during the 1920s (Chittenden and Fink 1922, Davis and Luginbill 1921).

History and Distribution

The GJB, a native insect, has a wide distribution in eastern United States, extending from Connecticut and southeastern New York to Florida and westward into Texas, Oklahoma, and Kansas (Figure 40). It naturally prefers rich sandy soils and loam high in organic matter (Chittenden and Fink 1922).

Figure 40. Distribution of the green June beetle in the United States. (Adapted from Chittenden and Fink 1922, fig. 4, drawn by R. McMillen-Sticht, NYSAES.)

Host Plants and Damage

Adult Feeding

Beetles injure many ripening fruits, including apricots, nectarines, peaches, plums, prunes, apples, pears, grapes, figs, blackberries, and raspberries (Plate 29). Thin-skinned fruits, especially figs, peaches, and grapes, are the most damaged.

Larval Damage

Grubs damage turf through constant burrowing and tunneling rather than by feeding on roots. As they crawl to the surface at night to feed on the dead and decaying organic matter, they sometimes make small mounds of soil 5–8 cm (2–3 in.) in diameter in addition to leaving distinct open, vertical burrows averaging 15–30 cm (6–12 in.) in depth. These burrows have about the same diameter as a thumb. In addition to decaying organic matter, grubs do feed on the fine roots of succulent plants, especially when decayed matter is scarce (Davis and Luginbill 1921).

The grubs cause similar mechanical damage to other crops. Young corn, oats, sorghum, and alfalfa growing on heavily manured fields are sometimes damaged. Also, many vegetables, strawberries, ornamental plants, and tobacco seedbeds are damaged as the grubs disrupt the surface soil, loosen roots, and feed on roots and crowns. The combined injury causes plants to wilt and die.

Description of Stages

Adult

The beetle is usually a beautiful velvet green dorsally but is somewhat variable in color with yellow-orange margins on the elytra (Plate 29). Ventrally it is shiny metallic green and orange yellow. The hornlike process of the clypeus is more prominent in males. Beetles are 2.0–2.5 cm in length and half as wide (Davis and Luginbill 1921).

Egg

When first deposited, eggs are pearly white and 1.5 mm wide by 2.1+ mm long. As the embryo develops, the eggs become nearly spherical and grow to 2.5–2.8 mm wide by 2.8–3.1+ mm long (Plate 29). One to two days before hatching, the embryo is plainly visible through the chorion. Like all other scarabaeid eggs, these will bounce from a hard surface without injury.

Larva

Recently hatched larvae are 7–8 mm long when outstretched and 2 mm wide and resemble older larvae except for the relatively longer hairs projecting laterally and posteriorly from the last segment (Plate 29). Full-grown third instars are 4–5 cm long by 1.3 cm in diameter and thicker at the posterior end (Plate 29). Head capsule widths are 1.7, 3.0, and 5.5 mm for the three larval instars, respectively.

The three pairs of legs are very small for the size of the grub and are not used for locomotion (Plate 29). The dorsal surface is transversely corrugated, with three distinct ridges per segment. Each ridge is covered with short, stiff hairs posteriorly directed that are used for its unusual dorsal locomotion.

Prepupa and Pupa

As in other scarabaeids, the prepupa ends the third-instar period. The pupa changes from white to light brown (Plate 30). Shortly before emergence, the metallic green and brownish tints of the adult become apparent (Plate 30). Pupae are 24–26 mm long and 13.0–14.5 mm wide through the thoracic region (Davis and Luginbill 1921).

Seasonal History and Habits

Seasonal Cycle

The GJB has one generation a year. It spends about 10 months of the year as a grub and overwinters as a third instar (Figure 41). Grubs resume feeding early in the spring and by late May and the 1st week in June form cells in which to pupate. After about a 3-week pupal period, the adults emerge, from mid-June onward. In Virginia, beetles appear about mid-June, continue through July and August, and disappear

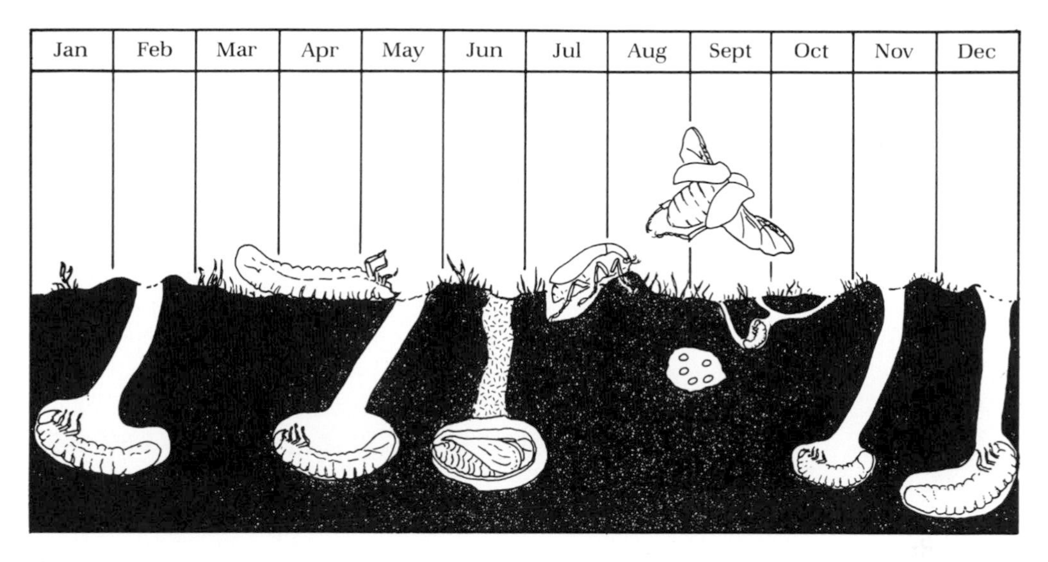

Figure 41. Life cycle of the green June beetle (diagrammatic). (Adapted from Chittenden and Fink 1922, fig. 5, drawn by R. McMillen-Sticht, NYSAES.)

by the 1st week in September. Oviposition occurs from mid-July through August. Eggs hatch in 10–15 days. Grubs molt twice by fall and become about 75% full-grown thirds before winter. They are either inactive or active during winter, depending on the temperature (Chittenden and Fink 1922).

Adult Activity

Daily Activity. During periods of abundance, the GJB females appear about daybreak. Shortly after sunrise, females begin to settle in the grass. Males appear and rapidly increase in numbers. By 7:00 AM, males dominate and fly rapidly 15–46 cm (6–18 in.) above the ground, with a buzzing sound resembling that of bumblebees. Males are attracted to the females by a milky fluid, which secretes a strong odor. Males easily find females even when the latter are hidden in the grass. Visible beetle activity gradually diminishes as the morning progresses, and by afternoon only an occasional beetle is observed (Davis and Luginbill 1921).

In Virginia, during a 4- to 6-week period when beetles are abundant, they are also active during late afternoon as they fly close to the ground or soar high among the treetops. Upon alighting, beetles gnaw into young twigs, causing them to break off. The beetles are also attracted to ripe fruit. After their daily flights and mating, the males burrow just beneath the grass, including that on putting greens, and make small mounds resembling miniature mole burrows. Most beetles spend the night in the soil or under debris, but an occasional beetle spends the night resting in shrubbery (Chittenden and Fink 1922, Davis and Luginbill 1921).

Adult Attractants. GJB adults were attracted to caproic acid while baits were being exposed for Japanese beetles. Weekly catches of about 25–90 adults with a 1:1 sex ratio indicated a definite attraction to both sexes (Muma 1944).

Trapping for oriental fruit moths with a mixture of malt extract and terpinyl acetate in Georgia peach orchards showed that the bait was attractive to GJB adults; a bait pail caught more than 400 beetles per week. A bait consisting of malt extract, benzoate of soda, and water was equally attractive (Beckhan and Dupree 1952).

Oviposition. After mating and before oviposition, females fly close to the ground, buzzing like bees to select a place to enter the soil. Upon alighting, they disappear rapidly into the soil and burrow 10–20 cm (4–8 in.) deep to deposit eggs. Under golf course turf, eggs have been found <5 cm (2 in.) below the surface (J. L. Hellman and R. Salvaggio, University of Maryland, personal communication, 1985). A sandy, moist soil containing well-rotted manure or other decomposing vegetation is most attractive for oviposition.

The GJB places its eggs in the soil in a way that is unusual for a scarabaeid. Eggs are deposited in balls of soil as large as a walnut and held together by a glutinous secretion (Plate 30). The numbers of eggs in each soil ball may vary from about 10 to 30; each egg has an individual cell within the large ball (Plate 30). Eggs in each soil ball are deposited at one continuous period of oviposition. A single beetle probably deposits as many as 60–75 eggs during about 2 weeks under normal conditions. It is not known how the female makes the soil ball once each egg has been placed in its individual cell (Plate 30; Davis and Luginbill 1921).

Larval Activity

Growth, Feeding, and Damage. Immediately upon hatching the young grub becomes active and crawls on its back, as do the older grubs (Plate 29). Grubs feed on animal manure and other decomposing vegetation and grow rapidly. The first grubs hatch about August 1 and become noticeable in lawns and gardens by the middle of the month.

The grubs' mode of locomotion is most distinctive, unlike that of any other turfgrass-infesting scarabaeid grub. They crawl on their backs by alternate contraction and expansion of their body segments, making use of their transverse dorsal corrugations and posteriorly directed spines. They are unable to walk with their legs, which are small and somewhat aborted. A grub enters the soil by lying on its back, bending the head and forepart of its body backward and downward, and burrowing in, using its head and feet (Davis and Luginbill 1921).

Grubs have distinct burrows that are mostly vertical, averaging 15–30 cm (6–12 in.) deep (Plate 29). By late fall, these burrows may reach depths of 46 cm (1.5 ft). In the sandy soils of Maryland golf courses, grubs have been found as deep as 0.9–1.1 m (3.0–3.5 ft; Hellman, personal communication, 1985). Their burrowing activities produce mounds of earth of 5–8 cm (2–3 in.) in diameter that closely resemble anthills. During the day, grubs remain at the bottom of their burrows. At night, they leave their burrows and crawl about, especially on warm, wet nights. On turfgrass soil, examinations should be made to a depth of about 10 cm (4 in.) to determine populations. If a mean of 6–8 grubs per 0.1 m^2 (per ft^2) is present, insecticidal treatments are usually considered necessary (Baker 1982, Chittenden and Fink 1922).

Overwintering and Spring Activity. During cold spells in late fall and winter, grubs burrow deeper and remain at the bottom of their burrows but do not go into true diapause. Warm days of midwinter make them active and bring them to the surface. As spring arrives, grubs again become active, feed for a short period, mature, and prepare an earthen cell at depths of about 20 cm (8 in.) in which to pupate following the prepupal period.

A cell is constructed of soil glued together by a fluid secreted by the grub (Plate 30). The cell is thin, about 1.0–1.5 mm thick, and may be crushed with moderate pressure between the thumb and forefinger. On one side of the cell there is a hard protuberance filled with sand rather than with silt or clay. This protuberance apparently receives the final discharge of excrement by the grub before being sealed off and smoothed like the rest of the cell (Davis and Luginbill 1921).

Pupation occurs during late May or early June in the latitude of Kentucky-Tennessee. After a pupal period of some 16–18 days, the insect becomes an adult. It remains in the cell for an additional 7–14 days before leaving the cell and emerging to the soil surface (Plate 30; Davis and Luginbill 1921).

Natural Enemies

Microorganisms

Grubs are sometimes infected by the green muscardine fungus, *Metarrhizium anisopliae.* Infection of grubs by the bacterium *Micrococcus nigrofaciens* Northrup has been reported but causes no appreciable population reductions (Davis and Luginbill 1921).

Parasitic and Predatory Insects

The digger wasp, *Scolia dubia* Say, is a common parasite of GJB grubs. The wasp enters a burrow, stings the grub, paralyzing it, and then attaches an egg to its ventral side. Upon hatching, the parasite larva feeds on the paralyzed GJB grub. The wasp larva completes its growth, spins a cocoon, pupates, and remains in its cocoon until the following year (Davis and Luginbill 1921).

Three sarcophagid flies reared from adults or pupae of the GJB were *Sarcophaga sarraceniae* Riley, *S. helicis* Towns, and *S. utilis* Aldrich (Davis and Luginbill 1921).

Vertebrate Predators

Birds are the most useful natural enemies of the GJB. No fewer than a dozen insectivorous birds feed on adults or grubs or both. Blackbirds and robins are the most active feeders.

Mammals known to feed on the GJB are the mole, opossum, chipmunk, and skunk (Davis and Luginbill 1921).

13

Scarabaeid Pests: Subfamily Dynastinae

Northern and Southern Masked Chafers

Taxonomy

The northern masked chafer, NMC, *Cyclocephala borealis* Arrow, and the southern masked chafer, SMC, *C. immaculata* (Olivier), belong to the order Coleoptera, family Scarabaeidae, subfamily Dynastinae, tribe Cyclocephalini. The NMC is called *Ochrosidia villosa* Burm. in some of the literature prior to 1940.

Importance

Both the NMC and the SMC are important turfgrass-infesting species as grubs in their geographic ranges, causing extensive damage to cultivated turf during late summer and early fall. Many of the grub problems in the midwestern and north central states are caused by masked chafers. Unlike other important species, such as the Japanese beetle and the European chafer, the masked chafer grubs are not usually considered serious turfgrass pests in the spring. Masked chafer grubs feed on the organic matter in the soil more than do other common turfgrass scarabaeids. As a result, larger populations are needed to cause turf damage. Since adults do not feed, they are not a pest of foliage (Potter 1981a).

History and Distribution

Both the NMC and the SMC are native to the United States and are distributed over a wide area east of the Rocky Mountains. The two occupy much of the same region (Figure 42). The NMC is troublesome as a turfgrass pest from Connecticut west to Ohio and Missouri, while the SMC is especially abundant in Kentucky, Indiana, Illinois, Missouri, and west to Texas (Potter 1981a). *Cyclocephala pasadenae*, the third species of importance in turfgrass, is probably the most common species in California (Ritcher 1966).

During the 1930s, the NMC was recognized as a turfgrass pest nearly simul-

Figure 42. States reporting three turfgrass-infesting masked chafers, *Cyclocephala* spp. Species not mentioned are considered to be either the northern or the southern masked chafer. (Data from U.S. Department of Agriculture 1951–1980; drawn by R. McMillen-Sticht, NYSAES.)

taneously in Connecticut, New York, and Ohio. In Connecticut, turf injury occurred from Hartford southwestward in many locations. In New York, damage was found only in an area within 56 km (35 mi) of New York City, where it was erroneously blamed on a resurgence of the Japanese beetle (Adams 1949b).

In more recent years, the SMC has been identified as a major, previously unrecognized pest of turfgrass in Texas. An overall mean of 39% of the grubs collected were the SMC, with the balance consisting of *Phyllophaga crinita,* the species considered to be the major scarabaeid injurious to turfgrass in Texas (Crocker et al. 1982).

Host Plants and Damage

Grubs of the NMC and the SMC feed on the roots of grasses present in each species' geographic range. In the midwestern regions of the United States, grubs of the SMC are particularly injurious to Kentucky bluegrass turf. The NMC has been reported as injurious to winter wheat in Nebraska (Johnson 1941, Potter and Gordon 1984).

The damage caused to turfgrass is typical of that caused by other scarabaeids. During late summer and early fall, third-instar grubs devour grass roots at or just below the soil-thatch interface, causing large patches of turfgrass to die through lack of roots.

Description of Stages

The NMC and the SMC are similar in appearance and size in all stages. Adults can be separated, but grubs are not distinguishable. Since the two are similar, the general description of the stages given by Johnson (1941) for the NMC can serve for both species.

Adults

The yellow-brown adult of NMC averages 11.8 mm in length and 6.8 mm in breadth for males and 11.0 mm by 6.7 mm for females. The sexes can be separated in both species by the distinctly longer lamellae of the antennae in the male, which are as long as or longer than all the other segments combined (Figure 43). The prothoracic legs of males have heavier tarsi, the claws are distinctly larger, and proximal segments 1–4 are as wide as or wider than they are long (Figure 43). In females these tarsal segments are longer than wide (Johnson 1941).

Both males and females of the NMC can be distinguished from those of the SMC;

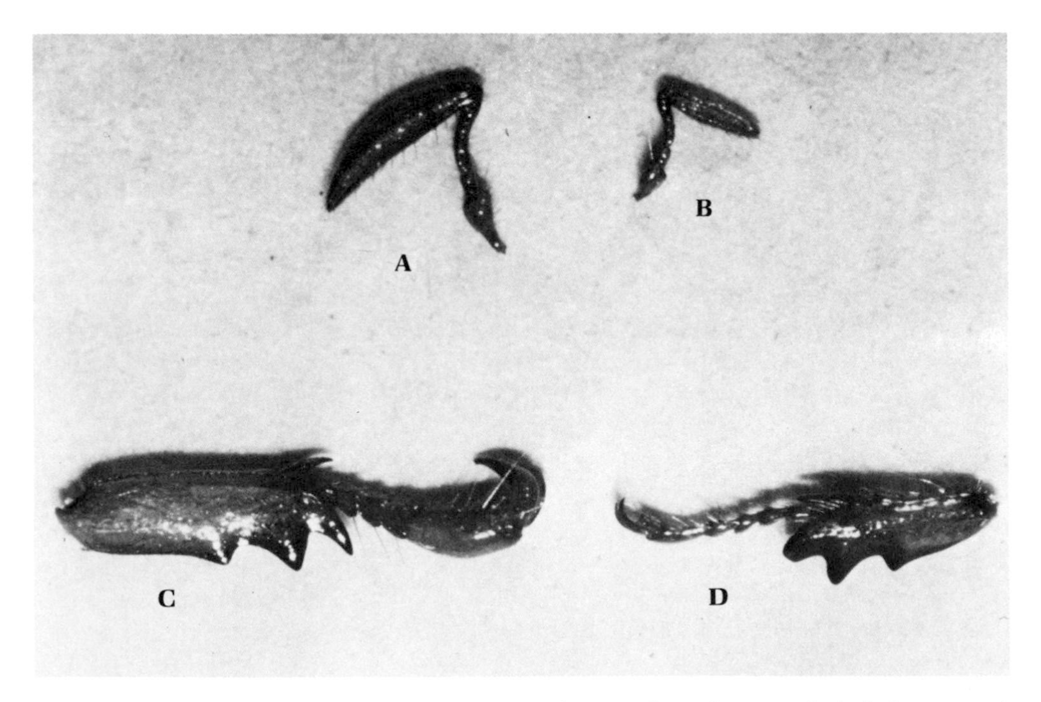

Figure 43. Sexual differences common to the northern and southern masked chafers. **A.** Male antenna. **B.** Female antenna. **C.** Male prothoracic leg. **D.** Female prothoracic leg. (Photos by B. Aldwinckle, NYSAES.)

the male characters are more distinct than those in the female. Males of the NMC are slightly larger, have many erect hairs on the elytra, and have much longer pygidial pubescence (Plate 31). Females of the NMC have dense hairs on the metasternum and a row of stout bristles on the outer edge of the elytra (Plate 31). Both of these characters are lacking in the female SMC (Potter 1981b).

Both sexes of both species have dark chocolate brown heads that shade to a lighter brown clypeus, a character that distinguishes them from other scarabaeid beetles of similar size and general coloration, such as several *Phyllophaga* species and even the European chafer.

Eggs

The pearly white, ovoid eggs of the NMC are delicately reticulate. Fresh eggs have a mean length and breadth of 1.7 mm by 1.3 mm and expand to 1.7 mm by 1.6 mm just prior to hatching (Johnson 1941).

A microscopic study of the eggs of the SMC by Potter (1983) revealed interesting features. The chorion of the youngest egg is only 1.25–1.60 μm thick exclusive of tubercles. By 8 days a thick serosal cuticle with two distinct layers, an outer portion measuring 7.0–7.5 μm and an inner layer that is 12.0–12.5 μm thick, forms. At maturity the serosal cuticle is resorbed, which allows the chorion to press against the egg-bursting spines on the thorax of the first-instar grub. Like other scarabaeid eggs, that of the SMC has a chorion that is noticeably elastic, so that eggs bounce without injury when dropped on a hard surface.

Larvae

The NMC larvae have mean head capsule widths of 1.6 mm, 2.3 mm, and 4.1 mm for the first, second, and third instars, respectively. The latter averaged 22.7 mm in total body length (Johnson 1941). NMC larvae are slightly stouter but are similar in length to the European chafer larvae (Tashiro et al. 1969). Morphologically, larvae of the NMC and the SMC are so similar that they cannot be separated. Both have a raster of coarse, long, hamate (hooked) spines showing no distinct arrangement. Spines become larger as they approach the anal slit, which is transverse and arcuate (Plate 31; Johnson 1941).

Prepupae and Pupae

The soil particles that form the earthen cell used for pupation are firmly cemented together, apparently by a larval secretion (Plate 31). In appearance the prepupa in masked chafers is similar to that of other scarabaeids. As in other scarabaeids, the newly formed creamy white pupa turns reddish brown as it matures. Mean dorsal length is 16.8 mm. The sex of the pupa can be distinguished by a pair of conspicuous lobes (the aedeagal protuberance) on the venter of the ninth segment of the male. The sex of older pupae can also be determined by sexual differences in the prothoracic tarsi visible through the cuticle and described in connection with adult differences (Johnson 1941).

Seasonal History and Habits

Seasonal Cycle

Masked chafers have a 1-year life cycle, spending 14–21 days as eggs, 10–11 months as larvae, 4–5 days as prepupae, 11–16 days as pupae, and 5–25 days as adults. They overwinter as third instars deep in the soil. They migrate upward during March and April and resume feeding until May, when they move downward for pupation. The nocturnal nonfeeding adults are present for nearly a month during June and July. Oviposition begins a day or two after emergence. Third instars are present by early September. They feed vigorously, causing most of their damage to turfgrass before they move downward in October for winter hibernation (Johnson 1941).

Adult Activity

Seasonal Presence. The most detailed study comparing the seasonal occurrence of the NMC and the SMC was made by Potter (1981b). Black-light trap catches in Kentucky revealed that the flights of the NMC begin in early to mid-June, peak in late June, and end by early August. SMC flights begin at least a week later than those of the NMC, peak 1–2 weeks later than those of the NMC, and terminate in early August in Kentucky (Figure 44). Flight activity is greater after a heavy rain (Figure 45).

Thermal unit (TU) accumulations (also called degree-day accumulations) in air and soil with a base temperature of 10°C (50°F) correlated well with the beetles' first emergence but were less useful for predicting the date of 50% and 90% flight. The first emergence of the NMC can be expected after TU accumulations of about 500 and 540 (in degrees Celsius) in air and soil, respectively. TU accumulations of about 585 (air) and 660 (soil) were necessary for SMC emergence (Potter 1981b).

Adult Response to Black-Light Traps. Even though the ratio of *Cyclocephala* males to females is reported to be about 1.5:1, male catches in light traps (incandescent) exceeded catches of females by 7:1 (Neiswander 1938). Black-light trap catches were 89% males for the NMC and 82% males for the SMC in Kentucky. The low proportion of females may be related to their behavior. Unlike males, which actively skim the surface of the turf in search of females, the latter usually burrow beneath the sod soon after mating (Potter 1981b).

Nightly Activity of NMC and SMC. The nightly flights differ for the two species, with the SMC flying earlier in the evening than the NMC. Black-light trap catches in Kentucky showed that 90% of the SMC were caught between 9:00 and 11:00 PM, while more than 90% of the nightly catch of the NMC came after 11:00 PM, with the heaviest catch between midnight and 4:00 AM (Figure 46).

SMC adults emerge from the soil on warm, humid evenings at about dusk. Swarms of males are commonly observed skimming over the turf in search of females. Females begin emerging shortly after the first male flights and climb upward on grass blades.

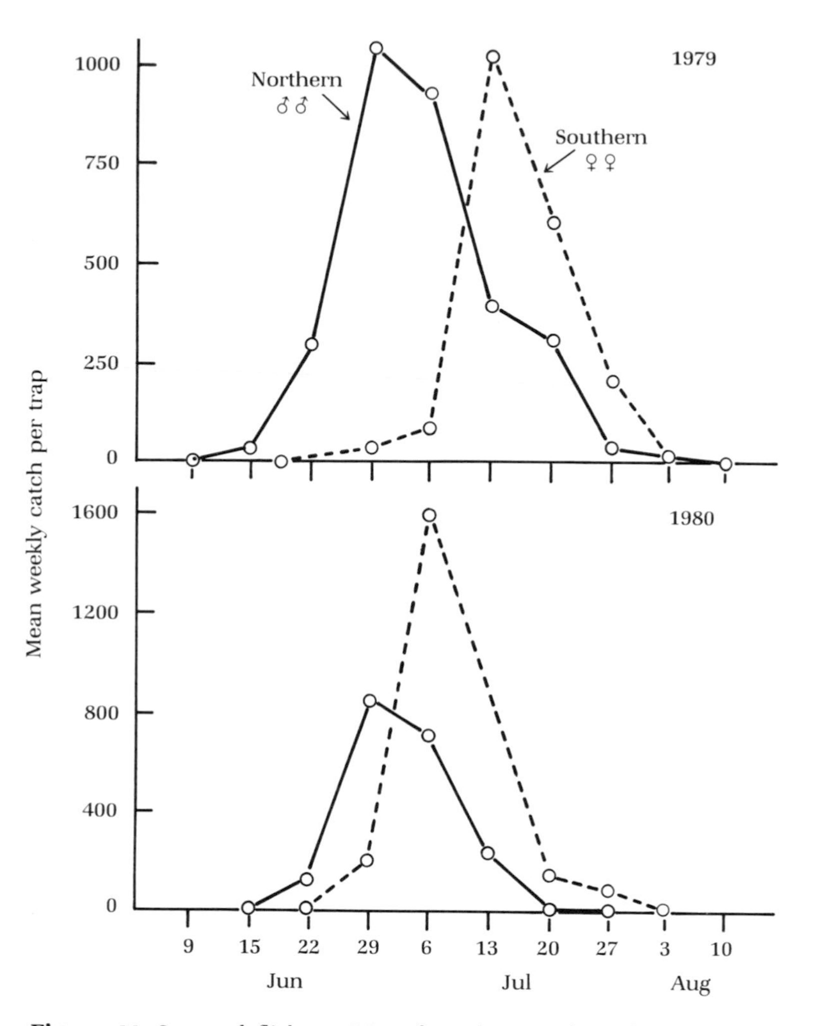

Figure 44. Seasonal flight activity of northern and southern masked chafer males as determined by black-light trap catches. (Adapted from Potter 1981b, fig. 2, courtesy of the Entomological Society of America.)

Searching males generally land downwind nearby and crawl upwind toward the female (Plate 31). Copulation takes place almost immediately. Mating pairs are often surrounded by 1–7 additional males, suggesting that a sex pheromone may be present (Potter 1980). NMC males were observed at about midnight, emerging and skimming over the sod surface in search of females. The sexual response of the two species is identical except for the temporal differences.

Traps baited with live females of either species made it evident that females are cross-attractive. Males in the laboratory responded to and mated with females of either species. Even though there appears to be a common airborne sex attractant, the two species have apparently remained reproductively isolated (Potter 1980).

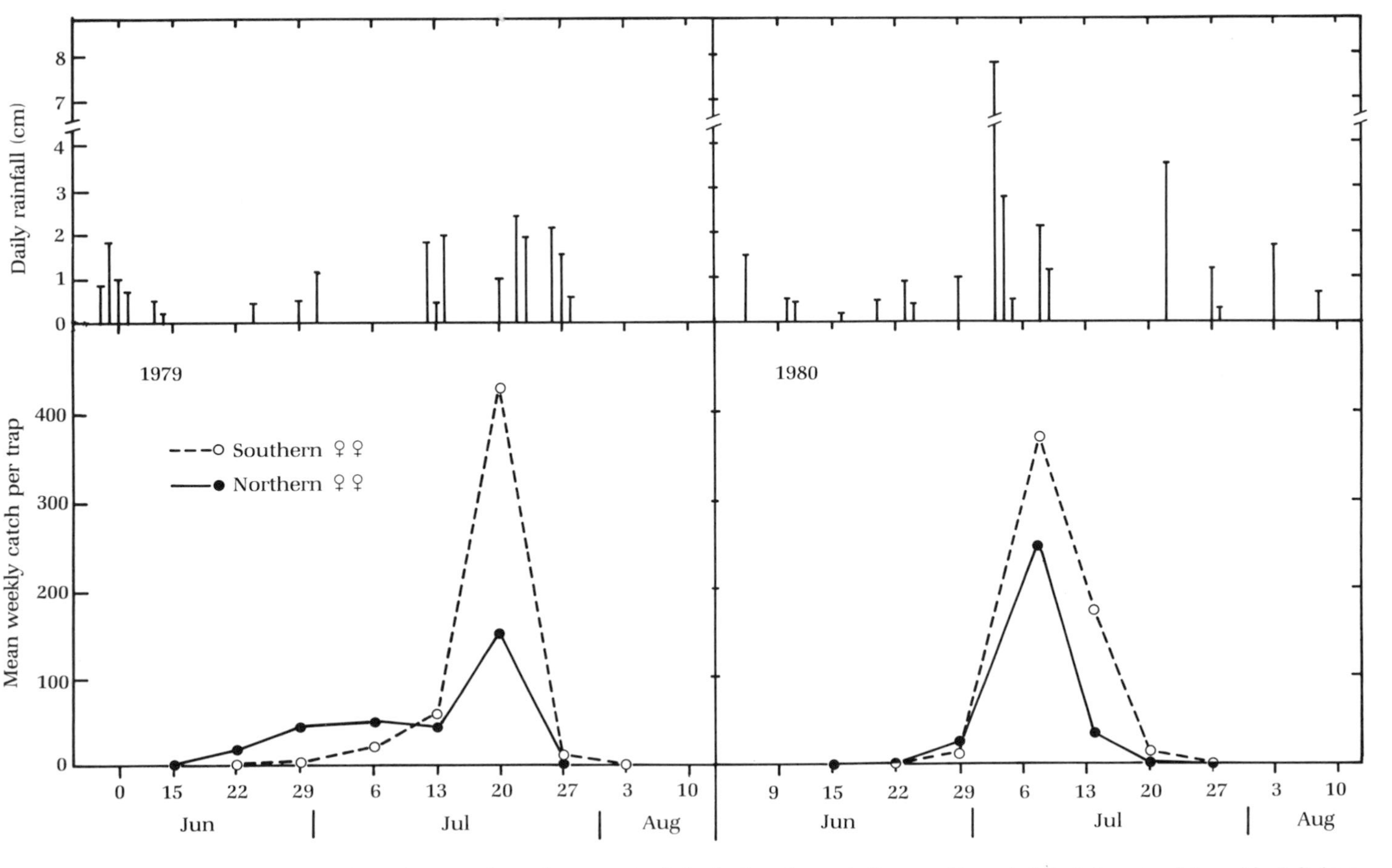

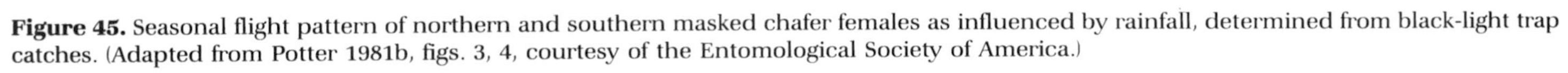

Figure 45. Seasonal flight pattern of northern and southern masked chafer females as influenced by rainfall, determined from black-light trap catches. (Adapted from Potter 1981b, figs. 3, 4, courtesy of the Entomological Society of America.)

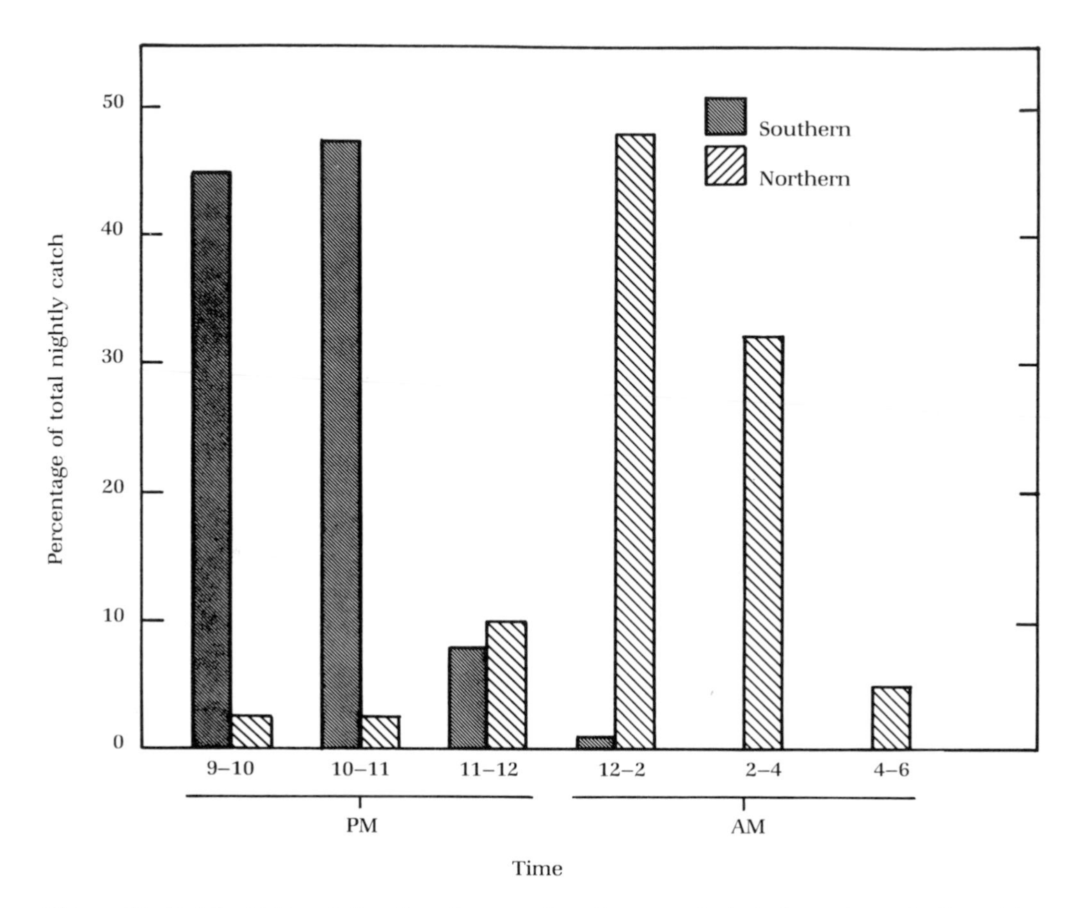

Figure 46. Nightly flight activity of southern and northern masked chafer males as indicated by black-light trap catches. (Adapted from Potter 1980, fig. 2, courtesy of the Entomological Society of America.)

Oviposition. In captivity, beetles of the NMC mate readily a day after emergence, and oviposition begins the next day. Females deposit about 11–12 eggs each in the field, but maximum oviposition in the insectary was 29. Hatching occurs in 14–21 days. In Kentucky, the SMC deposits most of its eggs in the upper 3 cm (1.2 in.) of soil (Johnson 1941, Potter and Gordon 1984).

Influence of Soil Moisture and Temperature. Soil moisture significantly affected oviposition. No eggs were deposited in air dry soil, and only a few eggs were deposited in soil of 5.0%–12.5% moisture. Oviposition was greatest at 25.5% soil moisture, averaging 2.5 eggs per female per day (Potter 1983).

Eggs developed normally at soil moistures of 10.3%–12.3% and above but shriveled and died in drier soils. The ability of eggs to withstand dessication changed markedly as they became older. Fully swollen 8-day-old eggs survived dessication much better than newly laid eggs or eggs that were close to hatching (Potter 1983). In the field, no eggs survived in dessicated turf where afternoon soil temperatures exceeded 40°C (104°F) and where soil moisture dropped to <8%. Egg survival under irrigated turf ranged from 55% to 75% (Potter and Gordon 1984).

Larval Activity

Hatching and Growth. Feeding begins almost immediately after hatching, as is evidenced by day-old first instars, which are discolored by food in the alimentary tract. Larvae can develop with a diet of organic matter, but their root-feeding habits make them destructive to turf. Since eggs of the NMC are deposited 11–15 cm deep (4.5–6.0 in.), newly hatched grubs must migrate upward to near the soil surface for their major root-feeding activities (Johnson 1941).

Grubs grow quickly and by late August reach their full size of about 2.3 cm (<1 in.) in length. Damage to turf usually occurs in September and October. Feeding continues until mid to late October, when cold temperatures force them to migrate downward to spend the winter at a depth of 36–41 cm (14–16 in.). In Connecticut, they descended as deep as 46 cm (18 in.), but about 70% hibernated at depths of 13–25 cm (5–10 in.). Winter mortality may be heavy, with as many as 50% dying in hibernation (Johnson 1941, Neiswander 1938).

Spring Activity. In Kentucky, grubs migrate upward to near the soil surface in late March and feed on grass roots until late May. Grubs move downward again to about 15 cm (6 in.) to become prepupae and pupae in June (Johnson 1941, Potter 1981a).

Miscellaneous Features

Threshold Population

Grub densities of 32–65 per 0.1 m^2 (30–60 per ft^2) occurred in Connecticut. In Ohio, 11–16 grubs per 0.1 m^2 (10–15 per ft^2) practically destroyed the grass (Johnson 1941, Neiswander 1938).

Potter (1982b) studied damage thresholds for the SMC in relation to soil moisture. SMC populations of 0, 3, 6, 12, 24, and 48 grubs per 0.1 m^2 (0, 2.7, 5.5, 11.0, 22.0, and 45.0 per ft^2), consisting of late second and third instars, were released on Kentucky bluegrass turf within enclosures. Half the enclosures with each population were irrigated twice a week, while the remaining half received only natural rainfall. This test, conducted during September and October, indicated that the economic threshold for masked chafer grubs is considerably higher than the usual rule-of-thumb estimate of 6–8 grubs per 0.1 m^2 (5.6–7.4 per ft^2).

After necessary adjustments had been made for loss of grubs, this study indicated that at least 9–10 grubs per 0.1 m^2 (8–9 per ft^2) were required to damage moisture-stressed Kentucky bluegrass turf. Well-watered Kentucky bluegrass turf tolerated

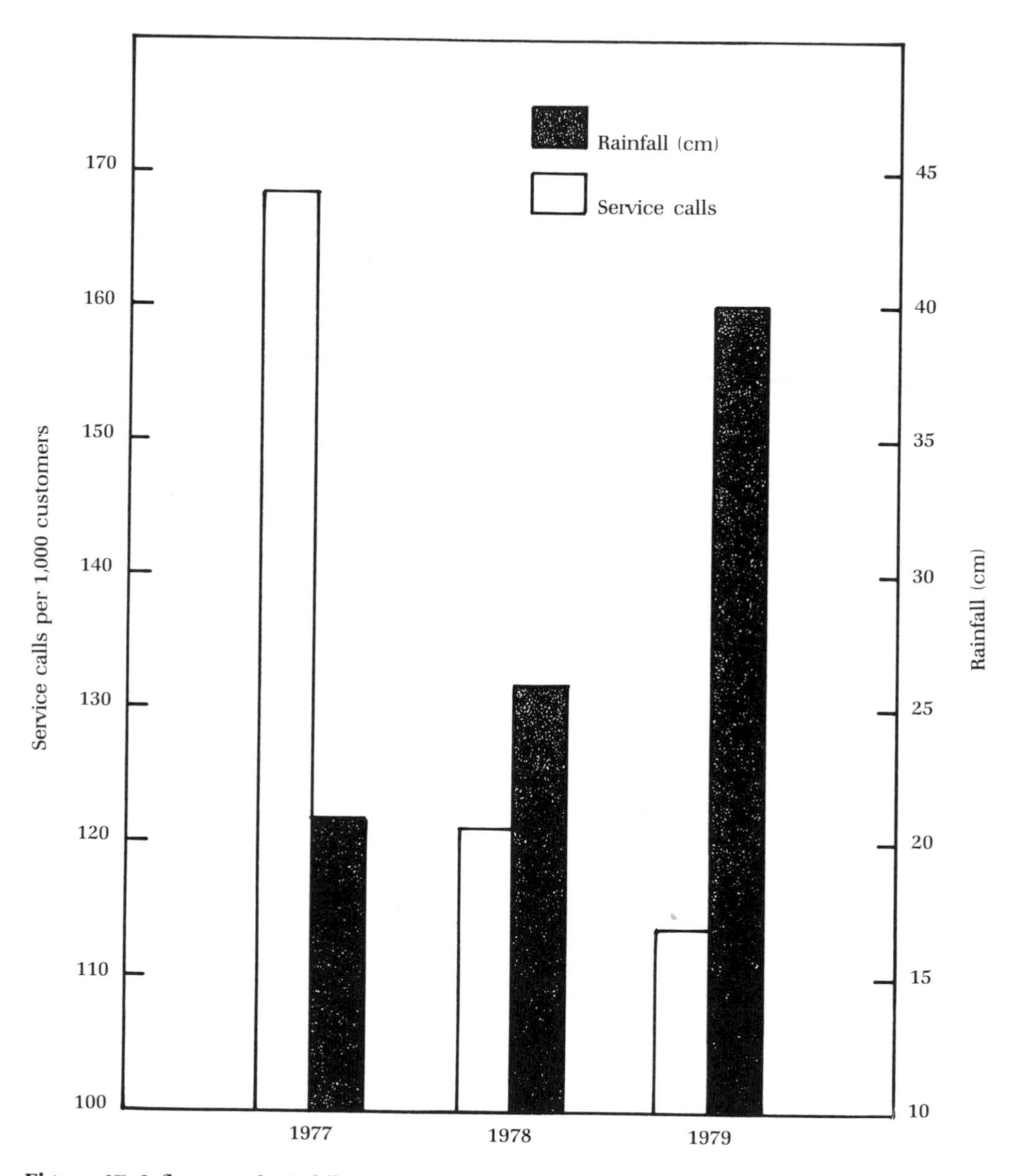

Figure 47. Influence of rainfall on numbers of grub-related service calls (per 1,000 customers) received by a Kentucky lawn care company during August 15–October 15 for 3 successive years. (Adapted from Potter 1981a, fig. 6, courtesy of *American Lawn Applicator*.)

15–20 grubs per 0.1 m^2 (14–18 per ft^2) before showing noticeable damage. Threshold populations of the SMC may be significantly higher than those of other species of similar grub size, such as the European chafer, since the former may feed on dead and decaying organic matter more than the latter do, especially as third instars (Ritcher 1966).

The influence of rainfall on the numbers of grub-related service calls made by a Kentucky lawn care company is a good indicator of the role of adequate moisture. Such a relationship is evident in Figure 47.

Natural Enemies

Microorganisms

Milky Disease. Japanese beetle larval surveys for milky disease in the Northeast revealed many instances of infected *Cyclocephala* grubs in at least eight states and the District of Columbia. The bacterium infecting *Cyclocephala* grubs was designated as type A *Cyclocephala* strain of *Bacillus popilliae* (White 1947).

More recent studies made in Kentucky demonstrated that the commercial *B. popilliae* (type A) spore talc mixture prepared with spores from milky Japanese beetle grubs had no infectivity on SMC grubs. Spore dusts prepared from diseased SMC grubs, however, were as ineffective on the SMC as commercial type A spore dusts are on the Japanese beetle. The fact that diseased grubs are commonly found in the field would indicate that milky disease has some value as a biological control agent for masked chafer grubs (Potter and Gordon 1984).

Parasites

In New Jersey, a native wasp of the genus *Tiphia* was reported to have a marked preference for parasitizing grubs of the SMC rather than those of the Japanese beetle in the same locality. Parasitism occurred in late August, when most Japanese beetle grubs were third instars, while the SMC grubs were still second instars (Jaynes and Gardner 1924).

Vertebrate Predators

Predators of other, similar scarabaeid grubs are undoubtedly natural enemies. These include birds, such as starlings, grackles, and crows, and mammals, such as moles, skunks, and raccoons.

14

Scarabaeid Pests: Subfamily Melolonthinae

Asiatic Garden Beetle

Taxonomy

The Asiatic garden beetle, AGB, *Maladera castanea* (Arrow), order Coleoptera, family Scarabaeidae, subfamily Melolonthinae, tribe Sericini, was first called the oriental garden beetle, *Aserica castanea* Arrow (Hallock 1929). It was later named the Asiatic garden beetle, *Autocerica castanea* Arrow (Hallock and Hawley 1936). *Maladera* is the currently accepted genus. Hallock and Hawley's (1936) studies have been the most complete on this insect. I base my account on their work and cite other material only when it supplements theirs.

Importance

The AGB, like the oriental beetle, is a relatively minor turfgrass and ornamentals pest when compared with the Japanese beetle. Adults are pests of some ornamental plants and vegetable gardens. While the insect breeds in greater abundance in weedy, abandoned areas than in lawns, the grubs are often destructive to turfgrass. Turf injury has been most prevalent in large metropolitan areas, where abandoned weedy lots and turfgrass areas exist side by side.

Because it is attracted to bright lights on warm nights of July and August, the beetle becomes a nuisance at all types of nighttime amusement and recreational parks, open air restaurants, and well-lit storefronts (Hallock 1936).

The AGB adult can create a medical problem through its habit of invading human ears. Most cases have involved people who slept on the ground in sleeping bags. Severe pain is caused when the tibial spurs pierce the delicate lining of the external auditory meatus. Medical attention is generally required for removal of the beetle, and most requests came between 11:00 PM and 1:30 AM (Maddock and Fehn 1958, Wills et al. 1969).

History and Distribution

Adults were first found in the United States in Rutherford, New Jersey, in 1921, but the beetle was not identified as an introduced insect until 1926. It is a native of Japan and China. Through 1936 it was distributed along the eastern seaboard, from Massachusetts into Maryland, and an isolated infestation appeared in South Carolina. The heaviest concentrations have been in the metropolitan New Jersey–New York area, including Long Island, and in Philadelphia, with very little spread since the 1930s (Hallock 1936).

Since there has been no concerted effort to determine its spread, very little is known of the insect's present distribution, but it is probably found in most northeastern states (Figure 48). In the late 1970s and 1980s, it was found in fair numbers infesting turfgrass in Rochester, New York, in mixed populations with the European chafer and the Japanese beetle. On Long Island during the same period grub populations were found that approached those of the oriental beetle and the Japanese beetle in mixed populations.

Host Plants and Damage

Adult Feeding

Beetles begin feeding at the margins of leaves, and when feeding is heavy, the entire leaf may be eaten, leaving only the midrib. Beetles may feed on more than 100 different plants, but the preferred food plants number about 30. The shrubs most commonly attacked include box elder, butterfly bush, Japanese barberry, rose sumac, and viburnum. The most favored flowers include aster, chrysanthemum, dahlia, goldenrod, and strawflower. Common fruits whose foliage is eaten are peach, cherry, and strawberry. Favored vegetables include carrot, beet, eggplant, pepper, and turnip (Hallock 1933, 1936).

Larval Feeding

The feeding habits of the larvae are less spectacular than those of adults, but larvae can be a serious turf problem. As many as 140 grubs per 0.1 m^2 (126 per ft^2) have been found in lawns, where only about 20 grubs per 0.1 m^2 (18 per ft^2) are needed to destroy a lawn.

Grubs of the AGB are normally less destructive to turf than larvae of the Japanese beetle or the oriental beetle. The reason is in part the AGB's tendency to feed at greater depths, which causes less damage. Most of the feeding occurs at 5.0–7.6 cm (2–3 in.) below the surface.

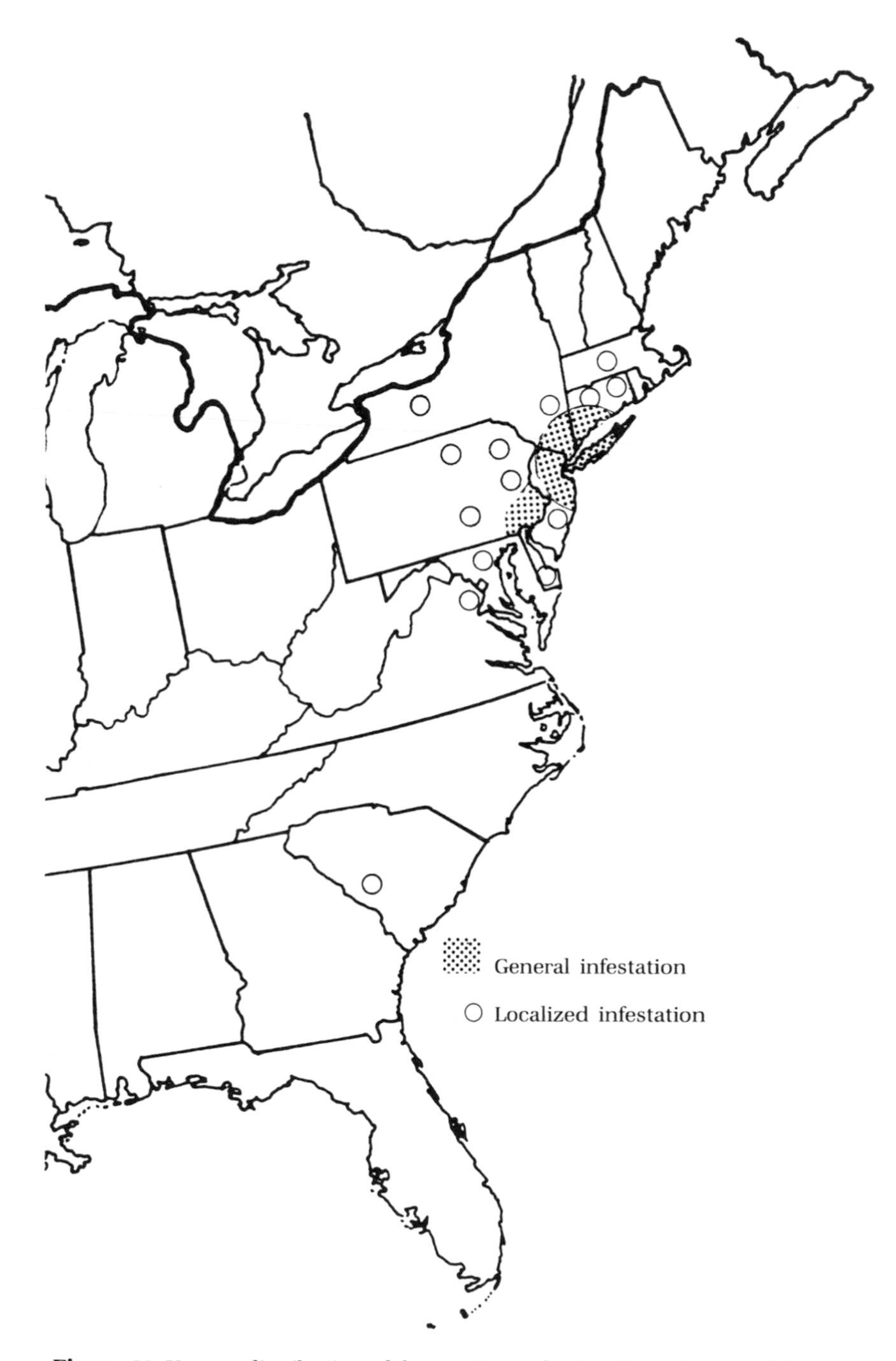

Figure 48. Known distribution of the Asiatic garden beetle in the United States. (Data from Hallock 1936, fig. 1, and U.S. Department of Agriculture 1951–1980; drawn by H. Tashiro, NYSAES.)

Description of Stages

Adult

The beetle is dull chestnut brown, with a slight iridescent velvety sheen, and 8–11 mm long (Plate 32). A few erect irregularly arranged, backward-projecting hairs are present on the top of the head. The surface of the elytra is essentially glabrous, but the outer lateral edge of the elytra has a row of fine hairs. Ventrally the exposed abdominal segments (5, 6, and 7) have a transverse row of spines extending the entire width of each segment (Plate 32).

Egg

The eggs are white, ovoid, and about 1.0 mm in diameter, deposited in clusters of as many as 19 eggs. They are loosely held together by a glutinous material (Plate 32). The eggs increase in size and become nearly spherical at maturity.

Larva

A newly hatched first instar is about 3.2 mm long. A full-grown third instar is about 19 mm long. Three distinct characters identify the larva of the AGB. The raster, consisting of a single transverse row of spines in a crescent shape, is the most noticeable character (Plate 32). Metathoracic legs have very small claws (they are practically nonexistent) as compared with the prothoracic and mesothoracic legs. Another prominent larval feature is the light-colored, enlarged bulbous stipe of the maxilla that is lateral to the mandibles (Plate 32). The palpi are usually in constant motion and at first sight make the larva appear to be chewing on a piece of vegetation.

Prepupa and Pupa

After feeding, the third instar constructs an earthen cell about its body, voids accumulated excrement, and becomes first a prepupa, then a pupa that is about 8.4 mm in length. The change to adulthood occurs as the pupal exuviae split to allow the beetle to emerge.

Seasonal History and Habits

Seasonal Cycle

In the known areas of infestation in the Northeast, the AGB has a 1-year life cycle. In the latitude of New York City, adults may be present from the last third of June until the end of October but are most abundant between July 15 and August 15. The average longevity of individuals is about 1 month. Eggs hatch in about 10 days. Grubs rapidly pass through the three larval instars, with about 75% becoming third instars before winter. Since first instars have never been found during spring, they are considered unable to survive the winter. Overwintering grubs migrate upward about mid-April to resume feeding on grass roots. During late May to early June, the mature larva constructs its earthen cell, where it transforms and remains as a prepupa for about 4 days before becoming a pupa for 8–15 days. The adult remains in the earthen cell until it turns from white to chestnut brown and hardens, before burrowing upward to emerge. Figure 49 illustrates the seasonal life cycle.

Should this insect spread to southern states, it may have two broods each year, since laboratory rearing has shown that it does not require a diapause. At a constant temperature of 27°C (81°F), three broods were reared between September and the end of the following June (Hallock 1936).

Adult Activity

Since adults are nocturnal, they are seldom seen on host plants by the general public. At dusk, beetles leave the ground, where they have been in hiding during the day. Beetles do not fly at temperatures below 21°C (70°F) but feed largely on plants near their daylight hiding areas and may remain quite unnoticed through the entire season. Large-scale emergence begins when temperatures are 18°–21°C (65°–70°F). During uncomfortably warm evenings when the temperature is above 21°C (70°F), beetles become very active and fly in all directions in search of favorite host plants (Allen 1944).

Oviposition. Well-kept lawns and other short-grass areas near preferred host plants are the likely oviposition sites. Each female deposits about 60 eggs that hatch in about 10 days. Eggs are normally deposited in the greatest numbers in overgrown, uncultivated areas where such weeds as daisy fleabane, goldenrod, plantain, orange hawkweed, ragweed, wild aster, and wild carrots abound.

Although orange hawkweed, *Hieracium aurantiacum* (Compositae), is not a food plant, the environment that it creates appears responsible for maximum oviposition. The leaves of this weed form a basal rosette that shades the ground and keeps the soil cool and moist. This creates a highly acceptable place for beetles to hide during the day, depositing their eggs in nearby turf soil. Larvae are usually 5–10 times more abundant in the vicinity of this weed.

Attraction to Lights. During warm nights, beetles are highly attracted to electric lights, especially to blue mercury-vapor lights. They may be found resting on the poles supporting streetlamps. Flight distances range from about 183 m to 457 m (200–500 yd) when the insects are attracted by lights. A trap illuminated by a 100-watt incandescent lamp has captured as many as 2,000 beetles an hour. A trap illuminated with a G-5 mercury-vapor lamp has captured as many as 4,700 beetles an hour.

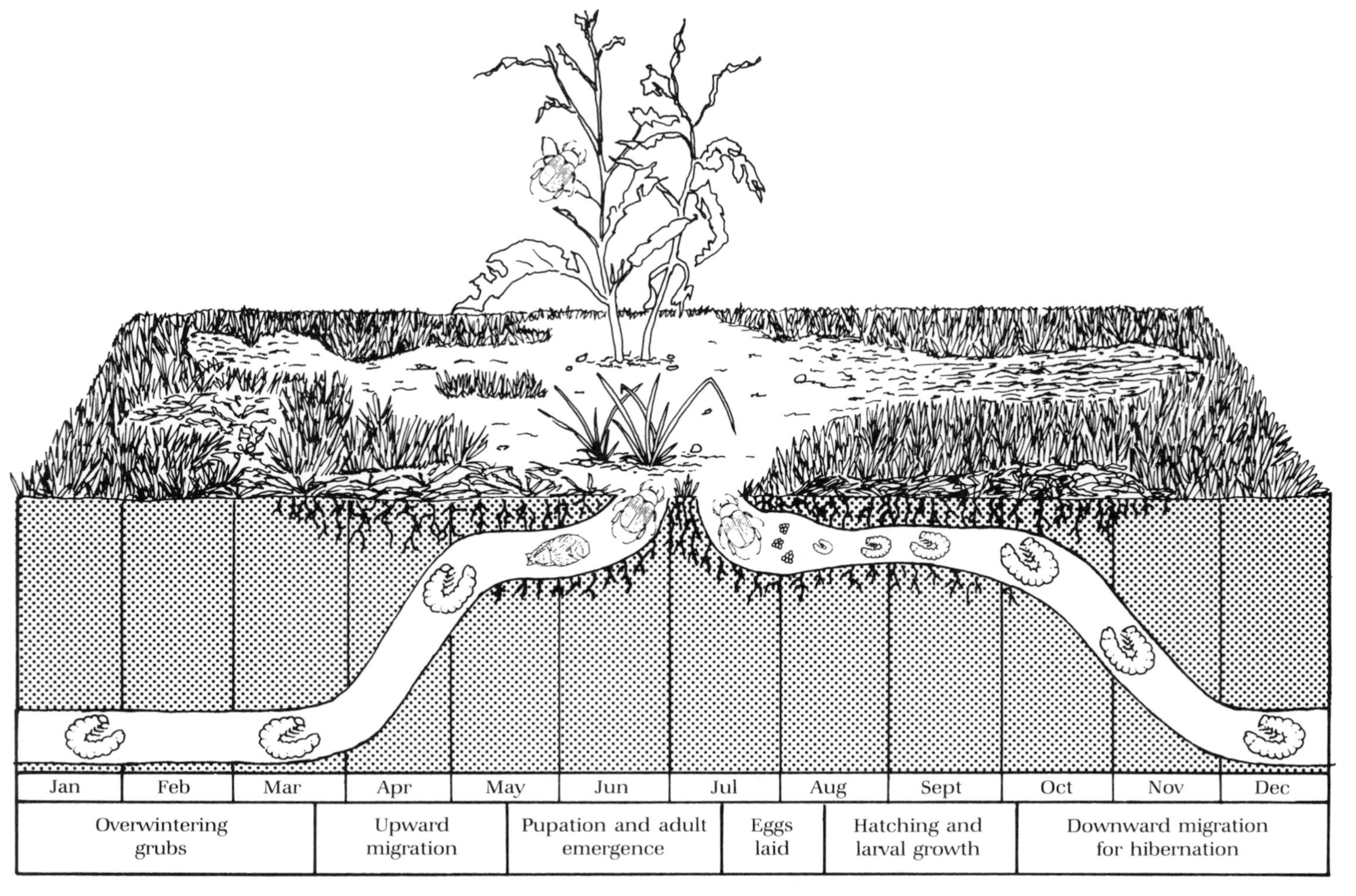

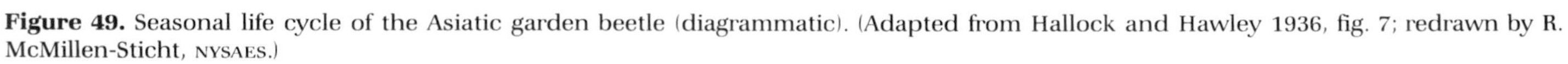
Figure 49. Seasonal life cycle of the Asiatic garden beetle (diagrammatic). (Adapted from Hallock and Hawley 1936, fig. 7; redrawn by R. McMillen-Sticht, NYSAES.)

Larval Activity

Grubs of the AGB occur in all types of sod, in weed patches, and to some extent in cultivated land such as gardens and nurseries (Hallock 1933). Larvae feed in the top 13 cm (5 in.) of soil with major feeding at a depth of 5–8 cm (2–3 in.). Because of the loose texture of cultivated soil, beetles go deeper into it than in turf. They are known to migrate as far as 0.9 m (3 ft) horizontally through turfgrass soils. Most plants are susceptible to larval feeding, but those with succulent roots are preferred.

Winter and Spring Activity. Beginning in October as the soil temperatures drop, larvae gradually descend 15.0–30.5 cm (6–12 in.) for overwintering; most are at depths of 2.3–28.0 cm (0.9–11.0 in.). Upward migration occurs during April, and by late April and early May the larvae are actively feeding on grass roots within the top 13 cm (5 in.) of soil. Pupation occurs at a depth of 4–10 cm (1.5–4.0 in.) during late May and June, with adult emergence beginning in late June.

Influence of Soil Moisture. Summer rainfall deficiencies markedly influence the survival of eggs and young larvae. Mortality may be so great during a dry summer that beetles are significantly reduced the following year.

Natural Enemies

The AGB is surprisingly free of natural enemies. Both adult and larval stages are free from parasitism by other insects and are not susceptible to disease organisms such as the milky disease bacteria. A wasp of the genus *Tiphia* is reported to parasitize the larvae in China and Japan. Starlings have been known to feed on beetles in their daylight hiding places.

European Chafer

Taxonomy

The European chafer, EC, *Rhizotrogus majalis* (Razoumowsky), order Coleoptera, family Scarabaeidae, subfamily Melolonthinae, tribe Melolonthini, was called *Amphimallon majalis* in all publications in the United States prior to 1978 (Sutherland 1978, Tashiro et al. 1969).

Importance

Wherever the European chafer establishes itself, the grubs become a more serious pest than Japanese beetle grubs, the other species most commonly found in mixed populations. Although EC population levels are generally lower, grubs are larger, come to the soil surface to feed almost a month earlier in the spring, and feed nearly a month later in late fall than the Japanese beetle. The European chafer damages all turfgrasses, whether fine turf or pastures, and many field and forage crops, including all grasses grown for hay, wheat, and barley. It has also been a serious nursery pest. Since adults do very little or no feeding, the European chafer is a less serious overall pest than the Japanese beetle.

History and Distribution

The European chafer is known to occur in western and central Europe but not in England or Holland. It was first discovered in the United States as a grub in 1940 in Newark, Wayne County, New York, then an active nursery-growing area (Gambrell et al. 1942).

For the first 10 years, the insect's spread appeared to be natural and contiguous, covering an area of about 997 km^2 (385 mi^2). During the period 1950–1966, numerous infestations were discovered in Connecticut, Massachusetts, New Jersey, Ohio, Pennsylvania, and West Virginia and in adjoining and nearby states, as well as in Ontario, Canada, along the Niagara frontier. Many of these infestations have been directly traced to shipments of infested nursery stock from Newark. Other localized infestations in the same states appeared quite certain to be the results of hitchhiking beetles, since they were found along major interstate highways and on railroad right-of-ways (Gambrell et al. 1942, Regnier 1939, Tashiro et al. 1969).

During the 1970s, infestations in New York became fairly contiguous, until the insect was present in all areas of the state except in the Adirondack Mountains and in eastern Long Island. It has also spread from localized infestations in other states except in West Virginia. This localized infestation, traced to plantings of grub-infested spruce, was successfully eradicated during the 1950s by the application of dieldrin over a 81-ha (200-acre) area surrounding the initial 10-ha (25-acre) infestation. The success of this single eradication effort can be attributed to the early use of an insecticide that had years of effective residual life (Tashiro et al. 1969, U.S. Department of Agriculture 1951–1980).

The results of the 1983 survey in Ontario, Canada, show that the generally infested area filled the Niagara peninsula, with the western boundary encompassing Burlington and Brantford and extending nearly to Simcoe. At least five isolated infestations stretch from Toronto to Windsor across the river from Detroit (Reid 1983).

Since the federal European chafer quarantine was rescinded in 1971 and scouting efforts ceased, information on the insect's later spread has been less defined in the United States. Figure 50 shows distribution in the United States and Canada as it is currently known.

Host Plants and Damage

Adult Feeding

Adults feed to such a limited extent that they are not considered an injurious insect in this stage. Literally thousands may congregate in a single small tree on a

Figure 50. Known distribution of the European chafer in North America through 1985. (Data from Reid 1983 and U.S. Department of Agriculture 1951–1980; drawn by H. Tashiro, NYSAES.)

given night without any visible damage evident to the naked eye. Adults feed little, nibbling only slightly along the margins of leaves (Plate 33).

Larval Feeding

Turfgrass damage by EC grubs is most serious, affecting many home lawns, golf course turf on fairways and roughs, parks, cemeteries, and other turfgrass areas. Like many other scarabaeids, grubs prefer to feed on fibrous roots, and therefore rooting habits and ground cover during the oviposition period largely determine which plants become hosts. Many annual crops escape damage because they are planted just as the grubs have completed their feeding in the spring and are harvested before the next brood of grubs becomes destructive. Pastures, winter wheat, and hay crops are subject to injury.

By early fall grubs are mostly third instars. Fall rains coupled with cooler temperature keep the soil moisture nearly continuously at field capacity. Under these conditions practically all the grubs are feeding within 2.5–5.0 cm (1–2 in.) of the soil surface, pruning off the fibrous roots. In degree the damage would be identical to that caused by other similar species, as noted in Chapter 10 (Plate 23).

Unlike the Japanese beetle grubs that often feed predominantly in the soil-thatch interface during fall and spring, chafer grubs are seldom observed in such a narrow horizon. Much damage occurs during the fall feeding period but often passes unnoticed; the above-ground portions appear healthy because of an abundance of moisture and cooler weather. As a result of root loss, even a momentary dry spell may cause large patches of turf to die. Under favorable fall conditions, even though the sod can easily be picked up because it lacks sufficient roots, it may appear healthy. In the spring, however, it may die as the grubs resume feeding on the previously weakened grass (Tashiro et al. 1969).

The density of grubs necessary to show damage is difficult to predict because so much depends on the vigor of the grass and on the moisture supply. In general, 10 or more third-stage grubs per 0.1 m^2 (per ft^2) in a high-maintenance turf, or 4–5 grubs per 0.1 m^2 (per ft^2) in a low-maintenance turf, are sufficient numbers to cause noticeable damage (Tashiro et al. 1969).

Larval feeding on ornamentals and nursery plants can be severe at times because it greatly reduces the fibrous roots, leaving only the larger tap roots. The lining-out stock of spruce and similar conifers may be seriously damaged or killed outright where larval populations are high (Figure 51; Tashiro et al. 1969).

Description of Stages

Adult

Mature adults are medium-sized, fawn-colored beetles with a rufous-yellow head and pronotum (Plate 33). A narrow band of light yellow hairs extends from under the caudal margin of the pronotum and slightly overlaps the proximal margin of the

Figure 51. Spruce seedling lining-out stock. **A.** Normal seedling. **B.** Seedling with fibrous roots destroyed by European chafer grubs. (From Tashiro et al. 1969, fig. 7, courtesy of the NYSAES.)

elytra. The heavily striated and lightly punctated elytra extend posteriorly to the middle of the penultimate abdominal segment, leaving the pygidium exposed (Gyrisco et al. 1954).

Males are slightly smaller than females, averaging 13.7 mm and 14.3 mm in length, respectively. Antennae are nine segmented and lamellate, with the club comprising three segments. Males have longer lamellae than females (Plate 33). Two June beetle adults similar in size that have been mistaken for the European chafer are *Phyllophaga tristis* (Fabricius) and *P. gracilis* (Burmeister). European chafer adults can be differentiated from *Phyllophaga* by the tooth on the tarsal claws, which are absent in the European chafer but present and distinct in *Phyllophaga* spp. (Figure 52; Gyrisco et al. 1954, Tashiro et al. 1969).

Egg

Freshly deposited eggs resemble other scarabaeid eggs, described in Chapter 10. Young EC eggs are about 0.49 mm wide by 0.73 mm long and become more spherical, measuring about 2.30 mm by 2.75 mm just before hatching (Plate 33). As in the case of other scarabaeid eggs, the resiliency of the chorion enables the egg to bounce from a hard surface without injury. As in other scarabaeid eggs, too, the tan mandibles of the embryo are clearly visible through the chorion prior to hatching (Tashiro et al. 1969).

Larva

European chafer grubs, like other scarabaeids, normally retain a C-shaped contour and develop brown to black posteriors as a result of ingested soil. The brown head capsules have average widths of 1.17 mm, 2.10 mm, and 3.00 mm, respectively, for the three instars. Full-grown third instars fully stretched out are about 6 mm wide in the thoracic region, the widest area, and about 23 mm long (Gyrisco et al. 1954).

Grubs of the European chafer can be distinguished from all other North American scarabaeid grubs essentially by a combination of two characters, the Y-shaped anal slit, with a stem half the length of each arm, and the raster, which has two subparallel rows of palidia converging toward the anterior and outwardly curved toward the posterior (Plate 33; Gyrisco et al. 1954).

Prepupa

In its terminal period, the larva changes from a C-shaped contour, becoming nearly straight, with a slight crook on the posterior end (Plate 26). Prepupae average 6 mm in width and about 19 mm in length and are extremely flaccid and tender (Tashiro et al. 1969).

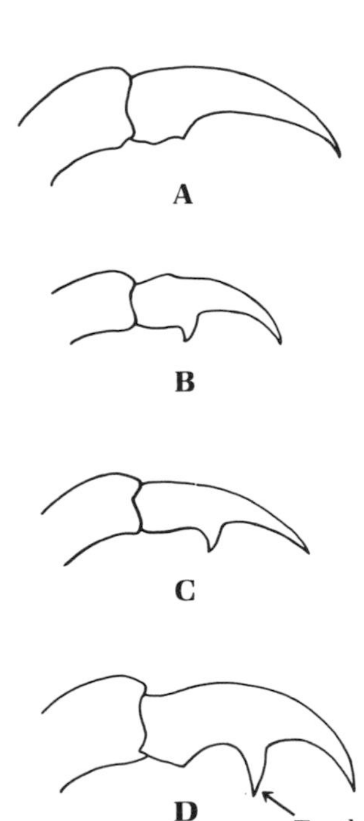

Figure 52. Differences in the tarsal claws between adults of the European chafer and native May or June beetles. **A.** European chafer. **B.** *Phyllophaga tristis.* **C.** *P. gracilis.* **D.** *P. hirticula.* (From Tashiro et al. 1969, fig. 10, courtesy of the NYSAES.)

Pupa

The European chafer pupa resembles that of any other scarabaeid as it lies in the earthen cell (Plate 26). Like other scarabaeid pupae, it turns from creamy white to the general adult color as it matures, but it remains very tender. Unlike the young Japanese and oriental beetle pupae, which are encased in the larval and prepupal exuviae, the European chafer sloughs off the exuviae to its posterior end. Pupae have an average width of 6.9 mm through the anterior abdominal region and an average length of 16 mm (Tashiro et al. 1969).

Differentiation of Sexes

Adults. Males can always be distinguished from females by antennal clubs that are twice as long as those in the female (Plate 33). Gravid females can also be easily separated from males by a bright yellow-tan ventral abdominal surface distended by more than 40–50 eggs (Plate 33). The color and distension disappear as eggs are expelled until the abdominal color resembles that on the rest of the body, as in the males (Tashiro et al. 1969).

Larvae. The sex of grubs can be determined in all three instars, in most cases by the appearance of the terminal ampullae in the males, as described in Chapter 10. Body fat sometimes obscures their appearance. When the ampullae are visible, the insect is certain to be a male. The ampullae appear more distinct in the EC males than in several other scarabaeid grubs.

Pupae. The sex of the pupae can be determined by differences in the ventral portion of the 10th abdominal segment (Figure 32). Male genital organs are much more pronounced. As in adults, the longer antennal club of the male is visible through the pupal integument (Tashiro et al. 1969).

Seasonal History and Habits

Seasonal Cycle

The EC has predominantly a 1-year life cycle, spends about 9 months of its year as a third instar, and overwinters in this stage. From 0.5% to 1.0% of the population require 2 years, overwintering the first winter as a second instar and the second winter as a third instar, emerging as an adult along with the regular 1-year-cycle beetles. Individuals requiring 2 years are found mostly in areas of tall grass and dense sod, presumably because of shading and cooler temperatures or in very gravelly soils with sparse vegetation, so that the food supply is limited. Figure 53 represents the life cycles of both the 1- and 2-year beetles on the basis of observations made during 1949–1961 on about 70,000 specimens found in more than 17,000 soil examinations in western New York (Tashiro and Gambrell 1963).

Adult Activity

Emergence and Seasonal Occurrence. After transformation from a pupa, the teneral adult remains in the earthen pupal cell for 3–5 days before it is ready to emerge and fly. It digs its way through the soil with its mandibles and fore tarsi. In western New York, adult European chafers start to emerge during the 2d or 3d week in June, reach peak populations during the last week in June or the 1st week in July, and then diminish rapidly. Flights are practically terminated by the end of July (Tashiro et al. 1969).

Phenological Relationships. Observations of the seasonal development of the European chafer over the years revealed that the flowering period of certain common plants could be correlated with the development of the European chafer. Observations in 1949–1961 revealed that the earliest seasonal development occurred in 1955 and the latest in 1961, with 1957 being a year of normal development. Weather

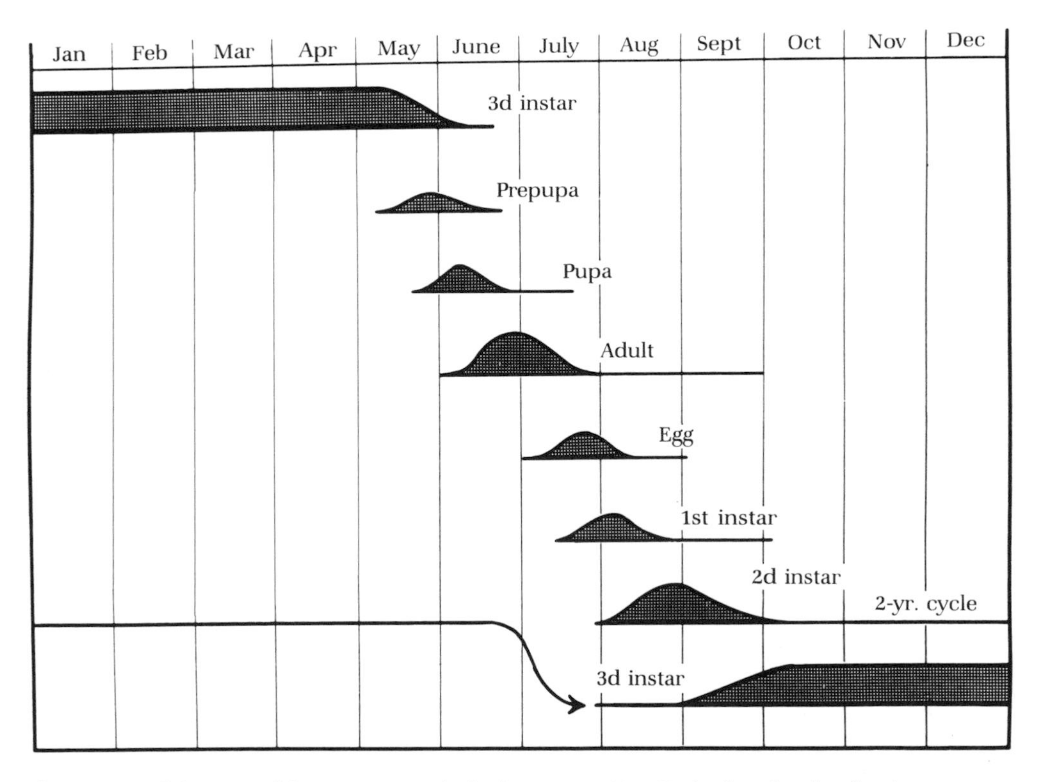

Figure 53. Life history of the European chafer in western New York, showing the dominant 1-year and minor 2-year life cycles. (From Tashiro and Gambrell 1963, fig. 3, courtesy of the Entomological Society of America.)

records (Figure 54) indicated that the spring of 1955 was the warmest, 1961 the coldest, and 1957 intermediate (Tashiro and Gambrell 1963).

Of the phenological relationships shown in Figure 55, the three most useful relationships have been the full-bloom period of Vanhoutte spirea, *Spiraea vanhouttei* (Briot) Zabel, indicating the presence of pupae, the early bloom of hybrid tea and floribunda roses, signalling early beetle flights, and the full-bloom period of the common catalpa, *Catalpa bignonioides* Walt., which correlated with peak flights of adults (Plate 34; Tashiro and Gambrell 1963).

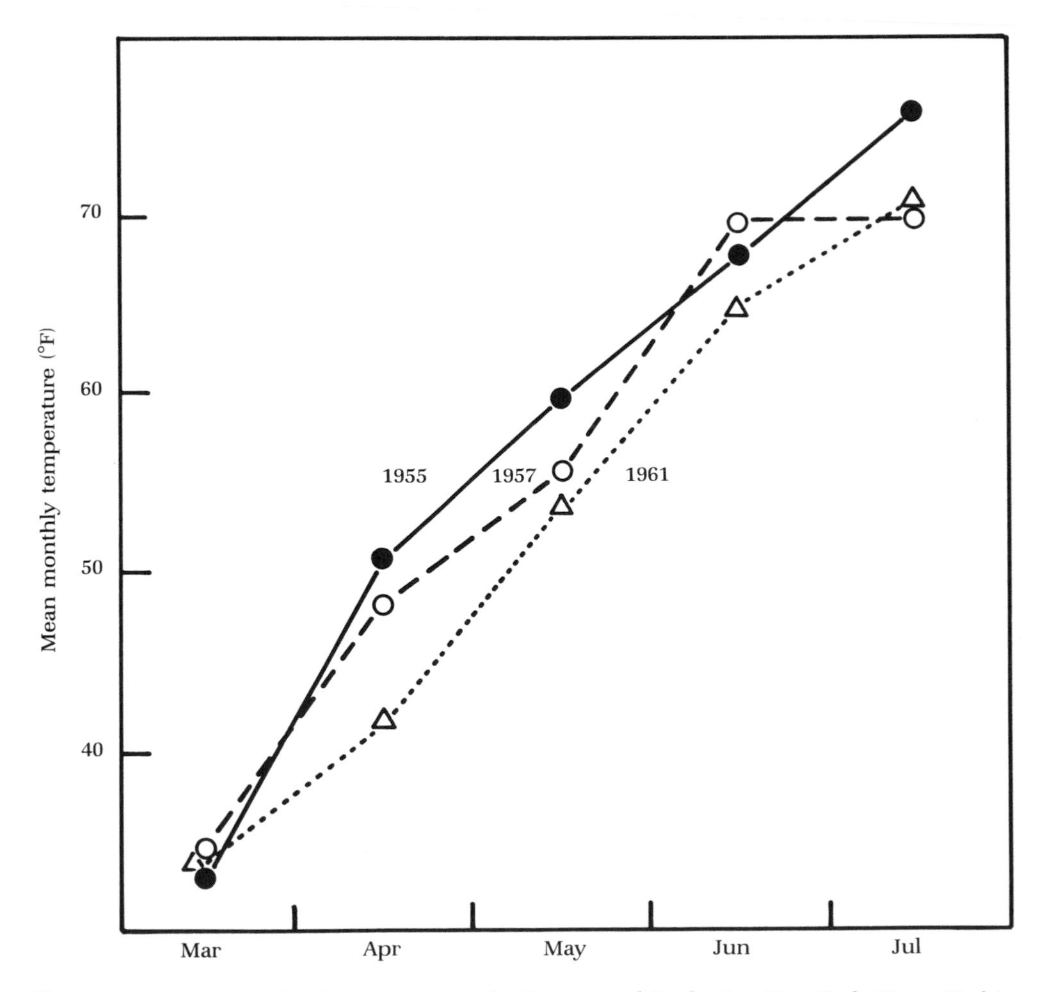

Figure 54. Mean monthly air temperatures for Geneva and Rochester, New York. (From Tashiro and Gambrell 1963, fig. 6, courtesy of the Entomological Society of America.)

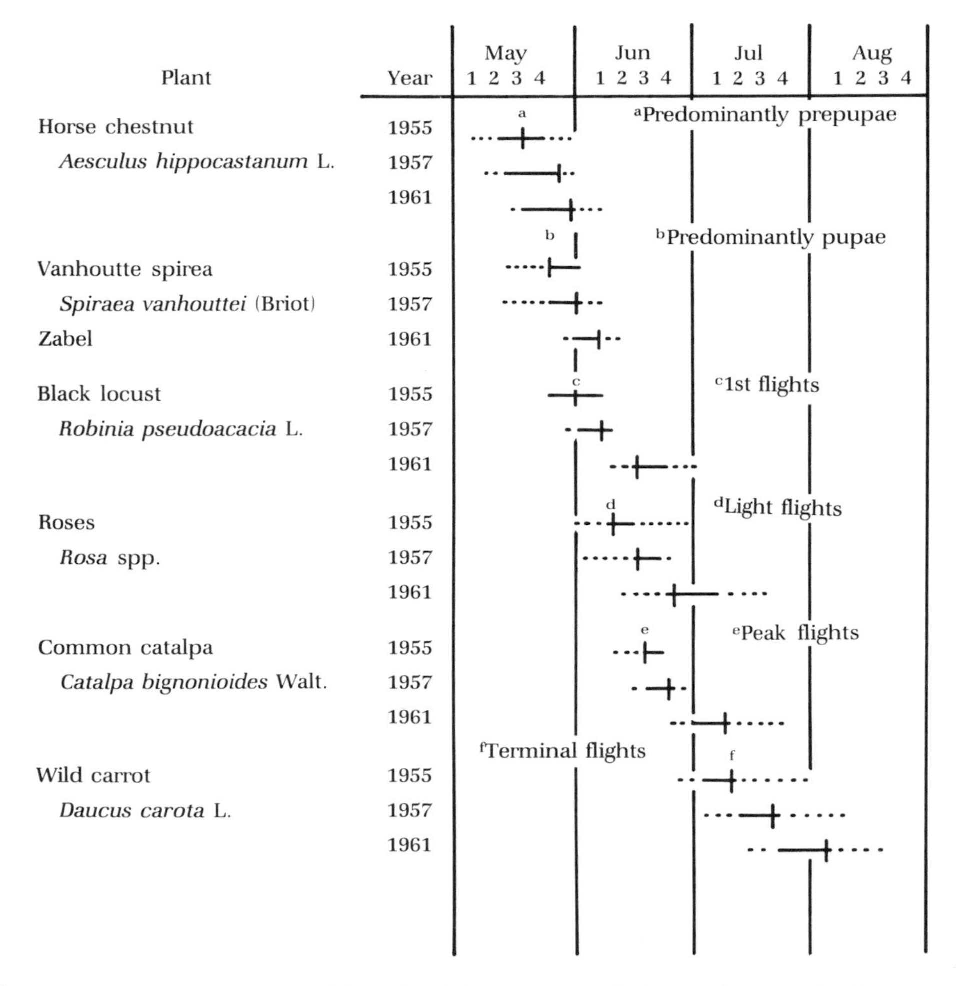

Figure 55. Various events in the life cycle of the European chafer in relation to the flowering of common plants during seasons that came early (1955), with average timing (1957), and late (1961) in western New York. For each year, the first broken rule indicates early bloom, the solid rule indicates full bloom, and the second broken rule indicates petal fall; the vertical line indicates chafer development. (From Tashiro and Gambrell 1963, fig. 5, courtesy of the Entomological Society of America.)

Daily Emergence and Mating Flights. During evenings of warm sunny days, adults begin emerging from the ground at about 8:30 PM EDT at a light intensity of about 1400 lux (130 ft-c), a level generally coincident with the disappearance of the sun below the horizon. Beetles crawl up grass stems to fly (Plate 34). They are very clumsy in their efforts to become airborne and may fall to the ground several times before they finally do so. Flight to nearby trees or other plants becomes apparent about 8:45 PM at a light intensity of about 323 lux (30 ft-c). Once airborne, the insects are strong fliers and fly to nearby or distant trees, even to the tops of trees as tall as 18–24 m (60–80 ft). The numbers approaching a given tree increase rapidly during the next 10–15 min, and peak swarming occurs about at 9:00 to 9:05 PM (Plate 34). On calm evenings, beetles hover completely around a tree, the majority in the central peripheral region. On windy evenings they hover on the leeward side. In the absence of trees or shrubs, they may swarm around such objects as light poles or chimneys. The primary attraction is a silhouetted object. There is no evidence that a given tree is preferred in the United States, but Regnier (1939) indicates that, in France, beetles have a strong preference for poplar, *Populus* spp. With several thousand in flight around a single tree, the noise produced by wing beats resembles that of swarming honeybees. This sound apparently has no physiological function (Gyrisco et al. 1954, Tashiro et al. 1969).

Weather conditions govern flight activities. Flights are greatly reduced when temperatures are below 19°C (66°F) or after an afternoon rainstorm. No flights take place at air temperatures below 11°C (52°F) (Gyrisco et al. 1954).

Mating and Oviposition. Mating takes place as soon as beetles have come to rest (Plate 34). There is no evidence of courtship before copulation. When several males attempt to copulate with a single female, each succeeding male attaches to the dorsum of the previous male until a string of six or more beetles has formed. Shortly after 10:00 PM, mating pairs and individuals start falling from the trees. At about this time, literally thousands of beetles can be dislodged by shaking the tree (Plate 34). Flights back to the ground continue throughout the night, and at dawn the few remaining beetles return to the soil (Gyrisco et al. 1954, Tashiro et al. 1969).

Oviposition occurs in moist soil as females deposit eggs singly in an earthen cell, generally at depths of 5–10 cm (2–4 in.). Formation of the cell and oviposition are described and illustrated in Chapter 10. Total eggs per female, determined through dissection, varied from 0 to 52 and, in captivity, from 2 to 46 eggs, with an average of 22. Ground cover has a pronounced effect on oviposition. Soil kept fallow during the entire flight period averaged only 0.3 grubs per 0.1 m^2 in the fall, compared with 14 grubs per 0.1 m^2 (13 per ft^2) in soil covered with turf during the flight period (Gyrisco et al. 1954).

The height of turf also has a pronounced effect on oviposition. Turf that was mowed during the entire flight season averaged 22 grubs per 0.1 m^2 (20 per ft^2) in the fall, compared with 13 grubs per 0.1 m^2 (12 per ft^2) in grass maintained at a height of a foot or more (Gyrisco et al. 1954).

Frequency of Flight and Longevity. Beetles feed so little that they depend entirely on the fat accumulated during the larval stage for their energy. Males had an average flight period of 6 days, and females 6.5 days, with an average adult life not exceeding 2 weeks. Within 2–3 days of natural death, abdomens become partially transparent and a mere shell, reflecting complete utilization of the stored fat. Many insects die on the soil surface under trees to which they last flew (Gyrisco et al. 1954, Tashiro et al. 1969).

Response to Chemicals. Exposure of more than 2,000 separate chemicals or combinations in traps originally used for the Japanese beetle revealed only mild attraction to a few compounds. Traps baited with java citronella oil–eugenol (3:1 vol) captured only about four times as many beetles as unbaited traps. Butyl sorbate was slightly more attractive. Both were used for scouting to a limited extent while the federal quarantine on the chafer was in force (Tashiro and Fleming 1954, Tashiro et al. 1964).

Response to Colors. Trap color had some influence on beetle response. A high-gloss finish was more attractive than a dull finish. Trap catches with red > black > yellow > blue > white were obtained, with only a 3.6-fold difference in catches in the two extremes (Tashiro and Fleming 1954).

Response to Fluorescent Lamps. When beetles were exposed at night to 15-watt fluorescent lamps ranging from peak emission of 2857 Å for the germicidal lamp to 6450 Å for the red lamp, they indicated distinct differences in degree of attraction. From the highest to the lowest, beetle catches were BL black light (3650 Å) > BLB black light (3650 Å) > erythemal (3100 Å) > blue (4400 Å) > germicidal (2857 Å) > green (5300 Å) = gold (5900 Å) = pink (6200 Å) = red (6450 Å). The BLB lamp filters out some of the 3650 Å radiation as it filters out most of the visible radiation with the blue-violet tubing, so that it was significantly less attractive than the BL lamp. The latter is the standard lamp used for attracting and capturing most night-flying pest insects (Tashiro et al. 1967).

Unlike the chemical baits that attract the beetles during their initial twilight flights, the black light attracts beetles after they have flown to a tree and have settled down. Therefore the most effective location was under a tree at which beetles congregate. Black-light traps far exceed chemicals in their ability to attract beetles, but because AC or DC current is needed to operate the lamp, they had a serious limitation for general scouting operations (Tashiro et al. 1967).

Larval Activity

Hatching and Development. Eggs hatch in about 2 weeks under normal field temperatures, and grubs are first instars for about 3 weeks, second instars for about 4–5 weeks, and third instars for about 9 months. In western New York, first instars are the dominant stage during the latter half of July, seconds during the latter half of August, and thirds from the latter half of September through the first half of May (Gyrisco et al. 1954).

Populations and Densities. After an adult infestation has been discovered in a new area, the buildup of larval populations follows a predictable trend. Grubs are not found in soil surveys during the first 2–3 years after discovery of the first beetles. Grubs subsequently increase steadily and reach peak populations during the next 4–6 years. Thereafter populations decline, and a moderate level is maintained indefinitely; peak populations are never again reached (Tashiro et al. 1969).

During years of peak grub populations, 22–32 third-stage grubs per 0.1 m^2 (20–30 per ft^2) are common. The largest concentrations are usually found at the junction of green and dead turf as a result of larval feeding. One such high-density record gives averages per 0.1 m^2 (per ft^2) of 60 (56) where grass was still green, 50 (47) where grass was about half dead, and 13 (12) where grass was entirely dead. Another high-density record shows 65 third instars per 0.1 m^2 (60 per ft^2) around a large dandelion, the only living plant, with its fibrous roots entirely consumed and its tap root more than half consumed (Gambrell 1943, Tashiro et al. 1969).

Environment Affecting Density. The highest populations of grubs are found out to a radius of about 23 m (75 ft) from the trees to which adult beetles have flown. Fewer grubs are found at locations farther from the trees. Soil moisture has a definite effect. Populations tend to be the highest where the average soil moisture during the entire larval season is below field capacity. In one comparison only 0.17 grubs were found per 0.1 m^2 (2 per yd^2) where the soil moisture averaged above 90% of field capacity, in contrast to 1.9–7.1 grubs per 0.1 m^2 (23–85 per yd^2) where the soil moisture averaged 41% and 65% of field capacity (Shorey et al. 1960).

Larval Distribution. Although eggs are deposited singly in earthen cells, they tend to be deposited in aggregates. As they hatch, and as larvae increase in size, there is a tendency toward more uniform distribution. Even third instars, however, are not distributed at random. They tend to be scattered in clumps and produce high and low counts. Lateral movement of third instars averaged 0.3 m (1 ft) a day in fallow ground. Horizontal movement of this magnitude can be expected when destructive populations must continue to advance into available sources of food supply as the grass is killed (Gyrisco et al. 1954).

Vertical distribution is governed during the growing season predominantly by moisture. Any time that moisture is adequate, grubs occupy the upper 5 cm (2 in.), where fibrous roots are most abundant. Drought drives grubs deeper into the soil in search of moisture and may be as deep as 20 cm (8 in.). Following a heavy rainfall or irrigation that moistens the soil continuously to the depth of the grubs, the grubs will migrate to within 5 cm (2 in.) of the surface within 24 hr. The greatest amount of vertical movement occurs during the winter (Tashiro et al. 1969).

Winter Hibernation

Vertical Distribution. During late fall or early winter, as the surface soil starts to freeze, grubs migrate downward to remain below the frost line. They may do so a month or more after cold temperatures have forced Japanese beetle grubs deeper in the soil. During early March, with 5 cm (2 in.) of frozen soil, 91% of the EC grubs were found 20–36 cm (8–14 in.) deep. In areas of heavy sod and thick snow cover, grubs may remain well within the upper 2.5–5.0 cm (1–2 in.) of the soil surface. Second instars migrate downward earlier in the fall, reach greater depths, and migrate upward later in the spring (Gyrisco et al. 1954, Tashiro et al. 1969).

Winter Mortality. A definite and fairly consistent winter mortality of grubs occurs (Table 12); 15 of 16 determinations showed population reductions from fall to spring ranging from 6% to 53%, with an average net loss of 24% (Burrage and Gyrisco 1954).

Spring Larval Activity. During late winter or early spring, grubs migrate upward following the frost line and may even be found in honeycomb ice at the soil surface during March in western New York. Grubs feed actively within 2.5–5.0 cm (1–2 in.) of the soil surface throughout April and into mid-May in western New York. Most of the spring grub damage appears during this period. This may be the first evidence of a grub population, as feeding commences on turf that overwintered in a weakened condition, its roots having been previously consumed in the fall. This is also a vigorous feeding period preparatory to pupation (Tashiro et al. 1969).

Upon completion of feeding, the grub changes from white and gray to cream with the accumulation of fat. The grub moves downward about 5–10 cm (2–4 in.) in heavy

Table 12. European chafer: Third-instar population trends between fall and spring in a permanent pasture, Wayne County, New York, 1952–1953

Grubs per unit area		Gain (+) or loss (−) in populations from fall to spring (%)
Thirds, fall	Thirds, spring	
51	31	−39
50	43	−14
71	61	−14
78	39	−50
43	50	+14
66	42	−36
50	45	−10
130	48	−53
57	41	−28
47	44	−6
46	37	−20
59	44	−25
75	51	−32
65	49	−24
61	49	−20
61	49	−20
Average net gain or loss (%)		−23.6

Source: Adapted from Burrage and Gyrisco 1954.

soil but as deep as 15–25 cm (6–10 in.) in sandy soil. Its body oscillates to form an earthen cell. The metamorphosis of fully mature grubs to adults follows the same sequence as that for other scarabaeids, as described in Chapter 10, and involves prepupal and pupal activities.

The prepupal period lasts 2–4 days before the prepupa becomes a pupa. The last larval exuviae, also the prepupal, are immediately sloughed off to the posterior.

During roughly a 2-week pupal period, oscillation of the body as it lies in the cell is the only movement. Being very tender, pupae are susceptible to mechanical injury. Dry soil conditions are not detrimental, but saturated soils produce high pupal mortality (Tashiro et al. 1969).

Natural Enemies

Microorganisms

Milky Disease. EC grubs are susceptible to both *Bacillus popilliae* and *B. lentimorbus* (Wheeler 1946), which show the same infectivity pattern and gross symptoms as in the Japanese beetle grubs (Plate 41). Infection by the former is also known as type A infection and that by the latter as type B. Bacterial strains naturally infecting EC grubs differ from those most infective to the Japanese beetle (Table 13). The Japanese beetles were highly resistant to strains most infective to EC grubs (Tashiro et al. 1969).

The regular strain of *B. popilliae* has been the most commonly used for large-scale colonization in the Japanese beetle program. The DeBryne strain of *B. popilliae*, selected from a farm by that name where milky disease bacterial spores were first distributed for EC grubs, has consistently been the most infective. The *Amphimallon* strain of *B. lentimorbus* is commonly found in many areas where EC grub populations have persisted for 10–15 years. It is also the dominant strain at the Eckert pasture (Table 14; Tashiro et al. 1969, Wheeler 1946).

Table 13. Virulence of different strains of milky disease bacteria to the third-instar European chafer and Japanese beetle

	Percentage of grubs infected[a]	
Species and strains of bacteria	E. chafer	J. beetle
Bacillus popilliae, regular	65	51
Bacillus popilliae, DeBryne	95	1
Bacillus lentimorbus, Amphimallon	92	9
Control (spore-free soil)	0	0

Note: Insects were incubated at 27°C (81°F) for 3 weeks in soil containing 10^9 spores per kilogram.

[a]100 grubs of each species exposed to each strain.

Source: Adapted from Tashiro et al. 1969.

Table 14. Incidence of milky disease in European chafer grubs in pastures inoculated in 1945 with *B. popilliae* spores, Port Gibson, Wayne County, New York

Brood	Season[a]	DeBryne pasture		Eckert pasture	
		Grub pop. per 0.1 m^2	Diseased (%)[b]	Grub pop. per 0.1 m^2	Diseased (%)[c]
1951–52	F	18	30	41	7
	S	13	5	18	2
1952–53	F	27	21	80	4
	S	17	8	11	0
1953–54	F	40	28	43	1
	S	5	0	0	0
1954–55	F	80	11	47	12
	S	38	21	62	18
1955–56	F	42	31	51	11
	S	23	21	26	18
1956–57	F	69	15	48	7
	S	24	15	14	7
Diseased grubs	F	—	22.7	—	7.0
(av. percentage)	S	—	11.7	—	7.5

[a]F = fall surveys made August–November 8; S = spring surveys made April–June 15.
[b]DeBryne pasture: 98.6% type A infections; 1.4% type B.
[c]Eckert pasture: 55.3% type A infections; 44.7% type B.
Source: Adapted from Tashiro et al. 1969.

When temperatures are in the optimum range of 20°–30°C (68°–86°F), EC grubs are more susceptible to milky disease organisms than Japanese beetle grubs when spores are ingested. In the field, however, infectivity is relatively low, and milky disease has never been a major factor in providing effective biological control. The highest incidence has approached 30%–35% infections (Table 14). Beyond this immediate area, infections of generally less than 5% have been the most common. Second instars have shown the highest incidence of disease in late August to early September. The need for high temperatures for consistent infectivity is apparent (Table 15; Tashiro et al. 1969).

As in the Japanese beetles, north of the latitude of Long Island, EC infections are low because of insufficient heat units in the soil during the critical feeding period when grubs are near the soil surface in early fall and late spring (Table 16). Average soil temperatures of 21°C (70°F) or higher are considered essential for consistent milky disease infections. At best, during the second-instar period mid-August to mid-September, soil temperature is generally at the minimum level for infection. During the third-instar period, temperatures are generally below minimum levels (Tashiro et al. 1969).

The etiology of milky disease infection in any species of grubs has never been clear. The assumption has always been that the spores germinated in the gut and passed through the gut wall into the coelomic cavity, where multiplication and

Table 15. Incidence of milky disease in the third-instar European chafer incubated in soil collected from sites yielding the highest known type A and type B infections

Incubation temperature			Percentage diseased, weeks incubation[b]			
°C	°F	Organism[a]	1	2	3	5
14	60	A	0	0	1	25
		B	0	0	0	7
21	70	A	0	0	57	86
		B	0	0	15	57
27	80	A	0	17	79	98
		B	0	7	38	83
32	90	A	9	31	83	90
		B	4	22	41	59
14–32	60–90	Spore-free soil	0	0	0	0

[a]Soil from DeBryne pasture yielding predominantly type A infections; soil from Bedette farm yielding predominantly type B infections.

[b]100 grubs incubated in soil from each pasture at each temperature.

Source: Adapted from Tashiro et al. 1969.

sporulation occurred until billions of spores were produced before the death of grubs. Histological studies on EC milky disease infections clearly demonstrated that vegetative cells penetrate the midgut tissues where vegetative growth, multiplication, and some sporulation occur. Sporulation continues after passage into the coelomic cavity (Figure 56; Splittstoesser et al. 1973, Splittstoesser et al. 1978).

Table 16. Larval development in the European chafer at various temperatures in 5 cm (2 in.) of soil, Newark, New York, in 1950–1959

		Mean weekly soil temperature	
Dominant stage	Period	°C	°F
First instar	July 15–31	21.6	71
	August 1–15	21.0	70
Second instar	August 16–31	20.0	68
	September 1–15	17.8	64
Third instar	September 16–30	15.0	59
	May 1–15	12.8	55
Prepupae	May 16–31	15.0	59
Pupae	June 1–15	19.0	66

Source: Adapted from Tashiro et al. 1969.

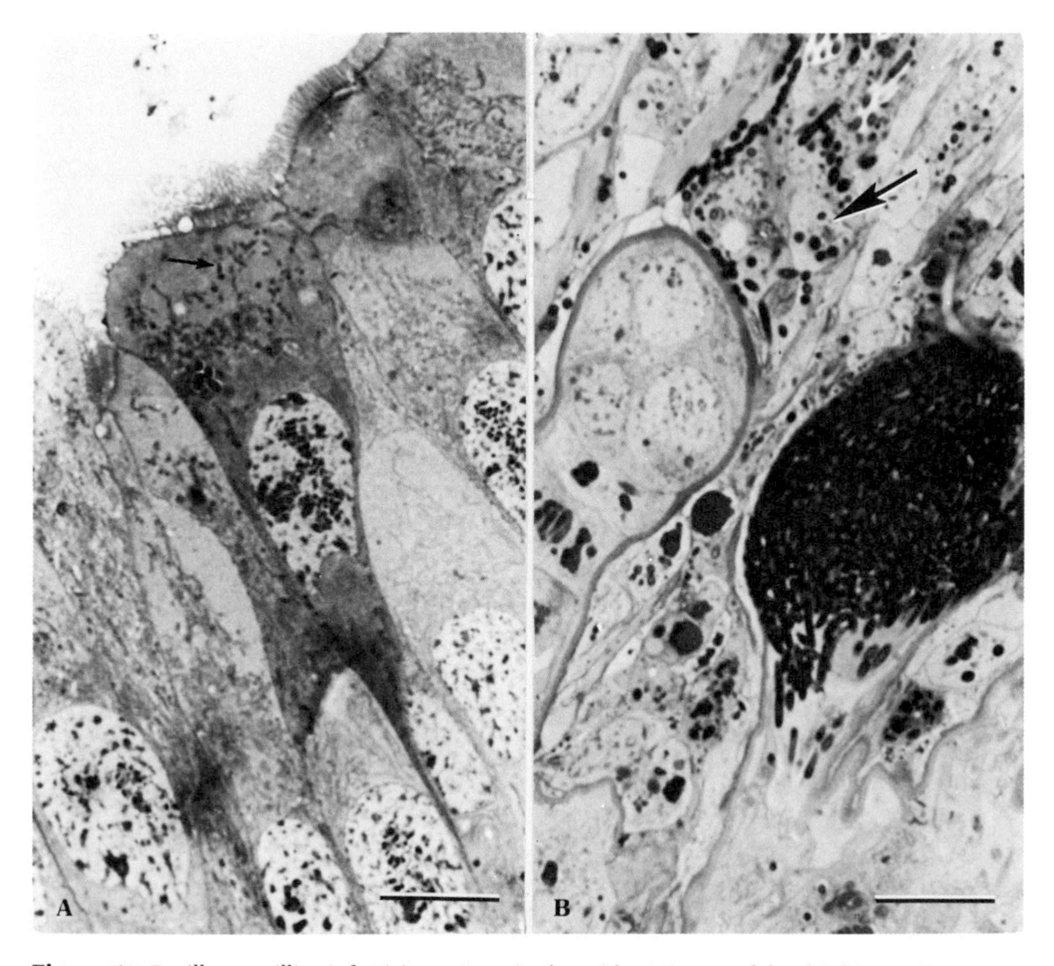

Figure 56. *Bacillus popilliae* infectivity pattern in the midgut tissues of the third-instar European chafer after ingestion of spore-inoculated peat. **A.** Vegetative rods (arrow) in distal portion of midgut columnar cell. × 1,360. Bar = 10 μm. **B.** Two adjacent nidi, one containing a dark mass of vegetative rods; bacteria also in the hemocytic capsule surrounding the nidi (arrow). × 1,360. Bar = 10 μm. (From Splittstoesser et al. 1978, figs. 1, 4, courtesy of Academic Press.)

Miscellaneous Microorganisms. Grubs infected by the green muscardine fungus, *Metarrhizium anisopliae,* are occasionally found, but the incidence of infection is so low that it is not considered an important factor (Tashiro et al. 1969).

The nematode *Neoaplectana glaseri* Steiner and the rickettsia *Coxiella popilliae* Dutky, causing blue disease in Japanese beetle grubs, are both infective to EC grubs under controlled laboratory conditions. Natural infections, however, have never been observed as they have been in Japanese beetle grubs in the field (Fleming 1962).

Insect Parasites and Predators

Seven parasitic insects contributing toward EC population reductions in Europe were introduced and released in New York over a 13-year period. There were four species of the family Tachinidae; two were larvaevorid internal grub parasites, and two were *Hyperecteina* species that deposit eggs on EC adults similar to *H. aldrichi* Mesnil (*Centeter cinerea* Ald.) on Japanese beetles. The latter has had an upsurge of activity in recent years. Two wasps of the family Tiphiidae included *Tiphia femorata* (F.) and *T. morio* (F.) (Parker 1959).

After many releases, there has been no evidence of establishment of any parasite species in New York. The success of parasites of both families in Europe is attributed to the presence of sufficient populations of an alternate host with a 3-year life cycle. During the period of releases, a similar alternate host in the release areas of New York varied from slightly present to practically nonexistent (Tashiro et al. 1969).

Two ground beetles of the family Carabidae considered responsible for some EC reductions are *Harpalus pennsylvanicus*, a predator on grubs, and *H. erraticus*, an egg predator (Gyrisco et al. 1954).

May or June Beetles

Taxonomy

The May or June beetles, M-JB, belong to the genus *Phyllophaga* Harris, order Coleoptera, family Scarabaeidae, subfamily Melolonthinae, tribe Melolonthini. The most complete taxonomic work on adults and larvae of *Phyllophaga* was prepared by Boving (1942) and contains larval keys for the differentiation of 61 species. Other major taxonomic contributions to this genus include the work of Ritcher (1966), covering larvae, and Luginbill and Painter (1953), covering adults. Early literature refers to this group as genus *Lachnosterna*. In spite of the importance of M-JB to American agriculture, not a single species has been given a common name (Blatchley 1910, Ritcher 1966, Werner 1982).

Importance

White grubs, as the larvae are commonly called, have been recognized as major pests of agriculture for many decades in the northeastern quarter of the United States, in Texas, and in the Canadian provinces of Ontario and Quebec. Larval damage to turfgrass, whether short turf or pasture sod, resembles that caused by other scarabaeid grubs. Because many species are relatively large, a few grubs per unit area can be most devastating (Luginbill and Painter 1953).

Adults of some species are serious pests of shade trees, and where these are abundant they can defoliate trees with young tender leaves. Favorite host trees include the oak, hickory, walnut, elm, and poplar.

History and Distribution

Members of the genus *Phyllophaga* that are native to North America include 152 valid species that occur in the United States and Canada. The greatest numbers of these species occur in the eastern half of the United States; only an occasional species appears on the West Coast (Figure 57). M-JBs are rather rare in the extreme southwestern United States. Texas has almost 100 species, but only one, *P. crinita,* is a serious turfgrass pest. In the north central and eastern states, where M-JB are the most abundant both as to species and populations, the seven most common and destructive species include *P. anxia* (LeConte), *P. fervida* (Fabricius), *P. fusca* (Froelich), *P. hirticula* (Knoch), *P. implicita* (Horn), *P. inversa* (Horn), and *P. rugosa* (Melsheimer). The most widespread throughout the country is *P. anxia*. Even in areas where a major, economically important species occurs, relatively few studies have been published (Frankie et al. 1973, Luginbill and Painter 1953, Ritcher 1966). Major species in several of these areas are listed below.

Northeastern United States and Canada

In the New England states and eastern Canada, the most common and injurious species is generally *P. anxia.* The principal species in southwestern Ontario, Canada, are *P. futilis* (LeConte), *P. fusca,* and *P. rugosa.* Twenty-six species are known to occur in New York. When the insects were studied as strawberry pests, the more common species were *P. rugosa, P. anxia, P. fusca,* and *P. longispina* (Smith). As pests of seedbeds of a forest tree nursery near Albany, *P. gracilis* (Burmeister), *P. tristis* (Fabricius), *P. crenulata* (Froelich), *P. fusca, P. fraterna* Harris, and *P. anxia* were the most numerous (Hammond 1940, Heit and Henry 1940, Henry and Heit 1940, Kerr 1941).

Ohio and Kentucky

Surveys on the dominance and abundance of M-JB species in the Midwest have been the most comprehensive in the above two states. In Ohio during the years 1935–1955, the most abundant and widely distributed species, as determined by light trap collections, were *P. fusca, P. futilis, P. hirticula,* and *P. rugosa* (Neiswander 1963).

The most comprehensive study on species dominance was undertaken in the bluegrass region of north central Kentucky centered on Lexington. Adult collections from host plants and larval collections during 1936–1939 revealed 36 species present in Kentucky. In the inner bluegrass region, where most of the fine horse farms are located, *P. hirticula* constitutes more than 80% of the population. Other dominant species are *P. fraterna,* which represents about 60% of the population in the intermediate bluegrass region, and *P. ephilida* (Say), which accounts for about 55% of the population in the outer bluegrass region (Ritcher 1940).

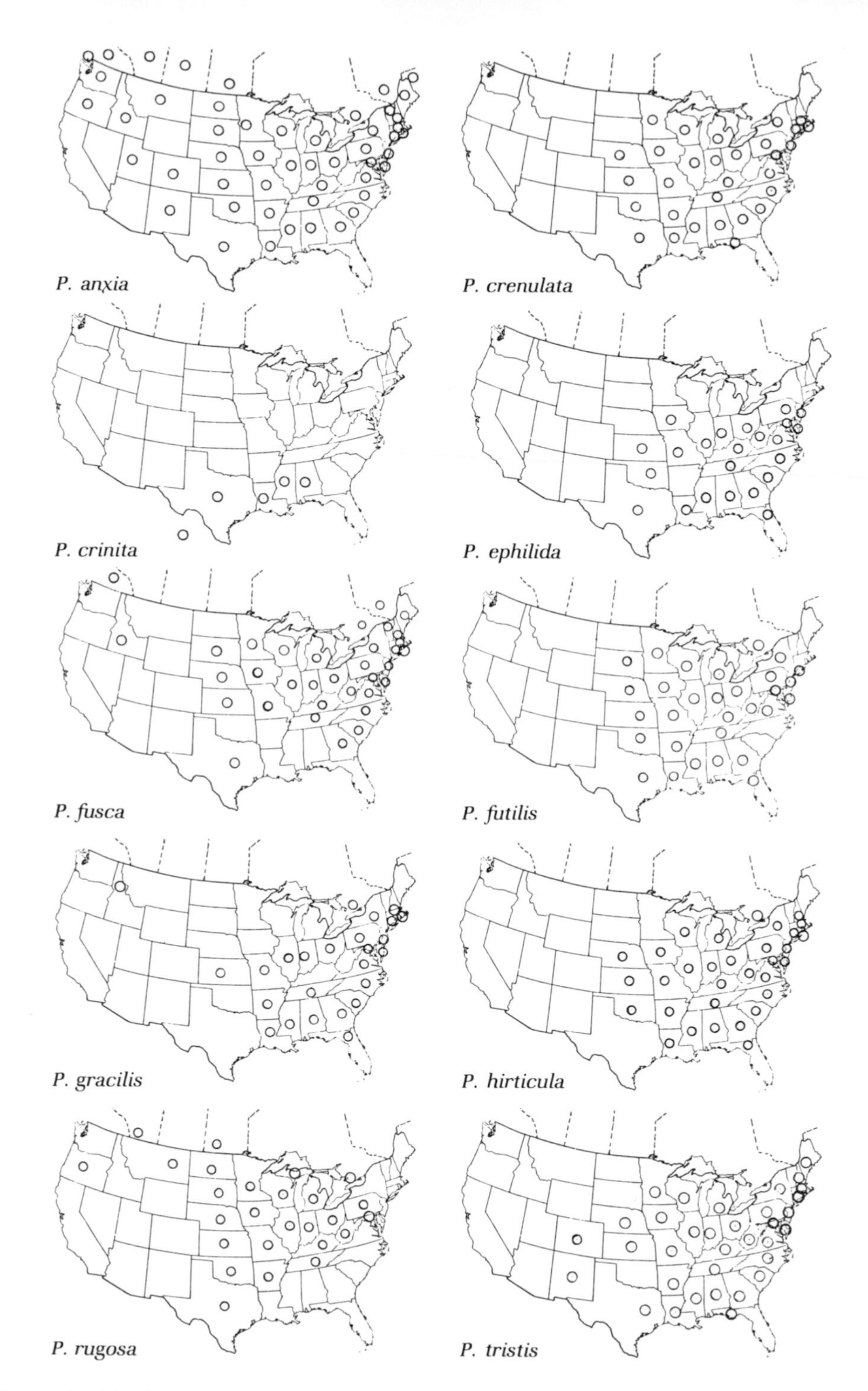

Figure 57. Distribution of major *Phyllophaga* spp. in the United States and southern Canada. (Adapted from Luginbill and Painter 1953; redrawn by H. Tashiro, NYSAES.)

Wisconsin and Nebraska

Surveys were made in Wisconsin during 1935–1937 by collecting adults from host plants. Of the 16 species collected, *P. rugosa* and *P. hirticula* were equally abundant and constituted 67% of the total M-JB population. They were followed by *P. tristis* and *P. fusca*. *Phyllophaga anxia* is the dominant species in Nebraska and is a perennial occurrence in the sand hill areas of north central Nebraska (Chamberlin et al. 1938, Rivers et al. 1977).

Southern States

The most abundant turfgrass-infesting species in Texas, especially abundant in the lower Rio Grand Valley, is *P. crinita* Burmeister. This species is recorded in only three more states, Louisiana, Alabama, and Mississippi, and in Mexico. It resembles *P. tristis* in size but can be differentiated by its much less hairy body (Frankie et al. 1973, Luginbill and Painter 1953, Reinhard 1940).

Host Plants and Damage

Adult Feeding

Feeding damage by M-JB is most common on certain shade and forest trees, although ornamental shrubs and even a few fruit trees are occasionally damaged. The degree of damage correlates with the emergence of beetles and the presence of young tender foliage. Oak trees are the preferred hosts, and bare or partly stripped trees are a common sight in the inner bluegrass region of Kentucky (Plate 36). Adults of some species feed regularly on grass blades. The peak emergence of *P. hirticula*, the dominant species, often comes during the 1st week in May, when the leaves of the pin, red, white, bur, and chinquapin oaks are young and tender. Other trees preferred and often stripped of leaves by various species of M-JB include persimmon, hickory, walnut, elm, and birch. Trees that often provide food but are never stripped include willow, plum, ash, sycamore, locust, sassafras, redbud, blackberry, and rose. Cherry and raspberry and some flower petals also provide food (Luginbill and Painter 1953, Ritcher 1940).

Larval Feeding

Bluegrass sod, whether short turf or pasture, is commonly destroyed by many species of *Phyllophaga*. Damage is caused primarily during the 2d year of both the 3-year cycle and the 2-year cycle as the larvae become third instars. Young forest and shade trees may be seriously injured when the grubs eat the fibrous roots and even girdle larger tap roots (Hammond 1940).

Since the early 1970s, *P. crinita* has been the most serious scarabaeid encountered in Texas turf, damaging bermudagrass, buffalograss, and St. Augustinegrass (Frankie et al. 1973, Reinhard 1940).

Description of Stages

Adults

The beetles are heavy, clumsy-looking insects, ranging in color from light or dark brown to mahogany brown and resembling several other turf-infesting scarabaeids in general appearance (Plate 35). The smallest and the largest species differ considerably in size. One of the smaller species, *P. gracilis,* varies in length from 10.5 mm to 13.0 mm, while *P. fusca,* one of the larger, varies in length from 17.5 mm to 23.5 mm. The adult of *P. gracilis* rather closely resembles the European chafer, both in size and in color. Body pubescence varies considerably in the M-JB. Among the more common species, the elytra of *P. tristis* are heavily pubescent, while those of *P. anxia* are nearly glabrous. As with many scarabaeids, the sexes can be distinguished by the longer lamellae of the antennae in males (Plate 36; Luginbill and Painter 1953).

Eggs

Like the eggs of many other turf-infesting scarabaeids, M-JB eggs are deposited singly in earthen cells. Eggs are pearly white and elliptical when they are freshly deposited and increase in size and become nearly spherical at maturity. For *P. hirticula,* a species of average size, young eggs averaged 2.38 mm by 1.49 mm and increased to 2.96 mm by 2.45 mm in length and width at maturity (Ritcher 1940).

Larvae

As is typical of other scarabaeids, newly hatched first-instar larvae are translucent white, but as soon as feeding begins, darkened areas form in the posterior regions. The larvae grow from less than 6.3 mm in length as firsts to 25.0–38.0 mm as fully grown third instars for the medium-sized species (Plate 36; Hammond 1940).

It is characteristic of M-JB larvae that the anal slit is V- or Y-shaped, with the stem of the Y much shorter than the arms (Plate 36). The lower anal lobe is usually divided by a sagittal cleft and is sometimes divided by a sagittal groove. The claws on the metathoracic legs are very small (Ritcher 1966).

Rastral patterns vary considerably, but the palidia of many are essentially two parallel rows of pali, or spines, and the palidia show a tendency toward convergence at both ends (Figure 58). In many larvae, the palidia are nearly straight, while in some they are sinuate or curved (Plate 36). The numbers of pali in each row vary considerably among species, from about 10 to about 30 (Boving 1942).

Pupae

As is typical of all turfgrass-infesting scarabaeids, the pupation of M-JB occurs in an earthen cell formed by the mature third-instar grub. Upon transformation from a prepupa, the pupa sloughs off the third-instar exuviae to the posterior end, as does the European chafer and the Asiatic garden beetle, which belong to the same subfamily.

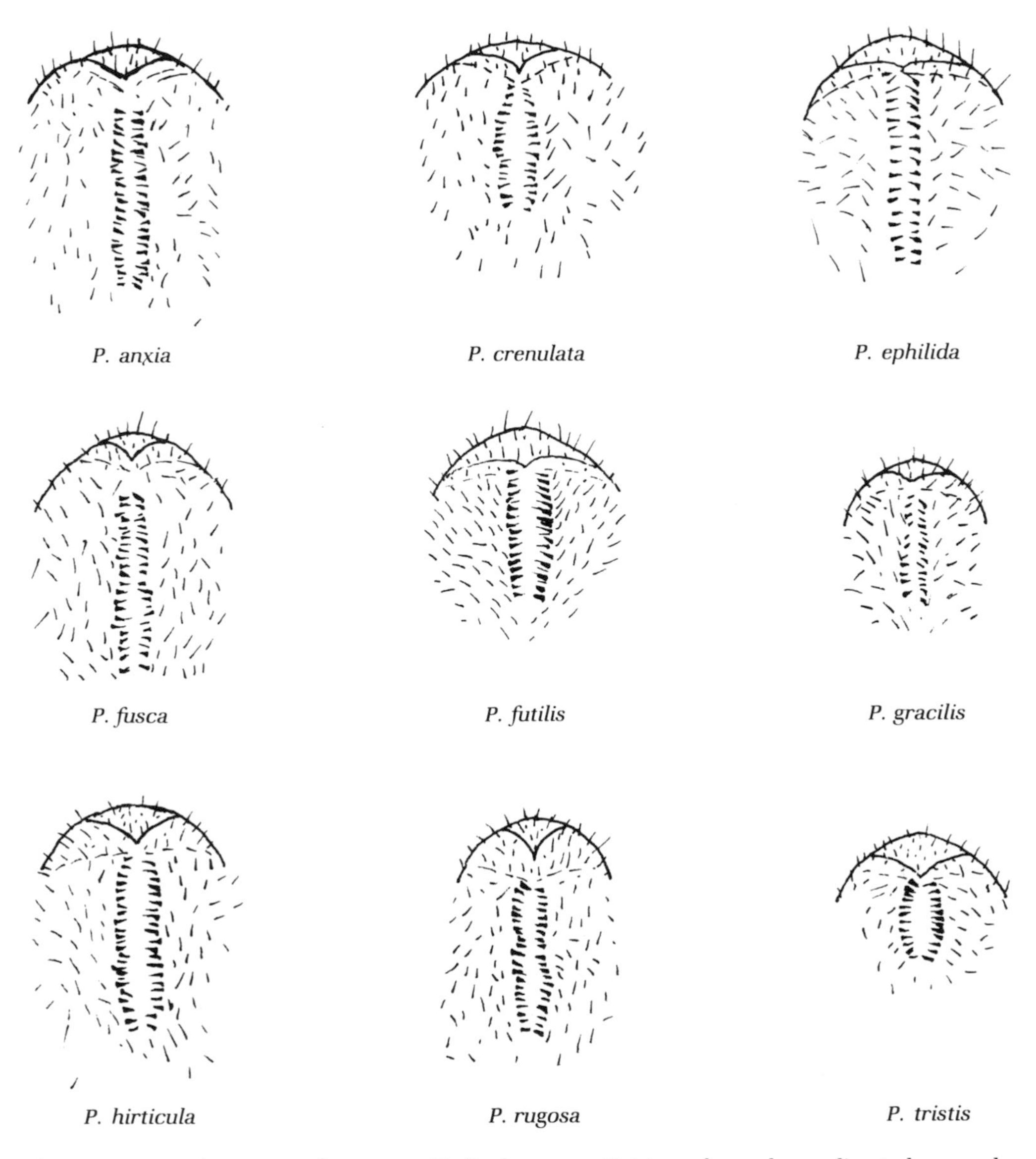

Figure 58. Rastral patterns of common *Phyllophaga* spp. Not to scale; anal area directed upward. (Adapted from Ritcher 1949, 1966 and Boving 1942; redrawn by H. Tashiro, NYSAES.)

Seasonal History and Habits

Seasonal Cycles

The life cycles of M-JB vary from 1 to 4 years for one generation, depending on the species and latitude. The most common in the geographic areas of greatest faunal presence are 3-year cycles and a few species of 2-year cycles. Those that have a 1-year life cycle are exceptions. In Kentucky, where the most comprehensive studies on biology of M-JB have been made, those with a 3-year life cycle include *P. hirticula,* the most dominant species, plus *P. rugosa, P. fusca,* and *P. crenulata,* all fairly common (Ritcher 1940).

Three-Year Life Cycle. A typical 3-year life cycle is illustrated by *P. hirticula* in Kentucky (Figure 59). During the 1st year, adults that have overwintered in the soil migrate to within 2.5–5.0 cm (1–2 in.) of the surface in March. They emerge as adults in May to feed on foliage and deposit eggs in June about 8–10 cm (3–4 in.) deep in the soil. Eggs hatch from June into July, and larvae feed within 2.5–5.0 cm (1–2 in.) of the surface during July into September. They become second instars in August–September and feed into October before migrating to lower depths for overwintering (Ritcher 1940).

During the 2d year, overwintering second instars migrate upward in April to within 2.5–5.0 cm (1–2 in.) of the surface, feed, and become third instars in June, doing most of their feeding thereafter, during July and August. They migrate downward again in late September and October to overwinter.

During the third year, overwintering third instars migrate upward in late March and feed during April and May to complete their larval development. During June they migrate to the lower depths again and form an earthen cell, where they become prepupae in late June, transform to pupae in July–August and become adults in August–September. These young adults remain in the same earthen cell until the next spring. Movement up from the earthen cell begins as early as February, and the first beetles arrive beneath the sod in late February and early March to complete a 3-year life cycle (Luginbill 1938, Ritcher 1940).

Life cycles studied in other areas of the midwest, in Ohio and southern Wisconsin, indicate that, with only minor exceptions, the M-JB have 3-year life cycles. With few exceptions, species in the Northeast and in the eastern Canadian provinces of Ontario and Quebec also have 3-year life cycles. *P. tristis* was found to have a 2-year life cycle in eastern New York (Chamberlin et al. 1938, Hammond 1940, Heit and Henry 1940, Neiswander 1963).

In Kentucky, some species that often have a 2-year life cycle occasionally have a 3-year life cycle in the same locality. Three of these include *P. futilis, P. bipartita* (Horn), and *P. inversa,* all present throughout the Midwest (Ritcher 1940).

Two-Year Life Cycle. Those M-JB with a 2-year life cycle molt either twice during the 1st year, to overwinter as third instars, or molt once the 1st year, overwintering as seconds, and molt again the 2d year and overwinter as thirds (Figure 60). Kentucky species, invariably of 2-year life cycles, include *P. ephilida* and *P. tristis.* Unlike *P. hirticula, P. ephilida* overwinters at a shallow depth and pupates in June, with adult emergence the same season.

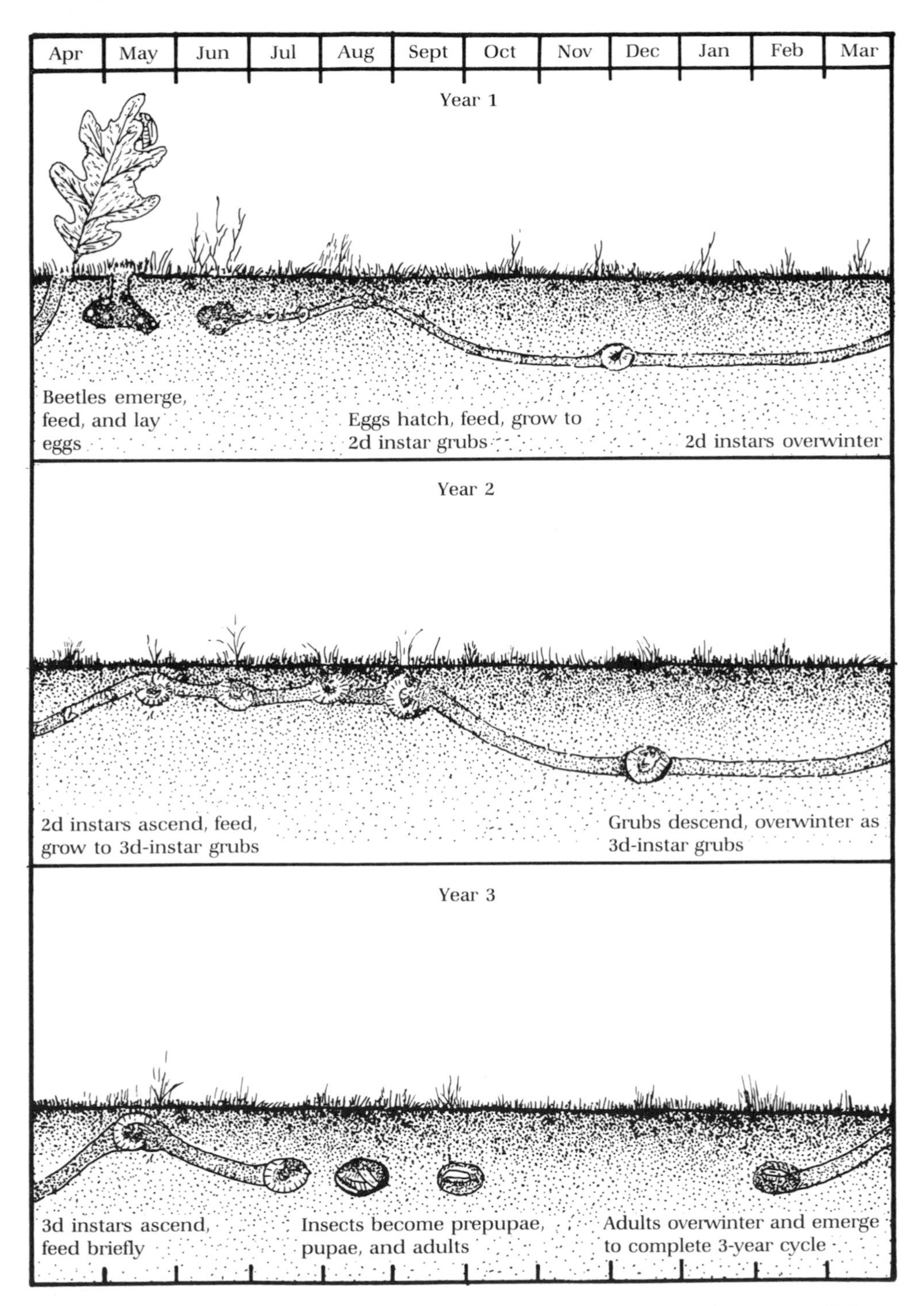

Figure 59. Three-year life cycle of *Phyllophaga* spp. based on midwestern observations. (Adapted from Luginbill 1938, fig. 13; redrawn by H. Tashiro, NYSAES.)

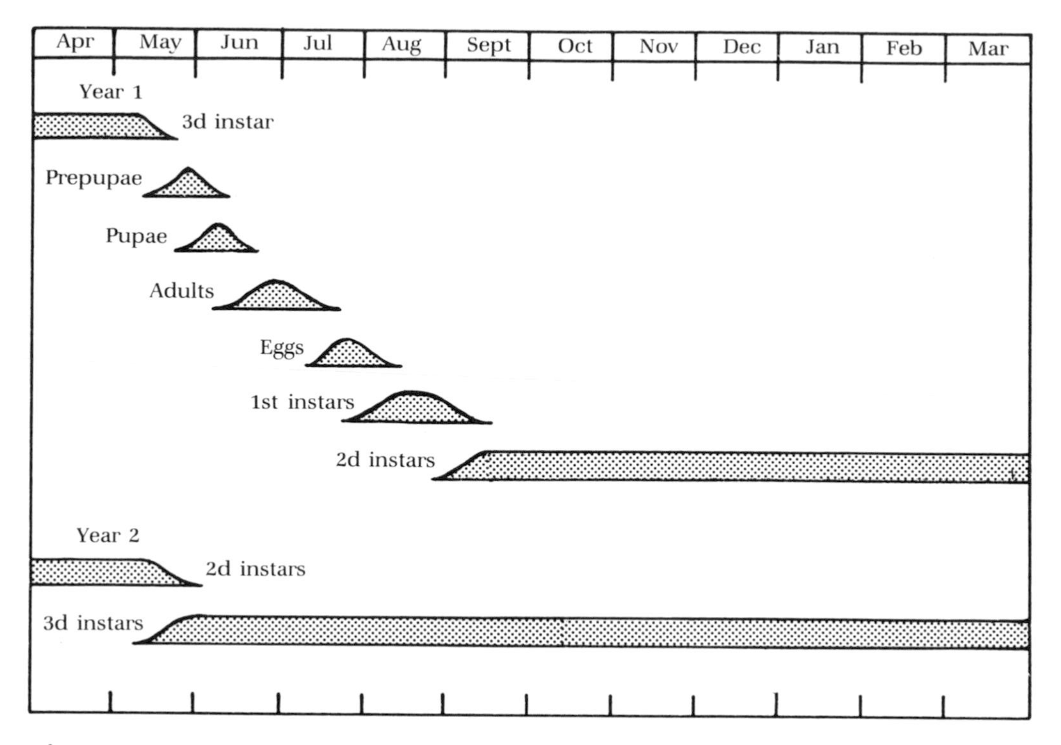

Figure 60. Two-year life cycle of *Phyllophaga ephilida* as indicated by observations made in Kentucky. (Adapted from Ritcher 1940, fig. 17; redrawn by H. Tashiro, NYSAES.)

After overwintering, the third instars migrate to very near the surface in April to feed a while before returning to form their earthen cells at the depth at which they hibernate. They are prepupae in late May, pupae in early June, and adults in late June, with emergence in July. Eggs are deposited about 8 cm (3 in.) deep in late July. They hatch in August, become second instars in September, and migrate downward in October to overwinter (Ritcher 1940).

During the 2d year, the second instars leave their overwintering depth and migrate to near the surface, where they feed on roots from April through July. They become third instars and are the most destructive grubs until October, when they migrate to lower depths for their 2d winter hibernation (Ritcher 1940).

One-Year Life Cycle. The most devastating of turfgrass scarabaeids in Texas is *P. crinita*. In the southern part of the state, it has a 1-year life cycle, while in northern Texas a small portion of the population is suspected of having a 2-year life cycle. Third instars overwinter. The beetle flight period is mainly from June into August. The beetle is no longer considered to have both a major and a minor flight period each year. Rather, adult flights are governed by moisture, with drought inhibiting flight nearly completely. Flights are reactivated within a week following rainfall of 1 cm or more. Most of the damage caused by third instars occurs during the summer and fall (Robert Crocker, personal communication 1985; Frankie et al. 1973; Gaylor and Frankie 1979).

Adult Activity

Daily Flights. During the day, beetles hid in soil or sod. Emergence from the ground and flight occur at dusk or shortly before, particularly during warm cloudy days when the temperature is above 15.5°C (60°F). Under these conditions, beetles fly to the tops of trees to feed. They sound like a swarm of bees during heavy flights. They return to the soil a little after daybreak. Because of these insects' nocturnal habits, most people see them only when they are attracted to the light and fly into screens and windows. On cool nights adults feed near the ground (Hammond 1940; Ritcher 1940, 1949).

Feeding Habits. Beetles of most species feed only at night. Trees with young, tender leaves are the most attractive food plants. The soft leaf tissues between veins is consumed, leaving the veins intact (Ritcher 1949). In Kentucky, *P. hirticula,* the dominant species, heavily damages oaks; the peak emergence of beetles coincides with the presence of young, tender leaves with only 2.5–5.0 cm (1–2 in.) of growth. Also feeding exclusively on oak is *P. tristis.* Trees that are not attacked include maple, basswood, wild cherry, and evergreens (Hammond 1940, Ritcher 1940).

Mating and Oviposition. Mating occurs only at night. While females are feeding, males search them out for copulation. Oviposition commences 9–10 days after mating. Each female deposits about 50 eggs during a 1–3-week period. Preferred sites for oviposition, which occurs during the day or night, include loose sod growing in sandy soil in the immediate vicinity of their food plants. Bare fallow ground is also found attractive for oviposition. Most eggs are found 8–18 cm (3–7 in.) deep in small, compact balls of soil. The earthen cell is held together by a secretion of the female (Hammond 1940, Ritcher 1940).

In Texas, *P. crinita* females have a preoviposition period of just over 7 days and an oviposition period of 3–5 days. From 30 to 40 eggs are deposited by each female at a depth of 5–10 cm (2–4 in.) in turfgrass soil. Beetles avoid heavily watered turf for oviposition (Frankie et al. 1973, Reinhard 1940).

Seasonal Flights. Although the name *May or June beetles* implies presence during these months, some beetles begin emerging from the soil in April. These include *P. fusca* and *P. tristis* in Kentucky. Most others begin in May; *P. fusca* is abundant through May and continues into early June. Species of a 3-year life cycle have a major flight year, but a few are also in flight the other 2 years, regardless of cycle (Hammond 1940, Ritcher 1940).

Artificial Stimuli. M-JB adults are attracted to incandescent lights in direct relation to the brilliance of the illumination. The majority of beetles captured in light traps are males. In Kentucky, more than 85% of the M-JB adults are males and in Texas about 80% of the *P. crinita* adults attracted to streetlights are males (Frankie et al. 1973, Ritcher 1940).

Larval Activity

Hatching and Development. Eggs hatch in 3–4 weeks. The young larvae first subsist on organic matter in the soil but soon turn to grass roots, their preferred food. Molting to second instars occurs during August or September for species with a 3-year life cycle. By mid-October, second instars have migrated downward for winter hibernation. Depending on species, those with a 2-year life cycle molt once or twice before overwintering (Ritcher 1940).

During the 2d year of a 3-year life cycle, the grubs feed most ravenously on roots and cause the most damage. Most species feed from 4 to nearly 6 months during the 2d summer, compared with less than a month during the 3d summer. By late August of the 2d summer, some grubs burrow downward for overwintering, while others feed until late October before hibernation (Hammond 1940).

Third-year grubs of *P. anxia* within the upper 5 cm (2 in.) are actively feeding. Grubs below that depth are inactive and display little movement when dug from the soil. During late April and May, grubs come to the surface to feed for about a month. In June they migrate downward to form earthen cells, become prepupae, change to pupae for about a month, and become adults that remain in the ground until the following spring (Hammond 1940).

In Texas, *P. crinita* first and second instars are in each stage about 3 weeks during early summer and become third instars for the rest of the summer and fall, when they cause turfgrass damage (Frankie et al. 1973).

Larval Distribution. Whenever grubs are within the upper 5 cm (2 in.), they are considered to be actively feeding. Descent to lower levels is made for winter hibernation or for completion of larval life and preparation for pupation. Depths of penetration are similar for both hibernation and pupation and vary according to species. One that penetrates deeply, *P. hirticula,* has been found 18–58 cm (7–23 in.) deep for both purposes. A species that overwinters at a shallow depth, *P. ephilida,* has been found at 5–15 cm (2–6 in.) deep both in hibernation and in pupation (Ritcher 1940).

Population Densities. In Canada 63 or more grubs per m^2 (75 per yd^2) are common and are found most frequently on higher ground in sandy soils. The presence of 12–17 grubs in a single hill of corn, sufficient to cause a complete loss of crop, is common (Hammond 1940, Luginbill 1938).

Pupation

The pupal period lasts about 1 month, depending on the species and the season, and occurs in July of the 3d year of a 3-year life cycle. Pupation occurs during June of a 2-year life cycle. The pupation of *P. crinita* lasts about 3 weeks in cells 8–15 cm (3–6 in.) deep (Frankie et al. 1973, Ritcher 1940).

Natural Enemies

Microorganisms

Phyllophaga spp. grubs infected with milky disease bacteria have been found in the field from widely scattered locations. Grubs of *P. anxia* from Clinton County, New York, in 1944, *P. fusca* grubs from Wayne County, New York, in 1945, and *P. hirticula* grubs from Kentucky, all infected with *Bacillus popilliae,* are the original sources of *B. popilliae* strains bearing the same specific names as grubs. Grubs of *P. anxia* and *P. fusca* infected with *B. popilliae* were found on numerous occasions in western New York during 1950 to 1982. The degree to which milky disease biologically controls M-JB grubs is not well known, but it is considered to be a relatively minor influence (Wheeler 1946).

The fungus *Cordyceps ravenelii* Berk and the green muscardine fungus *Metarrhizium anisopliae* are reported to be naturally infective to field populations of various species of grubs (Hammond 1940).

Insect Parasites and Predators

The M-JB complex has a large complement of parasites and predators. The adult *Pyrgota undata* Wied. (Pyrgotidae: Diptera) alights on the dorsum of M-JB adults that are feeding, causing the latter to take flight. Flight allows the parasite to insert her ovipositor into the thin integument of the abdominal dorsum. Death of the beetle occurs in 10–14 days. The parasite puparium is formed within the dead host (Davis 1919).

Grubs of M-JB are attacked by several wasps of the family Tiphiidae. These are external parasites. Stings by *Myzine* spp. females cause permanent paralysis, while stings by *Tiphia* spp. females cause only temporary paralysis of host grubs. Both parasites attach an egg to the integument of the grub while it is paralyzed. After hatching, the externally attached parasite larva feeds on the liquid contents of the grub (Plate 37). At the death of the grub, when only the integument and heavily chitinized portions remain (Plate 37), the parasite is ready for pupation. The parasite pupa forms within its fibrous cocoon (Plate 37). *Myzine quinquecincta* (Fab.) is a common parasite in Kentucky (Ritcher 1940).

The predators of M-JB pupae include larvae of a bee fly, *Exoprosopa fasciata* Macq., and larvae of a robber fly, *Diogmites discolor* Loew (Asilidae). In Kentucky, *D. discolor* has been found attacking pupae of *P. hirticula, P. fusca, P. inversa, P. rugosa,* and *P. tristis,* with an estimated 12% pupal destruction. Larvae of E. *fasciata* were found in pupal cells of *P. bipartita* (Ritcher 1940).

Vertebrate Predators

Vertebrate predators common to other turfgrass-infesting scarabaeids are also common to M-JB grubs. These include numerous insectivorous birds, skunks, raccoons, and armadillos that search for grubs as a source of food.

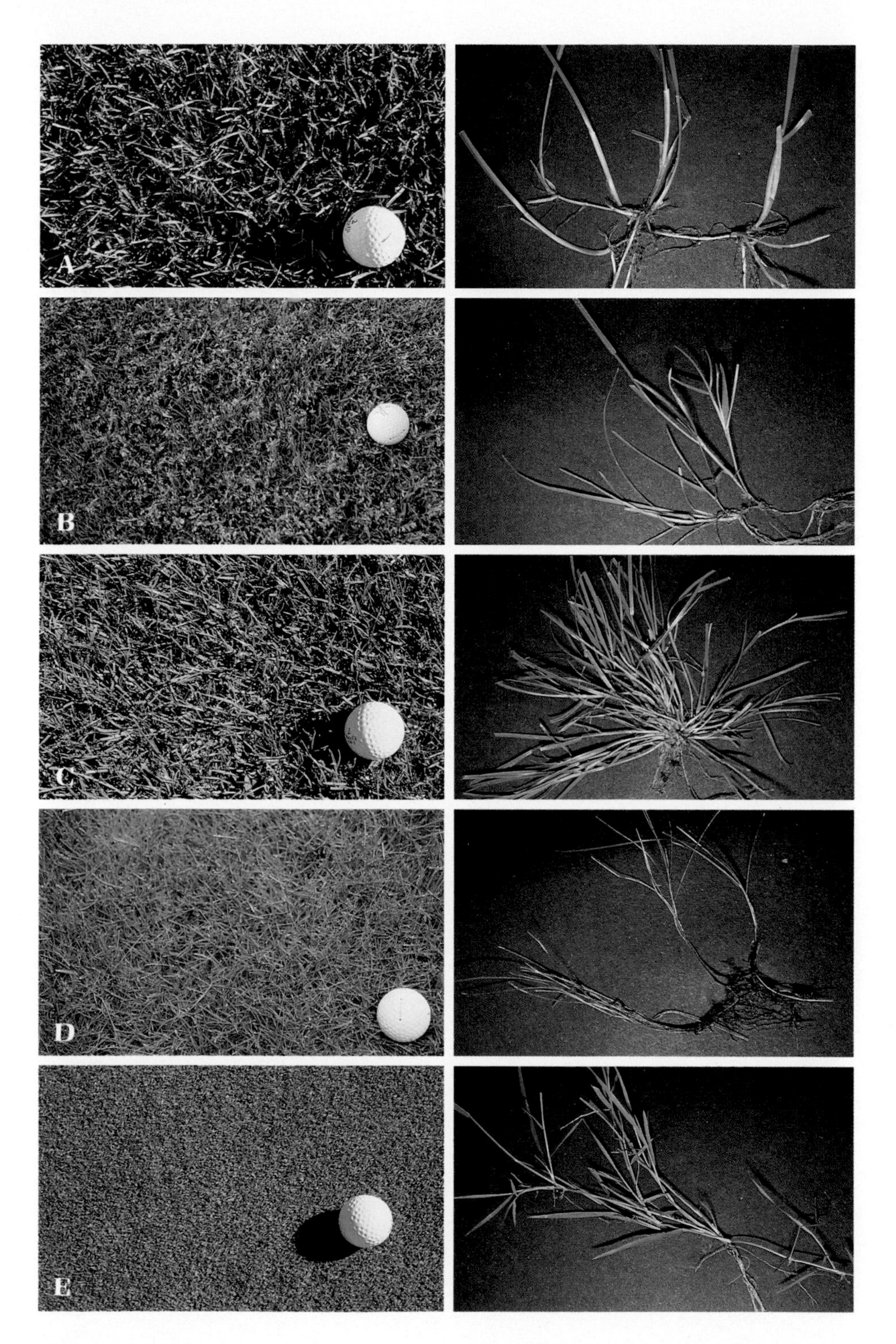

Plate 1. Turf (left) and vegetative growth characteristics (right) of cool-season grasses. **A.** Kentucky bluegrass. **B.** Annual bluegrass. **C.** Perennial ryegrass. **D.** Red fescue. **E.** Creeping bentgrass. (Photos courtesy of the New York State Agricultural Experiment Station, Geneva [NYSAES].)

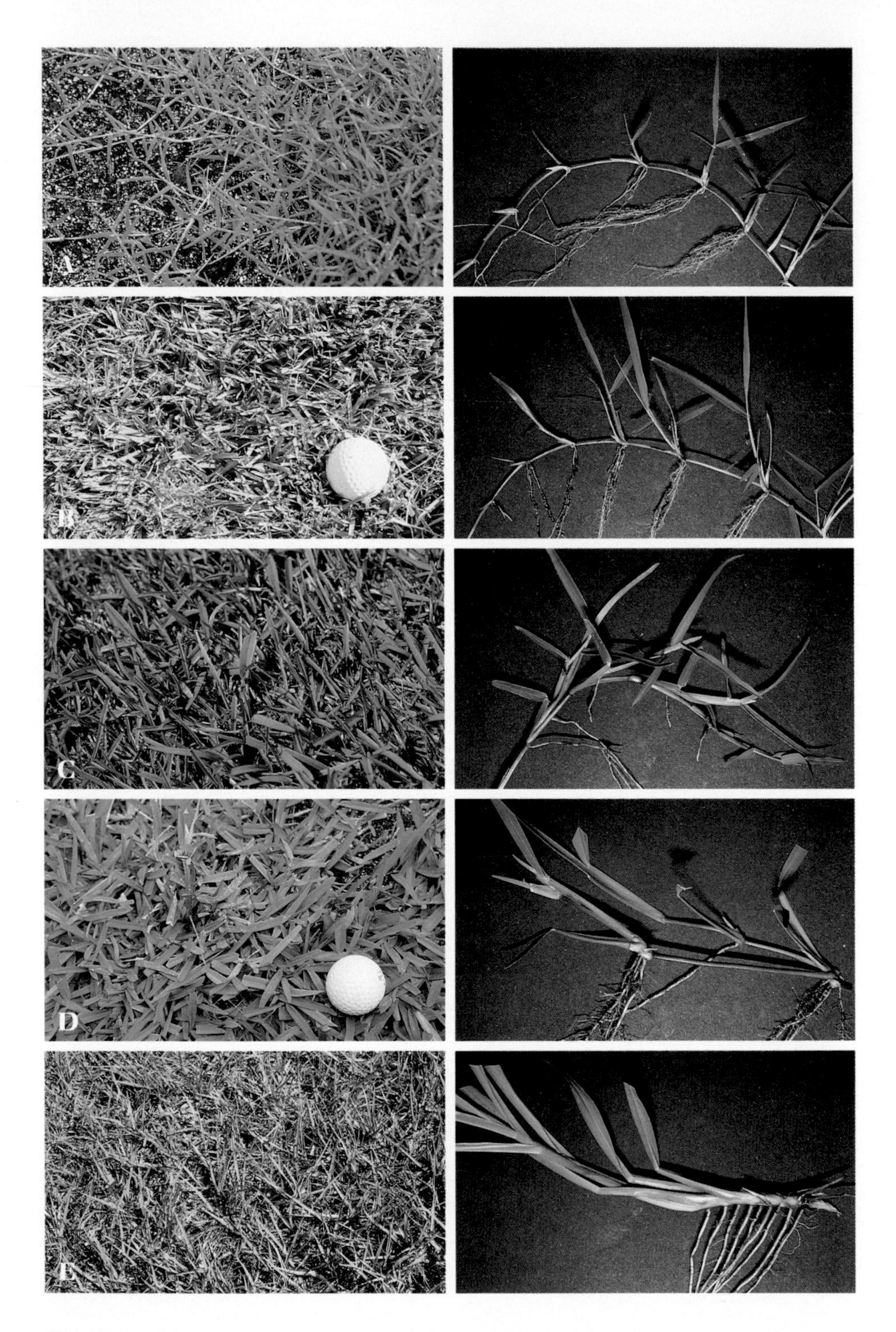

Plate 2. Turf (left) and vegetative growth characteristics (right) of warm-season grasses. **A.** Bermudagrass. **B.** Zoysiagrass. **C.** Centipedegrass. **D.** St. Augustinegrass. **E.** Bahiagrass. (Photos: NYSAES.)

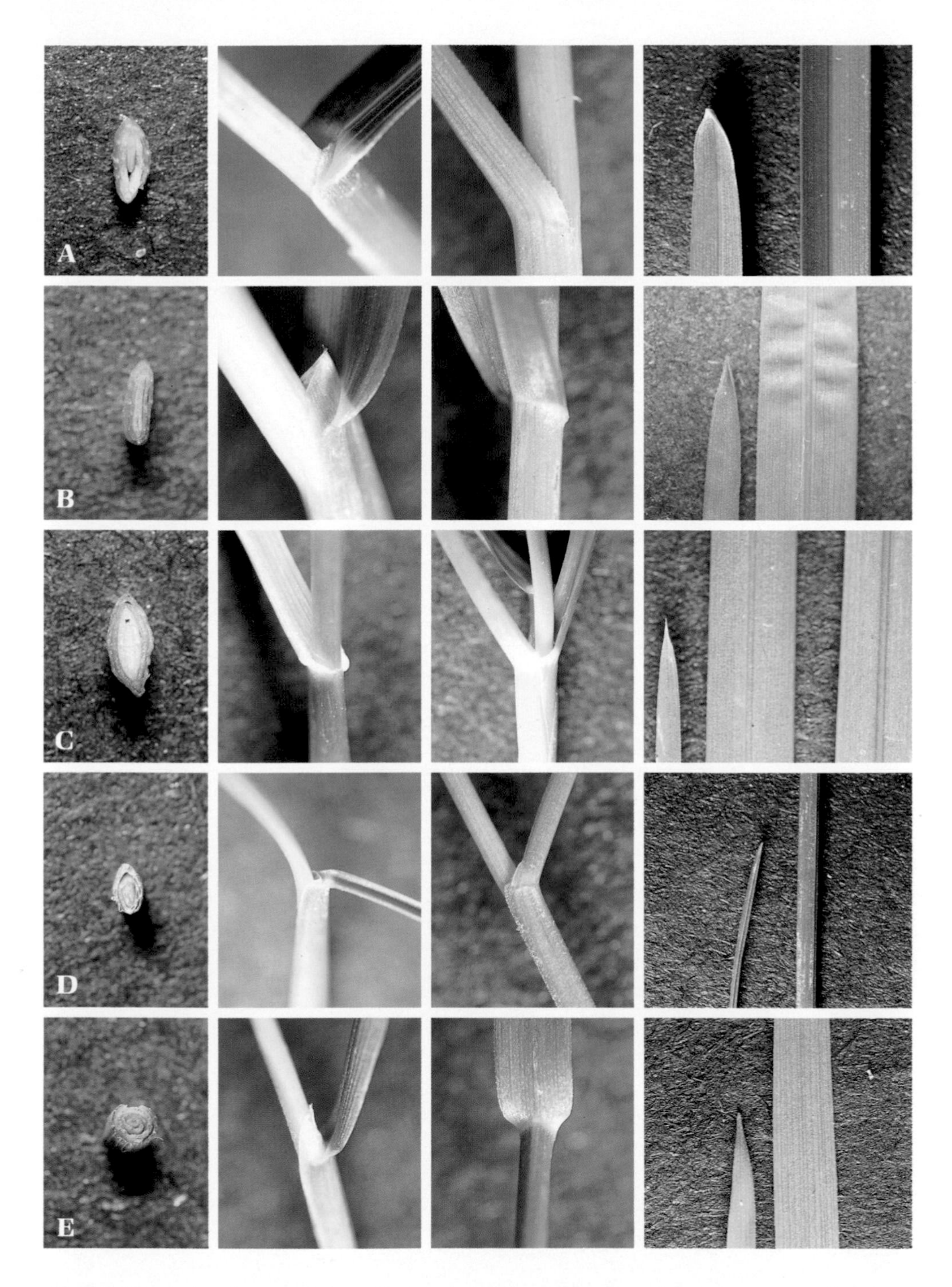

Plate 3. Vegetative characteristics of five cool-season grasses viewed at the same magnification. **A.** Kentucky bluegrass. **B.** Annual bluegrass. **C.** Perennial ryegrass. **D.** Red fescue. **E.** Creeping bentgrass. (See Table 1 for descriptions of characteristics; photos: NYSAES.)

Plate 4. Vegetative characteristics of five warm-season grasses viewed at the same magnification. **A.** Bermudagrass. **B.** Zoysiagrass. **C.** Centipedegrass. **D.** St. Augustinegrass. **E.** Bahiagrass. (See Table 1 for descriptions of characteristics; photos: NYSAES.)

Plate 5. Drought dormancy of Kentucky bluegrass-fescue lawn on poor soil and southern slope, with partial rejuvenation following rainfall. **A.** Dormant except for weeds and tall fescue. **B.** Start of recovery <1 week. **C.** Regrowth in <2 weeks. (Photos: NYSAES.)

Dichondra. **D.** Residential lawn. **E.** The same lawn viewed close up. **F.** Vegetative growth characteristics. (Photos: NYSAES.)

Plate 6. Bermudagrass mite. **A.** Normal (left) and mite-injured bermudagrass stem (right). **B.** Mites present under leaf sheath. **C.** Mites magnified about 55 times.

Winter grain mite. **D.** Injury to home lawn. **E.** Feeding injury to stem and leaf (right) normal. **F.** Adult female. **G.** Eggs. (Photos A–C, E–G: NYSAES; photo D courtesy of H. D. Niemczyk.)

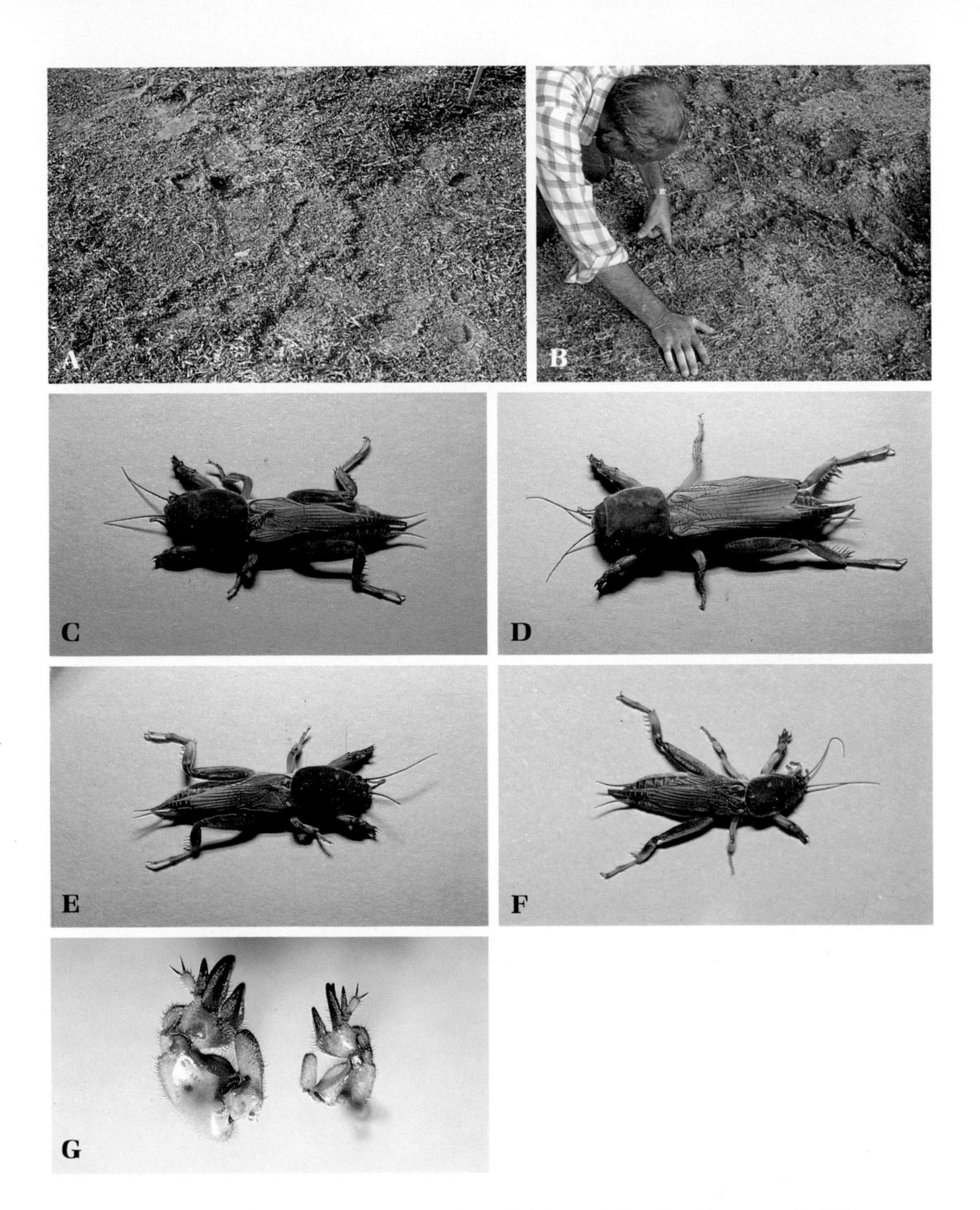

Plate 7. Mole crickets. **A.** Burrowing mounds and ridges. **B.** Depth of burrows. **C.** Male tawny mole cricket. **D.** Female. **E.** Male southern mole cricket. **F.** Female. **G.** Tibial dactyls of tawny (left) with narrow V-shaped separation and southern (right) with wide U-shaped separation. (Photos A, B courtesy of R. L. Crocker; photos C–G: NYSAES.)

Plate 8. Mole crickets. **A.** Short-winged mole cricket. **B.** Northern mole cricket. **C.** Cannibalism by southern mole cricket. **D.** Pitfall trap for capturing mole crickets designed by K. Lawrence. **E.** Sound traps for attracting and capturing the tawny and southern mole crickets. **F.** Tawny mole cricket killed by *Metarrhizium* sp. fungus. (Photos: NYSAES.)

Plate 9. Hairy chinch bug. **A.** Lawn damage. **B.** Adult feeding. **C.** Fescue tufts harboring high populations of adults. **D.** Adult (long winged). **E.** Adult (short winged). **F.** Male (left) and female (right) with keeled sternum. **G.** Young eggs. **H.** Mature egg. (Photos A–C, E, F: NYSAES; photos D, G, H courtesy of H. T. Streu.)

Plate 10. Hairy chinch bug. **A.** First-instar nymph just hatched. **B.** Late second-instar nymph. **C.** Fifth-instar nymph. **D.** *Beauveria* fungus–infected chinch bug. **E.** Big-eyed bug male (left) and female (right), a predator. **F.** Big-eyed bug feeding on early instar nymph. (Photos A, B, F courtesy of H. T. Streu; photos C, E: NYSAES; photo D courtesy of M. K. Sears.)

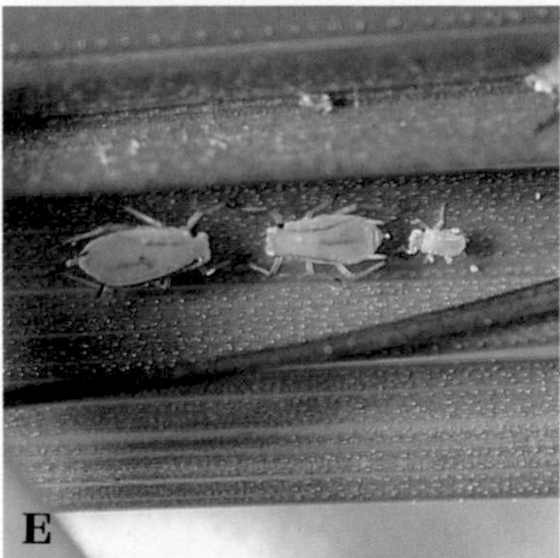

Plate 11. Greenbug. **A.** Feeding injury to Kentucky bluegrass leaf. **B.** Lawn damage. **C.** Maryland biotype winged female. **D.** Ohio biotype wingless adults. **E.** Maryland biotype wingless adults and nymph. (Photo A courtesy of H. D. Niemczyk; photo B by J. Meyers, ChemLawn Services Corporation, courtesy of H. D. Niemczyk; photos C–E: NYSAES.)

Rhodesgrass mealybug. **F.** Young adult female. **G.** Young larva. **H.** Nymph. (Photos: NYSAES.)

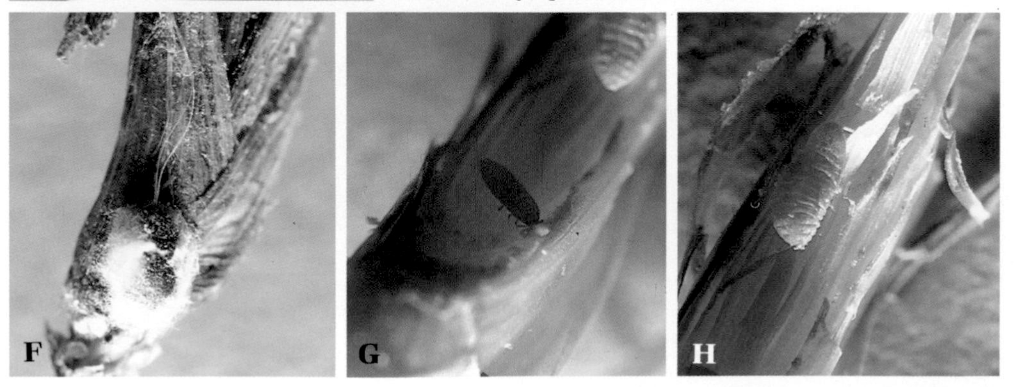

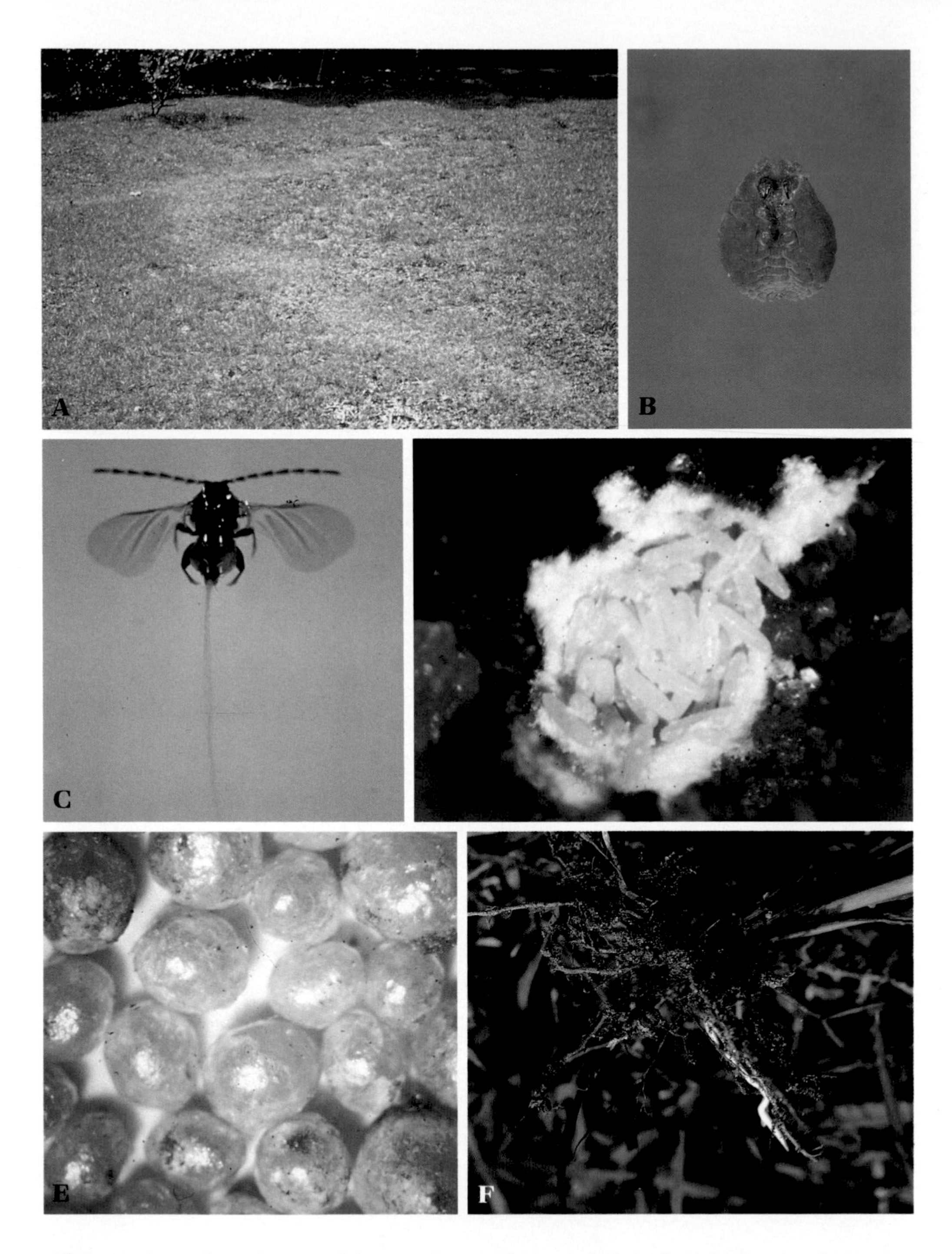

Plate 12. Ground pearl. **A.** Turf damage. **B.** Female, ventral view. **C.** Male. **D.** Egg cluster. **E.** Nymphal cysts. **F.** Female cyst attached to root. (Photos A, C, D, E by C. Kouskolekas, courtesy of P. P. Cobb; photos B, F courtesy of K. Kennedy.)

Plate 13. Major temperate-region sod webworm moths. **A.** *Parapediasia teterrella.* **B.** *Fissicrambus mutabilis.* **C.** *Pediasia trisecta.* **D.** *Chrysoteuchia topiaria.* **E.** *Tehama bonifatella.* **F.** *Crambus sperryellus.* **G.** *Crambus laqueatellus.* **H.** *Microcrambus elegans.* All photos of pinned specimens from the Cornell University collection, at same magnification. (Photos: NYSAES.)

Plate 14. Sod webworm damage. **A.** First instar feeding on epidermis. **B.** Early stages of turf damage. **C.** Spread of turf damage. **D.** More extensive damage, with starling peck holes. **E.** Larval burrow in thatch and fresh fecal pellets. **F.** Extensive, irreversible lawn damage, with much of the green areas consisting of clover and weeds. (Photos: NYSAES.)

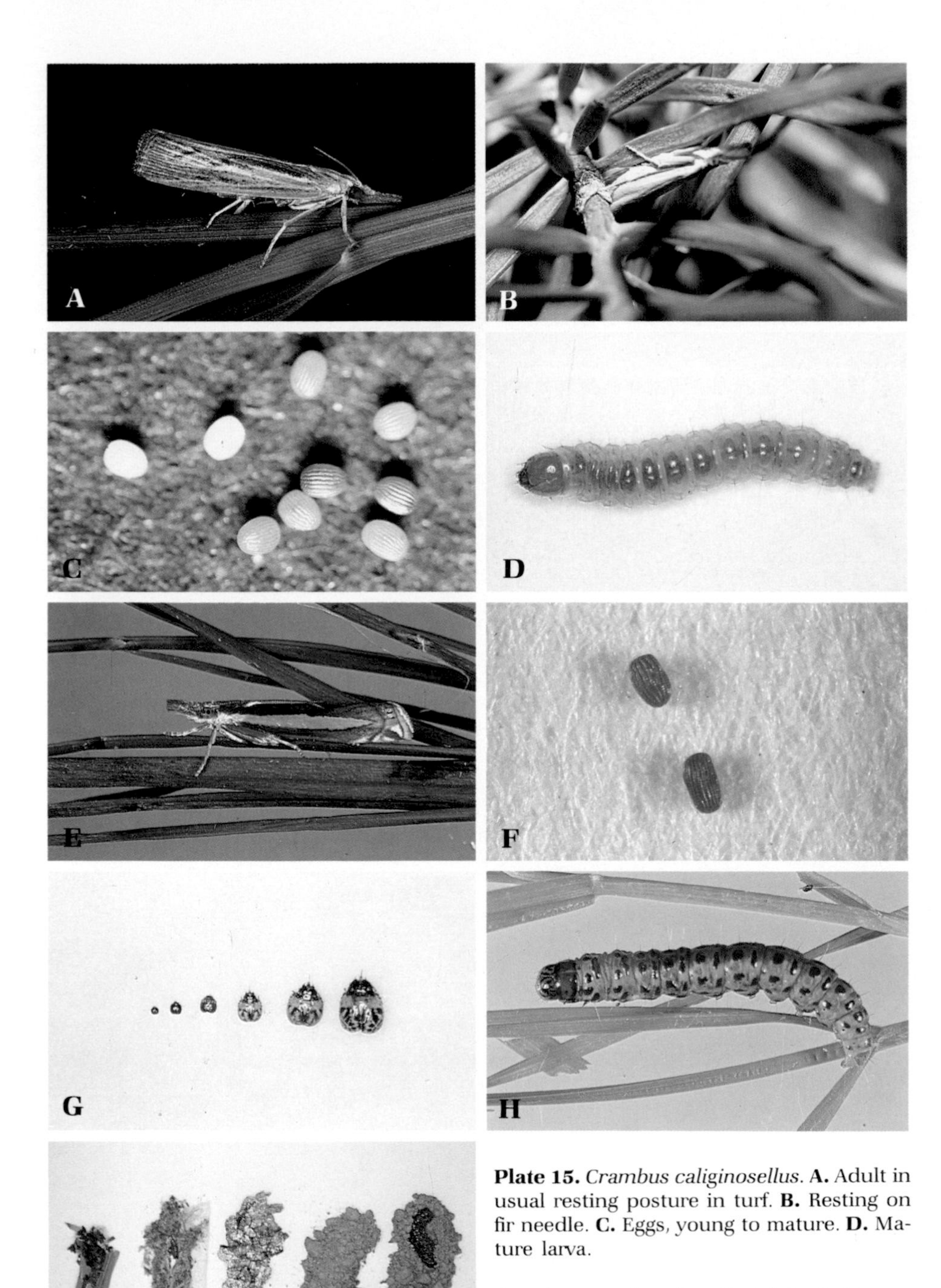

Plate 15. *Crambus caliginosellus.* **A.** Adult in usual resting posture in turf. **B.** Resting on fir needle. **C.** Eggs, young to mature. **D.** Mature larva.

Crambus praefectellus. **E.** Adult. **F.** Mature eggs. **G.** Head capsules of instars 1–6. **H.** Mature larva, resembling larvae of many species. **I.** Burrows and cocoons. (Photos: NYSAES.)

Plate 16. Grass webworm. **A.** Damage to bermudagrass. **B.** Male. **C.** Female. **D.** Male and female, ventral view. **E.** Mass of eggs on bermudagrass turf. **F.** Eggs, young (left) and mature (right). **G.** First-instar larva (feeds only on epidermis). **H.** Feeding habit of older larvae. (Photo A courtesy of C. L. Murdoch; photos B–G courtesy of D. M. Tsuda; photo H: NYSAES.)

Plate 17. Grass webworm. **A.** Mature fifth-instar larva. **B.** Cocoon of debris and fecal pellets. **C.** Pupa. **D.** Pupal terminalia: male (left) with undivided eighth segment and female (right) with divided eighth segment (arrow). **E.** Myna bird. **F.** Giant toad, predator of grass webworm larvae. (Photos A, B, D by D. M. Tsuda; photos C, E, F: NYSAES.)
Tropical sod webworm. **G.** Adult. **H.** Eggs showing head capsules. (Photos courtesy of D. J. Shetlar.)

Plate 18. Cutworm and armyworm moths. **A.** Black cutworm male. **B.** Black cutworm female. **C.** Variegated cutworm. **D.** Armyworm. **E.** Bronzed cutworm male. **F.** Bronzed cutworm female. **G.** Fall armyworm male. **H.** Fall armyworm female. (Pinned specimens collected and prepared by P. J. Chapman and S. E. Lienk, NYSAES; photos: NYSAES.)

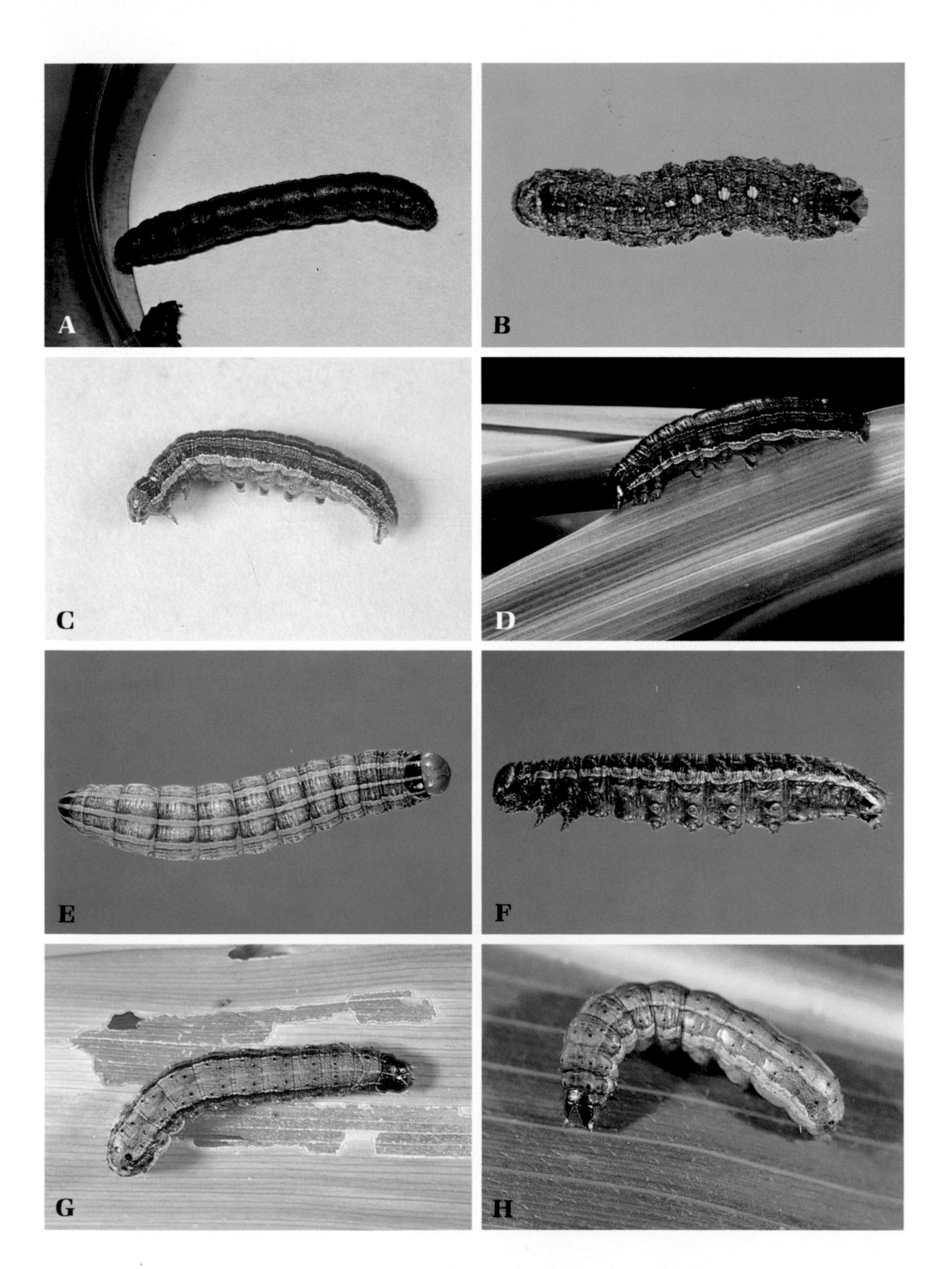

Plate 19. Cutworm and armyworm larvae. **A.** Black cutworm. **B.** Variegated cutworm. **C** and **D.** Armyworm color variations. **E** and **F.** Bronzed cutworm dorsal and lateral views. **G** and **H.** Fall armyworm color variations. (Photos A, C, G, H: NYSAES; photos B, D, F courtesy of R. W. Rings; photo E courtesy of H. D. Niemczyk.)

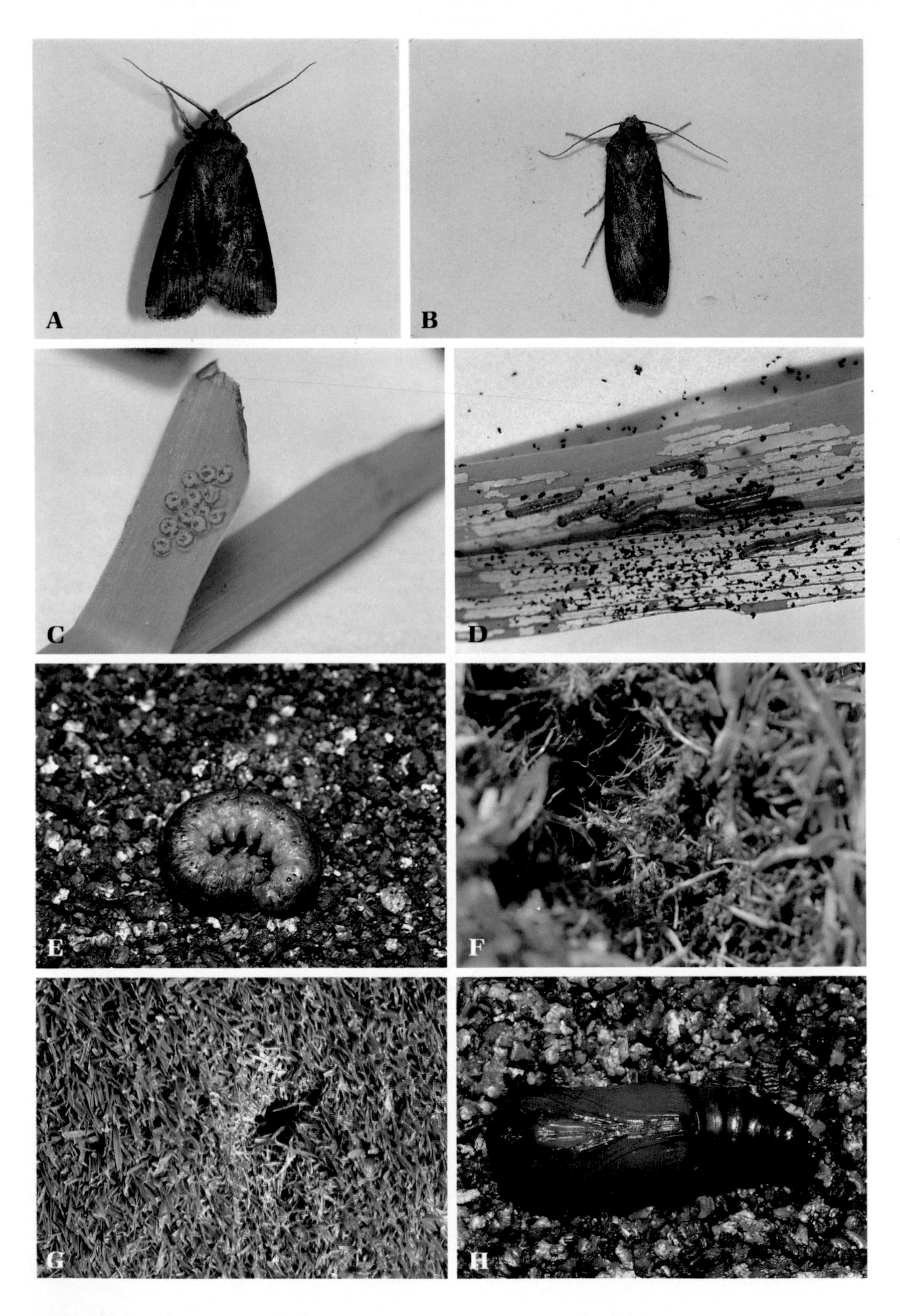

Plate 20. Black cutworm. **A.** Male (pectinate antennae). **B.** Female (filiform antennae). **C.** Eggs. **D.** First-instar larvae feeding only on epidermis. **E.** Mature larva. **F.** Frass-filled tunnel in bentgrass turf. **G.** Characteristic larval feeding around aeration hole on golf green. **H.** Pupa. (Photos: NYSAES.)

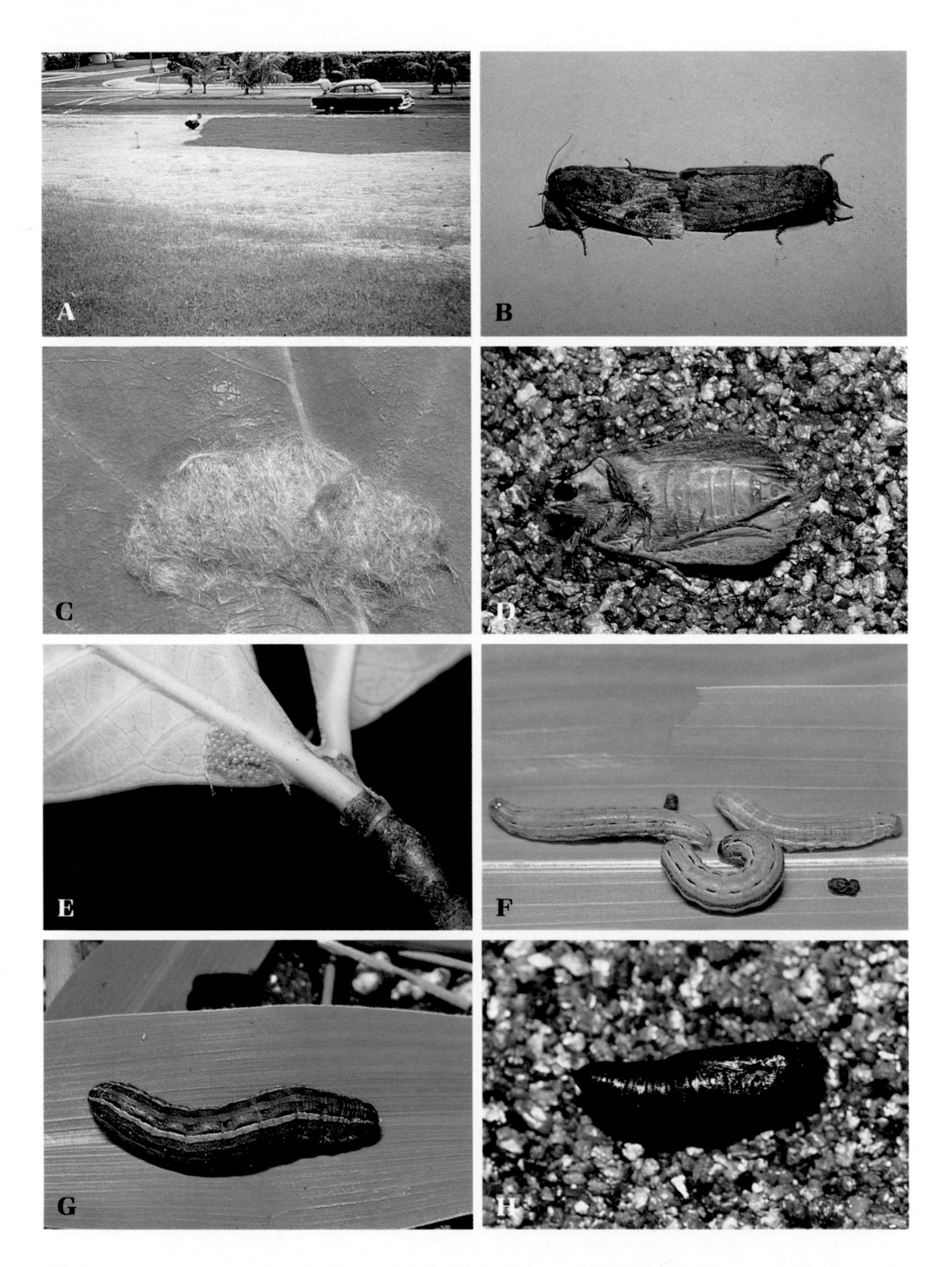

Plate 21. Lawn armyworm. **A.** Characteristic damage to common bermudagrass. **B.** Mating pair (male left). **C.** Egg mass by young female. **D.** Aging female with abdominal hairs nearly spent. **E.** Naked egg mass deposited by aging female. **F.** Young larvae. **G.** Mature larva. **H.** Pupa. (Photo A courtesy of A. K. Ota; photos B–H: NYSAES.)

Plate 22. Fiery skipper stages and habits. **A.** Male. **B.** Female feeding on lantana nectar. **C.** Eggs. **D.** First-instar larvae. **E.** Second-instar larva. **F.** Mature larva. **G.** Young pupa. **H.** Male (left) and female (right) pupal terminalia. **I.** Pupa in loosely woven cocoon. (Photos: NYSAES.)

Plate 23. Progression of turf damage by scarabaeid grubs. **A.** Slight thinning (lower left) in fall. **B.** Noticeable thinning during early spring. **C.** Extensive thinning but much still recoverable with fertilization. **D.** Irreversible damage in need of reseeding. **E.** Grubs at thatch-soil interface as many as half hidden in the soil. **F.** Typical renovation of small areas by removal of thatch and dead grass. (Photos: NYSAES.)

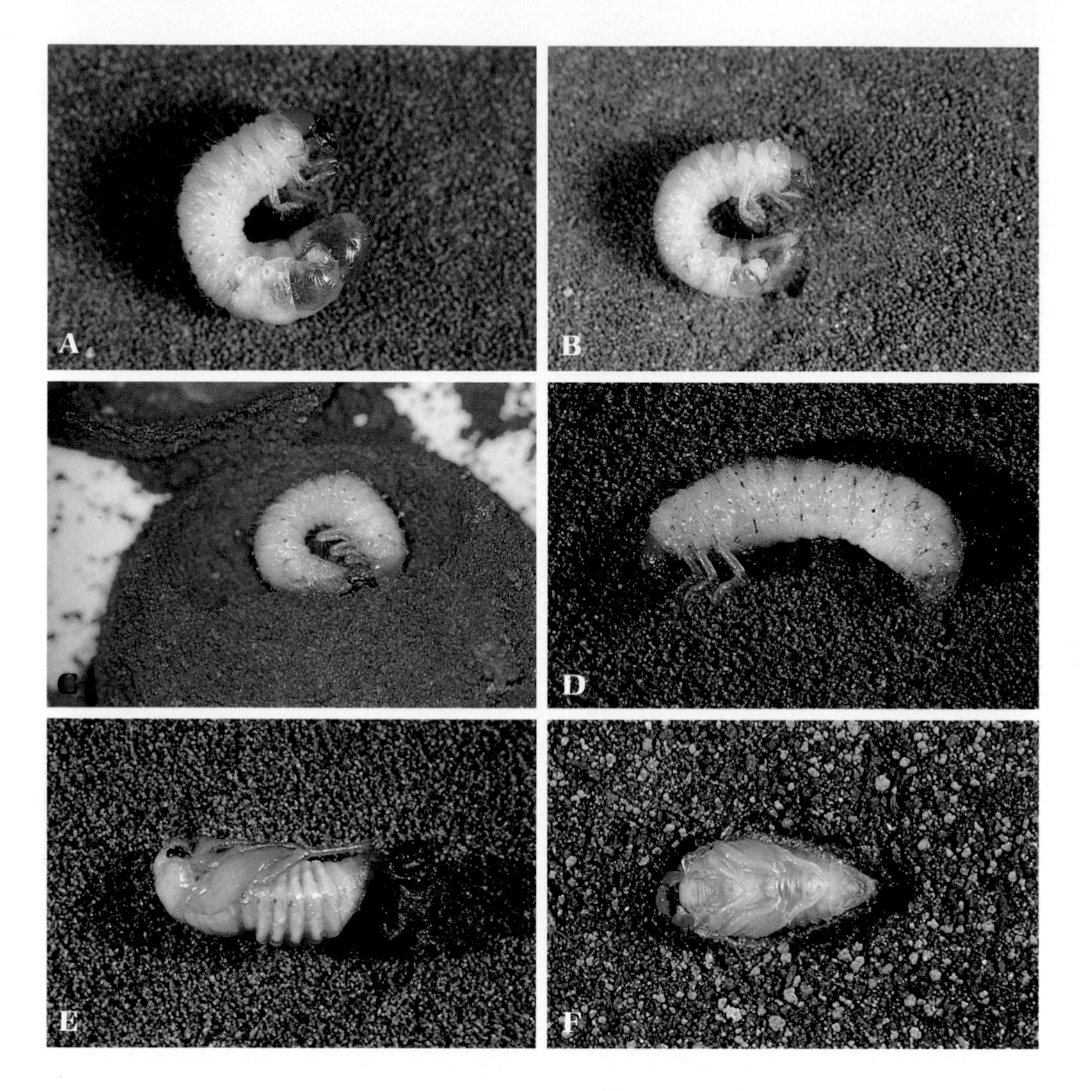

Plate 24. General differences in scarabaeid larvae, prepupae, and pupae. **A.** Midabdominal constriction in larva of masked chafer. **B.** Without constriction in oriental beetle larva. **C.** C-shaped oriental beetle prepupa. **D.** Relatively straight European chafer prepupa. **E.** Larval exuviae sloughed off in the European chafer pupa. **F.** Exuviae retained in the Japanese beetle. (Photos: NYSAES.)

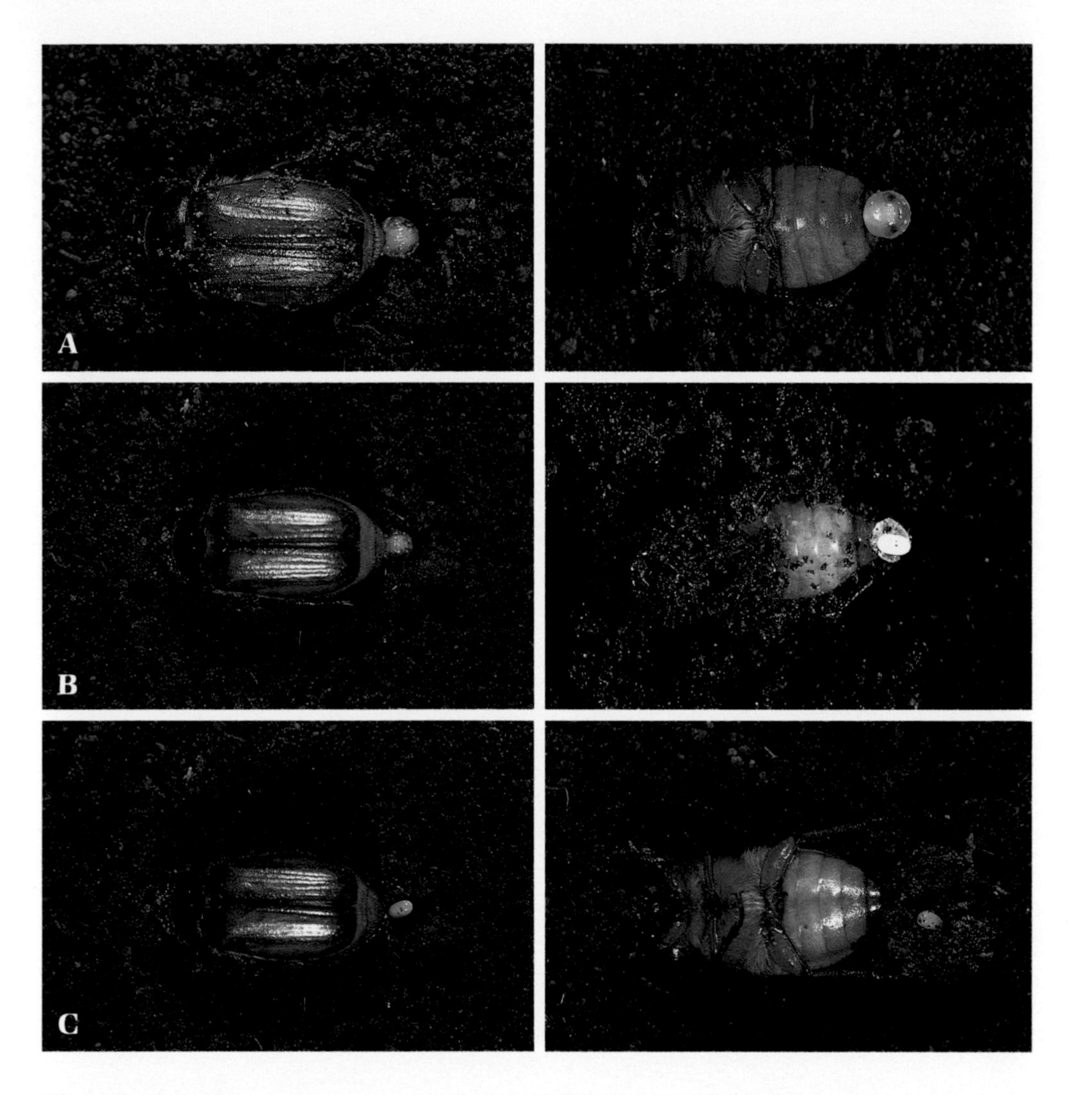

Plate 25. Oviposition sequence in scarabaeid adults, illustrated by the European chafer, dorsal views (left) and ventral views (right). **A.** Evagination of vagina as bulbous organ to compress soil to form earthen cell. **B.** Egg deposition, with simultaneous invagination of vagina. **C.** Egg in cell. (Photos: NYSAES.)

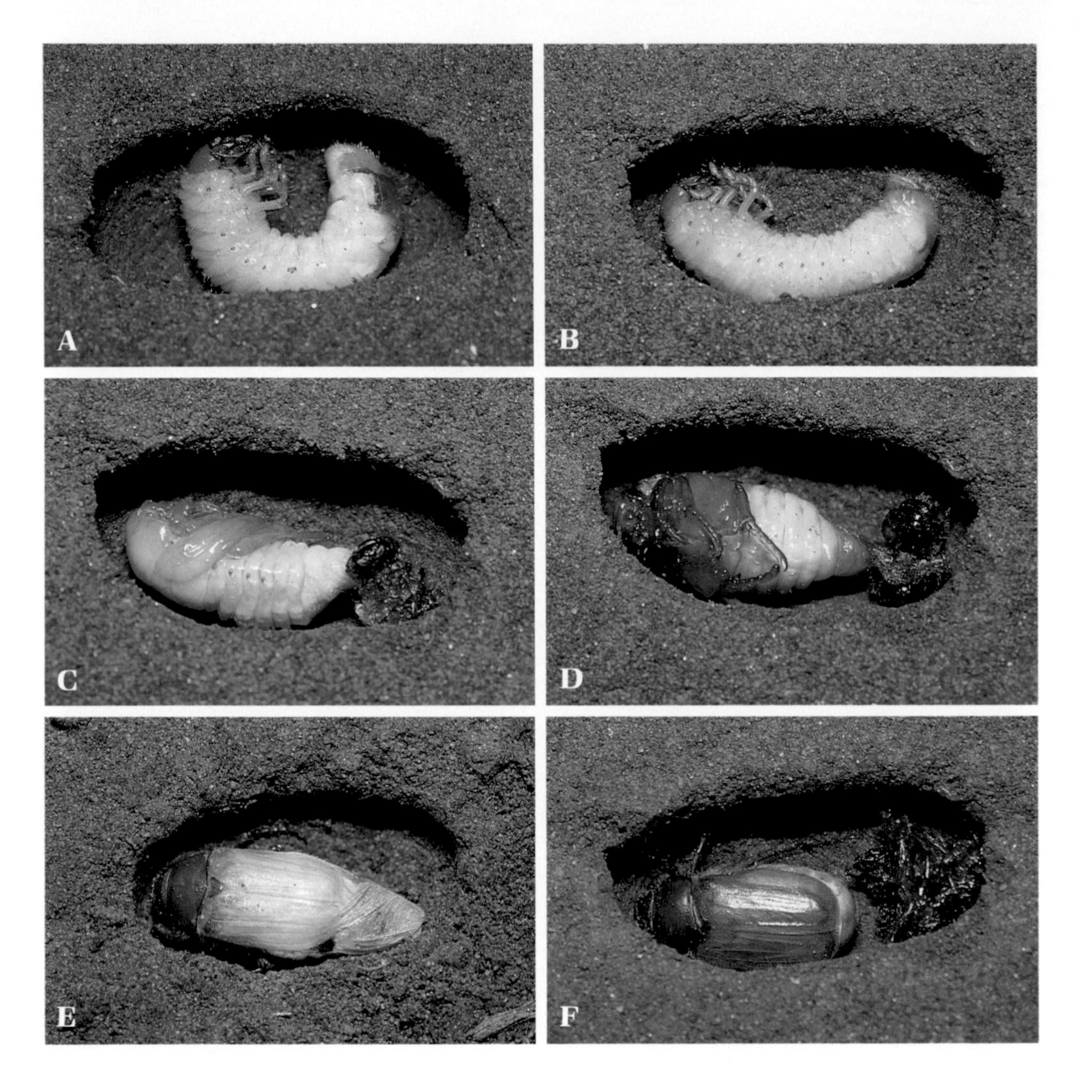

Plate 26. Scarabaeid larva-to-adult sequence in the same earthen cell, illustrated by the European chafer. **A.** Larva in own cell. **B.** Prepupa. **C.** Young pupa. **D.** Mature pupa. **E.** Callow adult with transparent elytra and unfolded hindwings. **F.** Mature adult ready to leave earthen cell. (Photos: NYSAES.)

Plate 27. Black turfgrass ataenius. **A.** Turf damage. **B.** Grubs in 0.1 m² (1 ft²) of soil-thatch interface with about 50% of grubs visible and remaining grubs hidden in soil. **C.** Callow to mature adults. **D.** Egg cluster. **E.** Healthy (left) and milky-diseased (right) third-instar grubs. **F.** Rasters of third-instar grubs. **G.** Pupae of different ages. **H.** *Aphodius granarius* raster. (Photos A–C, E–H: NYSAES; photo D courtesy of H. D. Niemczyk.)

Plate 28. Black turfgrass ataenius, preoviposition and blooming periods. **A.** Vanhoutte spirea in full bloom. **B.** Flowers. **C.** Horse chestnut in full bloom. **D.** Flower cluster. **E.** Black locust in early bloom. **F.** Flower cluster, showing development of plants when overwintering adults start laying eggs. **G.** Rose of sharon in full bloom. **H.** Flower, showing development of plant at start of egg laying by first-generation adults. (Photos: NYSAES.)

Plate 29. Green June beetle. **A.** Adult. **B.** Feeding on apricot. **C.** Eggs, young to mature. **D.** First-instar grub crawling on its back. **E.** Third-instar posterior segments and raster. **F.** Third-instar grub crawling on its back. **G.** Grub at bottom of vertical burrow, with Japanese beetle grubs on either side. (Photos A–F: NYSAES; photo G courtesy of D. J. Shetlar.)

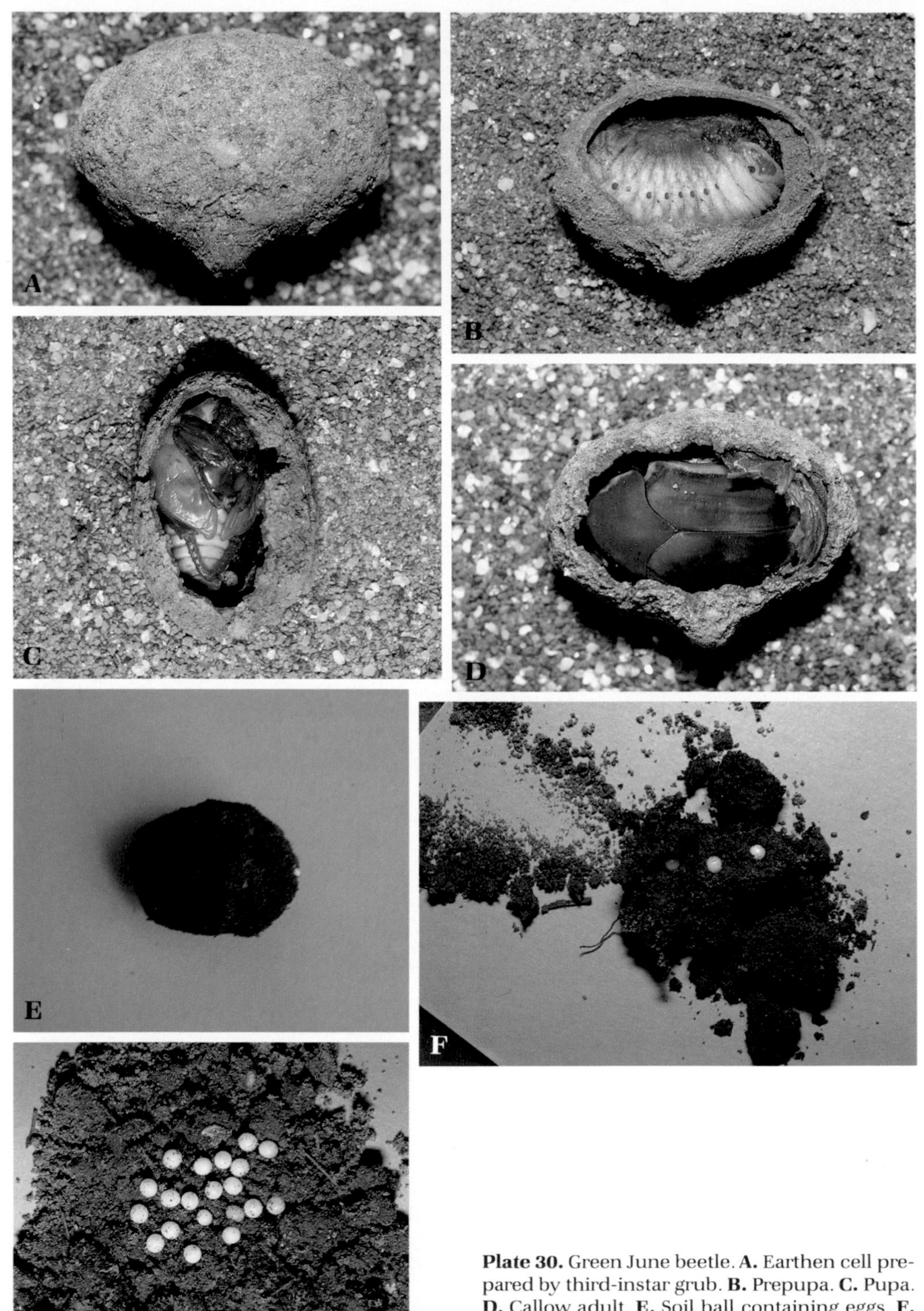

Plate 30. Green June beetle. **A.** Earthen cell prepared by third-instar grub. **B.** Prepupa. **C.** Pupa. **D.** Callow adult. **E.** Soil ball containing eggs. **F.** Same soil ball broken apart, showing eggs in cells. **G.** Twenty eggs from the same soil ball. (Photos: NYSAES.)

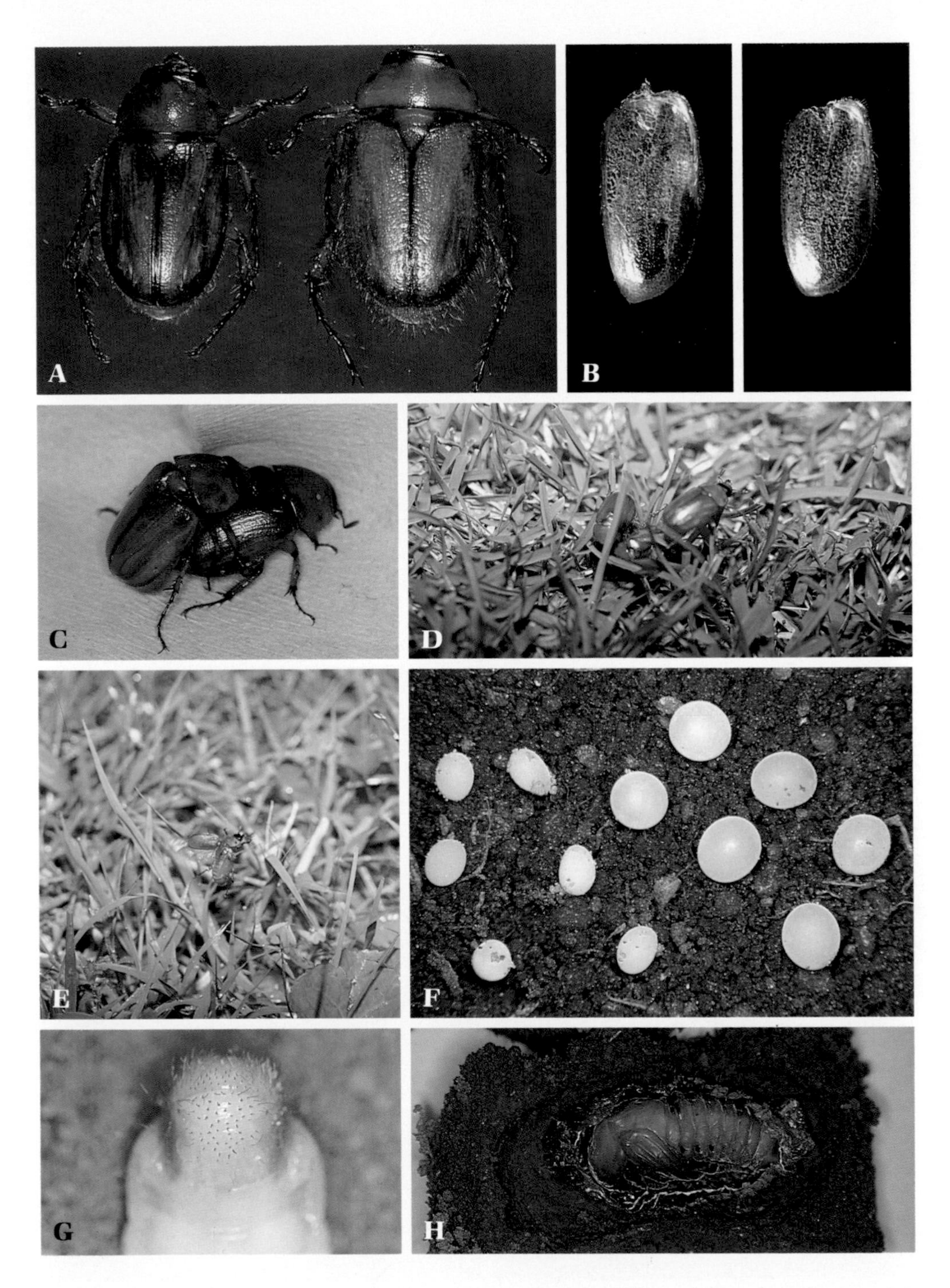

Plate 31. Southern and northern masked chafers. **A.** Adults of southern (left) and northern (right) masked chafer. **B.** Elytra of female southern (left) and northern (right) masked chafer, showing bristles along edge. **C.** Mating pair of southern masked chafer. **D.** Southern males clustering around single female. **E.** Start of flight. **F.** Young to mature eggs of southern masked chafer. **G.** Raster of southern masked chafer. **H.** Pupa of masked chafer in cell cemented by secretion. (Photos A, B courtesy of D. A. Potter; photos C–H: NYSAES.)

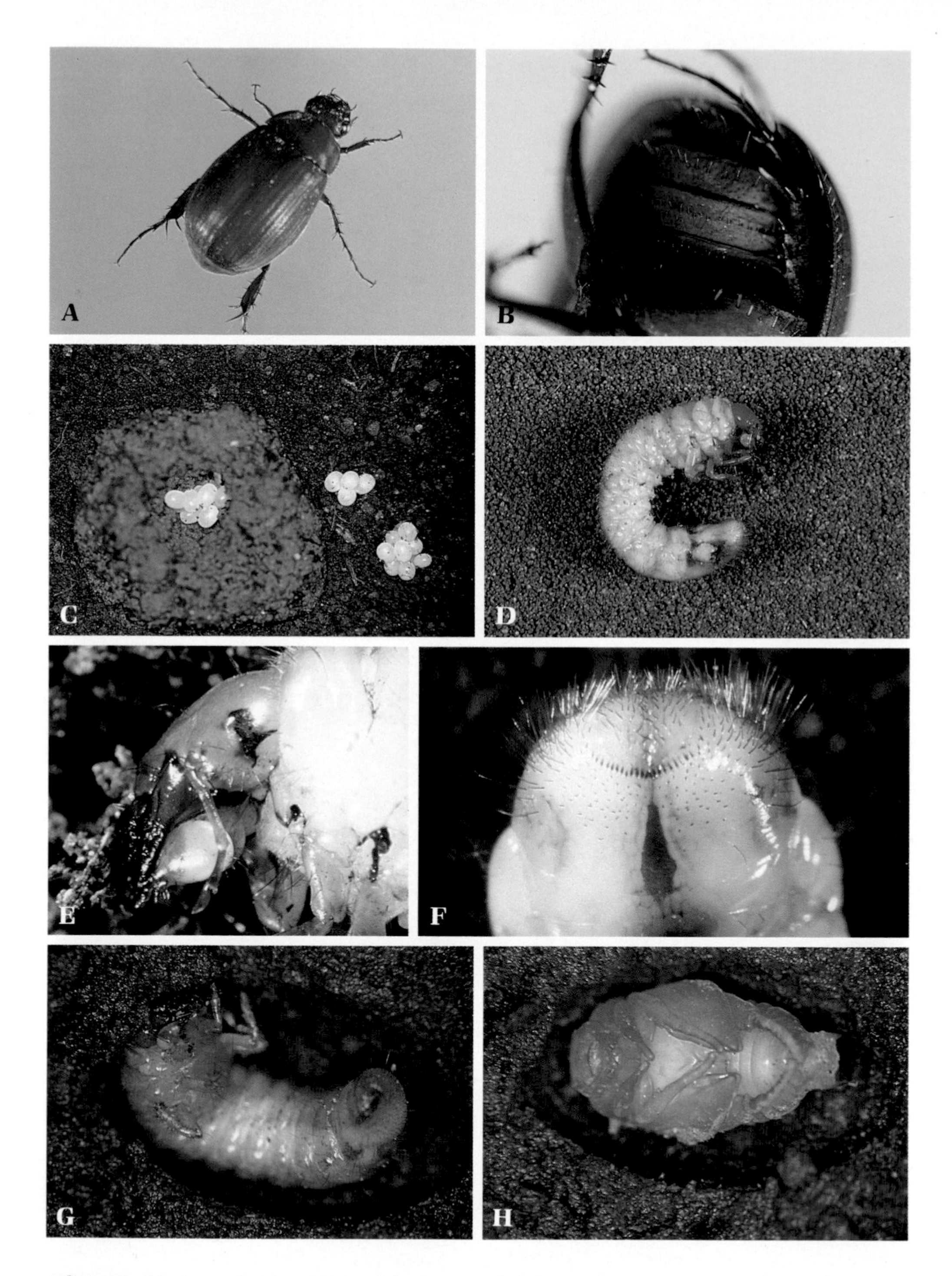

Plate 32. Asiatic garden beetle. **A.** Adult. **B.** Ventral abdominal segments 5, 6, and 7, each with a row of spines. **C.** Egg clusters and cemented cell with egg cluster. **D.** Third-instar grub. **E.** Bulbous stipe of left maxilla. **F.** Raster of third instar. **G.** Prepupa. **H.** Pupa in cell cemented by secretion. (Photos: NYSAES.)

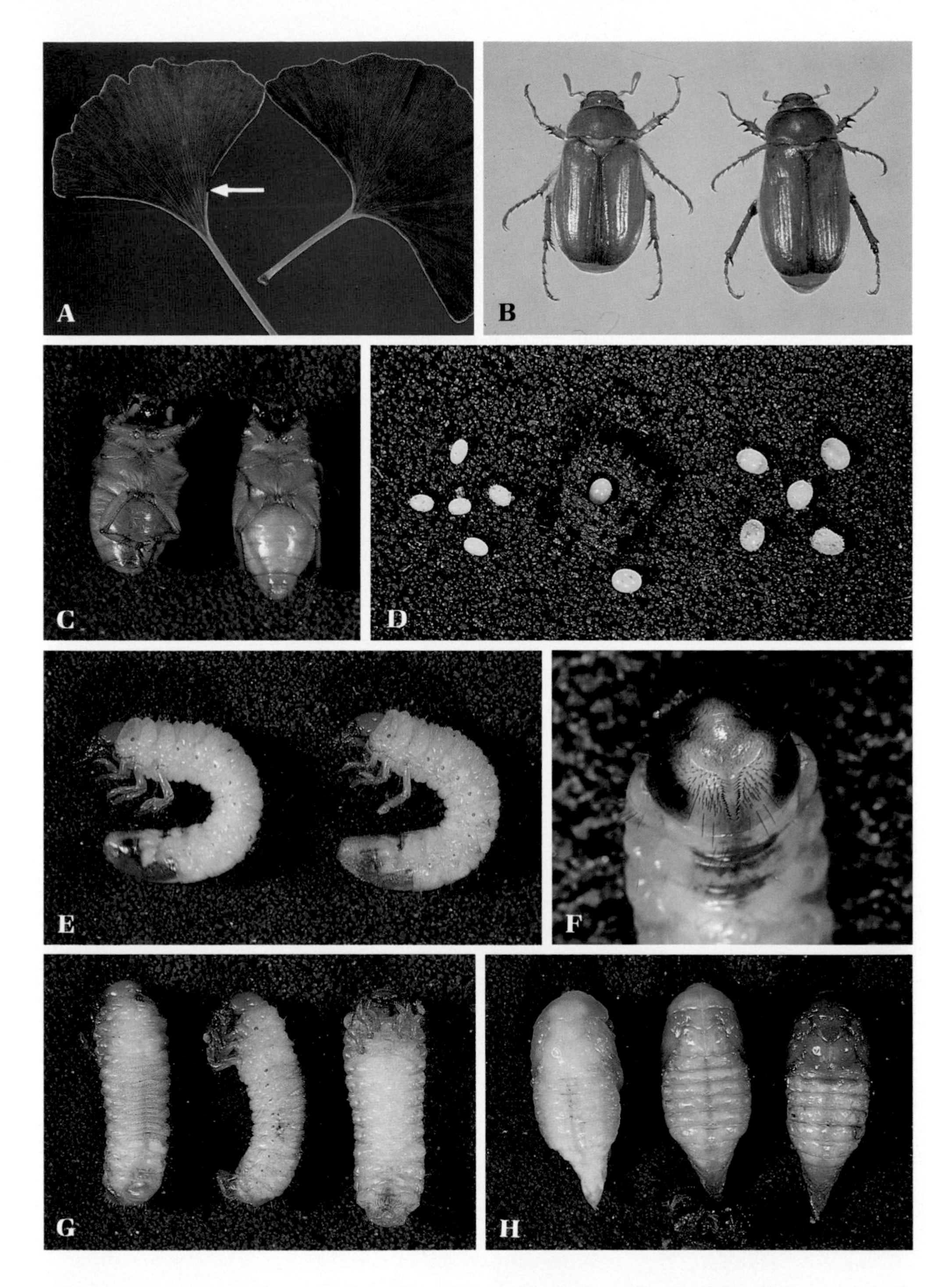

Plate 33. European chafer. **A.** Adult feeding injury (left leaf). **B.** Adult male (left) with longer club of antennae. **C.** Abdominal differences in male (left) and female (right). **D.** Eggs, young to mature, each originally deposited in an individual cell. **E.** Third-instar grub actively feeding (left, whitish) and mature grub that has completed feeding (right, yellowish). **F.** Raster of third-instar grub. **G.** Prepupae, three views. **H.** Pupae, young (left) to mature (right). (Photos: NYSAES.)

Plate 34. European chafer. **A.** The common catalpa in full bloom, the period of peak beetle flights. **B.** Catalpa flower cluster. **C.** Individual flower. **D.** Beetle climbing grass stem to take flight. **E.** Beetle mating flight to a locust. **F.** Mating pair. **G.** Mass of beetles shaken from tree. (Photos: NYSAES.)

Plate 35. Adults of the most common species of May or June beetle, *Phyllophaga*. **A.** *P. anxia*. **B.** *P. crinita*. **C.** *P. crenulata*. **D.** *P. fraterna*. **E.** *P. fusca*. **F.** *P. gracilis*. **G.** *P. hirticula*. **H.** *P. rugosa*. **I.** *P. tristis*. All photos of pinned specimens from the Cornell University collection, at same magnification. (Photos: NYSAES.)

Plate 36. May-June beetle. **A.** Defoliation of trees by adult feeding. **B.** Differences in club of antennae on male (left) and female (right). **C.** Small first-instar grubs. **D.** Mature third-instar grub. **E–G.** Modifications of rastral patterns and anal slit. In photo E, stem of Y-shaped anal slit so short it appears transverse. (Photo A courtesy of D. L. Potter; photos B–G: NYSAES.)

Plate 37. *Tiphia* wasp, a natural enemy of May or June beetles. **A.** Adult wasp. **B.** Young *Tiphia* larva feeding on liquid contents of grub. **C.** Mature *Tiphia* larva after the grub has been nearly completely consumed. **D.** *Tiphia* cocoon. (Photos: NYSAES.)

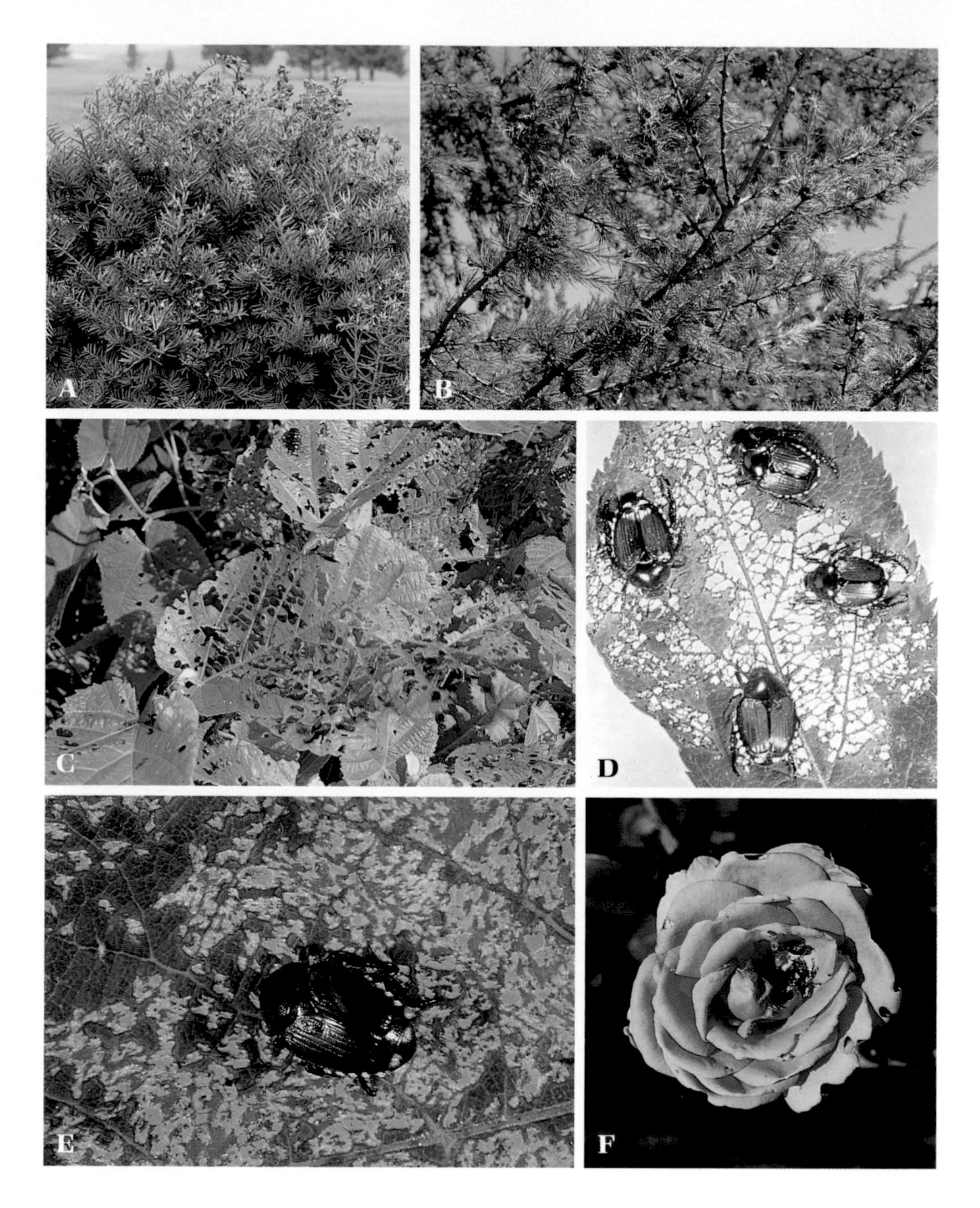

Plate 38. Japanese beetle adult feeding damage to favorite host plants. **A.** Yew. **B.** Larch. **C.** Linden. **D.** Apple. **E.** Grape. **F.** Rose. (Photos: NYSAES.)

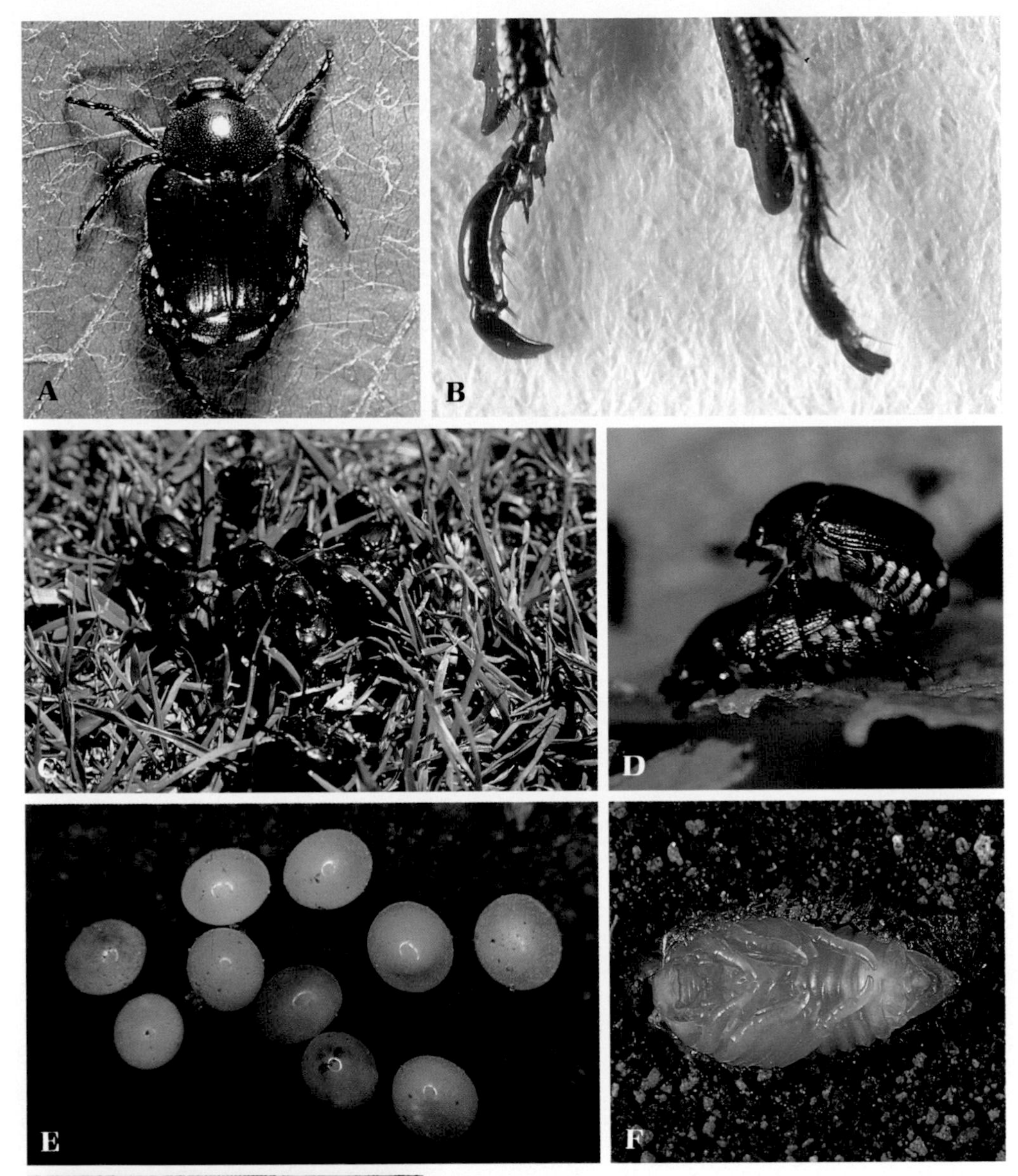

Plate 39. Japanese beetle. **A.** Adult. **B.** Forelegs of the male (left), with pointed tibial spur and first tarsal segment wider than long, and the female (right), with spatulate tibial spur and first tarsal segment longer than wide. **C.** Males attracted to single female in turf. **D.** Mating pair. **E.** Eggs, young to mature. **F.** Pupa with meshlike exuviae. **G.** Raster of third-instar grub. (Photos: NYSAES.)

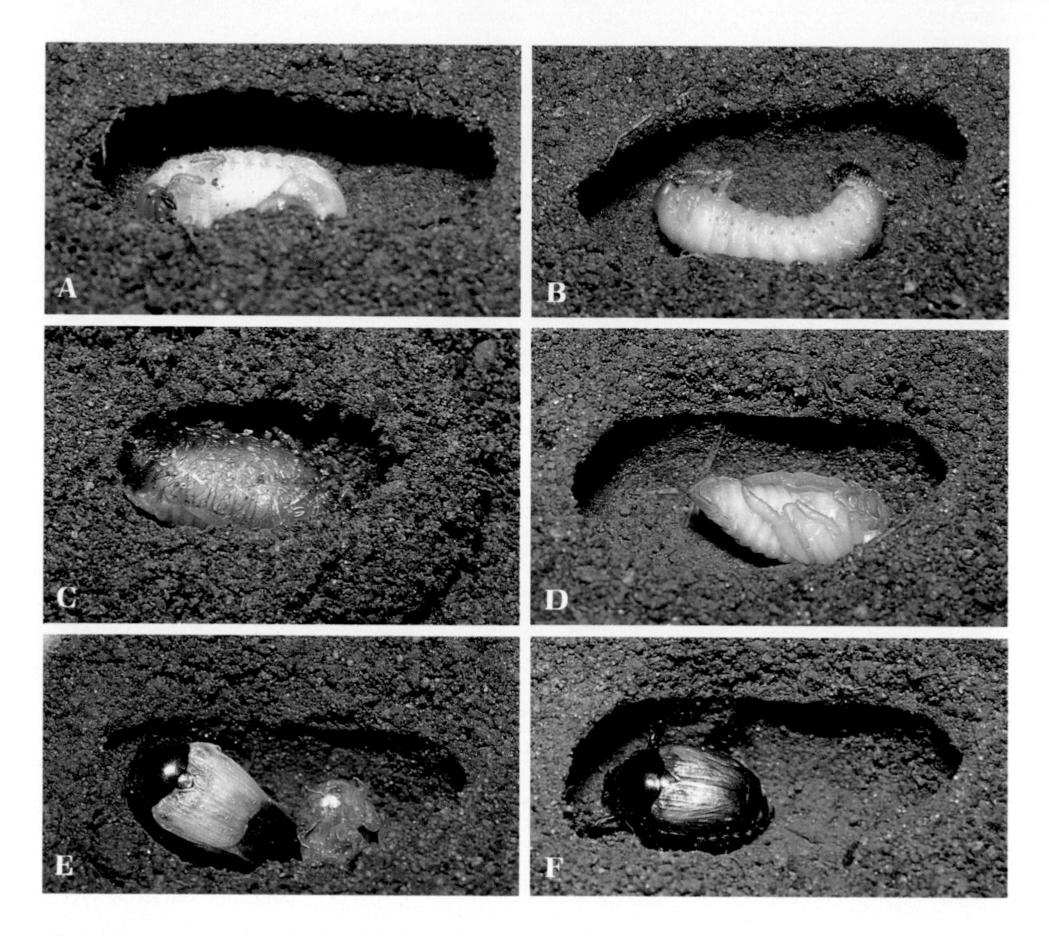

Plate 40. Japanese beetle development in earthen cell. **A.** Mature grub. **B.** Prepupa. **C.** Young pupa enclosed in larval exuviae. **D.** Older pupa free from exuviae. **E.** Callow adult. **F.** Mature adult ready to emerge and take flight. (Photos: NYSAES.)

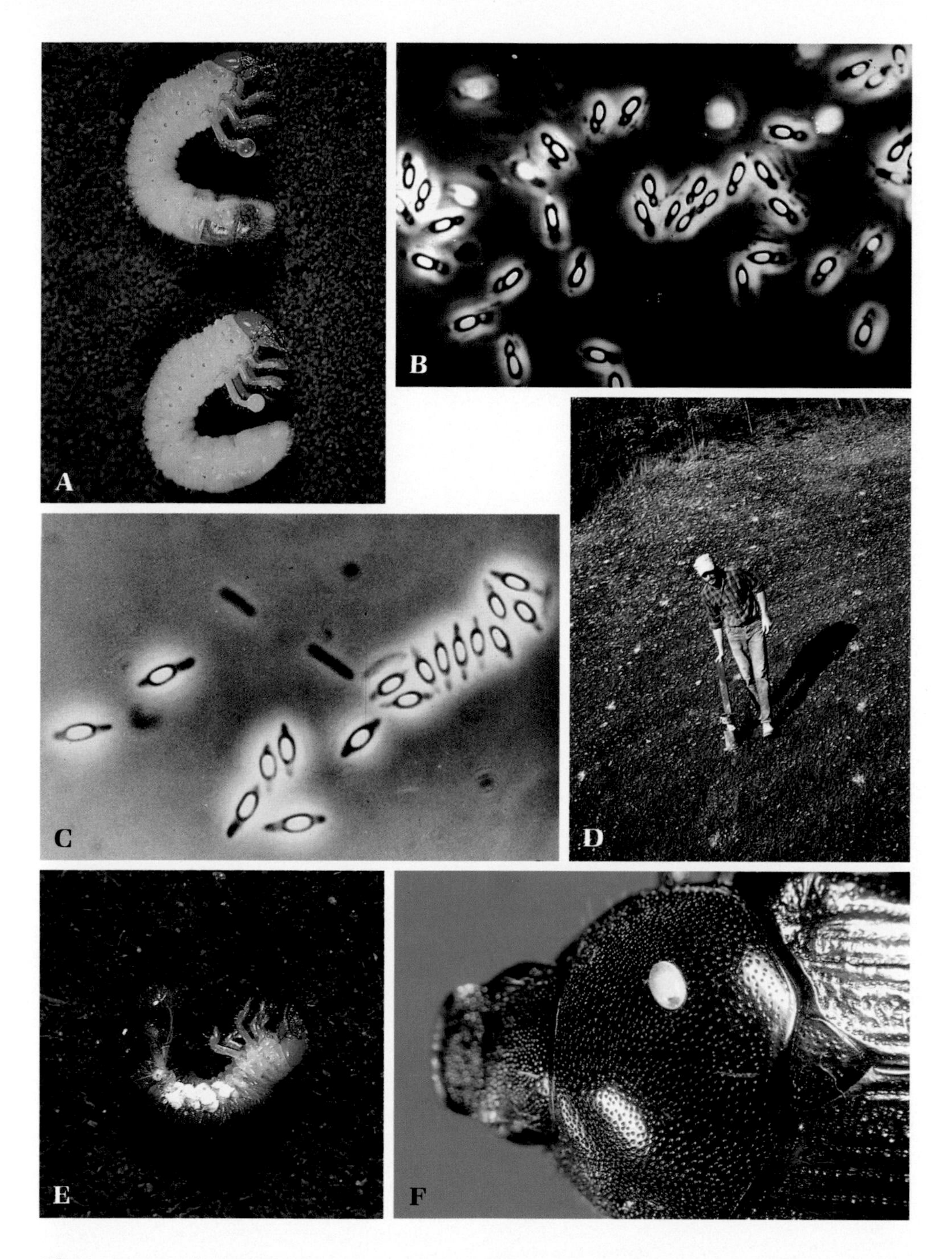

Plate 41. Natural enemies of Japanese beetle. **A.** Normal grub (above) and milky diseased grub (below). **B.** *Bacillus popilliae* spores, major milky disease bacteria. **C.** *B. lentimorbus* spores, minor milky disease bacteria. **D.** Artificial distribution of spore powder. **E.** Early *Metarrhizium*-infected grub. **F.** Beetle with attached egg of *Hyperecteina* spp., a tachinid fly. (Photos A–C, E: NYSAES; photo D courtesy of Fairfax Biological Laboratory; photo F courtesy of M. G. Klein.)

Plate 42. Oriental beetle. **A.** Canada hemlock with heavy grub infestations to soil depths of >30 cm (12 in.). **B.** Adult of the most common color. **C.** Antennal differences of the male (left) and female (right). **D.** Adults with color extremes. **E.** Mating (occurs readily in captivity). **F.** Eggs, young to mature. **G.** Raster (with fine palidia most visible in milky-diseased grub). **H.** Prepupa. **I.** Pupa (surrounded by exuviae). (Photos: NYSAES.)

Plate 43. Dichondra flea beetle. **A.** Normal dichondra leaves. **B.** Dichondra lawn destroyed by beetles and larvae. **C.** Crescent-shaped adult feeding scars on leaf. **D.** Adult beetle. **E.** Unusually large femur of dichondra beetle. **F.** Eggs. **G.** Larva (unusually long). (Photos A, C–G: NYSAES; photo B courtesy of L. R. Brown.)

Plate 44. Bluegrass billbug damage. **A.** Adults piercing stems. **B.** Mature larvae feeding in crown. **C.** Sawdustlike fecal pellets, evidence of billbug larval feeding. **D.** Lawn damage mainly along driveway. **E.** Spotty lawn damage. **F.** Uniform, extensive lawn damage. **G.** Insecticide-protected lawn shown in the foreground at left. (Photos A, C–G: NYSAES; photo B courtesy of H. D. Niemczyk.)

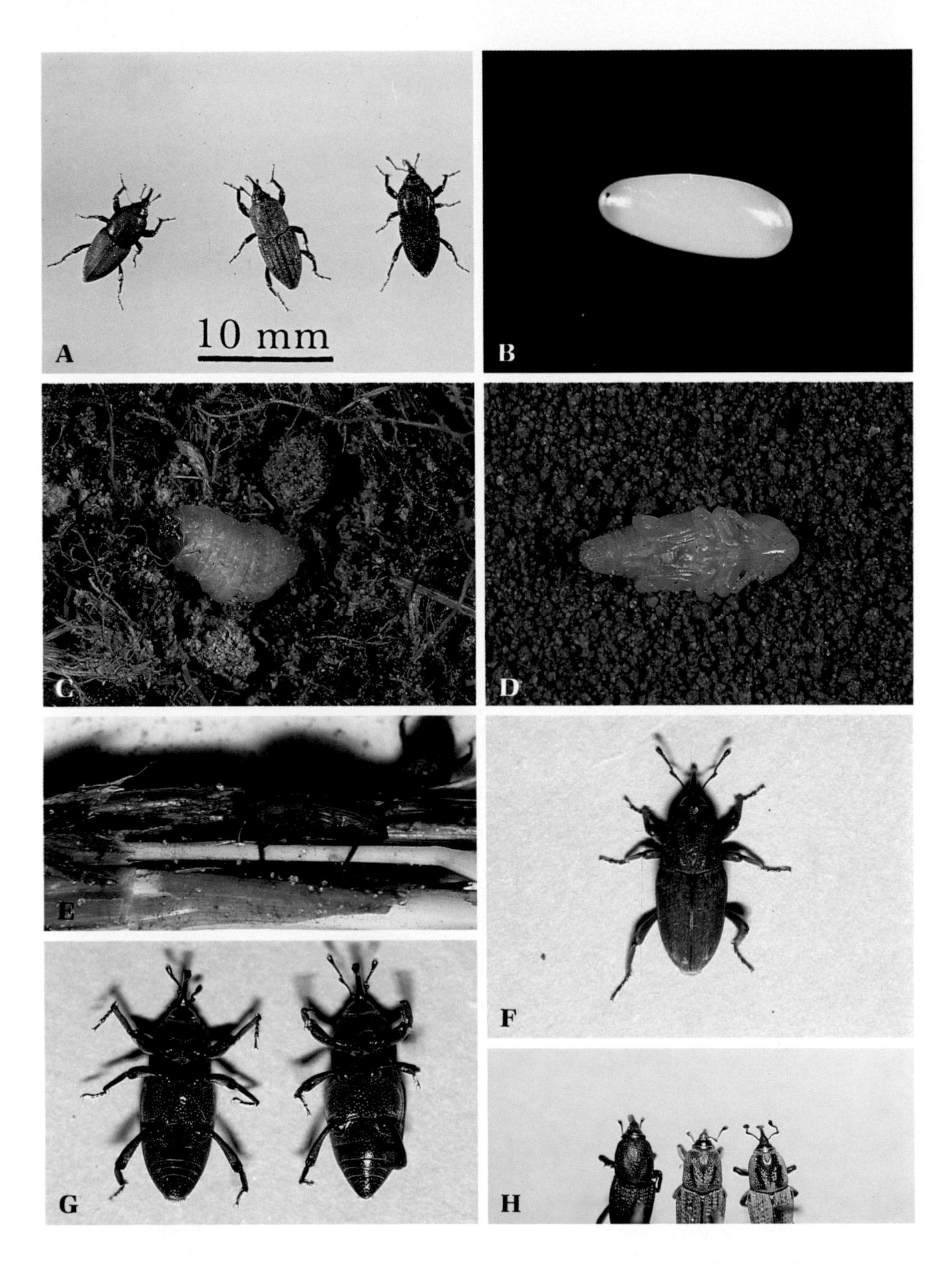

Plate 45. Bluegrass billbug. **A.** Callow adult (left), mature male (center), and mature female (right). **B.** Egg. **C.** Mature larva. **D.** Pupa. (Photos: NYSAES.)
Hunting billbug. **E.** Adult puncturing stem. **F.** Adult. **G.** Male (left), with rounded tip of abdomen, and female (right), with pointed tip of abdomen most visible ventrally. **H.** Pronotal markings of three billbug species (left to right): bluegrass, hunting, phoenix. (Photos: NYSAES.)

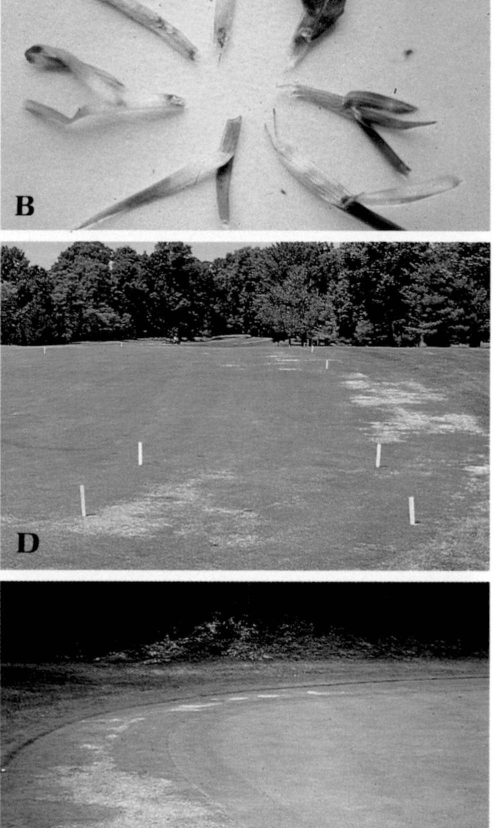

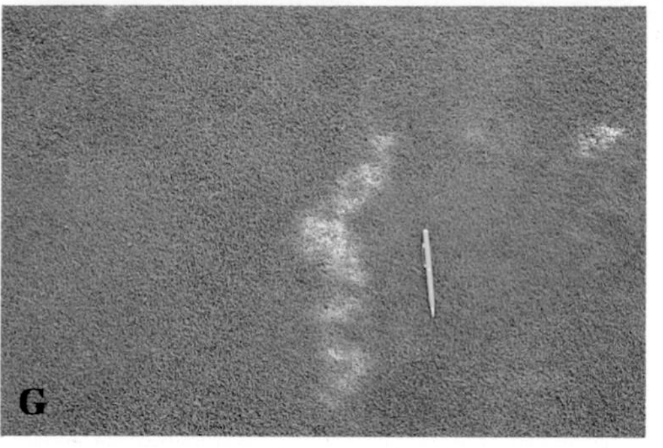

Plate 46. Annual bluegrass weevil damage to annual bluegrass. **A.** Adult feeding on leaves and stems. **B.** Larval feeding on leaves and stems. **C.** Fifth-instar larva feeding on crown. **D.** Early summer damage predominantly at edge of fairways. **E.** Complete destruction of annual bluegrass while bentgrass remains normal. **F.** Damage to collar. **G.** Damage to annual bluegrass on green. (Photos A, D, G: NYSAES; photos B, C, E, F courtesy of R. S. Cameron.)

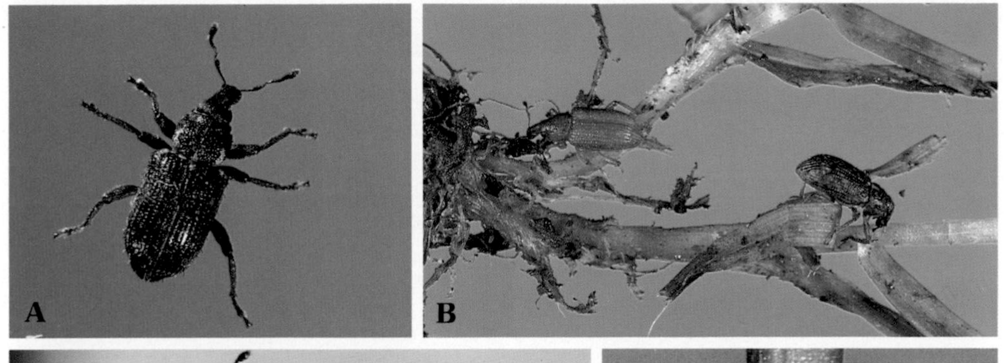

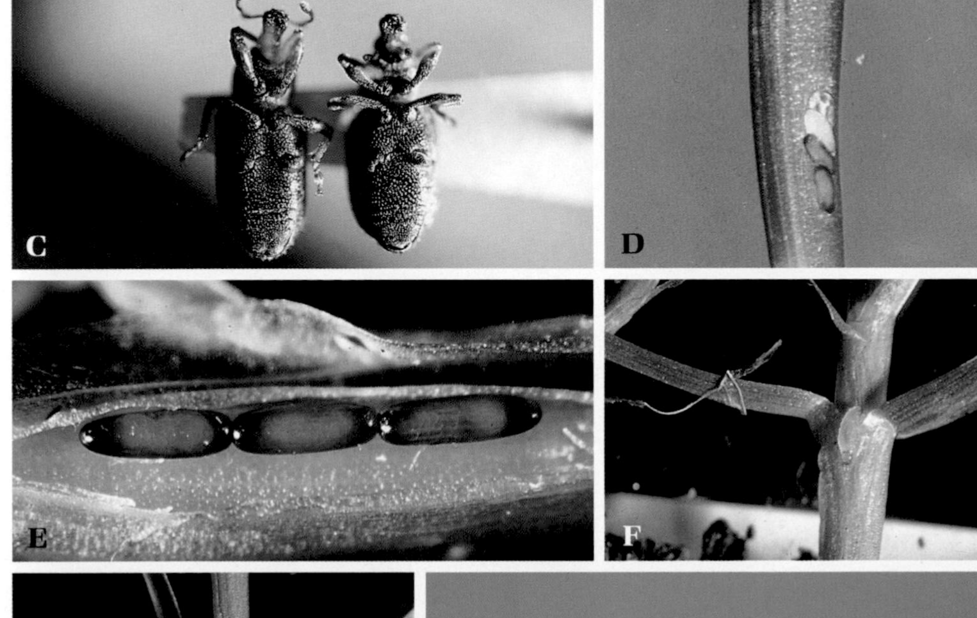

Plate 47. Annual bluegrass weevil. **A.** Adult that has freshly emerged, with full complement of scales and hairs. **B.** Callow and mature adults. **C.** Male (left), with median depression in first visible abdominal sternum, and female (right) without depression. **D.** Fresh (light) and mature (dark) eggs in stem. **E.** Mature eggs with leaf sheath removed. **F.** Second instar on stem. **G.** Third instar completely occupying stem. **H.** Mature larva. **I.** Pupae. (Photos A–E, H, I: NYSAES; photos F, G courtesy of R. S. Cameron.)

Plate 48. Annual bluegrass weevil plant relationships. **A.** White pines in the rough, the situation in which the largest populations of hibernating weevils are found. **B.** White pine litter containing a maximum of 440 weevils per 0.1 m^2 (per ft^2) of litter. **C.** Forsythia in full bloom, the time when weevil movement out of hibernation begins. **D.** Flowers. **E.** Pink flowering dogwood. **F.** White in full bract color when nearly all weevils have left hibernation quarters. **G.** White bracts. (Photos: NYSAES.)

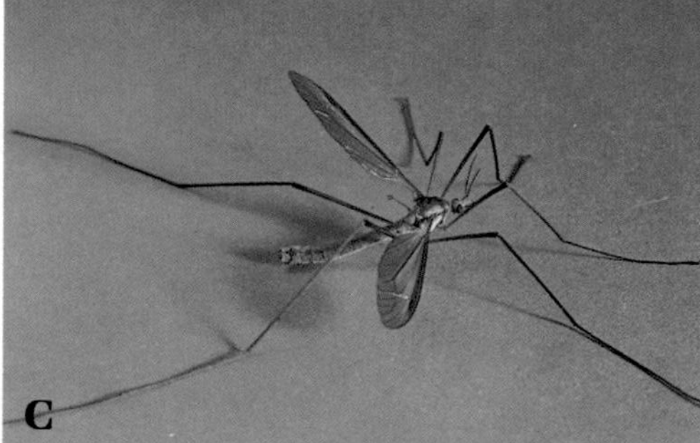

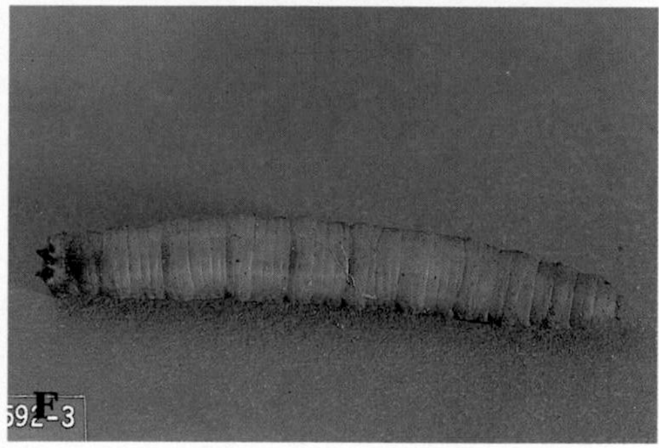

Plate 49. European crane fly. **A** and **B.** Light and severe injury to lawn. **C.** Male, showing wings longer than abdomen. **D.** Female, showing wings shorter than abdomen. **E.** Eggs. **F.** Larvae. **G.** Pupa. (Photos A, B courtesy of S. J. Collman; photos C–G obtained from the Ken Gray Collection, Oregon State University.)

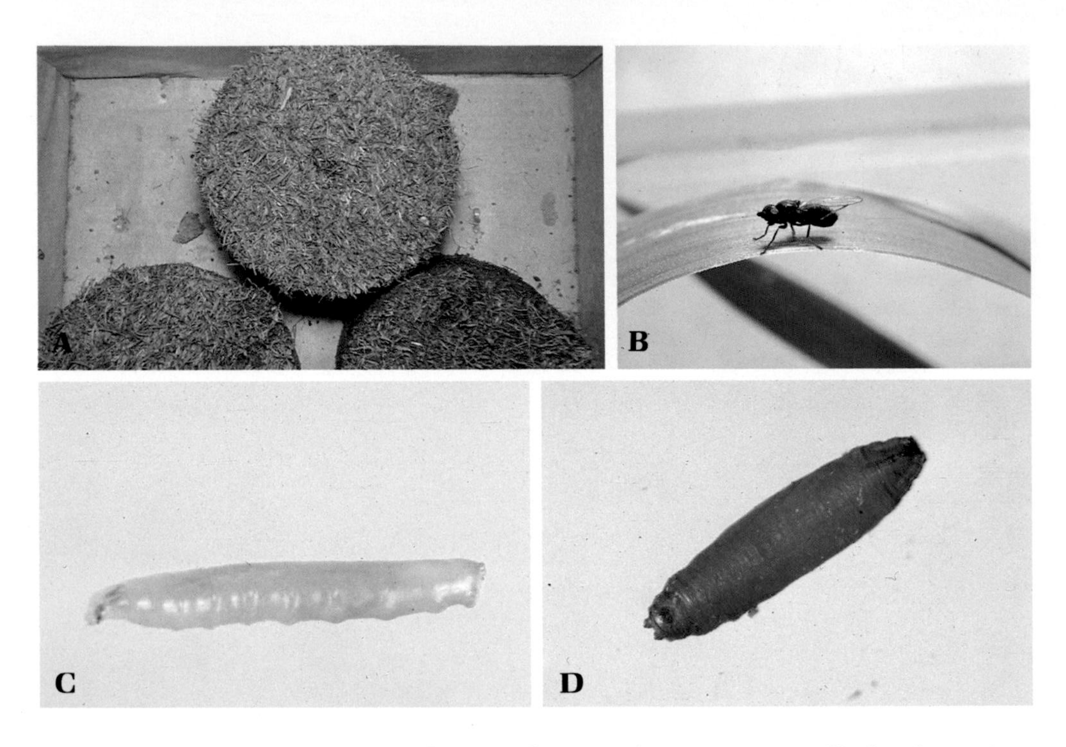

Plate 50. Frit fly. **A.** Normal to severe damage to bentgrass from a green caused by larval tunneling in stems. **B.** Adult fly. **C.** Third-instar larva. **D.** Puparium. (Photos courtesy of W. A. Allen.)

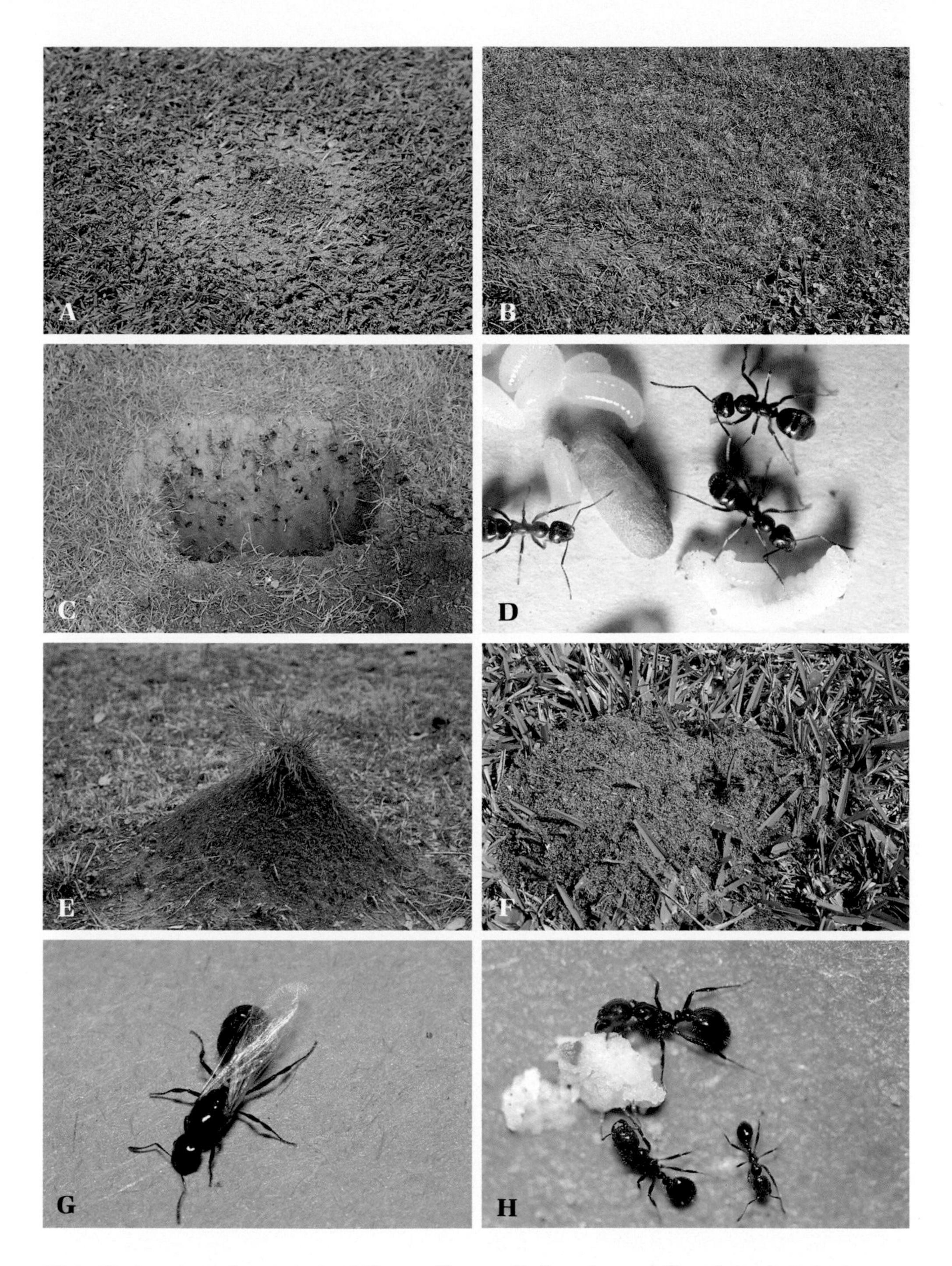

Plate 51. Ants in turfgrass. **A.** Ant hill on golf green. **B.** *Formica* sp. hill on lawn. **C.** Subterranean galleries below B. **D.** Workers, larvae, and pupa of B and C. **E.** Red imported fire ant mound around seedling pine. **F.** Fire ant mound in St. Augustinegrass turf. **G.** Fire ant winged female. **H.** Fire ant workers. (Photos A–D, F–H: NYSAES; photo E courtesy of P. P. Cobb.)

Plate 52. Cicada killer. **A.** Burrow and turf disturbance. **B.** Adult. **C.** Dog day cicada (prey). **D.** Cicada killer straddling cicada. (Photos A, D courtesy of D. J. Shetlar; photos B, C from pinned specimens in the Cornell University collection, NYSAES.)

Plate 53. Secondary turf mite and insect pests. **A.** Clover mite turf damage. **B.** Adult clover mite. **C.** Clover mite eggs. **D.** Short-tailed cricket. **E.** Grasshopper. **F.** Periodical cicada. **G.** Two-lined spittle bug adult. **H.** Nymph of two-lined spittlebug in spittle mass. (Photo A courtesy of H. D. Niemczyk; photos B, C, E, F: NYSAES; photo D courtesy of D. J. Shetlar; photos G, H courtesy of P. P. Cobb.)

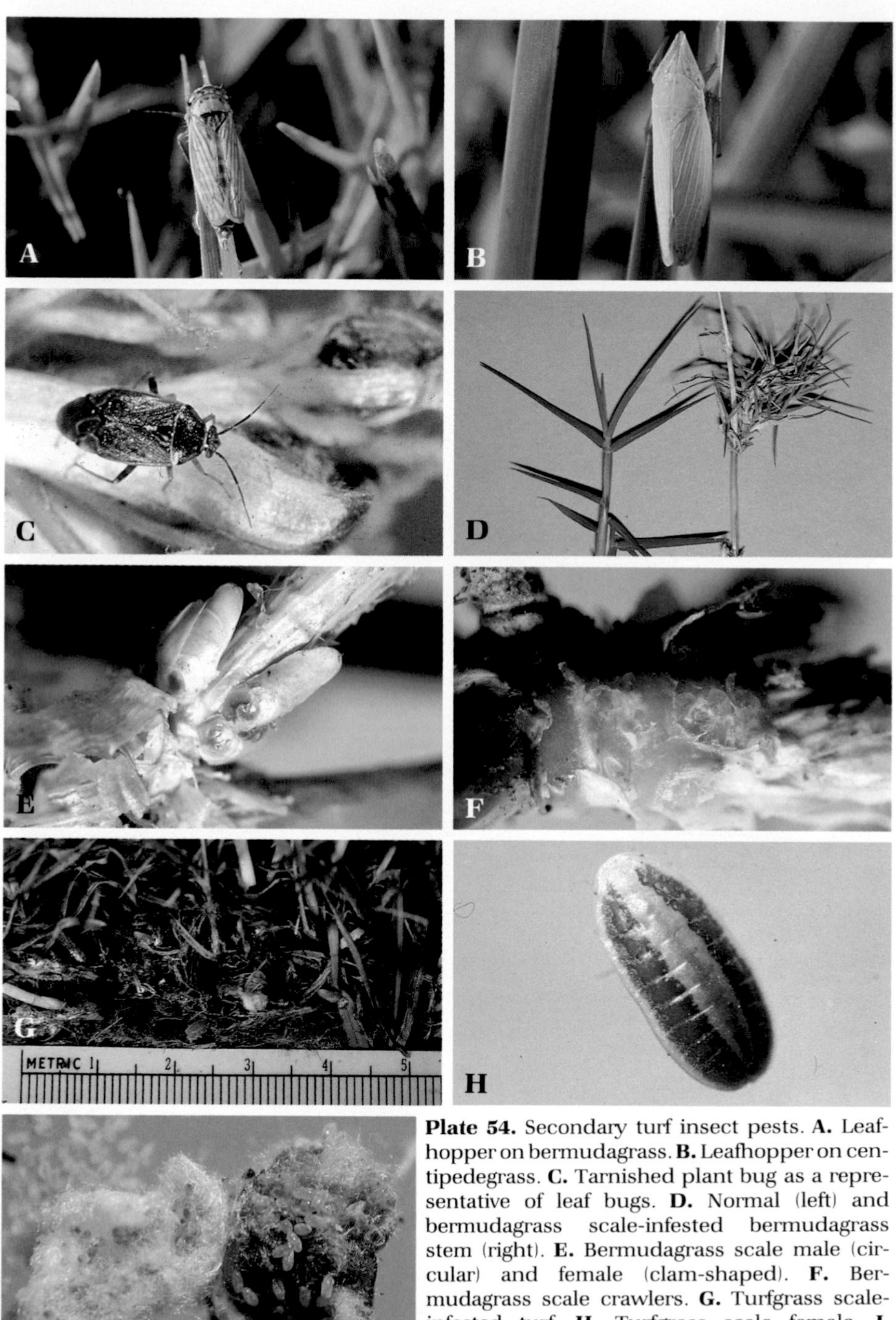

Plate 54. Secondary turf insect pests. **A.** Leafhopper on bermudagrass. **B.** Leafhopper on centipedegrass. **C.** Tarnished plant bug as a representative of leaf bugs. **D.** Normal (left) and bermudagrass scale-infested bermudagrass stem (right). **E.** Bermudagrass scale male (circular) and female (clam-shaped). **F.** Bermudagrass scale crawlers. **G.** Turfgrass scale-infested turf. **H.** Turfgrass scale female. **I.** Turfgrass scale eggs in cottony mass. (Photos A, B courtesy of D. J. Shetlar; photos C–F: NYSAES; photos G–I courtesy of M. K. Sears.)

Plate 55. Secondary turf insect pests. **A.** Cottony grass scale–injured home lawn. **B.** Cottony grass scale on stems. **C.** Cottony grass scale crawlers. **D.** Lucerne moth. **E.** Burrowing sod webworm male moth. **F.** Cigarette-paper sod webworm male moth. **G.** Cigarette-paper sod webworm larva. (Photos A, B courtesy of L. M. Vasvary; photo C courtesy of N. Hilton; photo D from pinned specimens in the Cornell University collection, NYSAES; photos E–G courtesy of D. J. Shetlar.)

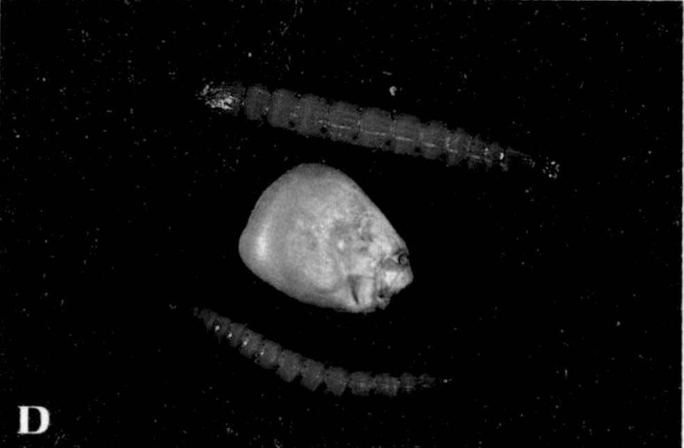

Plate 56. Secondary turf insect pests. **A.** Granulate cutworm moth. **B.** Striped grassworm moth. **C.** Corn wireworm adult. **D.** Corn wireworm larvae. **E.** *Polyphylla comes,* female (left) and male (right). **F.** Vegetable weevil adults. **G.** Vegetable weevil larva on dichondra. (Photos A, B, F from pinned specimens in the Cornell University collection, NYSAES; photos C, D courtesy of M. Villani; photo E courtesy of M. Klein; photo G courtesy of H. D. Niemczyk.)

Plate 57. Turf-associated earthworm and arthropods. **A.** Earthworm castings on golf fairway turf. **B.** Large, dried casting adding to roughness of fine turf areas. **C.** Earthworm with clitellum, showing the enlarged segments 31 or 32 to 37 that have a reproductive function. **D.** Snail. **E.** Garden slug. **F.** Sowbug. **G.** Millipede. **H.** Centipede. (Photos: NYSAES.)

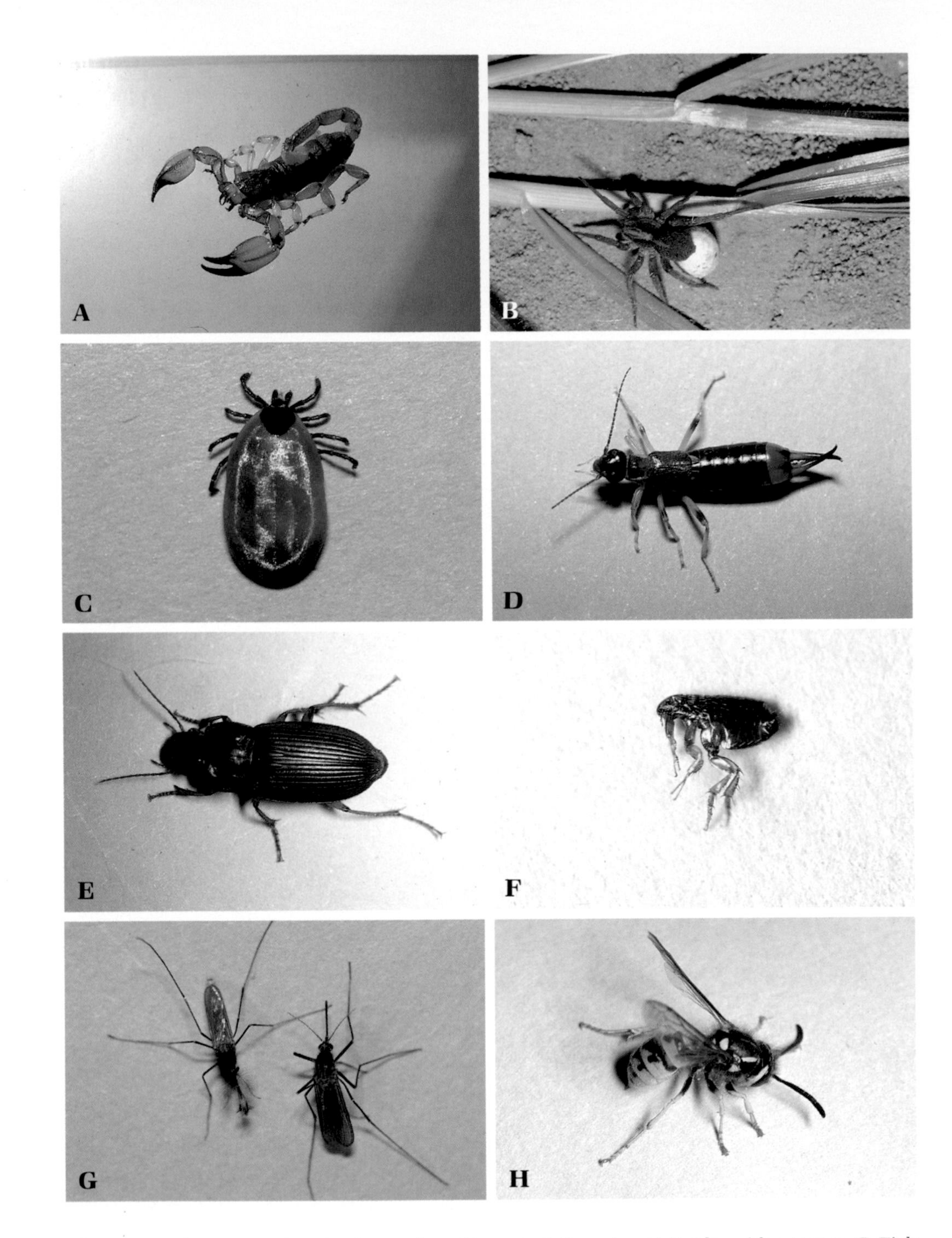

Plate 58. Turfgrass-associated arthropods and insects. **A.** Scorpion. **B.** Spider with egg case. **C.** Tick from dog. **D.** Earwig. **E.** Ground beetle. **F.** Flea from dog. **G.** Mosquito male with brushlike mouthparts (nonbiting) and female with styletlike mouthparts (biting). **H.** Eastern yellow jacket. (Photo A courtesy of L. R. Brown; photos B–H: NYSAES.)

Plate 59. Avian pests and beneficials of turfgrass. **A.** Starling peck holes. **B.** Flock of starlings attracted to grub-infested turf. **C.** European starling. **D.** Common grackle. **E.** Common grackle male (foreground) and female (background). **F.** Common crow. (Photos A, B, E: NYSAES; photos C, D, F by D. P. H. Watson, J. S. Dunning, and A. Cruikshank, respectively, obtained from the Cornell Laboratory of Ornithology.)

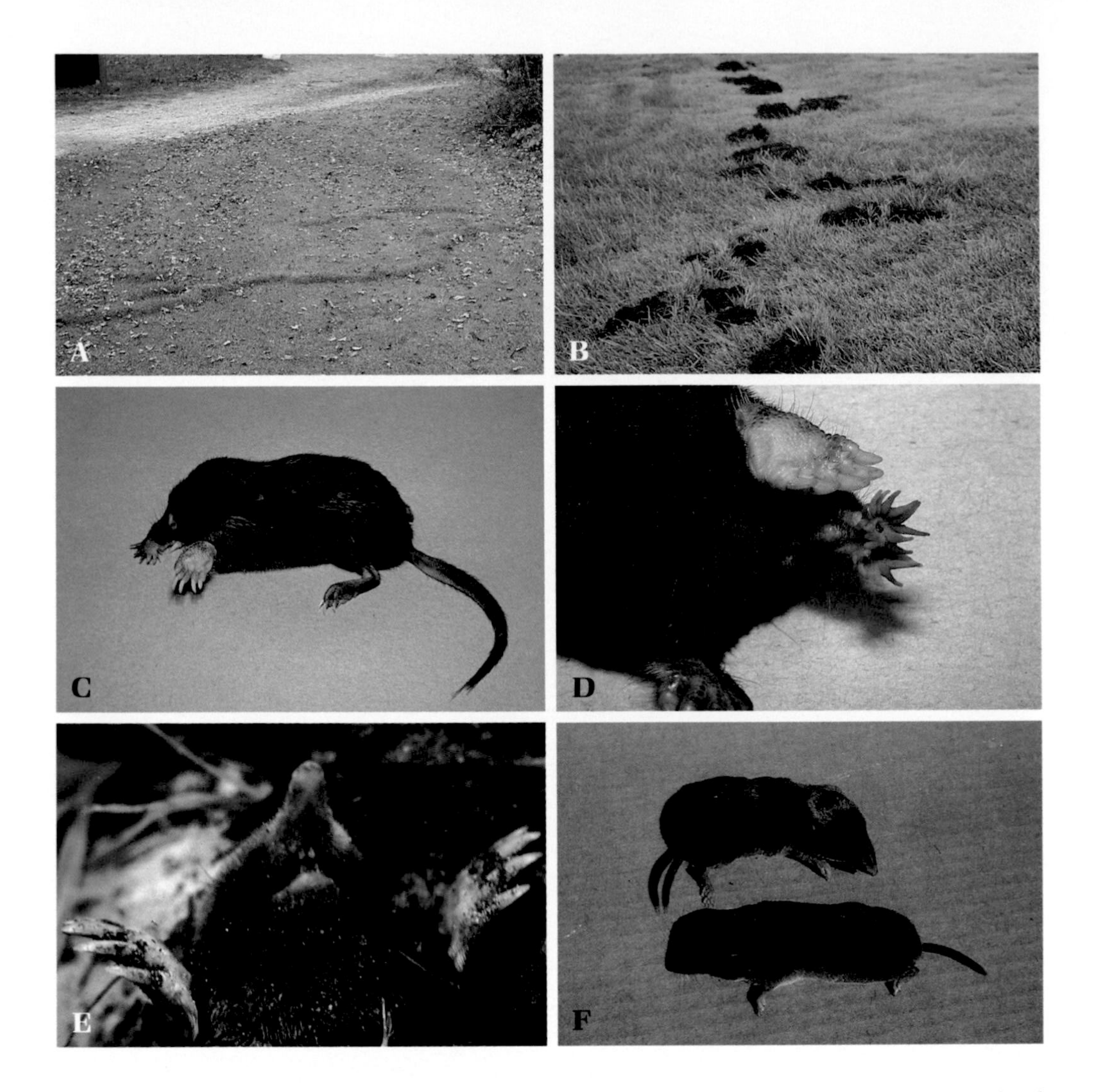

Plate 60. Moles and shrew. **A.** Characteristic ridges of eastern mole. **B.** Characteristic mounds of star-nosed mole. **C** and **D.** Star-nosed mole. **E.** Eastern mole. **F.** Short-tailed shrew. (Photo A courtesy of D. J. Shetlar; photos B–D, F: NYSAES; photo E by G. C. Hickman, obtained from the American Society of Mammalogists Mammal Slide Library, State University of New York, Oswego.)

Plate 61. Vole and chipmunk. **A.** Evidence of vole under snow on turf. **B.** Same lawn after snow melt. **C.** Meadow vole responsible for damage. **D.** Holes made by chipmunk in digging for grubs. **E.** Eastern chipmunk. (Photos A, B, D: NYSAES; photos C, E by L. L. Master and L. Elliot, respectively, obtained from the American Society of Mammalogists Mammal Slide Library.)

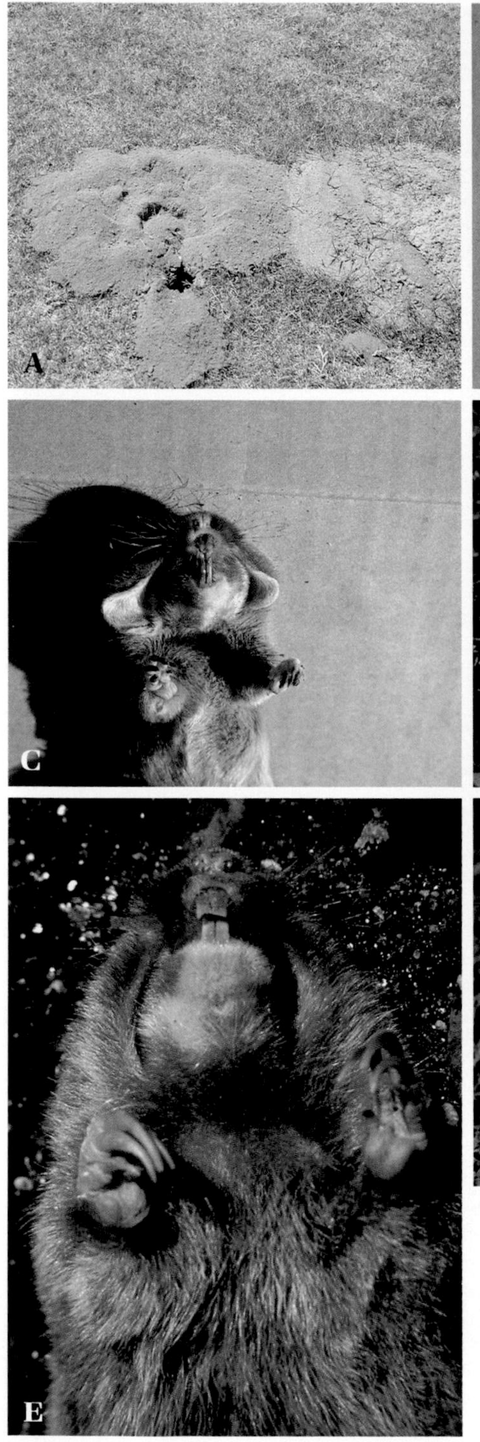

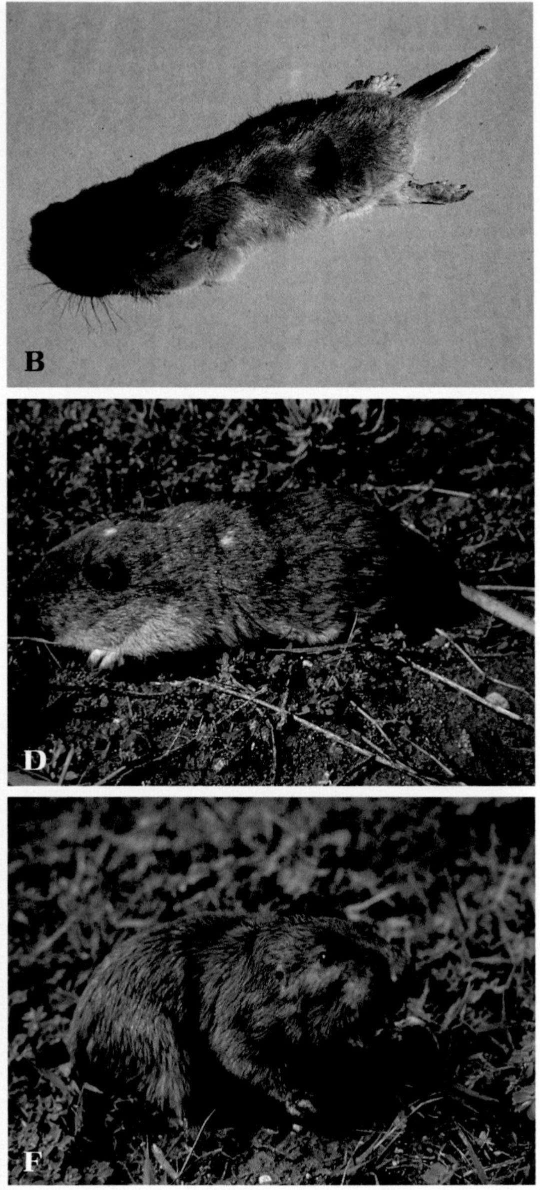

Plate 62. Pocket gophers. **A.** Old and fresh mounds made by a pocket gopher on a bermudagrass golf fairway; the tunnel opening (foreground) is atypical, but the plug of soil just to the rear, indicating the tunnel opening, is typical. **B** and **C.** Valley pocket gopher. **D** and **E.** Northern pocket gopher. **F.** Plains pocket gopher. (Photos A–C: NYSAES; photos D–F by G. C. Hickman, obtained from the American Society of Mammalogists Mammal Slide Library.)

Plate 63. Skunks, raccoon, and armadillo. **A.** Turf damage by skunks seeking scarabaeid grubs. **B.** Striped skunk. **C.** Hooded skunk. **D.** Turf damage by raccoon seeking scarabaeid grubs. **E.** Raccoon. **F.** Armadillo burrow in bermudagrass turf. **G.** Nine-banded armadillo. (Photos A, D, F: NYSAES; photos B, C, E, G by T. Yates, R. H. Barrett, W. F. Berliner, and L. L. Master, respectively, obtained from the American Society of Mammalogists Mammal Slide Library.)

Plate 64. Turfgrass thatch. **A.** Thatch of 8–10 cm (3–4 in.) thickness on home lawn. **B.** Golf fairway turf in which oriental beetle grubs remain exclusively in the heavy thatch.

15

Scarabaeid Pests: Subfamily Rutelinae

Japanese Beetle

Taxonomy

The Japanese beetle, JB, *Popillia japonica* Newman, order Coleoptera, family Scarabaeidae, subfamily Rutelinae (shiny leaf chafer), tribe Anomolini, is the single most important turfgrass-infesting scarabaeid in the United States.

Importance

The Japanese beetle, the most widespread and serious turfgrass insect pest in the Northeast when it is a grub, is also a major pest as an adult, when it feeds on nearly 300 species of plants, including fruits, vegetables, ornamentals, field and forage crops, and weeds. The grub is a major turfgrass pest of golf courses, recreational and industrial parks, school grounds, and home lawns. It has a wide geographic distribution in the Northeast and Midwest, where climatic conditions and large areas of permanent turf favor its development. The beetle's appetite for many ornamental plants greatly increases its pest status in landscape settings.

Insecticidal Resistance and Tolerance

With the advent and general use of the chlorinated cyclodiene insecticides (namely chlordane, dieldrin, aldrin, and heptachlor) in the early 1950s, the JB grubs and others that are closely related became insects that were easily controlled: a single application provided years of protection. The development of resistance to these insecticides in the late 1960s and early 1970s ended this era of easy control more than the legal banning of these compounds and initiated the use of short-term residual organophosphate insecticides (Tashiro and Neuhauser 1973).

During 1978, a New Jersey strain showed a 4.3-fold tolerance to chlorpyrifos, an organophosphate. Within the next 4 years, tolerance increased 6.8- to 11.8-fold in grubs from the same location. In addition, a population with a 42.2-fold resistance was present in Connecticut (Ahmad and Ng 1981, Ng and Ahmad 1979). Given time, it seems probable that resistance and cross-resistance to organophosphate and carbamate insecticides will develop as resistance developed to the chlorinated hydrocarbon insecticides. If it does, the control of these scarabaeid grubs will become increasingly difficult as the arsenal of effective insecticides continues to diminish.

History and Distribution

Introduction and Early Spread

The JB was first discovered in the United States in southern New Jersey during mid-August 1916. Prior to its accidental introduction, it was known to occur in Asia only on the four main islands of Japan; it is common but not abundant on the islands of Kyushu, Shikoku, and southern Honshu. Its greatest abundance occurs in northern Honshu and throughout Hokkaido, where grasslands abound. Little was known of its biology in Japan probably because it had little importance there as a pest (Fleming 1972).

Because of Federal Quarantines 40 and 48, established in 1920 against the Japanese beetle, accurate records have been kept on its spread. From its incipient infestation of 0.5 mi^2 in 1916, its spread has been rapid (Fleming 1972, 1976).

Year	Square miles infested	Additional states
1916	0.5	N.J.
1926	3,850.0	Pa., Del.
1937	13,850.0	N.Y., Md., Conn.
1946	37,500.0	Washington, D.C., Va., Mass., W.Va., R.I.
1952	76,500.0	N.H., Va., N.C., Ohio

Current Distribution and Eradication Efforts. Until 1972, the beetle occurred in every state east of the Mississippi River except Minnesota, Wisconsin, Mississippi, and Florida. By 1974 the beetle occupied much of the United States east of the 85th meridian. The capture of numbers of adults during 2 or more years in Cedar Rapids, Iowa, and Kenosha, Wisconsin, indicates the presence of an established infestation. The insect is now well established in Canada in the province of Ontario along the entire Niagara peninsula east of Hamilton and Simcoe, in Belleville, and near Gananoque. At least six other isolated infestations are present, extending from the easternmost at Farnham (near Montreal), Quebec, to the westernmost at Windsor, Ontario (Figure 61; Agriculture Canada 1983, Fleming 1976, U.S. Department of Agriculture 1972).

Of the several efforts to eradicate the JB east of the Mississippi River, none has been successful. Two infestations in California, the first in Sacramento, 1961–1964, and the second in San Diego, 1973–1975, have been successfully eradicated. At-

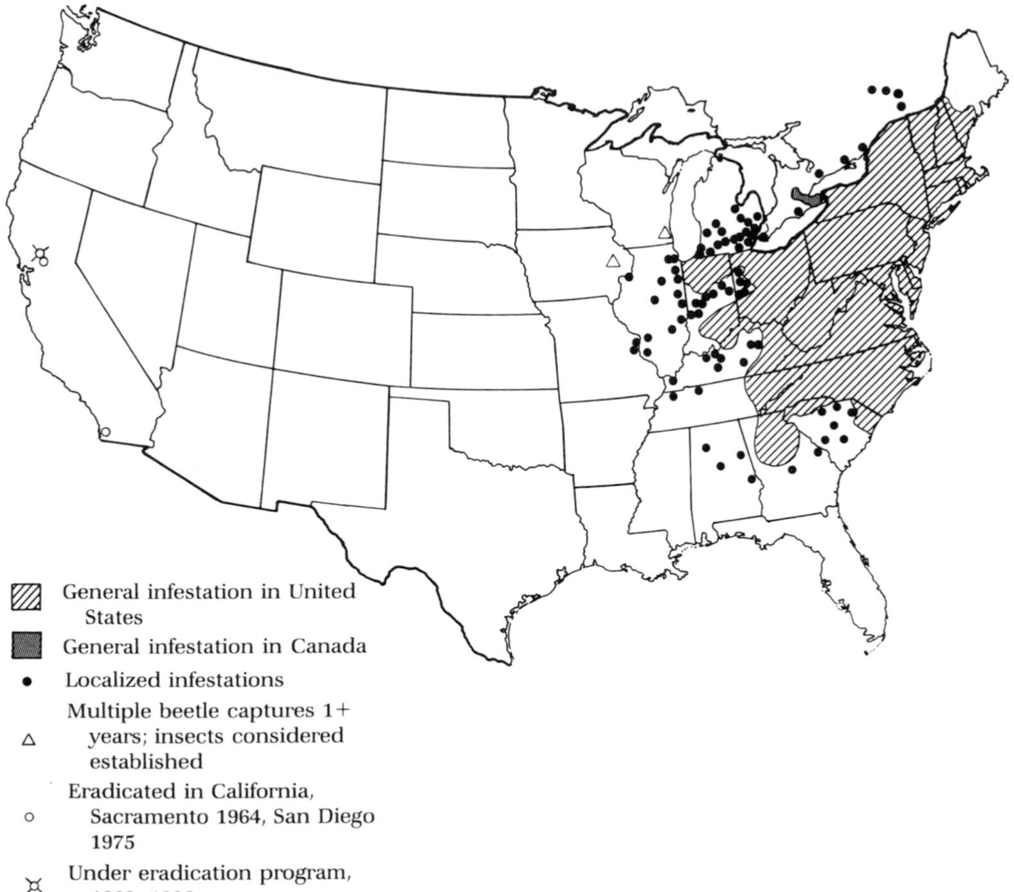

Figure 61. Known distribution of the Japanese beetle in North America through 1983. (Adapted from Agriculture Canada 1983 and U.S. Department of Agriculture 1951–1980; redrawn by R. McMillen-Sticht, NYSAES.)

tempts are currently being made to eradicate a third California infestation near Sacramento that was discovered in 1983. Environmental conditions that contribute to successful eradication exist in California. Cities surrounded by extensive areas of dry land during the period of beetle flights inhibit oviposition and the development of eggs and thus prevent spread. The lush, irrigated turf and ornamentals prevent the beetles from dispersing in search of food and oviposition sites (Fleming 1976, Gammon 1961).

Probable Ultimate North American Spread. The spread of the JB appears to be governed by temperature and precipitation. The beetle is adapted to a region where the mean soil temperature during the summer is between 17.5° and 27.5°C (between 63.5° and 81.5°F) and winter soil temperatures above −9.4°C (15°F), plus a region where precipitation is rather uniformly distributed throughout the year, averaging at least 25 cm (10 in.) during the summer. Since eggs must absorb water before and during embryonic development, soil moisture during the summer is essential for survival. The northern limits appear to be the elevated regions of the Northeast, extreme northern Michigan, and west of the Great Lakes to the Missouri River. The western limits are considered to be about the 100th meridian, where the semiarid region beyond is practically an insurmountable barrier. It has not become established in Florida and the Gulf Coast as predicted. The summer isotherm of 25°C appears to be the limit of its southern spread (Fleming 1972, Ludwig 1932).

Host Plants and Damage

Adult Feeding

Adults cause no injury to turf but are an important pest of many other plants. Of the nearly 300 species of host plants, the most favored include apples, cherries, grapes, peaches, plums, and blueberries in fruits; asparagus, beets, broccoli, rhubarb, and sweet corn in vegetables; maples, birch, crabapples, roses, sassafras, mountain ash, and linden in ornamentals; soybeans, alfalfa, clovers, and corn in field and forage crops; and smartweed, crabgrass, ragweed, and cattail in weeds (Fleming 1972).

Members of the family Rosaceae have a preponderance of the highly attractive host plants. More than 75 other families of plants have species that receive varying degrees of feeding injury. Since many ornamentals are favored host plants, the association of these ornamentals in a landscape situation makes the Japanese beetle a major pest of golf courses, parks, homes, and other well-landscaped areas (Plate 38; Fleming 1972).

Adults feed on the upper surface of the foliage of most plants, consuming soft mesophyll tissues between the veins, leaving a lacelike skeleton (Plate 38). Injured leaves eventually turn brown and die. Trees receiving extensive feeding turn brown and become partially defoliated. When fruit is attacked, the fleshy tissues are eaten. Beetles feed on the maturing silk of corn, so that kernels are malformed through lack of pollination.

A few plants are toxic to the beetles. They include the flowers and foliage of the cultivated geranium (*Pelargonium*) and the castorbean, *Ricinus communis,* and the flowers of the bottlebrush-buckeye, *Aesculus parviflora,* which cause paralysis and eventually death (Fleming 1972).

The presence of large numbers of dead beetles under food plants is sometimes mistakenly interpreted as evidence that such plants are toxic to beetles. After midsummer such an accumulation generally consists of aging beetles that lived their normal life span and died during their last feeding visit (Fleming 1972).

Larval Feeding

Grubs feed on the roots of a wide variety of plants, including most grasses, ornamental plants, and vegetables. Feeding by grubs on the underground stems and roots often passes unnoticed until the plants are badly damaged. Injury to well-kept turfgrass is usually not apparent when there are fewer than 10 grubs per 0.1 m^2 (per ft^2), but an unhealthy, poorly maintained turf may show injury, with 4 or 5 mature grubs per 0.1 m^2. Most of the severe turfgrass damage shows up in September–October, when third-stage grubs are feeding vigorously just before they go into hibernation, and again during April and May, as they feed heavily during third-instar maturation preparatory to pupation. During these two periods most grubs are feeding within the surface 5 cm (2 in.) of soil or in the soil-thatch interface. As many as 130 grubs have been found in 0.1 m^2 (122 per ft^2) of golf greens. Heavily damaged turf can easily be rolled back or lifted, since all of the fibrous roots have been eaten (Plate 23). Grubs may damage corn, bean, tomato, strawberry, and other field and garden plants by feeding on the roots. Nursery stock is often injured when grubs feed on and girdle the main root (Fleming 1972, 1976).

Description of Stages

Adult

Beetles are brilliant metallic green, with coppery brown elytra that do not entirely cover the abdomen. A row of 5 lateral brushes of white hairs on each side of the abdomen and a pair of these brushes on the dorsal surface of the last abdominal segment distinguish this beetle from all others with similar coloration (Plate 39). Beetles vary in length from 8 mm to 11 mm and in width from 5 mm to 7 mm, with the female slightly larger than the male (Fleming 1972, Hadley and Hawley 1934).

The sexes can be distinguished by small differences in the tibia and tarsus of the first pair of legs (Plate 39). The proximal four tarsal segments in the male are wider than they are long, while in the females they are as long as or longer than they are wide. The most pronounced difference is in the first tarsal segment. The apex of the female tibia is spatulate, while it is generally more pointed in the male. Differences in length of the lamellate club of the antennae, common in sexes of many scarabaeids, do not exist in the Japanese beetle (Hadley and Hawley 1934).

Egg

Young eggs are ellipsoidal and measure about 1.5 mm in diameter (Plate 39). As eggs mature, they double in size. In all other aspects they resemble the scarabaeid eggs described in Chapter 10 as they develop (Fleming 1972).

Larva

Newly hatched larvae are about 1.5 mm long and translucent creamy white. After feeding occurs, the accumulation of fecal matter in the hindgut makes them appear gray to black. In general larval appearance, shape, and C-shaped form, the larvae are no different from other common scarabaeid grubs found in turfgrass. Head capsule widths of the first, second, and third instars average 1.2 mm, 1.9 mm, and 3.1 mm, respectively. The middorsal length of the body, according to Boving (1939), averages 10.5 mm, 18.5 mm, and 32.0 mm for the three stadia, respectively. The figure for the length of third instars appears to be in error and conforms more to the findings of Hadley and Hawley (1934), who indicate that the insects are about an inch long (25.4 mm) when full grown.

The rastral pattern, ventrally located on the last abdominal segment, has characteristics that distinguish the JB from all other scarabaeid grubs in the United States. The palidia, two conspicuous rows of six or seven pali medially located to form a V, and the transverse anal opening distinguish this grub from all others (Plate 39; Fleming 1972).

Prepupa

At the prepupal stage the JB, like other turf-infesting scarabaeids, remains in the earthen cell constructed by the mature larva. It straightens out and forms only a slight crook on the posterior end and, again like other scarabaeids, is very flaccid and tender at this stage.

Pupa

The young pupa forms within the old larval and prepupal exuviae, which change in appearance to a fine, light tan meshlike tissue (Plate 39). This covering, similar to that found in the oriental beetle, splits along the middorsal line as the pupa matures. Within the cuticula, the pupa somewhat resembles the adult except that the wings, legs, and antennae are folded closely to the body. Its average dimensions are 7 mm wide by 14 mm long. The male pupa has a three-lobed eruption covering the genitalia on the ventral abdominal segment, which differentiates it from the female.

Seasonal History and Habits

Seasonal Cycle

The JB has a 1-year life cycle (Figure 62) throughout most of its range, but in the more northern latitudes or higher elevation a substantial portion of the population requires 2 years to complete a generation. These areas are southern Maine, New Hampshire, Vermont, and the Adirondack regions of New York. Even in western New York as many as about 5% of the overwintering third instars are from the previous year's generation (author's personal observations).

The normal first-beetle emergence and the peak emergence periods at increasing latitudes, represented by several states, are shown in Table 17 (Fleming 1962). In western New York, where mixed populations of the JB and the European chafer are common, the first emergence of the former occurs as adults of the latter are ending for the season.

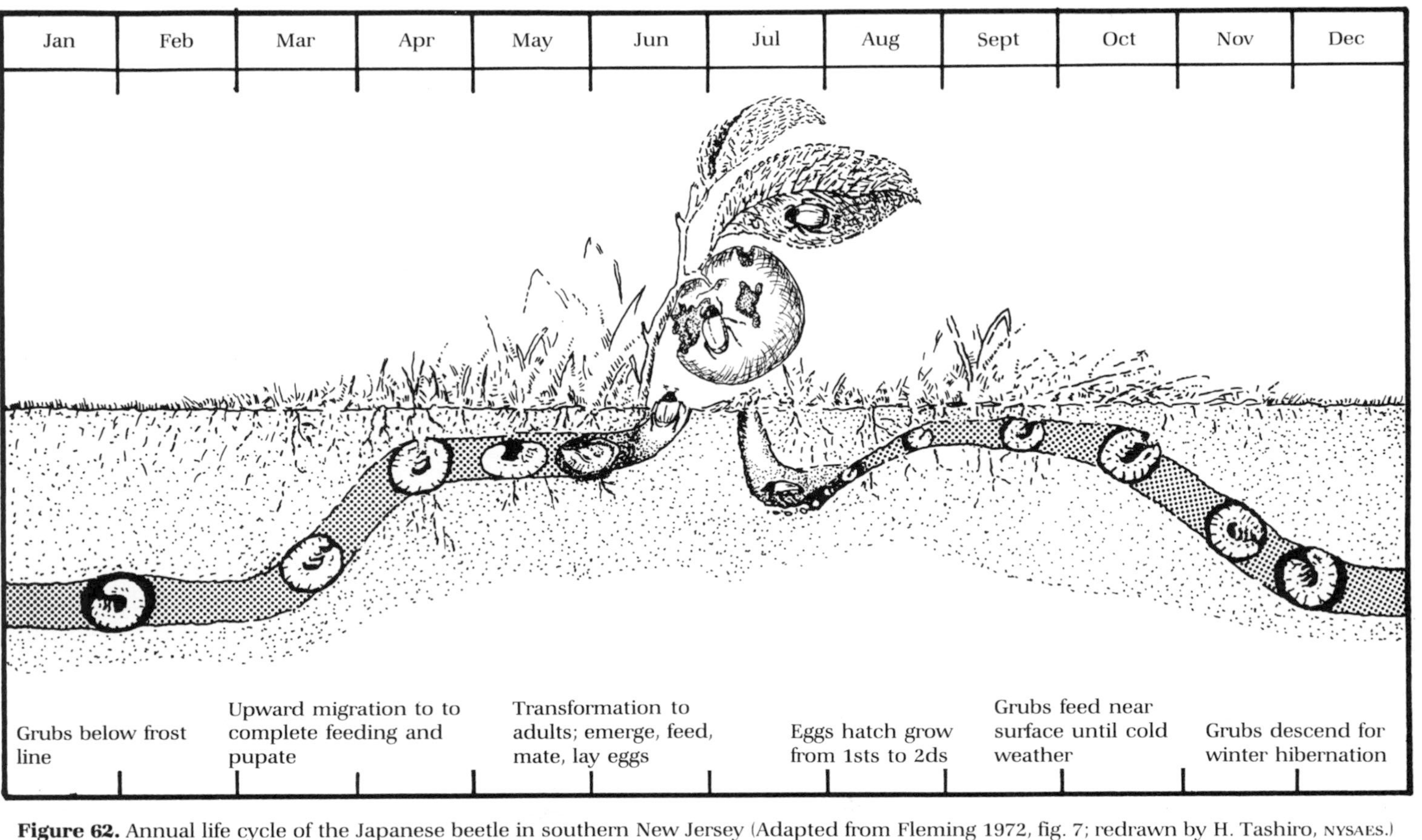

Figure 62. Annual life cycle of the Japanese beetle in southern New Jersey (Adapted from Fleming 1972, fig. 7; redrawn by H. Tashiro, NYSAES.)

Table 17. Initial emergence and maximum abundance of Japanese beetle adults relative to altitude and latitude

		Month and quarter[a]		
Location	Latitude	Initial emergence	Maximum abundance	Nearly gone
Cen. N. Carolina	35	May 3	Jun 2	Jul 2–3
Mts. N. Carolina	35	Jun 4	Aug 2–3	Sept 4
Mts. Tennessee	35–36	Jun 1	Jun 3	Sept 1
Cen. Virginia	38	May 4	Jul 2	Aug
Sacramento, Calif.	38	Jun 1	Jul 1	—
Cen. Maryland and Delaware	39	Jun 2–3	Jul 2–3	Sept 4
S. New Jersey and S.E. Pennsylvania	40	Jun 3	Jul 4	Sept
S.E. New York, Connecticut, Rhode Island, and S. Mass.	41–42	Jun 4	Jul 4	Sept
S. New Hampshire and Vermont and W. New York	43	Jul 1	Jul 4	Sept

[a]1, 2, 3, 4 denote quarters of months.
Source: Adapted from Fleming 1972.

Individual beetles live 4–6 weeks, with females depositing eggs most of their lives. Eggs hatch in about 2 weeks, become second instars in 2–3 weeks, and become third instars about 3–4 weeks later. During September and October most grubs feed vigorously and become nearly full grown.

Grubs begin to move downward during the first freezing air temperatures in the fall for hibernation. Grubs move up again during April and May to feed and reach maturity. The prepupal period lasts about 10 days in May before the insects become pupae. After 8–20 days they become adults in June in southern New Jersey and in July in Western New York to complete a 1-year life cycle.

Adult Activity

Emergence and Flight. Beetles emerge from the pupal exuviae by a dorsal split. The adult remains in the same earthen cell 2–14 days while its cuticula hardens, its wings expand to normal size, and the body becomes fully pigmented. When it is ready to leave its earthen cell, the beetle uses its mandible and fore tibia, adapted for digging, to rise to the soil surface, a process that may take a day or more (Fleming 1972).

After the initial emergence, most beetles leave their tunnels during the morning of clear days for feeding and mating. They return to the soil in the late afternoon or evening to spend the night. Beetles also remain in the soil on cold wet days (Fleming 1972).

Evidence of Sex Attractant. Virgin females emerging from the ground are very attractive to males, especially during early summer. Males fly low over turf in the early morning of clear, warm days, searching for emerging females. As a virgin female emerges, males alight a short distance on her leeward side and crawl toward her. When many males congregate around a single female, *balling* occurs, and she is known to have 25–200 males balling around her (Plate 39). Practically no actual mating takes place during the balling process, which ends about midday (Smith and Hadley 1926).

The balling phenomenon indicates the presence of a powerful volatile sex pheromone. Unmated females from balls were more attractive to males than the then best chemical lure, a 9:1 mixture of phenylethyl butyrate and eugenol (Ladd 1970).

Sex Pheromone. The pheromone isolated from virgin females was identified as (*R,Z*)-5-(1-decenyl)dihydro-2(3*H*)-furanone. The synthetic sex attractant developed is referred to as *R,Z*-furanone (Japonilure). To date the most attractive lure is phenethyl propionate (PEP) + eugenol + geraniol, 3:7:3 (PEG) exposed jointly with as little as 5 mg of *R,Z*-furanone in each trap. Bag-a-Bug is a plastic JB trap with a plastic bag to serve as a beetle holder. Other brands contain the same lures. The trap is baited with a dual-lure system of Japonilure and PEG. Even when this highly attractive trap is present, nearby favored host plants are more important in concentrating beetles (Klein 1981, Ladd et al. 1981, Tumlinson et al. 1977).

Mating and Oviposition. Copulation occurs frequently during the early morning or evening, usually on the adult food plant, where the female usually continues to feed (Plate 39). Mating occasionally occurs on the ground. There is no uniformity in the numbers or frequency of copulation, but the females usually mate between each oviposition period (Fleming 1972).

After burrowing to a depth of 5–10 cm (2–4 in.), the female deposits an egg in an earthen cell, presumably prepared in the same manner as that of the European chafer female (Chapter 10). There is apparently no secretion to cement and strengthen the walls of the cell. A female may enter the soil 16 or more times to deposit 40–60 eggs during her life.

Although most of the eggs are deposited in short-cut turfgrass and pastures, some are also deposited in cultivated fields of rye, corn, beans, tomatoes, and nursery stock, especially in adjacent turfgrass areas where the ground is dry and hard. During summers of deficient rainfall, females select poorly drained ground, irrigated areas, and fallow fields, where the soil is loose for oviposition (Fleming 1972, Smith and Hadley 1926).

Most of the eggs are deposited during the period of maximum adult abundance, which varies for different localities, depending on latitude and altitude. In the major infested areas, this occurs from early to mid-June in the more southern latitudes and lasts until late July and into August in the more northern latitudes (Fleming 1972).

Flights. During early morning, beetles usually rest quietly on plants. On clear days when the temperature reaches about 21°C (70°F) and relative humidity is below 60%, beetles fly in all directions. Flight usually ceases or is retarded above 35°C or when the relative humidity rises above 60%. Flights are also greatly diminished on cool windy days and cease completely on rainy days (Fleming 1972).

The average outward spread from a point of introduction is 16–24 km (10–15 mi) per year. Sustained flights of 8 km (5 mi) have been recorded during offshore winds from New Jersey to fishing boats in the Atlantic (Fleming 1958, Smith and Hadley 1926).

Beetles cause no harm to the turf but are a nuisance on the golf greens when they are numerous, since they must be brushed aside if they block a putt. When abundant, they can also be annoying when they fly, as they accidentally strike the head and face.

Larval Activities

Hatching and Growth. Eggs hatch in about 2 weeks. The grubs form a cell in the soil a little larger than their body and feed on fine rootlets that grow into the cell. The first molt occurs in 2–3 weeks. The second molt occurs 3–4 weeks later for the final instar.

Grubs show a positive thigmotactic response to living roots, stones, and other objects in the soil. While in its earthen cell, the grub assumes a C-shaped form and does not straighten out until it is burrowing horizontally through the soil by digging with its mandibles. The cell is enlarged as the grub grows, to permit free movement of the body. During this entire period most of the grubs are within the top 5 cm (2 in.) of soil during summer if there is adequate moisture (Table 18). As the surface soil dries, many will burrow below this level to seek moisture. By fall the third-instar

Table 18. Vertical distribution of immature Japanese beetles during a calendar year in southern New Jersey and southeastern Pennsylvania

	Percentage of total at indicated depths (cm)			
Month	0–5	5–10	10–15	15–25
Dec–Feb	5	62	31	2
Mar	11	64	23	2
Apr	63	28	9	<1
May	95	5	<1	—
Jun	76	24	<1	—
Jul	84	16	<1	—
Aug	94	6	<1	—
Sept	93	7	<1	—
Oct	63	30	7	—
Nov	10	65	24	1

Note: Data cover 11 years at eight sites.
Source: After Hawley 1944.

grub feeds and grows rapidly. With an abundance of moisture many are within the soil-thatch interface. While moisture affects their vertical movement, soil temperature affects vertical movement the most (Fleming 1972).

Winter Hibernation. As the surface soil cools to about 15°C (59°F), grubs begin to move downward and continue until the soil reaches 10°C (50°F). Thereafter the grub becomes inactive. By November in the latitude of southern New Jersey, the bulk of the population is at 5–15 cm (2–6 in.) depth. Less than 3% will move downward to a depth of 15–25 cm (6–10 in.). Grubs remain at their hibernational level all winter (Fleming 1972).

Most grubs overwinter as third instars, but in the more northern latitudes those with a 2-year life cycle overwinter as second instars their 1st winter and as third instars during their 2d winter.

Winter Mortality. With heavy snow and thick sod cover, grubs may be closer to the surface and may survive with only low mortality. With relatively barren soil they tend to move deeper. Lack of a snow cover on barren soil results in extremely high grub mortality.

Extremely high grub mortality occurred in the New York City area during the winter of 1980–1981. On December 25 and 26, 1980, there was a sudden drop in temperature, with only a trace of snow cover. Lows of −18.0° and −15.0°C (−1° and +5°F) and −20.5° and −19.0°C (−5° and −2°F) were recorded at LaGuardia and Westchester County airports, respectively. During the previous 2 weeks the maximum and minimum temperatures ranged from 0° to −12°C (32°–10°F); National Oceanic and Atmospheric Administration 1980).

During April 1981 no grubs were found in any thin turfgrass area that showed heavy turf damage and evidence of soil disturbance by skunks and other vertebrates as they dug for grubs during the previous fall. The presence of grub populations high enough to destroy turf in the fall, however, had been reported by several turfgrass managers. Only one or two grubs could be found, and they were in areas of heavy turfgrass cover, which presumably acted as an insulating barrier during the December freeze.

Spring Larval Activity. Migration upward usually commences during the latter half of March in southern New Jersey as soil temperatures rise above 10°C (50°F), and by late April and early May grubs are primarily within the surface 5 cm (2 in.) of soil. Most of the grubs feed very actively for 3–4 weeks at or very near the soil-thatch interface.

Prepupal and Pupal Activities. When feeding has been completed, the mature grub moves downward to a depth of 5–10 cm (2–4 in.) and constructs its horizontal earthen cell. Here the insect transforms from the larva to an adult in a sequence differing in only minor ways from that described and illustrated in Chapter 10 for the European chafer. Pupation occurs within the larval-prepupal exuviae, which later split to release the pupa. As with the European chafer, the callow adult has soft transparent elytra and still unfolded hind wings, while the rest of the body has turned the metallic green characteristic of the mature adult (Plate 40).

Influence of Soil pH and Liming. Work conducted in Ohio during the 1950s appeared to indicate that significantly more eggs were deposited in soil of pH 4.5 and lower than in soil of pH 7.6+. Also, larval survival was considered better in soils of 4.5–5.0 than in soils of 6.8–7.0 (Polivka 1960a, 1960b; Wessel and Polivka 1952).

More recent laboratory studies in New York indicated that there was no significant difference in larval survival in soil of pH 4.6, 5.9, 6.7, and 7.6 in either the Japanese beetle or the European chafer during the entire first-instar or second-instar periods when the larvae were held at the optimum temperatures of 25° and 30°C (77° and 86°F; Vittum and Tashiro 1980).

In Massachusetts, elemental sulfur or dolomitic limestone was added to Japanese beetle–infested turfgrass to adjust the soil pH to about 5, 6, 7, and 8. The existing and subsequent generations of grubs were counted. In a follow-up test a single application of either dolomitic limestone or hydrated lime was made to turf during April, May, July, and August at agronomically acceptable rates. In none of the cases was there any evidence that soil pH adjustments or lime applications influenced larval populations (Vittum 1984). Since most turfgrasses grow best at pH 6–7, manipulation of soil pH outside this range is not a practical method of reducing grub populations even if pH had an influence.

Natural Enemies

Like all other insects of major importance, the Japanese beetle is attacked by a large number of microorganisms, invertebrate parasites and predators, and vertebrate predators that help keep the beetle in check. The explosive populations that developed during the years immediately following the 1916 discovery no doubt reflected the arrival of a pest insect in a new environment where its more effective natural enemies were absent. The most effective parasites and predators from the Orient were introduced and have established themselves, and native organisms adapted to become enemies of the Japanese beetle, so that such explosive population growth no longer occurs.

Microorganisms

Japanese beetle grubs are infected by various microorganisms, including bacteria, rickettsiae, fungi, protozoa, and nematodes. Of these only the bacteria causing milky disease have had a significant influence in biological control.

Milky Disease. A few abnormally white grubs were found in New Jersey in 1933. The blood of these grubs teemed with bacterial spores. Two species of the bacteria that caused this condition were described and named, *Bacillus popilliae*, the dominant species, and *B. lentimorbus*, of relatively minor importance. The former has been called the type A milky disease organism and the latter the type B. Differences in the morphology of the spores, most readily apparent with about 400× magnification under phase contrast, distinguishes the two species. Spores of *B. popilliae* have a parasporal body at the end of the sporangium, while *B. lentimorbus* does not (Plate 41). The normally clear translucent blood starts to become turbid as vegetative cells multiply and eventually turns milky white as sporulation of the bacteria occurs, hence the name milky disease (Plate 41). Milky disease has been the most effective of all biological agents in controlling the Japanese beetle. Although the term *milky spore disease* is currently popular, the correct name of the disease caused by these bacteria is still *milky disease* (Dutky 1940, Fleming 1962).

The origin of these native organisms is unknown, but they are thought to be obligate parasites of native white grubs of the family Scarabaeidae. Eventually these bacteria became adapted to the Japanese beetle. They are noninfective to any insect outside this beetle family or to any other organism.

The spores normally occur in the soil and remain viable for years. When grubs ingest sufficient quantities of spores along with roots, organic matter, and soil, the spores lodge in the epithelial tissues of the midgut. Germination and multiplication occur (presumably as in the European chafer, Chapter 19) in these tissues and invade the coelomic cavity as vegetative cells and spores (Splittstoesser et al. 1973).

Unlike grubs infected with bacteria that cause septicemia and rapid death, milky-diseased grubs live for weeks, even months, and continue to feed as they gradually change in external appearance, become weaker, and eventually die. Infection is possible in all three instars, but infected grubs never transform to the next stage. As maximum sporulation occurs and grubs approach death, spore contents can vary from 2 billion to 5 billion per grub. Upon death and decay of the cadaver, a high concentration of spores is released into the soil, and any grub feeding in the area is subject to infection. Spores remain viable after passing through the digestive tracts of birds and mammals and are believed to spread from area to area primarily in this way.

Temperature range from growth and development of *B. popilliae* in a grub is about 16°–36°C (61°–97°F), and a temperature of about 21°C (70°F) is required for rapid buildup of spores in an area. This minimum temperature requirement prevents milky disease from becoming a highly effective biological control agent north of New York City and through central Pennsylvania (latitude 40°), where soil temperatures are seldom sustained at a high enough level over an extended period during the critical larval feeding period. This period encompasses the spring and early fall, when soil moisture is sufficiently high to allow grubs to feed actively near the surface. As Poughkeepsie, New York, less than 160 km (100 mi) north of New York City, soil inoculation at 2,000 lb of spore dust per acre was required to provide infectivity patterns similar to those that 10 lb per acre are expected to provide south of latitude 42°. The degree of infection that can be expected south of Long Island is shown in Table 19, where gradual declines in grub populations are accompanied by increases in incidence of disease (Adams 1946, White 1941).

For preparing spore powder for field innoculations, third-stage grubs are field collected, each grub is injected with a spore suspension using a hypodermic syringe, and grubs are incubated in moist soil with germinating seeds as food at approximately 30°C (86°F) for rapid disease development. After about 10–14 days, when grubs have reached maximum sporulation, they are ground, mixed with talc, and standardized at 100 million spores per gram of powder. Field colonization is accomplished by depositing 2 g of spore powder at intervals of 1.5 m or 3.0 m (5 or 10 ft) on turfgrass areas infested with a high population of grubs (Plate 41). Broadcast application of the spore powder over an entire area would dilute spore concentrations to levels too low to cause infections. During 1939–1951, nearly 90 tons of spore powder were applied to more than 40,500 ha (100,000 acres) at 132,000 sites in 14 states (Fleming 1968). Currently there are only two sources of milky disease spore powder, Fairfax Biological Laboratory, Inc., Clinton Corners, New York 12514, and Reuter Laboratories, Gainesville, Virginia 22065.

Evidence that Japanese beetle grubs may be developing resistance to milky disease (Dunbar and Beard 1975) in Connecticut prompted a study on Long Island during 1979–1980 and in Connecticut in 1981–1982. Of some 30+ sites of permanent turf where soil was collected and bioassayed for spore contents, better than 95% of the sites had spore contents judged to be from 1 billion to 10 billion spores per kg of

Table 19. Milky disease in Japanese beetle grubs in relation to season and population during 1939–1940

Month and year of sampling	Grub populations in numbers per 0.1 m²	Percentage of living grubs diseased
	New Jersey	
May 1939	36	<1
Jun 1939	5	18
Sept 1939	40	19
Jun 1940	5	60
	Maryland	
Aug 1939	37	4
Nov 1939	20	19
May 1940	18	2
Jun 1940	6	67
Aug 1940	14	30
Sept 1940	11	28

Source: Adapted from White 1941.

soil. No evidence of resistance to milky disease was found in any location via bioassay (Tashiro, O'Knefski, and Nelson, unpublished data).

Microorganisms of Secondary Importance. Japanese beetles have been found in the field infected with a host of various organisms. All contribute toward population reductions even though they play a minor role. Fleming (1962) listed the following. A rickettsia, *Coxiella popilliae,* produces a greenish blue discoloration of larval fat bodies and bears the name *blue disease. Metarrhizium anisopliae,* a widely distributed entomopthorous fungus, also infects Japanese beetle grubs (Plate 41); the incidence of infection is usually low. Spores of *Beauveria bassiana,* another common entomopthorous fungus, occasionally infect adults more than larvae. Japanese beetle grubs infected with *Neoaplectana glaseri* were first found in New Jersey. This nematode is a general parasite of insects. The second-stage nemas enter the grub orally and develop an infection. The nematode has successfully been colonized in New Jersey and in Maryland, but its influence as a biological control agent of Japanese beetle is considered to be very limited.

Parasites and Predators

During 1920–1933, some 49 species of parasites and predacious insects were imported from the Orient and Australia and were released into Japanese beetle–infested areas. Five species have become established and are a part of the permanent fauna. The most effective and widely distributed of the introduced insects were two larval parasites (Fleming 1962).

Larval Parasites. The spring, or Korean, tiphia, *Tiphia vernalis* Rohw., which attacks overwintering grubs in the spring, and the Japanese tiphia, *T. popilliavora* Rohw., which attack young grubs in late summer, have been the most effective. The female tiphia searches out a grub in the soil, stings it to paralyze the grub temporarily, then attaches a single egg to a particular ventral body fold. Upon hatching, the maggotlike larva attaches itself externally, pierces the cuticle with its mouthparts, and feeds on the liquid contents of the grub. When the parasite larva is fully developed, it will have consumed the entire grub cadaver except for the cuticle and the head capsule. Pupation takes place in a strong, water-tight, silken cocoon in the earthen cell formerly occupied by the grub. The parasite emerges as an adult the following year.

Under favorable conditions, *Tiphia vernalis* can parasitize more than 60% of the grubs and *T. popilliavora* somewhat less. They attack only Japanese beetle grubs (Fleming 1976). The adults feed on the flowers of the wild carrot and on the honeydew on leaves on warm sunny days. Many new colonies have been established by collecting feeding adults and releasing them in other areas with high grub populations. During the past 10–15 years, both *Tiphia* species have been very scarce, and in many areas they have not been seen. Very limited, if any, investigation has been conducted on either species in recent years.

Adult Parasite. Another introduced insect, *Hyperecteina aldrichi,* a tachinid fly, appeared from mid-June to mid-July in New Jersey and was seen feeding on honeydew and on the exudate of nectar glands. The female attaches an egg about 1 mm in diameter to the thorax of the beetle (Plate 41). The egg hatches in 24 hr, and the larva feeds internally on the vital organs as the beetle buries itself in the ground. The parasite kills the beetle in about 5 days, pupates, and overwinters (Rex 1931).

During the 1930s and 1940s, *H. aldrichi* was of only minor importance. During 1979, personnel of the Fairfax Biological Laboratory found that about 20% of the adult Japanese beetles in central Connecticut were parasitized, and they named it the *winsome fly* (not an approved common name; Fairfax Biological Laboratory, Inc., 1979).

Vertebrate Predators

Birds and mammals considerably reduce Japanese beetle grub populations just as they do those of other scarabaeid grubs but may also cause considerable turfgrass damage in the process of digging for their quarry. Starlings, grackles, robins, and other birds feed on large numbers of grubs especially during early spring and fall. When crows dig for grubs, they can cause considerable damage to the weakened turf. Mammals attracted to grub-infested turfgrass as a source of food include moles, skunks, and raccoons.

Oriental Beetle

Taxonomy

The oriental beetle, OB, *Anomala orientalis* Waterhouse, order Coleoptera, family Scarabaeidae, subfamily Rutelinae, in the United States was first named the Asiatic beetle (Friend 1929). This ambiguous term was replaced by the presently accepted common name in 1933 (Adams 1949a).

Importance

The OB has been considered a relatively minor turfgrass and ornamentals pest in comparison with the Japanese beetle. Adults are usually inconspicuous and are largely unnoticed by the general public. In localized sites in southeastern New York, New Jersey, and western Connecticut, this grub often exists in populations mixed with the Japanese beetle where the oriental beetle is the dominant species, accounting for 50% to 75% of the grubs present. Because the two are practically indistinguishable to the naked eye, many mixed populations are often mistaken as consisting solely of Japanese beetle grubs. Since the oriental beetle grubs are far less susceptible to milky disease infections, the presence of strictly normal grubs is often interpreted as meaning that Japanese beetles are no longer susceptible to milky disease (Adams 1949a). This confusion itself makes it important to distinguish between the grubs of the two species.

Currently, at least on Long Island, the OB has become a serious pest of field-grown nursery stock and potted plants maintained outdoors. Oviposition occurs in the soil around the plants. The third-instar grubs subsequently infest the entire soil mass of potted plants and reach soil depths exceeding 30 cm (1 ft) of field-grown stock (Plate 42). Certified treatments to eradicate the grubs in either situation do not exist, and infested nursery stock being shipped into noninfested areas, especially Canada, is subject to rejection.

Much of the nonreferenced information that I present below comes from studies made in New Haven, Connecticut, by Friend (1929). I have included additional references to update the information.

History and Distribution

The OB is probably a native of the Philippine Islands that was carried to Japan and was introduced from Japan to the United States. Sometime before 1908, it was introduced to the Hawaiian Island of Oahu, where it became a serious pest of sugarcane. On the mainland, adults were first collected in 1920 in a New Haven, Connecticut, nursery, having presumably been imported directly from Japan in infested balled nursery stock. Twelve years later, the OB was limited to an area within 145 km (90 mi) of New York City. It is currently present in six northeastern states and North Carolina (Figure 63). The natural spread of the OB has been slow, presumably because it is not a strong flier (Bianchi 1935, Britton 1925, Hallock 1933).

Host Plants and Damage

Adults occasionally cause a little damage by feeding on flowers of such plants as roses, hollyhock, phlox, and dahlias, but they are not considered a serious pest. Grubs, however, are a serious pest of turfgrass because they eat roots within 2.54 cm (1 in.) of the soil surface. From 43 to 65 grubs per 0.1 m^2 (40 to 60 per ft^2) are often present, causing complete destruction of turf. The presence of as many as 590 per 0.1 m^2 (550 per ft^2) has been reported. Severe injury has also occurred to strawberry beds. Severe damage to sugarcane in Hawaii ceased after the introduction of a scoliid wasp, *Scolia manilae* Ashmead, which is a highly effective parasite of the OB under Hawaiian conditions (Britton 1925, Hallock 1930).

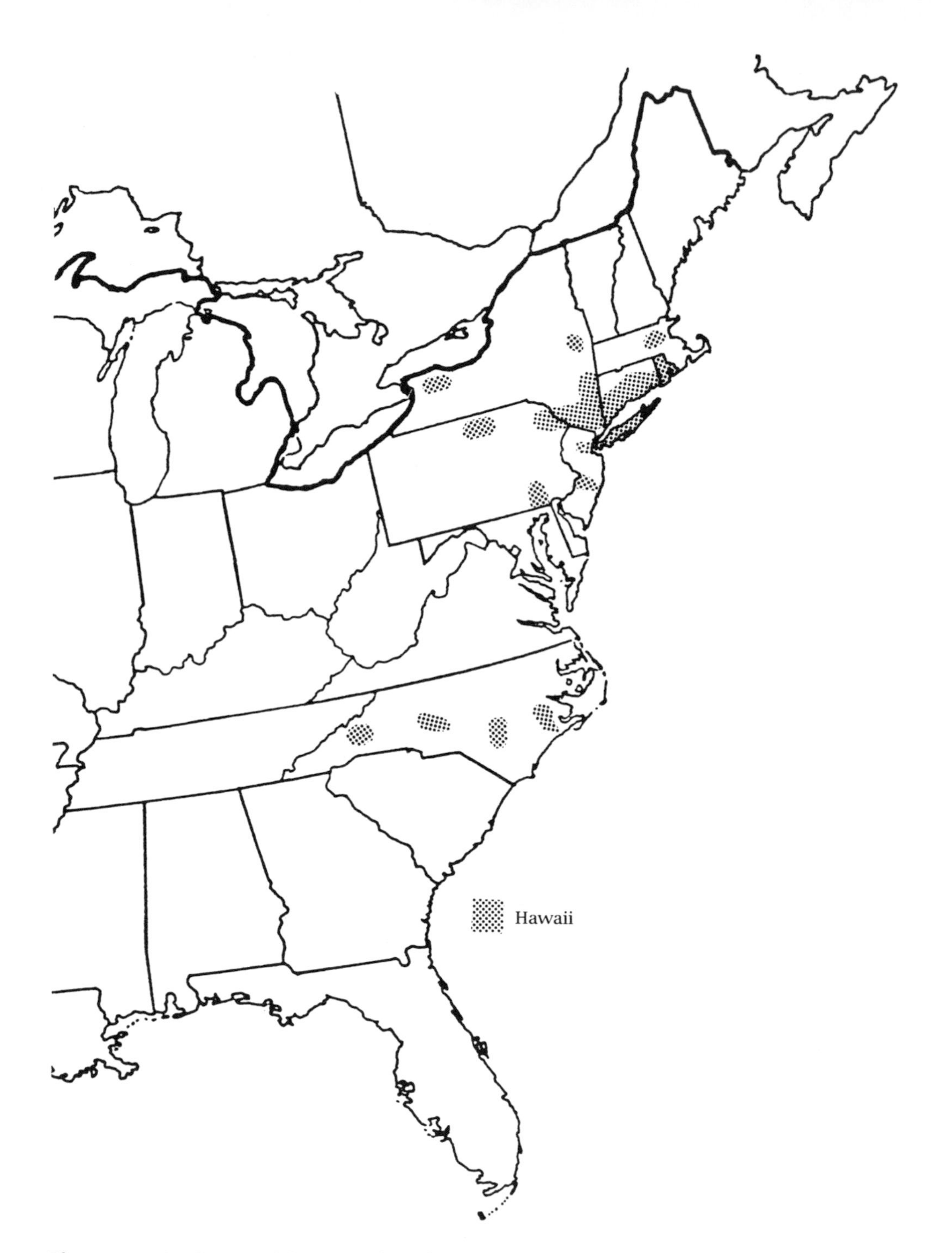

Figure 63. Distribution of the oriental beetle in the United States. (Data from U.S. Department of Agriculture 1951–1980; drawn by H. Tashiro, NYSAES.)

Description of Stages

Adult

The adult OB is broad bodied, spiny legged, and convex backed, characteristics of many scarabaeid beetles. Individuals are mostly straw colored, with black markings to a varying extent across the elytra (Plate 42). Males have a mean length of 9.0 mm; females are slightly larger, with a mean length of 10.3 mm. The sexes can be distinguished by the lamella of the female antenna, which is distinctly shorter than the rest of the antenna. In the male, the lamella is as long as the remainder of the antenna (Plate 42). In color adults range from black to straw, with intermediate stages having black elytral bands and thoracic markings (Plate 42). Eight stages having black elytral bands and thoracic markings (Plate 42). Eight variations are given (Figure 64). Patterns resembling those on the beetles labeled E, F, and G in Figure 64 are dominant, accounting for about 75% of the total.

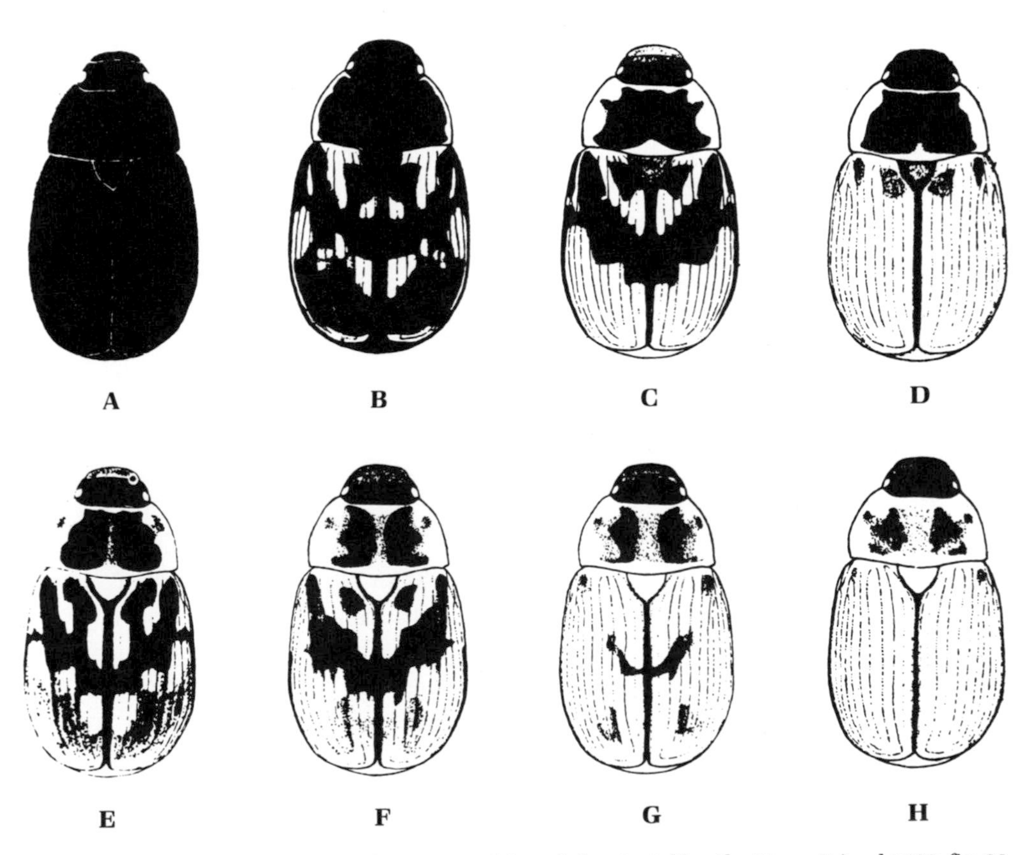

Figure 64. Variations in colors and patterns of the adult oriental beetle. (From Friend 1929, fig. 36, courtesy of the Connecticut Agricultural Experiment Station, New Haven.)

Egg

When first deposited, the egg is white, ovoid, and smooth and is about 1.2 mm × 1.5 mm long. At maturity it is more spherical and about 1.6 mm × 1.9 mm (Plate 42).

Larva

First-instar grubs range from about 4 mm in length on hatching to about 8 mm at maturity. Second-instar grubs attain a length of about 15 mm. Fully grown third-instar grubs are about 20–25 mm in length. Mean head capsule widths are 1.2 mm, 1.9 mm, and 2.9 mm for first-, second-, and third-instar grubs, respectively.

The characteristics of the rastral pattern on the ventral side of the 10th abdominal segment help distinguish the OB larva from most other turfgrass-infesting larvae. The palidia, a pair of subparallel rows of recumbent setae pointing toward the median line, are much narrower and finer than in most species (Plate 42). There are usually 11–14 pali, but the number may vary from 10 to 16 in either palidium and may differ by as many as 3 pali between palidia. This setal arrangement alone cannot be relied upon completely, since the scarabaeids *Phyllopertha horticola* and *Anomala* (*Pachystethus*) *lucicola* have similar arrangements. The anal slit in the OB is transverse (Ritcher 1966).

Prepupa and Pupa

The prepupa, as in other scarabaeids, is a quiescent stage that terminates the third-instar stage.

The young pupa develops within the old larval and prepupal exuviae, as does the Japanese beetle (Plate 42). Both belong to the subfamily Rutelinae. The exuviae split longitudinally to release the maturing pupa. The pupa upon maturity is about 10 mm long by 5 mm wide in greatest diameter. The ventral side of the abdomen differs in the two sexes. Posterior to the ninth segment of the male and ventrally, there are two lobes that are absent in the female.

Seasonal History and Habits

Seasonal Cycle

The OB has received particular attention as it occurs in turf areas in New Haven, Connecticut. Grubs are normally abundant only in well-kept lawns. Normally the OB overwinters as third-instar grubs, migrates to near the soil surface in April, and feeds for about 2 months before pupating in June. Adults are present during late June into August, with most of the eggs laid during July and early August. First instars are dominant in August, second instars in September, and third instars in October. Downward migration for hibernation begins after the middle of October. A generation a year is usual (Figure 65), but as many as 15% may fail to complete their development the 1st year and may overwinter as second instars. During the second season, these grubs become fully developed third instars. They move to depths of 20–36 cm (8–14 in.) in the soil for the second winter, where they remain until June of

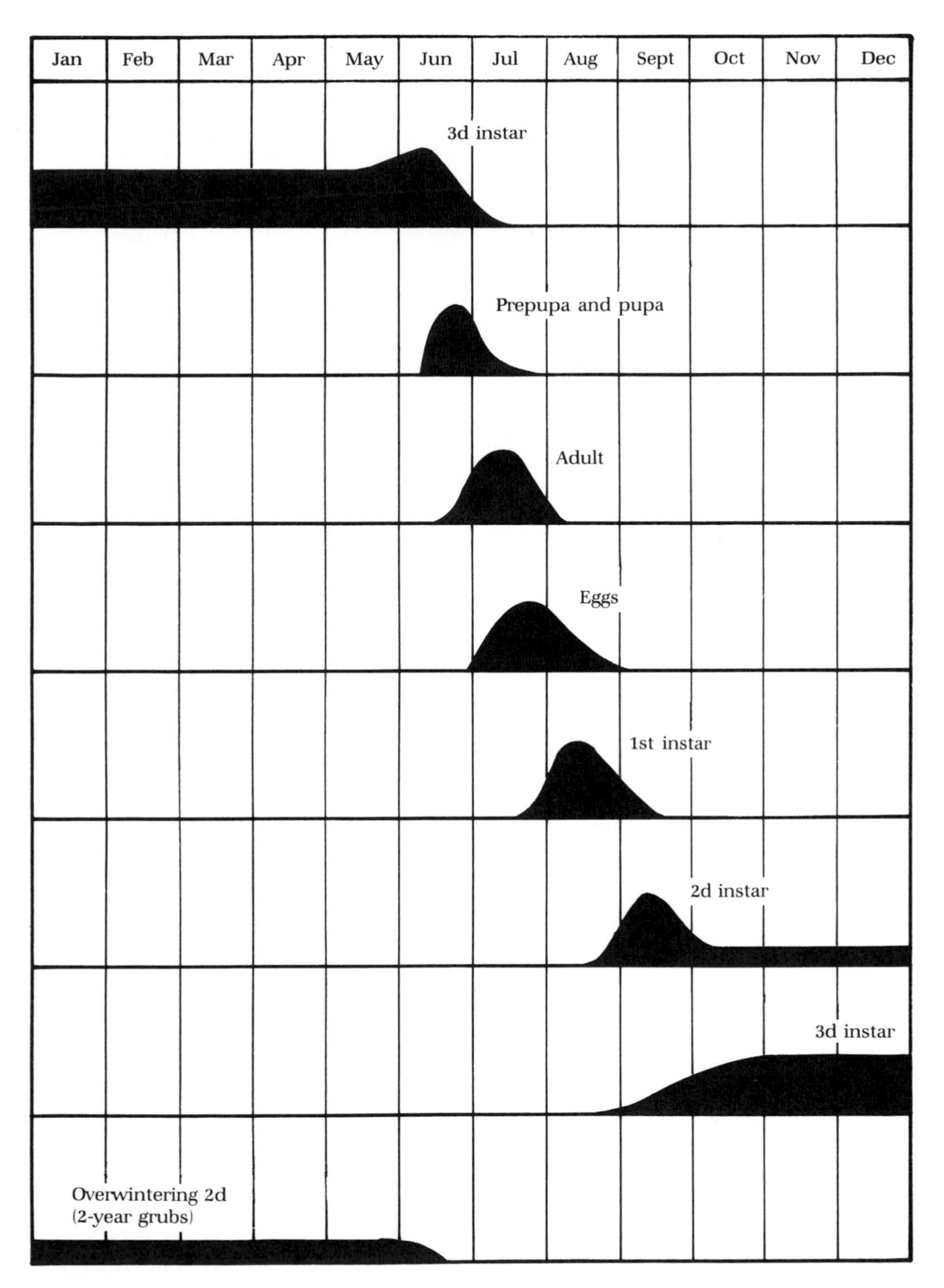

Figure 65. Life cycle of the oriental beetle in Connecticut. (Adapted from Hallock 1935, fig. 1; redrawn by R. McMillen-Sticht, NYSAES.)

the following year before becoming pupae and synchronizing their life cycles with the normal 1-year life cycles (Hallock 1935).

The days required for each stage at New Haven during 1926 and 1927 for individuals reared in the insectary under conditions as close as possible to normal were: adults (to oviposition) 4, eggs 25–28, larvae 311–317, prepupae 5–7, and pupae 14–15.

Adult Activity

Adult Emergence and Flight. Adults remain in their earthen pupal cell for about a day, until the chitin hardens. Emergence occurs from late June into August. Beetles are active during the day, flying only in bright sunlight on very warm days. Most flights take place between 8:00 AM and 4:00 PM. They rarely have more than 46 m (50 yd) of sustained flight. Beetles are also found at night, crawling on the ground under lights. Adults eat very little, mostly feeding in flowers, but are not considered injurious to flowers. Feeding activities may be primarily in search of moisture (Hallock 1930, 1933).

Mating and Oviposition. A few beetles have been found mating in the field. In captivity they mate very readily (Plate 42). The interval between mating and oviposition may be as short as 1 day but is normally about 5 days. Oviposition occurs in 4–20 days, with the median being about 7 days; eggs are deposited both day and night. Eggs are deposited singly at depths of 2.5–28.0 cm (1–11 in.). Adults will oviposit in ground covered with sod or in barren soil. About 25 eggs per female are considered the average, but as many as 63 eggs have been deposited by a single individual. The egg stage lasts 17–25 days in the field. As with other scarabaeids, moisture seems essential for development of the embryo. Just before hatching, the brown tips of the mandibles are visible (Hallock and Hawley 1936).

Larval Activity

As soon as hatching occurs, the young larvae move upward to within 2.5 cm (1 in.) of the soil surface to feed on grass roots and organic matter. Larvae will live and grow on organic matter in the absence of live roots, but growth is retarded and ultimate size is reduced.

Duration of the first instar is less than 30 days at 23°–25°C (73°–77°F) in living sod. Duration of the second instar is about the same. By the 1st week in September, the second instars have become common, and turf damage is apparent. By the last week in September, most of the larvae are third instars. About the middle of October, larvae begin to move deeper for hibernation.

Winter and Spring Larval Activities. Downward movement of grubs generally starts in October, when the mean soil temperature at the 8 cm (3 in.) depth falls to about 10°C (50°F). They spend the winter at depths of 20–43 cm (8–17 in.), penetrating more deeply in lighter, sandier soils. Winter mortality in the field is negligible;

there are as many larvae in April as in the previous November and December (Hallock 1935).

In spring, as the soil temperature at the 8-cm level (3 in.) reaches a mean of about 6°C (43°F) during late March or early April in New Haven, grubs start to move upward. As the soil temperature at 8 cm (3 in.) reaches 10°C (50°F), usually about mid-April, grubs are at their feeding position just below the surface. They remain at this level until the first part of June, when they migrate downward to pupate. A few larvae feed not on living grass roots in the spring but instead on organic matter, and they pupate. Optimum conditions for larval development appear to be an open lawn exposed to the sun and a good cover of grass on a soil of fairly rich, sandy loam. Horizontal movement of grubs has been recorded as about 1.2 m (4 ft) (Hallock 1935).

Pupation

About the first of June, larvae dig to a depth of 8–23 cm (3–9 in.) for pupation. Some may go as deep as 30 cm (12 in.).

The prepupal period lasts about 6–13 days, with an average of 8 days. Pupation occurs during late June and early July in most cases and lasts about 2 weeks.

Natural Enemies

Microorganisms

Milky Disease. OB grubs show only limited susceptibility to milky disease in the field. In areas where spore dust had been applied several years previously, milky disease was found in only 5% of the grubs during the fall. This very limited incidence in the OB has led turf management personnel to believe that milky disease is no longer effective on the Japanese beetle. Grubs of the two species often occur in mixed populations, with the OB being the dominant species. Often such a population is mistakenly identified as consisting only of Japanese beetle grubs (Adams 1949a).

In a more recent study, both Japanese beetle and OB grubs showed a remarkably low incidence of milky disease in Connecticut, and it was believed that resistance to milky disease was developing (Dunbar and Beard 1975).

Parasites

In Hawaii, *Scolia manilae* Ashmead has been successfully introduced and has been so effective in parasitizing OB grubs that the latter are no longer a serious problem there. This parasite was introduced near Philadelphia as a potential parasite of the Japanese beetle but did not survive the winters (Britton 1925).

Vertebrate Predators

All of the vertebrates that feed on other scarabaeid grubs also feed on many OB grubs. The birds include starlings, grackles, and crows. The mammals include moles, skunks, and raccoons.

16

Coleopteran Pests: Family Chrysomelidae

Dichondra Flea Beetle

Taxonomy

The dichondra flea beetle, DFB, *Chaetocnema repens* McCrea, family Chrysomelidae, subfamily Halticinae, is a member of a genus of minute flea beetles generally <2.5 mm and most 1.5–2.0 mm in length. A closely related species, *C. confinis* Crotch, feeds on bindweed and sweet potato plants in the morning-glory family, Convolvulaceae. Dichondra is also a member of this plant family.

Practically all the information we have on the DFB comes from McCrea (1972), supplemented by Bowen (1980) and V. A. Gibeault, Cooperative Extension, University of California, Riverside (personal communication, 1985). The account below draws mainly on information from McCrea (1972) and on the author's personal observations unless otherwise indicated.

Importance

The DFB is a very destructive beetle on lawns of dichondra, *Dichondra* spp., in California, often destroying many of them. Beetles are so small that they are difficult to detect readily. Once an infestation has become established, destruction of the dichondra lawn is very rapid, and many homeowners are unable to determine the cause until irreversible damage has taken place.

History and Distribution

Depradation of dichondra lawns in southern California by *Chaetocnema* spp. became serious during the mid 1960s and was erroneously identified as having been caused by *C. confinis* or *C. magnipunctata* Gentner. Since 1972, *C. repens* has been identified as a new species and has correctly been determined to be the cause of dichondra devastation. This beetle is the most serious pest of dichondra throughout the southern California coastal areas from San Diego into Santa Barbara County; into

Antelope Valley in the western Mojave Desert; in the interior southern California counties of Imperial, Riverside, and San Bernardino; and at least as far north as Fresno in central California. As far as is known, the flea beetle is associated with dichondra throughout its range into the San Jose and San Francisco Bay area and near Sacramento (Figure 66; Gibeault, personal communication, 1985; McCrea 1972; author's personal observations).

Host Plants and Damage

Only the broad-leaved plant dichondra is affected by the DFB. It does not feed on any of the grasses (Bowen 1980).

Damage appears first as crescent-shaped mandibular cuts made by adults on the upper surface of the leaves (Plate 43). Soon thereafter the leaves turn brown and die

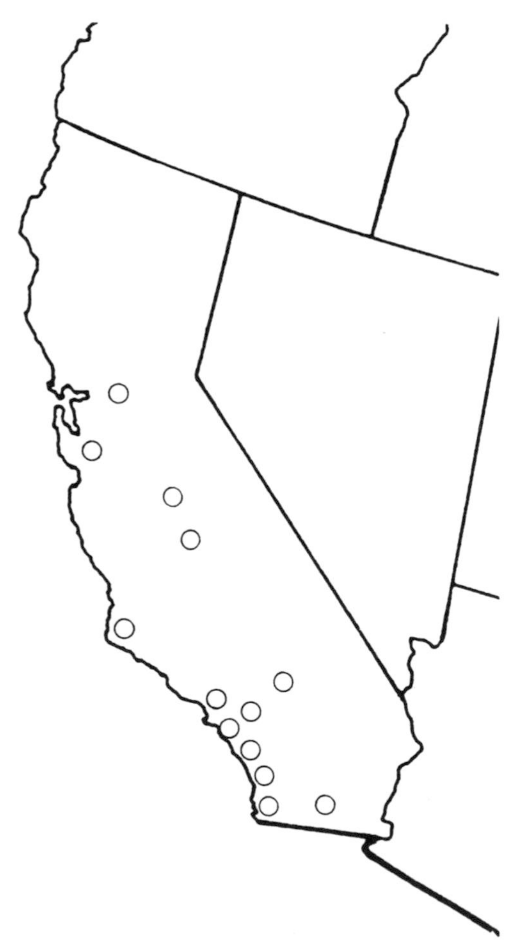

Figure 66. Distribution of dichondra lawns and the associated dichondra flea beetle in California. (Data from McCrea 1972 and author's personal observations; drawn by H. Tashiro, NYSAES.)

(Plate 43); dying strips of dichondra appear adjacent to sidewalks. Damage is thought to start near sidewalks for any of three reasons. Hitchhiking beetles may start an infestation. Temperatures are warmer near the walks and permit earlier seasonal activity. Increased moisture due to runoff from paved areas may create a more favorable habitat for the beetle.

Injury is often assumed to be caused by lack of water or by a fertilizer burn (Bowen 1980). In southern California, within 3 weeks following the first symptoms, as much as a third of a lawn of average size can be totally destroyed, and within 4–5 weeks the entire dichondra lawn is often destroyed.

If much damage by adults is evident, larvae are usually present, feeding on the roots and crown. Root damage appears to start at the tip of the small roots. Larvae work their way to the crown, leaving skeletonized roots with much frass. Sometimes the larvae burrow up roots and completely disappear from sight. McCrea (1972) is convinced that dichondra turf would survive the mandibular cuts on the leaves by adults if the larvae did not destroy the root system.

Description of Stages

Adult

Adults are black to reddish black and slightly bronzed. Their antennae and legs are reddish yellow, to reddish brown. The underside is reddish brown to reddish black and shiny. One of the more conspicuous features is the femur of the hindleg, which is enormously broad even for a flea beetle (Plate 43). Beetles are broadly oval, robust, and 1.5–1.8 mm long.

DFB adults have distinct punctures over the compound eyes and on the prothorax and can thereby be distinguished from the very similar *C. confinis.*

Egg

The minute egg is about 250 μm wide by about 480 μm long and is deposited on leaves and stems.

Larva and Pupa

The DFB has four larval instars. They grow from first instars with head capsule widths of 115–125 μm to fourth instars with head capsule widths of 237–265 μm. Larvae appear unusually long relative to their diameter (Plate 43). Fourth instars are about 6 mm long. Pupae are first whitish and about 1.4 mm long.

Life History and Habits

Life Cycle

Adults overwinter in the soil around the crown and in leaf trash near dichondra lawns. A warm spell in January may cause adults to surface and feed. Adults generally become active in May and continue through October in warm areas. A generation is completed in about a month.

Eggs hatch in three days. The four larval instars require 22–25 days, and the pupal period lasts about 5 days. Teneral adults require a day or two for maturation. A total of 32–35 days are thus required for a complete generation. Breeding is continuous through the warmer months. The last generation of pupae become adults to overwinter in November.

Adult Activity

If adults are present, moving the open palm lightly over the surface of the lawn will cause the beetles to jump and alight on the back of the hand. Beetles can also be seen on the upper surface of the leaves, with typical feeding damage (Bowen 1980). Adults can be captured by sweeping the lawn with an insect net.

Mating pairs are often observed on the upper surface of a leaf (author's personal observation). Copulation lasts about 30 min. Eggs are laid on leaves and stems and on moist blotters in a laboratory rearing cage.

Larval Activity

Larvae live in the soil and under heavy infestations completely destroy the dichondra stand. The largest numbers are found 2.5–5.0 cm (1–2 in.) deep, near small rootlets of dichondra. Some are found from 7.6–10.0 cm (3–4 in.) deep in the soil.

Young larvae begin feeding on roots soon after hatching, often burrowing completely inside the root. Upon completion of larval life, pupation occurs in the soil. Larval feeding on the roots is the most destructive activity of the DFB.

Miscellaneous Features

Larvae can be raised on damp blotters. Rearing through an entire life cycle was most readily accomplished by infesting flats of dichondra. Larvae from sod and soil samples have readily been retrieved by placing the samples in Berlese funnels.

Natural Enemies

None has been reported.

17

Coleopteran Pests: Family Curculionidae

Bluegrass Billbug

Taxonomy

The bluegrass billbug, BGB, *Sphenophorus parvulus* Gyllenhal, belongs to the order Coleoptera, family Curculionidae, subfamily Rhynchophorinae. Early literature placed the BGB and other closely related species under the genus *Calendra* (Vaurie 1951).

Importance

The BGB is an important pest of home lawns in many scattered locations. In Ohio, BGB injury to home lawns would rank second in importance to the hairy chinch bug. About 95 years ago the BGB was first reported damaging turfgrass in Nebraska. In recent years it has become so severe a threat that the breeding of turfgrasses for resistance to this insect was deemed fully warranted (Bruner 1890, Kindler and Kinbacher 1975, Mahr and Kachadoorian 1983, Niemczyk 1983).

Evidence in Ohio that the BGB has developed resistance to the chlorinated cyclodiene insecticides may be partially responsible for the upsurge of this insect in turfgrass. Prior to the development of resistance, these insecticides, used against other turf insects such as the scarabaeid grubs, may have held the BGB in check (Niemczyk and Frost 1978).

In Rochester, New York, and nearby areas, the BGB suddenly became a serious home lawn problem in 1967 at nearly the same time that European chafers manifested resistance to chlorinated cyclodienes. The eastern problem with the BGB appears to parallel the Ohio situation (author's personal observations).

History and Distribution

The BGB, a pest of cool-season grasses, is generally troublesome across the northern half of the nation in localized areas. It can be found as a pest wherever Kentucky bluegrass is grown. Serious outbreaks occurred in Salt Lake County, Utah, during the early to mid-1960s. During the late 1960s outbreaks were prevalent in Washington, Nebraska, Kansas, Wisconsin, South Carolina, Pennsylvania, New York, and Massachusetts (Niemczyk 1983, Tashiro and Personius 1970).

According to Satterthwait (1932), the BGB has a much wider distribution, occurring from Canada to Florida and Texas and extending from the Atlantic coast to South Dakota and probably throughout the United States.

Host Plants and Damage

Grasses Damaged

As its name implies, the primary host plant of the BGB is Kentucky bluegrass, but it also attacks perennial ryegrass and fescue. In Nebraska the BGB was first reported in the late 1800s as an important pest of timothy (Mahr and Kachadoorian 1983, Roselle 1975).

Nature of Damage

Adults feed on the grass stems by inserting their beaks into the center of the stem (Plate 44). Adult feeding damage is not considered to be of major concern.

Eggs are inserted singly between leaf sheaths just above the crowns, into the feeding punctures. As they hatch, young larvae feed within the stems and hollow them out. The larger larvae feed in the crown at the base of the stems, so that the stems can easily be pulled out by hand (Plate 44). After larvae have consumed most of the crown, the plants break off easily at that point. As the later-instar larvae approach maturity, they migrate below the stems and also feed on roots. A light tan, sawdustlike excrement accumulates in the area of heavy feeding (Plate 44). Turf damaged by the BGB appears wilted but does not respond to watering. Feeding by grubs and mole crickets makes the soil loose and soft, but turf killed by the BGB has a firm soil (Fushtey and Sears 1981, Mahr and Kachadoorian 1983, Niemczyk 1983).

Period of Damage

July to early August is the most critical period for damage. The turf is under moisture stress, and the rapidly growing larvae of the BGB are engaged in peak feeding. Damage occurs as spotty brown patches, first appearing along sidewalks and driveways or near trees and considered to be associated with the insect's hibernation quarters (Plate 44). Many infestations show brown patches scattered over the entire lawn that later coalesce to produce total turf destruction to nearly the entire lawn (Plate 44; Niemczyk and Frost 1978, author's personal observations).

Description of Stages

Adult

Adults of the BGB are gray, black, or brown weevils (Plate 45). Their color may be modified by the dried mud that often adheres to their bodies. The BGB has a heavily punctated thorax and coarsely striated elytra. The antennae are attached near the base of the snout in all *Sphenophorus* spp., while in *Hyperodes* spp., the only other major genus infesting turfgrass, the antennae are attached near the apex of the snout.

The BGB is 7–8 mm long, exclusive of the length of the snout. The tarsi of the legs and most of the antennae are generally reddish brown. This same coloration often exists at the anterior edge of the pronotum, the lateral edges of the elytra, and coxal area, and the apex of the snout, and gradually fades into adjacent dark gray to black areas. Teneral adults (Plate 45) are reddish brown and are often seen with mature adults in the soil-thatch interface or in the soil. Males and females can generally be distinguished by the tip of the abdomen, which is more rounded in the male and more pointed in the female (Plate 45). This difference is more apparent from the ventral view, as shown for the hunting billbug (Plate 45).

Egg

The eggs of the BGB are oblong and clear to creamy white. They are found in the stem and between leaf sheaths (Plate 45). Eggs are 0.6 mm wide by 1.5 mm long (Tashiro and Personius 1970).

Larva

The BGB larvae are white and legless, with brown head capsules (Plate 45). All instars resemble each other except in size. Mature larvae of the BGB are about 8 mm long (Niemczyk and Frost 1978).

Pupa

The pupae of the BGB are creamy white, with the bodies resembling those of adults in many ways (Plate 45).

Seasonal History and Habits

Seasonal Cycle

The BGB has a 1-year life cycle. Individuals overwinter as young adults produced during the summer and early fall.

During the spring, adults leave their hibernation quarters and are most frequently seen crawling over paved sidewalks and driveways on their way to turf areas. After a brief period of feeding, females deposit eggs in the stems. After hatching, young larvae feed in stems, but as they grow they start feeding in the crowns. By midsummer, larvae will have molted several times, and the ultimate-instar larvae produce the most damage as they continue to feed in the crowns and roots. Larval feeding is completed by mid- to late summer, with pupation taking place in the soil. Young adults of the new generation are present during early to midfall, seeking hibernation quarters.

Adult Activities

During September and October, adults of the BGB are often observed climbing screens, shingles, and masonry walls to a height of 1.2–2.0 m (4–6 ft) or attempting to fly in a light breeze. In a few days with cooler temperatures, the adults seek hibernating quarters in weeds and leaf litter, in hedgerows, in surface litter around the foundation of buildings, and in other protected areas. A favorite site for overwintering BGB adults is the junction of turf and sidewalk (Niemczyk 1983).

During the first warm sunny periods of spring, BGB adults leave hibernating quarters and are seen crawling over paved surfaces on their way to turf areas on which they can feed. In New York this phase occurs starting in mid-May, with peak adult movement observed during June.

Adults feed on grass stems just above the crown by inserting their mouth parts into the stems. Shortly after feeding begins, eggs are inserted between leaf sheaths 1–2 mm above the crown. Oviposition occurs during May and June in the latitude of Wisconsin and New York (Mahr and Kachadoorian 1983).

Larval Activity

Hatching, Growth, and Maturity. Eggs hatch within the stems in about 2 weeks, and young larvae feed there a short time before burrowing downward to feed in and on the crowns. The entire crowns may be consumed by the larger larvae. Before maturity, larvae move to the root zone and feed on roots and rhizomes (Niemczyk 1983).

Wherever the BGB larvae have been actively feeding, they leave an accumulation of fine, whitish, sawdustlike frass. The presence of this excrement is positive evidence of BGB damage. Mature larvae penetrate to a soil depth of 8–10 cm (3–4 in.) to transform to the pupal stage (Niemczyk 1983).

Miscellaneous Features

Sampling Techniques

The presence of BGB adults is most simply determined when adults are observed crawling over paved areas adjacent to turf areas. When two or more adults per minute are counted during the spring, damage to adjacent turfgrass can be expected. Turf treatment during April through mid-May in Ohio to kill adults before major oviposition occurs was found to prevent severe turf damage by larvae during late June and July (Figure 67; Fushtey and Sears 1981, Niemczyk 1983).

It is almost impossible to see adults in turfgrass, but a drench that will cause adults to surface has been developed. Turf is mowed to a height of 1.3 cm (0.5 in.). A pyrethrin drench of 2.5 ml of a 1.2% pyrethrin and 9.6% piperonyl butoxide in 3.8 l (1 gal) of water is applied to 9.3 dm^2 (1 ft^2) area of turf. On turf with 0.45–1.15 cm of thatch, adults surfaced for about 20 min. During a visual search, 1.0–1.9 adults could be found, but 5.5–7.3 adults were observed using the drench. Larvae did not respond to the drench (Kindler and Kinbacher 1982).

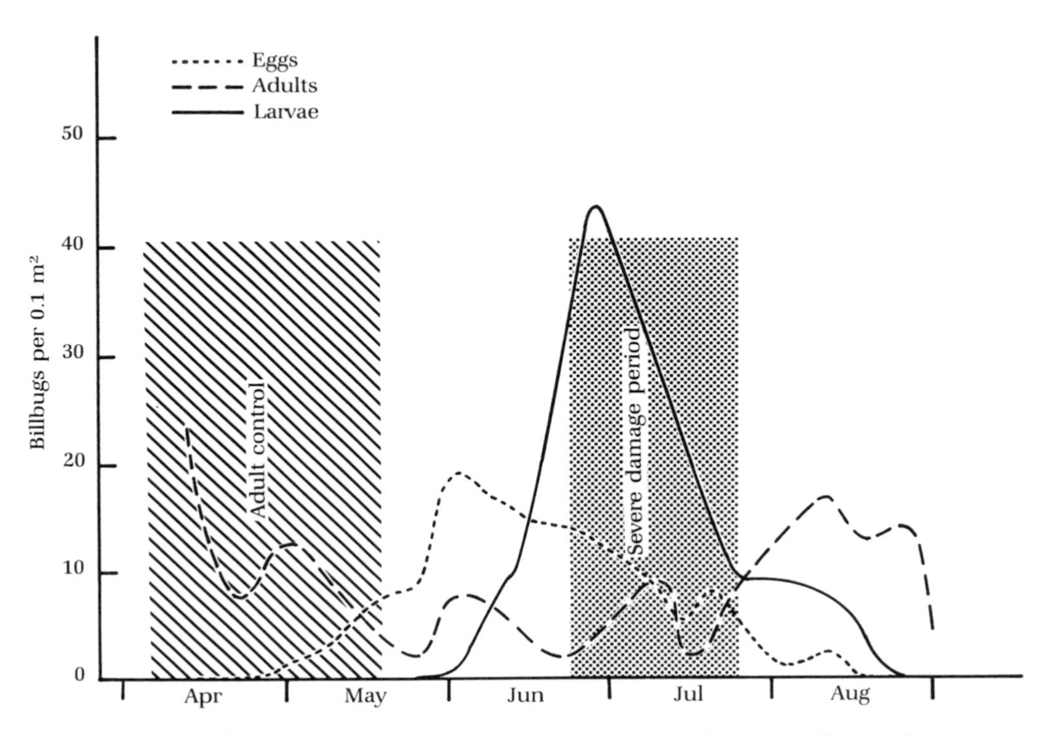

Figure 67. Seasonal populations of eggs, larvae, and adults of the bluegrass billbug and the spring treatment period for control of adults to prevent summer larval damage. (Adapted from Niemczyk 1983, p. 4, courtesy of *American Lawn Applicator*.)

For removal of eggs and larvae from stems, 3.8-cm (1.5 in.) lengths of stem above the crown were covered with tap water and placed in a blender for three 3-sec runs. The material was passed through stacked sieve screens (U.S. Standard Sieve Series No. 10, 20, 30, and 35). Contents of the 10 and 20 sieves were washed back into a blender and were blended a second time. Extraction produced an average of 2.08 eggs and 1.15 larvae for each 25 stems (Kindler and Kinbacher 1982).

Plant Resistance

In recent years, a search has been made to determine the presence of resistance to BGB in existing cultivars of turfgrass. Studies conducted in New Jersey indicate that the perennial ryegrass cultivars 'HS-1352', 'Pennant', and 'Regal' show resistance to the BGB in the amount of damage caused and in the numbers of billbugs present (Ahmad and Funk 1983).

Also in New Jersey, the resistance of perennial ryegrasses to billbugs appeared to be associated with the presence of the *Lolium* endophytic fungus. Resistance was observed to relate to high levels of the endophyte that is primarily transmitted by seed and vegetative propagation. The *Lolium* endophyte gradually loses its viability when seed is stored for prolonged periods (Funk and Hurley 1984).

Searches for Kentucky bluegrass cultivars with resistance to the BGB have been made in Nebraska. Of 15 cultivars evaluated for field resistance to BGB, none escaped injury, but there were varying degrees. 'Park', 'Nebraska Common', and 'South Dakota Certified' had the lowest numbers of billbugs per unit area, while cultivars showing the greatest amount of damage were 'Warren A-20' and 'Nugget' (Kindler and Kinbacher 1975).

Field studies made in New Jersey in the search for Kentucky bluegrass cultivars resistant to BGB corroborated some of the results of Nebraska. Plots of the cultivar 'Kenblue' had a mean of 1.3 larvae per ft^2, while 'Nugget' had a mean of 60.3 larvae per ft^2. The common types of Kentucky bluegrass, such as 'Arboretum', 'Delta', 'Park', and 'Nebraska Common', were among the more resistant (Funk and Ahmad 1983).

Natural Enemies

Microorganisms

There appears to be no evidence that any microorganisms infect the BGB.

Insect Parasites and Predators

At least two parasites have been reported from several billbugs, including the BGB. These include a hymenopterous egg parasite, *Anaphoidea calendrae* Gah., and a dipterous parasite, *Myiophasia metallica* Tns. (Tachinidae) (Satterthwait 1931, 1932).

Vertebrate Predators

Toads, reptiles, and numerous birds, including crows, blackbirds, cardinals, and other species, are listed as predators (Satterthwait 1932).

Hunting Billbug

Taxonomy

The hunting billbug, HB, *Sphenophorus venatus vestitus* Chittenden, order Coleoptera, family Curculionidae, subfamily Rhynchophorinae, is one of about five subspecies of *S. venatus.* Another subspecies, *S. v. confluens,* is destructive to orchard grass, *Dactylus glomeratus,* in Oregon and also treats bentgrass and Kentucky bluegrass as host plants. These subspecies too were placed in the genus *Calendra* in the early literature (Kamm 1969, Oliver 1984, Vaurie 1951).

Importance

The HB is an important pest of turfgrasses in the Southeast. The most severe injury occurs in zoysiagrass and in improved strains of bermudagrass. In Hawaii, the HB was second only to the lawn armyworm as a pest of lawn and turf, zoysiagrass again being the most severely damaged (LaPlante 1966b, Oliver 1984).

History and Distribution

The HB is probably the most abundant and widespread of the five subspecies. It occurs in the Atlantic coastal states from Virginia to Florida and along the Gulf to about Brownsville, Texas, and north to Kansas and Missouri. It is also present in some of the Caribbean islands. The HB was introduced into Hawaii in about 1960 and is presently found throughout most of the state (LaPlante 1966b). In southern California the HB has repeatedly been intercepted in zoysiagrass and centipedegrass shipped from Georgia and other eastern states, and at least one golf course infestation has been found. The Phoenix billbug, *S. phoeniciensis* Chitt., (Plate 45) is the established and commonly found species in southern California (LaPlante 1966b, Morishita et al. 1971, Vaurie 1951).

Host Plants and Damage

The HB most severely damages zoysiagrass and the improved strains of hybrid bermudagrasses. Other turfgrasses that serve as hosts include St. Augustinegrass and centipedegrass. Nonturf host plants include nutsedge, crabgrass, signal grass, barnyard grass, wheat, corn, sugarcane, Pensacola bahiagrass, and leatherleaf fern. In southern Louisiana, it has been the most troublesome in turfgrass nurseries, golf courses, and turf around commercial buildings (Oliver 1984, Woodruff 1966).

Damage by the HB shows up as irregular elongated or rounded areas of brown and dying grass. Grass in these areas lacks sufficient roots to obtain water and nutrients or to anchor itself to the soil. Extended dry weather makes the injury more pronounced. Unlike damage by scarabaeid grubs, mole crickets, and other turf insects, which cause the turf and soil to feel soft and spongy, under grass damaged by the HB the soil remains firm underfoot.

Adults and young larvae feed on stolons, crowns, and new leaf buds; older larvae attack roots and runners to a depth of about 8 cm (3 in.). The grass dies as a result of the combined onslaught. HB-injured zoysiagrass becomes chlorotic, and tufts are easily lifted from the sod mat (Brussell and Clark 1968, Oliver 1984).

Description of Stages

Adult

HB adults are nearly identical in appearance to BGB adults but are slightly larger, measuring 8–11 mm in length (Plate 45). The coarsely punctated pronotum, with its smooth, nonpunctated Y-shaped median area and parenthesislike curved markings on the sides, characterizes this species and differentiates it from the bluegrass billbug (Plate 45). Adults are generally black but may appear brown or gray when dried mud adheres to the punctated thorax and striated elytra. Some specimens are reddish and apparently callow. Adults usually feign death for short periods when they have been disturbed (Oliver 1984, Vaurie 1951, Woodruff 1966).

Egg

Eggs are oblong and clear to creamy white. They appear in slits in leaf petioles or in the stems that are made by adult females. They are also in leaf sheaths and at the top of the crown (Kelsheimer 1956, Woodruff 1966).

Larva

The larva of the HB, nearly identical to that of the BGB, is a white, legless insect with a brown head capsule. All instars resemble each other except in size. The last-instar mature larva is 7–10 mm long and is found in the crown or root area just below the thatch (Oliver 1984).

Seasonal History and Habits

Seasonal Cycle

The HB has one generation a year. It overwinters as an adult but may also be seen most months of the year. Adults emerge from hibernation during spring. After a short feeding period, eggs are deposited in the stems of host grasses. Larval development extends into summer. The average period from egg hatch through pupation is about 30 days (Oliver 1984).

Adult Activity

Like the BGB, adults of the HB that have emerged from overwintering sites are first seen crawling over driveways and walks and along curbs. Adults usually feign death for short periods when disturbed. They also cling tightly to a stem or leaf when attempts are made to collect them. Adult feeding and oviposition are most pronounced in the spring, with larval development extending into summer (Oliver 1984).

As adults of the new generation develop and start to migrate over paved areas, they become most noticeable in midday during late summer months. The adult can fly a short distance but cannot sustain flight for long. During fall, as temperatures decline, adults seek hibernation quarters in leaf litter, in weedy areas, and in other partially protected situations (LaPlante 1966b, Oliver 1984).

Larval Activity

Upon hatching, young larvae feed within stems, hollowing them out as well as filling them with frass. Older, larger larvae feed on the crown of the grass plants. Finally, as the ultimate-instar larvae approach maturity, they feed on the roots of plants. They penetrate the soil to a maximum depth of about 8 cm (3 in.).

Natural Enemies

The HB, like the bluegrass billbug, appears free from attacks by microorganisms. Apparently the same insect parasites attack both species of billbugs (Satterthwait

1931, 1932). In Hawaii, the giant toad of the genus *Bufo* consumes large numbers of adults. Examination of their scat reveals a diet consisting at times nearly 100% of weevils.

Annual Bluegrass Weevil

Taxonomy

A common name of the annual bluegrass weevil, the ABW, has been assigned tentatively to an insect currently classified as *Hyperodes* near *anthracinus* (Dietz) and belonging to the order Coleoptera, family Curculionidae, subfamily Cylindrorhininae. Species determination has not been resolved for the ABW. R. E. Warner of the U.S. National Museum believed that the ABW represented a species between *H. anthracinus* (Dietz) and *H. maculicollis* (Kirby) and therefore classified it as *H.* sp. near *anthracinus* (Cameron and Johnson 1971b). The common names *turfgrass weevil* and *annual bluegrass weevil* have been suggested. Since Schread's publication (1970b), which has priority, proposes the latter name and since the insect is a pest of only the annual bluegrass, *Poa annua* L., it appears appropriate to designate it the *annual bluegrass weevil.* To become official, however, a common name must be submitted to the Entomological Society of America for approval only after the insect has received a specific designation as to genus and species.

The present literature lists some 48 species of *Hyperodes* present in North America north of Mexico. These species are extremely difficult to classify because of their similarities in size and color. The only comprehensive key to adults indicates that *H. anthracinus* has sparse pubescence, a shiny body surface, and a strong median prothoracic carina, while *H. maculicollis* has dense pubescence, a dull body surface, and a short prothoracic carina abbreviated before and behind (Blatchley and Leng 1916, Kissinger 1964, Stockton 1963).

Charles O'Brien (Florida A & M University, Tallahassee) is making a thorough revision of the weevils belonging to *Hyperodes* and *Listronotus,* two very closely related and superficially separated genera. He has tentatively designated the ABW from New York State as *Listronotus maculicollis* (personal communication, 1985). Until final revision and to avoid confusion, however, I shall designate the ABW using the old binomial of *Hyperodes* near *anthracinus.*

Importance

The ABW is a turfgrass pest only of short-cut annual bluegrass maintained at cutting heights of 0.6–1.3 cm (0.25–0.50 in.) or less. It is therefore a major pest exclusively of golf course fairways, tees, greens, and aprons of greens and tennis court turf. Cutting heights of 4 cm (1.5 in.) or higher prevents this insect from becoming a pest of home lawns. It has not been injurious to any other species of turfgrass associated with these two recreational facilities.

Occasionally golf course superintendents claim that minor infestations occur on pure bentgrass greens. In most cases, close examination should reveal that the damage is occurring in small patches of annual bluegrass that have invaded the seemingly pure bentgrass greens. Such has been the situation in the peripheral areas of the major infestations that have occurred within an 80-km (50-mi) radius of New York City.

Within this 80-km radius, major efforts were made by many golf courses to eliminate annual bluegrass because it has the inherent weakness of dying out during the midsummer heat. Since elimination efforts have generally been unsuccessful and automatic irrigation systems have made it easier to maintain annual bluegrass through the summer, most courses make major efforts to maintain this grass through control of the ABW.

Our current understanding of the ABW suggests that the major obstacle to maintaining annual bluegrass during the summer has largely been infestations of this insect and not the turf's inherent inability to withstand summer stresses. The annual "spring die out" recognized on tennis courts and golf courses on Long Island for 40–50 years was no doubt largely caused by this insect (Cameron and Johnson 1971b, Vittum 1980).

Control is an expensive maintenance operation for golf courses; many treat the entire fairways, tees, and greens with an insecticide. Courses with fewer resources may spot treat the areas that are most severely and repeatedly infested year after year. Whether the entire playing area is treated or spot treatments are employed, once a course has received a treatment program, it normally becomes an annual operation.

History and Distribution

The earliest reports of the ABW as a turfgrass pest came from Connecticut in 1931. On Long Island it was first reported injuring turf in Nassau County in 1957 and again in 1961. By 1965–1967, additional injuries on Long Island and in Westchester County just north of New York City were being reported by golf course superintendents. In 1966–1967, *Hyperodes* sp. damage occurred on an Ithaca, New York, golf course some 400 km (250 mi) from New York City. Turf-damaging *Hyperodes* were collected also in the Pocono Mountains of northeastern Pennsylvania and from Connecticut courses (Britton 1932, Cameron and Johnson 1971b, Cameron et al. 1968).

More recently (during the late 1970s and early 1980s), turf-damaging *Hyperodes* have been present on many New York State golf courses, in Elmira, in the southern tier area of the state, near Buffalo, in the western edge, in Lake Placid, in the Adirondack Mountains, and in Cooperstown, in the center of the state. Several locations throughout New England have been affected. The most severe and perpetual problems with the ABW continue to occur on Long Island, in Westchester County, New York, and in Fairfield County, in southwestern Connecticut, where there is as high a concentration of golf courses as in any part of the country.

Host Plants and Damage

Host Plants

In the field, annual bluegrass is the only turfgrass that has suffered ABW feeding damage and only when golf course fairways, tees, greens, and aprons of greens are close cut (<0.6 cm). There has been no evidence of turf damage in roughs, presumably because of the much higher cut. Annual bluegrass maintained on tennis courts (also cut very low) has also suffered damage (Cameron and Johnson 1971a).

Adult weevils feed on the leaves and stems of annual bluegrass but cause no serious problems. When given a choice, weevils will feed on clover, plantain, dandelion, and even mulberry (Cameron and Johnson 1971b). In feeding preference studies with turfgrasses, adult weevils found tall fescue as attractive as annual bluegrass. Other plants fed upon but found less attractive than annual bluegrass were Canada bluegrass, rough bluegrass, Kentucky bluegrass, creeping bentgrass, perennial ryegrass, and red fescue. No feeding preference tests have been performed with larvae, but field observations indicate that only annual bluegrass has ever shown damage. Patches of bentgrass within an annual bluegrass stand remain completely healthy when the annual bluegrass is completely killed by larval feeding (Cameron and Johnson 1971a, 1971b; Vittum 1980).

Damage

Adult Feeding. Adult weevils feed mostly on leaves and stems at the base of leaf blades of annual bluegrass (Plate 46). Notches are chewed out of the edges of blades, or holes are eaten out of the centers. Adult feeding may weaken grass stems but rarely kills them (Cameron and Johnson 1971b).

Larval Feeding. Damage initially appears as small yellow-brown spots on annual bluegrass that are caused when larvae sever stems from the plant (Plate 46). One larva can kill as many as a dozen individual stems. The final, or fifth-instar, larva feeding on the crown of the grass causes the greatest damage (Plate 46). On fairways, damage is first evident near the edges and gradually spreads toward the centers. Since hibernating adults disperse primarily by crawling, they first feed and oviposit in close-cut annual bluegrass along the edge of fairways (Plate 46). As larvae grow, the spots coalesce, causing large areas of turf to die. Concentrations of larvae as great as 500 per 0.1 m^2 (465 per ft^2) can kill large areas of turf but may leave patches of bentgrass totally uninjured (Plate 46). Larvae severely damage the collars of greens when annual bluegrass is the dominant species (Plate 46). As previously noted, what appears to be bentgrass injury on greens is sometimes actually minute, undetected patches of annual bluegrass being killed by the larvae (Plate 46; Cameron and Johnson 1971c, Tashiro 1976b, author's personal observations).

Severe damage occurs during May and early June. Larvae, pupae, and adults can be found beneath the damaged turf by mid-June in southeastern New York (Cameron 1970).

Damage during midsummer by second-generation larvae is normally much less severe and more localized. Given areas of a particular golf course sometimes show repeated localized infestations year after year. This situation could relate directly to favorable adult overwintering sites. On fairways, while the spring damage is generally more severe along the edge, damage during midsummer is generally scattered.

Description of Stages

Figure 68 shows all stages of the ABW.

Adult

Description: **Hyperodes *near* anthricinus.** ABW adults average 3.6 mm in length and resemble those of the turf-infesting billbugs but are much smaller (Plate 47). The antennae are attached near the apex of the beak, with the scape directed back toward the base of the beak in a long groove called the *scrobe*. The scape is nearly as long as the pedicel and flagellum together. The beak bears three carinae dorsally, the median carina being the strongest. The prothorax bears a dorsal median carina and a postocular lobe that often partially covers the compound eye (Cameron and Johnson 1971b).

Mature adult weevils are generally black and are clothed with fine hairs and yellow-brown and grayish white scales. Scales on the elytra are in scattered groups that create a faint mottled appearance. As adults grow older, many of the hairs and scales are worn off, making the body appear more shiny and black. Callow adults, generally found only in the soil, are reddish brown to brown and gradually turn gray to black. The abdomen is the last part of the body to darken (Plate 47).

The male and female resemble each other, the latter being just slightly larger than the former. Sexes can be separated by differences in the ventral abdominal segments. In the male the third abdominal sternum (first visible) has a median depression, while in the female there is no such depression. Rather, this sternum tends to bulge slightly in the middle. The last sternum (seventh) of the male is flattened and is slightly impressed at the apex, while in the female this sternum has a large, shallow depression (Plate 47). The sex ratio is nearly 1:1 (Cameron and Johnson 1971b).

Differences in Adults Revealed by Scanning Electron Microscopy. In an attempt to clarify possible species differences among ABW individuals from different locations, adults from western New York were compared with those of southeastern New York via scanning electron microscopy (SEM). Adults from Westchester County were found to have distinctly fewer scales per unit area immediately above coxa 1 than adults from Elmira in western New York (Figure 69). SEM photos of other areas of the weevils from the two locations showed only slight differences. On the basis of these studies, the weevils from Elmira are considered distinct from weevils of Westchester County. The former were considered to be possibly *H. maculicollis* (Vittum 1979).

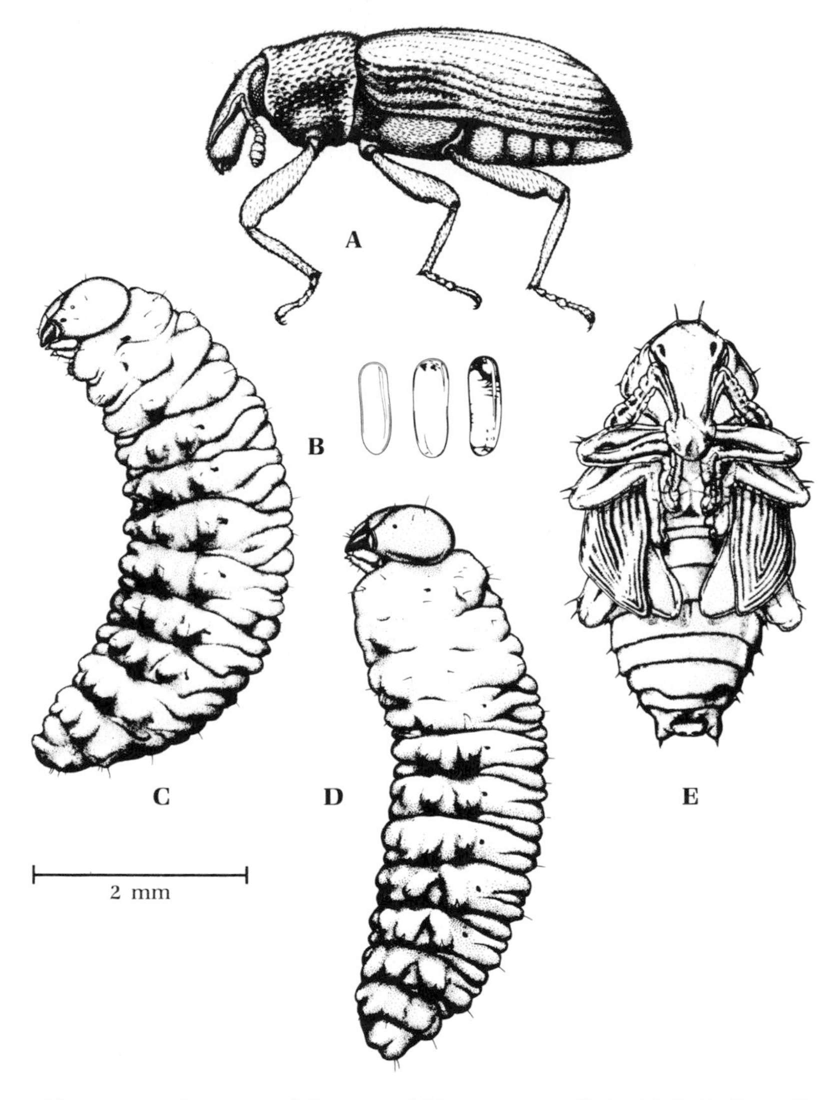

Figure 68. Life stages of the annual bluegrass weevil. **A.** Adult. **B.** Eggs. **C.** Ultimate-instar larva. **D.** Prepupa. **E.** Pupa. (From Cameron and Johnson 1971b, figs. 34–38, courtesy of the Cornell University Agricultural Experiment Station, Ithaca.)

Egg

Eggs of the ABW, deposited between leaf sheaths, are approximately three times longer than they are wide and average 0.8 mm by 0.25 mm. They are rounded at both ends and are generally straight, but some may be curved longitudinally. They are also generally round in cross section, but a few may be flattened. The chorion is

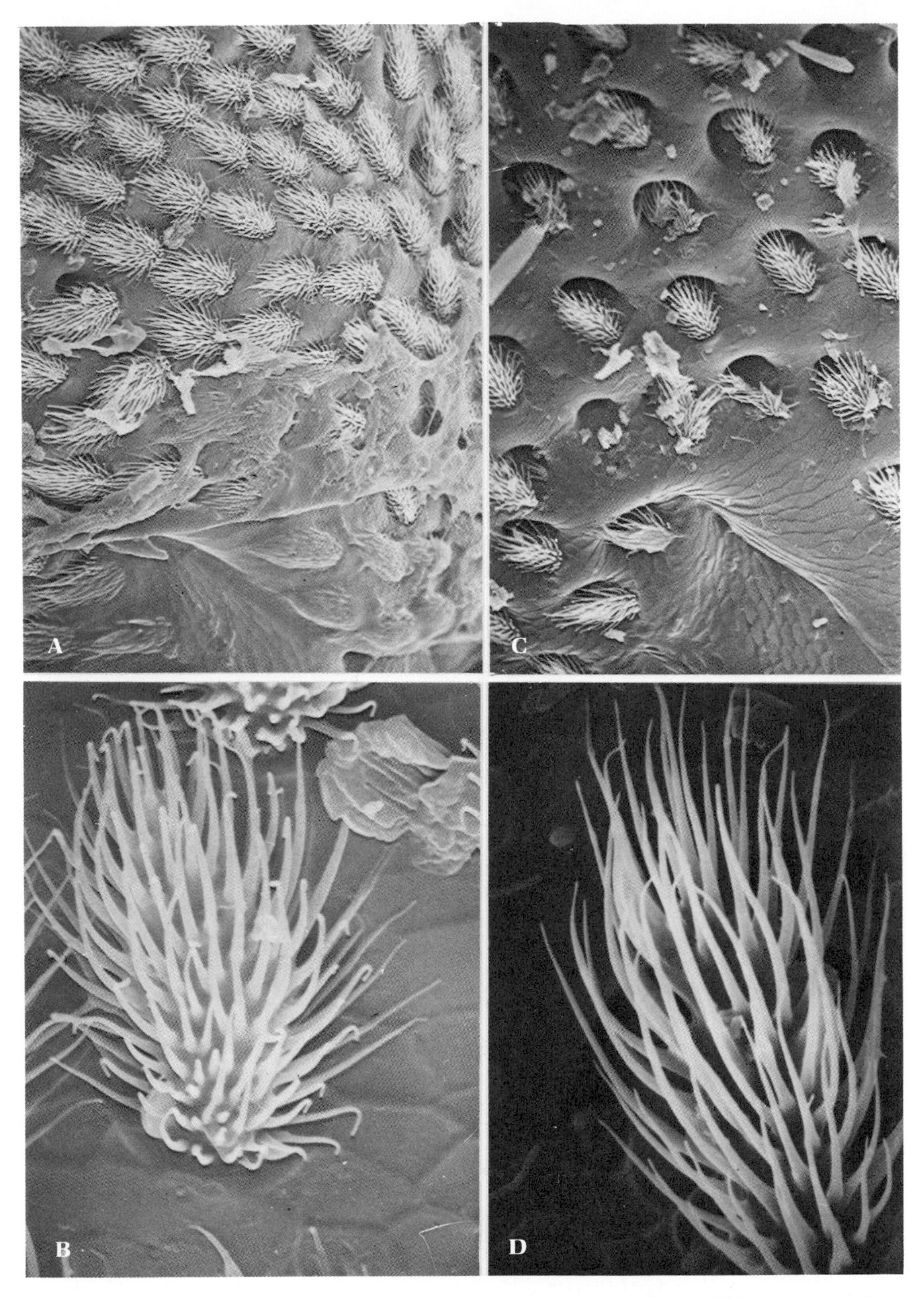

Figure 69. Scales and individual scale immediately dorsal to coxa 1 of annual bluegrass weevil. **A.** Weevil from Elmira (western New York), × 245. **B.** Weevil from Elmira, × 1,360. **C.** Weevil from Mamaroneck (southeastern New York), × 246. **D.** Weevil from Mamaroneck, × 1,860. (Adapted from Vittum 1979, figs. 9–12, courtesy of P. J. Vittum, University of Massachusetts.)

shiny, smooth, and usually unsculptured except for longitudinal ridges and grooves that were probably created by the ridges of the leaf sheath within which the eggs are deposited. Eggs are pale yellow to white, with a transparent clear chorion when they are first deposited. Within a day or two, the chorion becomes smoky black but remains transparent to reveal the developing embryo (Plate 47; Cameron and Johnson 1971b).

Larva

The larva is creamy white, legless, and wider in the middle than either end. It grows from a length of about 1 mm in a young first instar to about 4.5 mm in a mature fifth instar (Plate 47). First instars are nearly straight but gradually become more crescent shaped as they grow. The dorsum of each segment has three fleshy lobes, and the three legless thoracic segments have ventral swellings, each with long setae. The dark brown head capsule has a distinct epicranial suture that makes an inverted Y. Two pairs of ocelli are present, one located just above the mandibles and the other in the genal region (Cameron and Johnson 1971b).

Prepupa

The prepupa is nearly identical to the last instar externally except that all three thoracic segments lose many of their folds, become distended, and make the prepupa slightly longer than the last instar. Prepupae do not move about but are capable of performing a circular movement of the abdomen that is often the most distinguishing characteristic (Cameron and Johnson 1971b).

Pupa

The pupa, exhibiting many adult characteristics, is first creamy white throughout (Plate 47), but parts of the body gradually darken to a reddish brown, becoming black as it matures. First the eyes, then the mandibles, the tarsal claws, the beak, and the articulations of the legs change color, in this order (Cameron and Johnson 1971b).

Seasonal History and Habits

Seasonal Cycle

The ABW overwinters as an adult primarily in the litter under trees of golf course roughs, with the largest concentrations found in white pine, *Pinus strobus*, litter (Plate 48). Migration from hibernation quarters to fairways commences during early spring about the time that forsythia, *Forsythia* spp., is in full bloom (about mid-April in southeastern New York; Plate 48). During the period when flowering dogwood, *Cornus florida* L., is in full bloom (full bract color) and when redbud, *Cercis canadensis* L., is in full bloom, all adults will have left hibernating quarters and will have moved out into fairways, tees, and greens (Plate 48; Tashiro and Straub 1973, author's personal observations).

Oviposition occurs shortly after adult feeding begins. Larval development through five instars requires about a month, and development from egg to adult requires about 2 months. The generalized life cycle of the ABW in southeastern New York and southwestern Connecticut, found in fairways, tees, and greens during spring and summer, is shown in Figure 70 (Cameron 1970, Vittum 1980).

A good correlation exists between growing degree-day accumulations in Westchester County, New York, and maximum presence of all stages except eggs (which were not studied). The spring generation of the ABW shows a variation of about 2 weeks among years for any given stage (Figure 71). The maximum presence of small larvae extended from the last week in May to the first week in June, that of mature larvae occurred during the first half of June, that of pupae occurred during the second half of June, and that of callow adults came during the last week in June and the 1st week in July (Vittum 1980).

A second generation was clearly evident, with each stage spanning a longer period, depending on the year. Small larvae were present during the last 3 weeks of July, large larvae during the second half of July into early August, pupae during the last of July and the first half of August, and callow adults during the first 3 weeks of August.

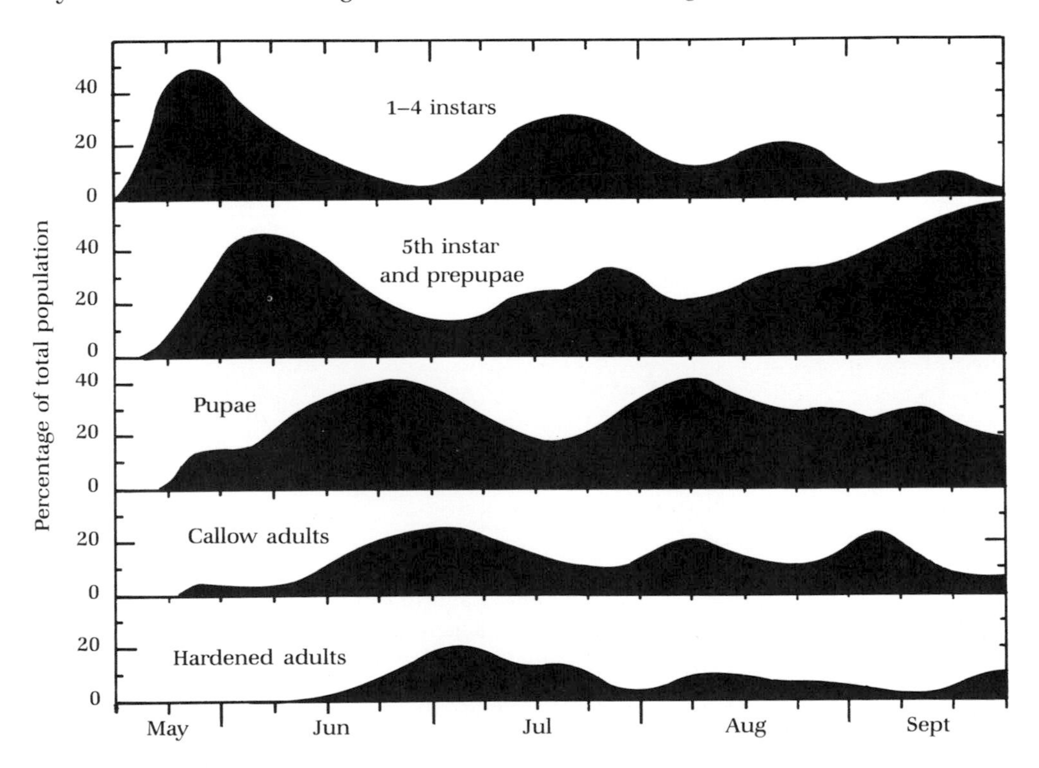

Figure 70. Annual bluegrass weevil fairway and tee populations exclusive of eggs, estimated on the basis of turf-soil plugs (10.8 cm in diameter) from Westchester County, New York, and Fairfield County, Connecticut, golf courses, 1976–1979. (Data from Vittum 1980; drawn by R. McMillen-Sticht, NYSAES.)

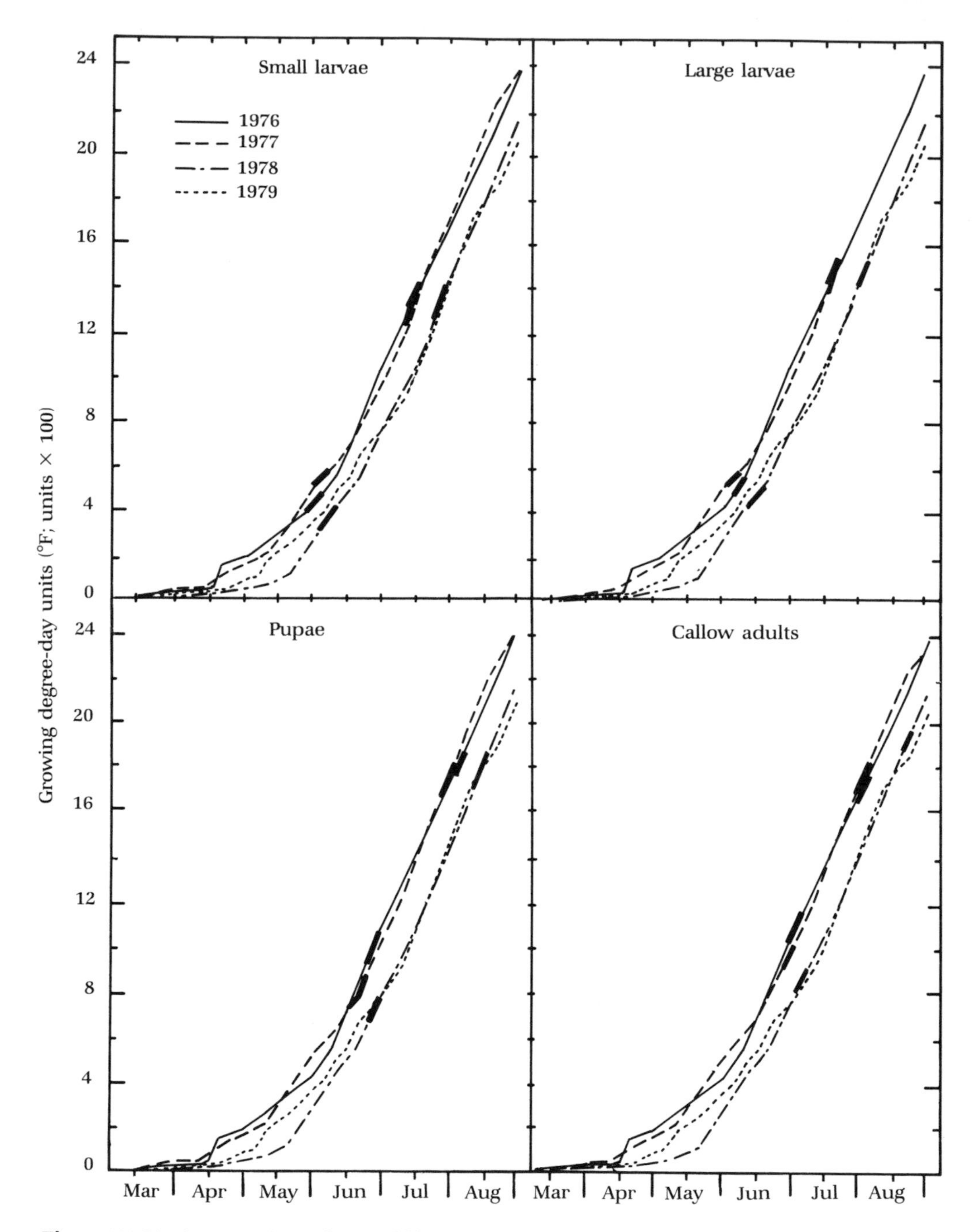

Figure 71. Maximum activity of annual bluegrass weevil small larvae (first four instars), large larvae (fifth instar and prepupae), pupae, and callow adults relative to growing degree-day accumulations with a base 11°C (52°F) in Westchester County, New York. Temperature records from Westchester County Airport, 1976–1979. (Adapted from Vittum 1980, figs. 38–41, courtesy of P. J. Vittum, University of Massachusetts.)

There was also evidence of a third generation of small larvae during 1976 and 1977 during the second half of August. These years had unusually warm springs, which apparently accelerated development throughout the entire growing season (Vittum 1980).

Adult Activity

Hibernation. The bulk of the population overwinters as adults, and migration to overwintering sites begins by late summer. Prior to 1970, the major hibernation sites of weevils was thought to be in tufts of fescue. Adults varying from about 15 to a high of 42 per 0.1 m^2 (per ft^2) were found in litter under hedges, in fescue under trees, and under leaves on tennis courts, but none were found in fairways, tees, greens, and aprons (Cameron and Johnson 1971b).

Dominant Overwintering Site. During the 1970s the primary overwintering sites were determined to be in white pine needle litter. Samples of 0.1 m^2 of litter to the soil surface collected from the same five white pines over the period of a year showed that a maximum of 443 adults were present in each 0.1 m^2 of litter (Table 20). There was a consistency of high populations under a given tree, but no environmental conditions such as grouping of trees, thickness of litter, tree size, and so forth appeared to influence populations. Soil samples removed from under white pine litter also demonstrated that roughly 40%–60% of the diapausing weevils were in the upper layer of soil (Table 21).

Litter samples taken from under trees of other species in the roughs of various golf courses showed weevils to be present under spruce, hickory, maple, birch, and

Table 20. Diapausing annual bluegrass weevil adults in white pine litter in the roughs of a Westchester County, New York, golf course, 1976–1977

Month or half month of litter collection	No. 0.1 m^2 samples per tree	Adults per 0.1 m^2 litter under 5 trees (av.)[a]					Av. per 0.1 m^2 each period	Population range per 0.1 m^2
		1	2	3	4	5		
Oct 1976	4	103	224	—	—	—	163	333[2]–59
Nov	2	113	317	67	59	—	139	443[2]–14
Feb 1977	6	71	127	57	43	17	63	215[2]–1
Mar	3	62	78	47	48	11	49	89[1]–4
April 1	2	27	50	26	15	7	25	99[2]–4
April 2	4	8	8	3	2	1	4	15[1]–0
May 1	4	<1	<1	1	0	<1	<1	2[3]–0
July	1	51	8	4	26	2	18	51[1]–2
Aug	1	9	20	5	18	5	11	20[2]–5
Sept	1	9	47	9	7	3	15	47[2]–3
Oct	1	12	66	16	27	6	25	66[2]–6

Note: A superscript in the cells indicates tree number.
[a]Weevils were extracted by submerging litter samples to force the insects to surface.

Table 21. Distribution of diapausing annual bluegrass weevil adults in white pine litter and soil on a Westchester County, New York, golf course, 1976–1977

		Percentage of total		
			Soil-depth (cm)	
Tree no.	Weevils per 0.1 m^2 (av.)[a]	Litter	0–1.0	1.0–2.5
	November 15, 1976			
1	112	31	69	—
2	346	58	42	—
3	67	67	33	—
4	60	76	24	—
Av. (%)	—	58	42	—
	March 20, 1977			
1	106	58	27	14
2	165	47	33	20
3	92	51	38	11
4	73	66	18	16
5	16	69	25	6
Av. (%)	—	58	28	14

[a]Weevils were extracted by submerging samples to force the insects to surface.

other trees. Where white pine and other trees occurred in close proximity, weevil densities were always the highest under white pine.

Extraction of Weevils. The extraction of adult weevils from litter and soil samples is readily accomplished by submerging the samples entirely under water, which forces adults to crawl to the surface. Paper toweling placed on the water causes the weevils to cling to the undersurface. The weevils can readily be removed from the paper and can be retained for further studies.

About 50% of the adults surface during the first 10 min of submergence, 70% in 30 min, 90% in 60 min, and virtually 100% in 2.0–2.5 hr of submergence (author's personal observations).

Emergence and Dispersal. White pine litter sampled at regular intervals during the spring of 1976 and 1977 revealed the period of weevil emergence and dispersal out of diapausing sites (Table 22). All weevils were in diapausing sites throughout March. Migration into fairways occurred during both years, beginning in early April, and by the end of the month the weevils were nearly all dispersed. Eggs were abundant in the stems during the last week in April and the 1st week in May. Dispersal from diapausing quarters is primarily by crawling, as evidenced by turf damage first at the edge of fairways, where oviposition first occurs.

Table 22. Annual bluegrass weevil seasonal development in Westchester County, New York, golf courses, in 1976 and 1977

Week (beginning Sunday)	Total insects observed	Percentage in each group[a]			
		Diapausing adults	Fairway adults	Eggs in stem	Larvae
		Spring 1976			
March 30	35	100	—	—	—
April 5	68	87	13	—	—
April 12	81	91	9	—	—
April 19	33	79	21	—	—
April 26	86	0	0	100	—
May 10	67	0	27	60	13
May 17	23	4	0	30	65
		Spring 1977			
March 20	1590	100	—	—	—
April 3	570	99	1	—	—
April 10	343	94	2	4	—
April 17	135	59	8	33	—
April 24	88	11	24	59	6
May 1	71	14	13	70	3
May 8	56	7	7	79	7
May 15	20	0	10	70	20

[a]0.1 m^2 white pine litter samples were collected and submerged for extracting diapausing adults; turf 10.8 cm in diameter and soil plugs were teased apart for fairway adults and larvae; leaf sheaths were removed for counting eggs.

Weevils also disperse by flight and are attracted to black light (Table 23). The highest catches occurred in July, indicating that the first-generation (summer) adults were most readily attracted. A few of the overwintering adults, however, were also attracted and were caught during early May. Weevils have also been observed flying onto fairways in full sunlight. Weevils have also been caught on 20–30 cm (8–12 in.) plywood painted yellow and coated with Tanglefoot or other sticky substances (Cameron and Johnson 1971b, Schread 1970b, author's personal observations).

Adult Feeding. Young adults are quite active during the day as they emerge from the soil. They are generally black, but some remain the reddish brown of callow adults for a week or longer after emergence. Mature adults are rather difficult to find during the day; they tend to remain inactive in the grass and thatch. At dusk, adults begin to crawl to the tips of the grass stems to feed. During midsummer nights, adults can often be swept into a net from grass more than 1.3 cm (0.5 in.) tall but not from greens because the grass is too short. With the aid of a strong light, adults can readily be seen on the surface of greens. Adults feed mostly on leaves and at the junction of leaves and stems. They chew holes in leaves and notches along the

Table 23. Annual bluegrass weevil adult catches in black-light (15-watt) traps at Westchester County, New York, golf courses, in 1976

Half-month period	Locations 1	2	3	4	Total	Percentage of total
April 2	0	0	0	0	0	0
May 1	0	90	0	0	90	5
May 2	0	0	0	0	0	0
June 1	23	69	0	0	92	5
June 2	226	112	15	20	373	21
July 1	800	50	135	0	985	55
July 2	174	16	30	0	220	12
August 1	14	3	3	0	20	1
August 2	0	0	3	0	3	<1
Total	1,237	340	186	20	1,783	—
Percentage of total	69	19	10	1	—	—

Note: Traps were emptied every Monday morning, and catches were grouped into half-month periods.

margin. Adult feeding is relatively insignificant; adults may weaken the grass, but they rarely cause it to die (Cameron and Johnson 1971a, 1971b).

Oviposition. After a brief period of feeding, the females are ready to begin oviposition. Eggs are deposited, usually in groups of 2 or 3 between leaf sheaths. Dissections reveal the development of more than 50 eggs in each female, and the full complement is presumably deposited in the field. Laboratory rearing has resulted in an average egg production of 11.4 eggs per female. Coincidentally, when seven pairs of diapausing adults were isolated and held for as much as 10 weeks with annual bluegrass, they produced an average oviposition of 11 eggs. One female deposited eggs for 9 weeks (Table 24; Cameron and Johnson 1971b, author's personal observations).

Manipulation of Adults for Laboratory Studies. Adult weevils are difficult to hold with forceps for sexing or for any other purpose. They may readily be handled individually by a suction device with a glass pipette drawn to a size that is large enough to accommodate the head but not the thorax (Figure 72; Tashiro et al. 1977).

Larval Activity

Hatching, Growth, and Feeding. Upon hatching, most first instars burrow into the stems almost immediately, feeding both upward and downward from the point where they hatch. The youngest and most tender central portion of the stems is eaten. Larvae may emerge and reenter and same stems several times or may move to

Table 24. Survival and oviposition from paired annual bluegrass weevils, in 1977

	Male and female pairs 1–6: Survival and total egg production per female[a]					
Weeks from pairing	1	2	3	4	5	6
1	0	0	0	0	0	0
2	0	6	0	0	0	0
3	4	9	0	0	0	1
4	4	10	6	7	0	0
5	0	0	7	0	4	0
6	2	0	0	0	0	0
7	8	0	0	0	0	0
8	5	0	0	0	0	0
9	5	0	0	0	0	0
10	0	0	0	0	0	0
Total eggs per female	28	25	13	7	4	1

[a]Weevils collected April 12 as diapausing adults.

other stems nearby. Larvae are mobile and do not shun light, so they probably move from stem to stem above ground. Tunnels made by larvae are filled with frass that gradually turns brown. Feeding by small larvae seldom kills stems, but the central leaves turn yellow and gradually die (Cameron and Johnson 1971b, Cameron et al. 1968).

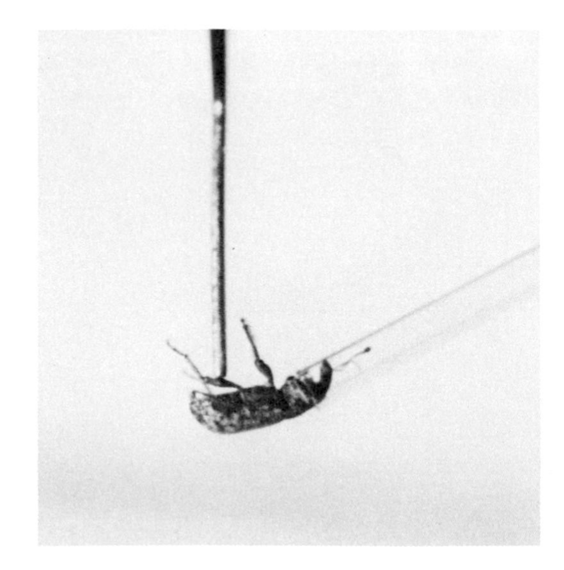

Figure 72. Method of holding weevils with a vacuum for individual handling. (From Tashiro et al. 1977, fig. 1, courtesy of the Entomological Society of America.)

As the larvae increase in size, they make larger tunnels, completely hollowing out the centers and leaving only the outer leaf sheaths. As the larvae become too large for grass stems, particularly the fifth (ultimate) instar, the insect begins feeding externally at the base or crowns (Table 25). Feeding damage occurs only on annual bluegrass; all other turfgrasses and especially bentgrasses in close association with annual bluegrass remain undamaged. Since annual bluegrass is likely to die under moisture stress, ABW-damaged stems in a weakened condition die out quickly with lack of moisture. Mature fifth instars are found in the thatch and surface soil among roots that are not damaged. Once in the soil, the mature larvae are very close to becoming prepupae and pupae (Cameron and Johnson 1971b).

Prepupal and Pupal Activity

When development of the fifth instar is completed, the larva tunnels about 0.6 cm (0.25 in.) into the soil and forms an earthen cell. Within the cell the pupa begins to form within the cuticula of the fifth-instar larva. The prepupa rotates within the earthen cell and moves its abdomen in a circular motion. This stage lasts 2–5 days.

Upon completion of the prepupal period, the fifth-instar exuviae split dorsally just behind the head to release the young, creamy white pupa. It also rotates within the earthen cell. The pupal period requires 3–9 days. Upon completion of the pupal development, the adult breaks the pupal exuviae dorsally along the head and thorax and frees itself. The light-colored, soft adult remains in the earthen cell 3–8 days, gradually turning from a reddish brown callow adult to a blackish mature adult. Some adults remain reddish brown for more than a week after emergence (Cameron and Johnson 1971b).

Natural Enemies

Few enemies of the ABW are known. A few larvae and pupae were found during 1977 that were flaccid with a coral to lavender hue. These were infected with a

Table 25. Location of annual bluegrass weevil larvae and pupae in 368 soil-plug samples from golf courses on Long Island, New York, in 1968

		Percentage of total		
Stage insect	Number	Inside stems	Above soil	In soil[a]
Small larvae	99	26	72	2
Medium larvae	214	2	84	14
Large larvae	245	0	49	51
Pupae	130	0	0	100

Note: Soil-plug samples were removed with a standard golf cup cutter 10.8 cm in diameter.

[a]Varied from av. 0.3 cm for small larvae to 0.8 cm soil depth for pupae.

Source: Adapted from Cameron and Johnson 1971b.

common entomogenous nematode of the genus *Neoaplectana.* Several species of carabid beetles freely feed on larvae and pupae of ABW in the laboratory. Nearctic *Hyperodes,* including *H. maculicollis,* have been taken from the stomachs of toads, *Bufo* spp. Blackbirds have been observed actively pecking the turf infested by the ABW (Cameron and Johnson 1971b, Johnson and Cameron 1969, Stockton 1956, Vittum 1980).

18

Dipteran Pests: Families Tipulidae and Chloropidae

European Crane Fly

Taxonomy

The European crane fly, ECF, *Tipula paludosa* Meigen, order Diptera, family Tipulidae, subfamily Tipulinae, has elongated maxillary palpi that distinguish members of this subfamily from other subfamilies. *Leatherjacket* is the name given to the larva of the ECF (Alexander 1919, Jackson and Campbell 1975).

Importance

While the ECF has a very limited distribution in North America, it is of considerable importance as a pest of lawns and golf courses in southern British Columbia and western Washington State. It has had an even greater impact on the dairy industry of this area by damaging pastures and hayfields. The largest populations occur in grasslands; the worst attacks are apparent usually in the spring (Jackson and Campbell 1975).

The great abundance of adults for a short period during late summer and their habit of collecting on sides of buildings create a nuisance to the general public (Jackson and Campbell 1975).

History and Distribution

The ECF is a native of northwestern Europe, where it occurs from the lower Scandinavian area to northern Italy and from Great Britain into the U.S.S.R. In North America it was positively identified in 1955 as it infested flower beds in Cape Breton Island of Nova Scotia. Its most probable source of introduction was soil that was dumped after being used as ship ballast (Fox 1957, Jackson and Campbell 1975).

On the West Coast it was first discovered in Vancouver, British Columbia, causing severe damage to lawns in 1965. The source of this infestation is not known. In the state of Washington, it was first detected in a light trap during late summer of 1966 at Blaine, bordering Canada. During the next 7 years it gradually spread southward in localized infestations to the vicinity of Tacoma (Jackson and Campbell 1975, Wilkinson and MacCarthy 1967).

Presently the ECF has a continuous southward distribution that includes Olympia, Washington, and at least an isolated infestation farther south along the Columbia River in Wahkiakum County, Washington (A. Antonelli, Washington State University, personal communication, 1985). In 1984, the presence of the ECF in three Oregon counties of Clatsop, Columbia, and Washington, west and northwest from Portland, was confirmed (Figure 73; J. Capizzi, Oregon State University, personal communication, 1985).

The maritime climate of coastal British Columbia and western Washington of relatively mild winters and abundant rainfall is an ideal climate for the ECF. A similar climate to the south extends the area of potential infestation as far south as northern California.

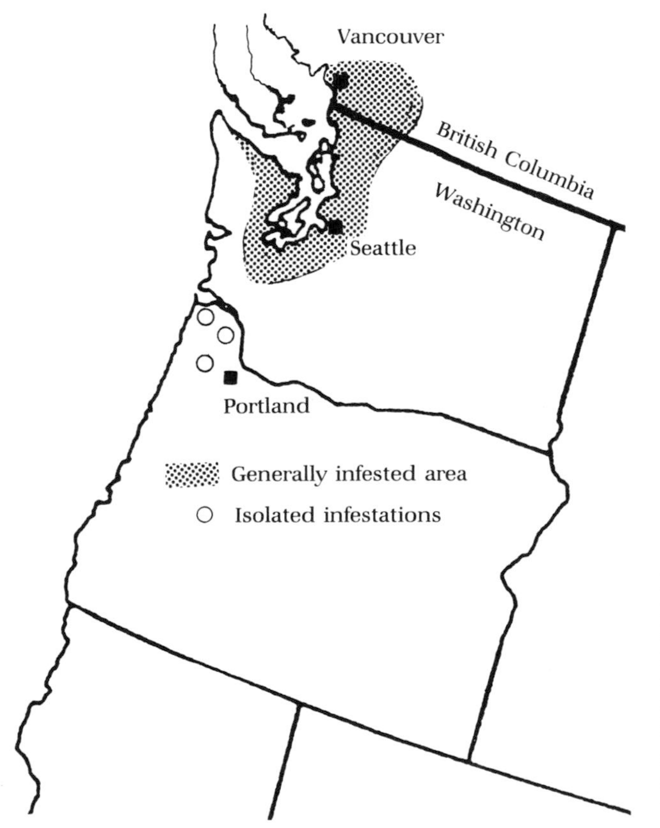

Figure 73. Distribution of the European crane fly in southwestern British Columbia to northwestern Oregon. (Data from Jackson and Campbell 1975, fig. 1, and 1985 personal communications with A. Antonelli, Washington State University, and J. Capizzi, Oregon State University; drawn by H. Tashiro, NYSAES.)

Host Plants and Damage

Larvae of the ECF feed primarily on the grasses of lawns and pastures, but strawberries, flowers, and vegetable crops have also been attacked (Wilkinson 1969).

When feeding below the surface, larvae mainly attack root hairs, roots, and crowns (Plate 49). The nightly migration of large larvae from the soil to the surface disrupts the soil surface and destroys young seedlings. They eat stems, grass blades, and leaves when they feed above ground (Wilkinson and MacCarthy 1967).

As few as 10–11 larvae per 0.1 m^2 (per ft^2) have caused serious injury to grassland. As many as 120 larvae per 0.1 m^2 (110 per ft^2) have been counted in April–May in Vancouver (Wilkinson 1969). While there are varying accounts of population density relative to economic damage, counts as high as 30 per 0.1 m^2 (27 per ft^2) have been noted in healthy fairway turf in Washington, with no appreciably noticeable damage (A. Antonelli, personal communication, 1985).

Description of Stages

Adult

The ECF adults are fairly large for crane flies. Males are 14–19 mm long and females 19–25 mm long. The wings of males are longer than the abdomen but shorter than the abdomen in females (Plate 49). Females have 10 clearly defined abdominal segments, with segments 8–10 modified into a functional ovipositor. The sex ratio was established as about 1.7 males to 1 female by Coulson (1962), but in Washington the male-female ratio ranged from about 1.8:1 to 1.2:1 (Jackson and Cambell 1975).

Callow adults are pale greenish with a transparent appearance that is retained for several hours because the meconium is not discharged until after the adult emerges. Callow females have an elongated, swollen abdomen (Byers 1961, Coulson 1962).

Egg

Eggs of the ECF are typically shiny black and elongate-oval, with one side flattened and one end more pointed than the other (Plate 49). They are 1.1 mm long by 0.4 mm wide (Wilkinson and MacCarthy 1967).

Larva

Tipulid larvae have four instars, the last three of which are similar in appearance (Plate 49). They are nearly cylindrical but taper slightly at both ends. Larvae are light gray to grayish, greenish brown, with irregular black specks of various sizes. The cuticle is slightly transparent, revealing two longitudinal tracheal trunks and the alimentary canal (Jackson and Campbell 1975).

The truncated end of the last abdominal segment consists of the upper spiracular area and the lower anal area. The dorsal area has the spiracular disc with two rounded spiracles and six tapering anal lobes, the characteristic numbers in Tipulinae (Brindle 1960).

Pupa

The pupa is formed inside the last larval cuticle, the puparium (Plate 49). The five posterior abdominal segments have protuberances bearing caudally directed spines, so that the pupa can move vertically in their burrows in response to various stimuli (Alexander 1919).

Seasonal History and Habits

Seasonal Cycle

The ECF is a univoltine species that overwinters as a third instar. The insects become fourth instars in April, feed most vigorously to maturity, and stop feeding about the middle of May. Then they remain inactive for a prolonged period until pupation, which begins as early as mid-July but peaks about the end of August. During late August and September, the ECF passes in rapid succession through the pupal, adult, egg, and first-instar stages (Figure 74; Coulson 1962, Wilkinson and MacCarthy 1967).

During 1972 and 1973 in Washington, a few flies emerged as early as July and a few as late as early October, but 95% emergence occurred during about 20 days in late August and early September (Jackson and Campbell 1975).

By the time peak flights occur, peak oviposition has already taken place, since eclosion, mating, and oviposition occur predominantly on the 1st night of adult life. Eggs hatch in 11–15 days. Larvae feed ravenously and usually complete the first two instars in less than 2 months. Typically, third instars overwinter from November into April. The ECF does not have a true winter diapause (Barnes 1937, Coulson 1962, Jackson and Campbell 1975).

Adult Activity

Emergence, Mating, and Flight. Adults typically emerge shortly after sunset, finish mating by midnight, and lay most of their eggs by dawn, all during the same night. Teneral males are seen occasionally in the afternoon and can be captured on sticky board traps above 1 m (3 ft) in height throughout the afternoon. In Washington, the greatest male movement and mating occurred between 7:00 and 12:00 PM PDT. Males are attracted to females even while the latter are emerging from their pupal cases, and union occurs as soon as the teneral female has emerged (Jackson and Campbell 1975).

After union takes place, the female drags the male up the stem, with the male hanging downward and nearly motionless. Normally they may remain *in copula* for about an hour, but they may remain coupled for three hours or more. Fully gravid young females are apparently incapable of flight; they move clumsily over the soil

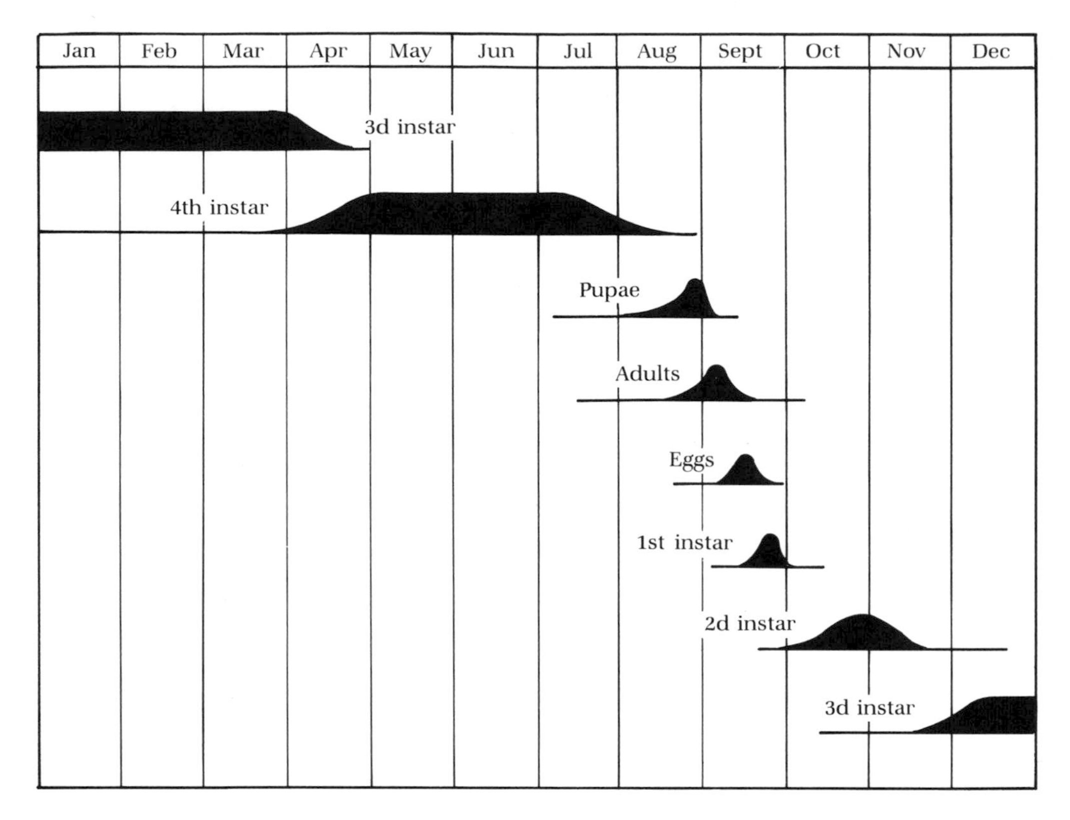

Figure 74. Life cycle of the European crane fly in northwestern Washington. (Adapted from Jackson and Campbell 1975, fig. 7; redrawn by R. McMillen-Sticht, NYSAES.)

surface, stopping occasionally to probe the soil for oviposition. Oviposition is mostly completed by dawn of the first night. Older females that have laid most of their eggs can fly. These females may take to the air while still in copula. These females alone can fly 1.8 m (6 ft) or more in height, as evidenced by catches on sticky board traps (Coulson 1962, Jackson and Campbell 1975).

Oviposition and Fecundity. Shortly after mating ceases, oviposition begins, and it continues until nearly all eggs are laid. From 50% to 75% or more of the eggs are laid by midnight of the day of emergence and about 10% on the next day. Females insert their terminal abdominal segments just below the soil surface (no more than 5 mm) to lay eggs. Nearly 70% of the eggs are within 0–1 cm (0–0.4 in.) in depth, and more than 90% are within the 1–3 cm (0.4–1.2 in.) in depth. Eggs are laid in rapid succession until from 200 to more than 330 have been deposited (Barnes 1937, Coulson 1962, Jackson and Campbell 1975).

Longevity. Adults take no food but do drink water, which lengthens longevity. Males live an average of 7 days, with a maximum longevity of 14 days. Females live an average of 4–5 days, with a maximum of 10 days (Barnes 1937).

Egg Development

The fresh eggs of *Tipula* have such high moisture requirements that they will collapse within 2–4 min unless they are in a saturated atmosphere. Fertile eggs absorb about half their weight in water when they are 2–3 days old. The optimum temperature for development and hatching is between 14° and 15°C (57°–59°F; Barnes 1937, Janisch 1941, Jackson and Campbell 1975, Laughlin 1958).

Larval Activity

The eggs hatch in 11–15 days, and larvae feed from the 1st day. First instars are about 2.7 mm long and grow to second instars about 4–5 mm long in about 2 weeks. They spend the winter as nondiapausing third instars, a stage that lasts as long as 4 months (Jackson and Campbell 1975).

Growth is rapid in spring, when most of the turf damage occurs. Upland soil holding three times its dry weight of water provides a minimum of larval mortality. Young larvae appear to prefer green leaves to roots and grow most rapidly, with the least mortality on white clover. Larvae surface to feed mainly at night but remain below the thatch to feed on roots and crowns during the day. On warm evenings and dark warm days, they can be seen feeding on the surface. Otherwise they are generally found within the upper 2.5 cm (1 in.) of sod. After mid-May, many are found as deep as 7.6 cm (3 in.) below the surface (Maercks 1939, Rennie 1917, Sellke 1937, Wilkinson 1969, Wilkinson and MacCarthy 1967).

Two periods of rapid growth occur, in the fall when they grow from first to third instars and again in the spring when they transform from third to fourth instar; they reach their peak weight of 300–500 mg in June (Jackson and Campbell 1975).

Pupal Activity

The pupae can move up and down in their earthen burrows in response to various stimuli, aided by curved spines on the abdomen. After adult emergence, the pupal cases can easily be found in the field, since half to two-thirds of their length protrudes from the soil. The sex of the individuals can be determined from their vacated genital sheaths (Jackson and Campbell 1975).

Miscellaneous Features

Factors Producing High Mortality or Survival

A very high mortality of eggs and first instars occurs when the surface soil atmosphere is even slightly less than 100% RH. Nearly 100% mortality of these stages has been observed in England, with death due to desiccation and infertility. Mortality from severe winter weather can be as high as 30% to 40% in third-instar larvae (Coulson 1962).

The ECF has been increasing in importance in the Pacific Northwest because of the mild winters and cool summers and rainfall in excess of 50 cm (20 in.). A 30-year average indicates that the wet coastal belt of British Columbia and northern Washington is practically ideal. Even the extremes of summer heat and dryness and the coldest of winters in this area have permitted more than 40 larvae per 0.1 m^2 (per ft^2) to develop (Wilkinson 1969).

Population Survey Techniques

Adults are best sampled with a lightweight sweep net. Immatures may be sampled in the fall with a soil corer 10.16 cm (4 in.) in diameter to a depth of 7.6 cm (3 in.). Larvae do not penetrate below this depth. The cores may be dried in a Berlese funnel to force larvae out. This method gave a more accurate count than the application of an aqueous solution of *o*-dichlorobenzene as an irritant to force larvae to the surface of the turf (Jackson and Campbell 1975, Shaw et al. 1974). Antonelli (personal communication, 1985), rather than using Berlese funnels, determines larval presence by merely breaking the samples apart. Spring sampling for larger larvae is generally more useful for the purpose of making decisions on the need for control.

Economic Threshold Population

In British Columbia, 20 or more larvae per 0.1 m^2 (per ft^2) in a lawn were considered the threshold population for a recommendation that insecticidal control be undertaken (Wilkinson and Gerber 1972).

In Washington, a survey can be made in March, or when temperatures become consistently warmer, by selecting several areas in the lawn and digging up patches of turf 0.1 m^2 (1 ft^2) and 2.5–5.0 cm (1–2 in.) in depth. If average counts exceed 25 per 0.1 m^2 (per ft^2) in healthy turf, an insecticidal treatment should be considered. Turf in poor condition can show damage with 15 larvae per 0.1 m^2 (per ft^2; Antonelli and Campbell 1984; Antonelli, personal communication, 1985).

Natural Enemies

Several microorganisms, parasites, and predators are associated with the ECF in Europe. They exert little influence on population reductions and are probably of even less effect in America. A tachinid parasite, *Siphona geniculata* (de Geer), considered to be the most promising, has been obtained from Germany and was released in Vancouver. The predation of larvae by starlings and native moles, *Scapanus* spp., exerts some influence (Wilkinson 1969, Wilkinson and MacCarthy 1967).

Frit Fly

Taxonomy

Frit fly, FF, *Oscinella frit* (L.), order Diptera, belongs to the family Chloropidae.

Importance

The FF is considered responsible for more damage to golf greens and collars (edges of greens) than is generally recognized. Damage occurs when larvae feed on the terminal shoots of grasses. The adults annoy golfers because the flies are attracted to white balls, on which they will land. The FF is a serious pest of cereals in Europe but not in the United States. In Virginia, the FF is a serious pest of reed canary grass, *Phalaris arundinacea* (L.), a hay crop, and much knowledge of this insect in this country comes from studies on this crop (Allen and Pienkowski 1974).

History and Distribution

The FF is widely distributed in the temperate Northern Hemisphere. It is reported from northern Europe into Russia. In the United States it is commonly found north of the latitude of Virginia and Kansas, wherever grass is green for a considerable part of the year. It is most abundant in regions where winter wheat is grown, from the Great Lakes to the Ohio River and westward to Missouri. In Virginia it is more abundant at higher elevations. Infestations in lawns in Connecticut and Rhode Island have been reported often (Aldrich 1920, Allen and Pienkowski 1974, Vance and App 1971).

Host Plants and Damage

Common cool-season turfgrasses are hosts, with bluegrass and bentgrass most susceptible, especially when these grasses are kept watered and short by mowing or grazing. Fly density tends to be higher where tiller density is the thickest. Larvae infest stems of barley, wheat, oat, and rye in Europe and feed also on immature kernels, producing empty kernels called *fritz* in Sweden, hence the common name. In the mountains of Virginia, the FF is an important pest of reed canary grass where the yields of the second hay harvest are significantly reduced (Aldrich 1920, Allen and Pienkowski 1974, Jepson and Heard 1959).

Larval feeding in the upper primordial leaves causes general yellowing and death of the central leaf, while surrounding shoots remain green. On golf courses, damage is often first apparent on collars and approaches the center of the green. Higher elevations of greens are usually the first to show symptoms. Greens with soil high in organic matter appear more susceptible. Tunneling of the stems near ground level causes the upper portion of the plants to become brown and die (Plate 50; Aldrich 1920, Bowen 1980).

Description of Stages

Adult

The flies are 1–2 mm long and black, with a shiny dorsum, yellow halteres, and yellow markings on the tibial segments of the legs (Plate 50; Aldrich 1920, Dahlsson 1974).

Egg

Eggs are pure white and 0.7–0.8 mm long, with a finely ridged surface. They are not easily detected because of their small size, and they can be jarred loose easily from leaves (Aldrich 1920, Allen and Pienkowski 1974).

Larva

There are three larval instars. Young larvae are yellow white, with two black mouth hooks and posterior spiracles. Mature larvae are about 3 mm long and yellow, with black curved mouth hooks and visible anterior and posterior spiracles (Plate 50; Aldrich 1920, Bowen 1980).

Pupa

Pupae are yellow at first, turn dark brown, and are <3 mm long (Plate 50; Bowen 1980).

Life History and Habits

Seasonal Cycle

Maggots overwinter in grass stems infested the previous fall. Pupation and adult emergence occurs early the following spring, about March in California and mid-April at Blacksburg, Virginia. Three generations a year are reported in the higher elevations of Virginia (Figure 75) and four each year in Indiana. At Blacksburg, peak flights of flies occurred from April to early June from overwintering maggots. Populations of first-generation flies from late June to mid-August are the smallest, and those of the second generation during mid-August to early October are intermediate. Peaks in larval population closely followed those of the adults. First-generation larvae complete development in about 5 weeks, second-generation larvae do so in 8–10 weeks, and the third generation overwinters. The average life cycle from adult to adult is 21–58 days (Aldrich 1920, Allen and Pienkowski 1974, Bowen 1980).

Adult Activity

Upon eclosion, adults climb up the inside of the dead and dried stems and escape through cracks and crevices. When flying, flies are attracted to and land on white

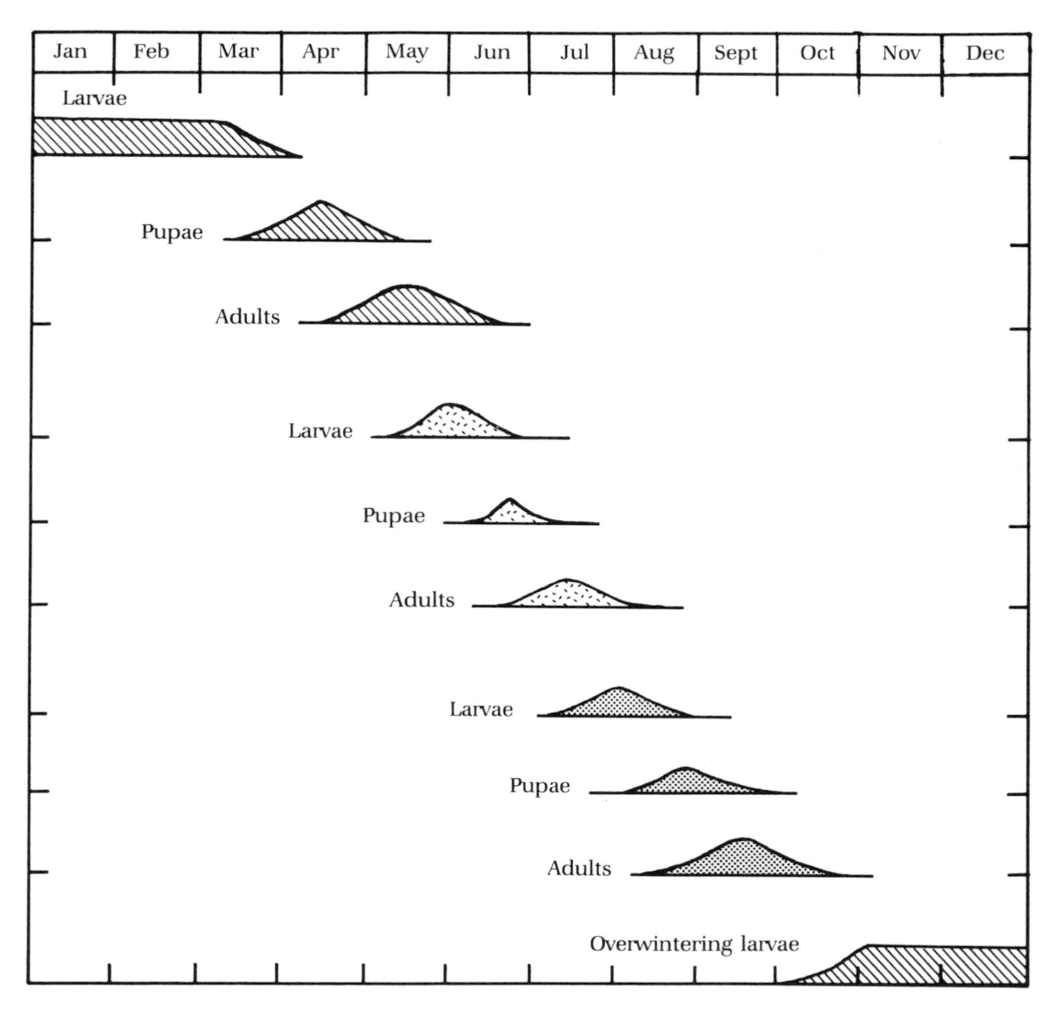

Figure 75. Life history of the frit fly in southwestern Virginia. (Data from Allen and Pienkowski 1974; drawn by H. Tashiro, NYSAES.)

objects. They are often seen hovering over golf greens and can be collected from midmorning onward throughout the summer (Baker 1982, Niemczyk 1981).

Mating and Oviposition. Males mate within 6 hours of emergence. Females do not mate until the second day. Mating pairs have been observed on leaves and stems of host plants from about 9:30 AM until sunset. Eggs are usually laid on leaves, in leaf sheaths, and in the brown stubble. Females live an average of 5.5, 18.0, and 26.0 days in the spring, summer, and fall generations, respectively. Egg production averaged 17, 59, and 32 eggs for the spring, summer, and fall generations, respectively (Allen and Pienkowski 1974, Southwood et al. 1961).

Egg Survival

The incubation period of eggs in the field is 3–7 days. Rainfree periods are most important for good survival when tall grasses are infested, because eggs are dislodged during rain. In Virginia, with less frequent rains in spring, egg mortality is <5%. During summer, with more frequent rains, mortalities often exceed 50% because eggs are dislodged (Allen and Pienkowski 1974).

Larval Activity

Upon hatching, the larvae immediately enter the grass stems, migrate to the crown, and girdle the embryonic growing tissue. Soon the whorl leaf begins to yellow, wither, and brown. As many as six larvae generally develop per tiller, with more developing per tiller in the spring than in the summer. Upon completion of larval development, the puparium is formed at the base of the whorl leaf. Overwintering larvae remain at the point of original feeding at the base of the leaf (Allen and Pienkowski 1974).

Miscellaneous Features

Populations have been surveyed with a sweep net. A water trap with a wetting agent has also been used. A blue- or violet-colored water trap gave higher catches and more balanced sex ratio catches than any other color. Adult flies have been successfully collected with a DeVac machine and and have been separated from grass litter using Berlese funnels (Allen and Pienkowski 1974, Oschmann 1979).

Natural Enemies

Several insects are reported to be parasites or predators of FF in Canada and the United States, but none causes high mortality (Allen and Pienkowski 1975).

19

Hymenopteran Pests: Families Formicidae and Sphecidae

Ants

Taxonomy

Many species of ants, order Hymenoptera, family Formicidae, invade turfgrass areas throughout the country. Members of the subfamily Formicinae that are commonly reported include the cornfield ant, *Lasius alienus* (Foerster), the mound-building ant, *Formica exsectoides* Forel, and the red ant, *F. pallidefulva* Latreille. All others mentioned herein belong to the subfamily Myrmicinae and include the pavement ant, *Tetramorium caespitum* L., harvester ants, *Pogonomyrmex* spp., and the complex of fire ants of the genus *Solenopsis.* There are four fire ants, two native to the United States and two introduced. Natives include the fire ant, *Solenopsis geminata* (Fabricius), and the southern fire ant, *S. xyloni* McCook. The two introduced are the black imported fire ant, *S. richteri* Forel, and the red imported fire ant, RIFA, *S. invicta* Buren. The RIFA is by far the most important of all the genera and species mentioned (Creighton 1950, Lofgren et al. 1975).

Importance

Ants are primarily troublesome in turfgrass areas because they build mounds as they form subterranean homes for their colonies. They seek out drier, well-drained sandy soils that have low water-holding capacity. The galleries they form, which damage roots, add to the desiccation of the soil, and the turf in the surrounding areas becomes thin and unsightly. Mounds of various sizes and shapes, formed according to the habits of the ant species, are detrimental to lawn mower blades. Some mounds are too high to mow over. Some ant species have vicious bites and stings, with the RIFA being the most serious (Baker 1982, Oliver 1982a, Schread 1964).

Probably because exaggerated claims have been made that RIFA is a pest serious enough to kill livestock and wildlife and to destroy crops, more public funds have been spent on this insect than on any other in history. Although $200 million have been used for eradication and control efforts, the ant continues to spread (Oliver 1982a).

History and Distribution

Ants have a cosmopolitan distribution from the Arctic to the most tropical areas of the world. Harvester ants occur west of the Mississippi River, the two most common being the Texas harvester ant and the western harvester ant. Connecticut lawns have been invaded by colonies of the cornfield ant, the red ant, the mound-building ant, and the pavement ant. In the Northeast, the cornfield ant is the most frequently encountered species on golf greens. All four species of fire ants are found only in the southern states (Baker 1982, Schread 1964).

Both imported fire ants were introduced into or near Mobile, Alabama, at different times. The black imported fire ant, probably introduced from Argentina, was first recorded in 1918 and still remains in a small area of northeastern Mississippi and northwestern Alabama (Figure 76). The RIFA, a native of Brazil, was introduced about 1933–1940. Because it is replacing the other three fire ants and is becoming the most common in nine states from North Carolina to Texas, it is usually the insect in question whenever fire ants are discussed. A narrow band passing from

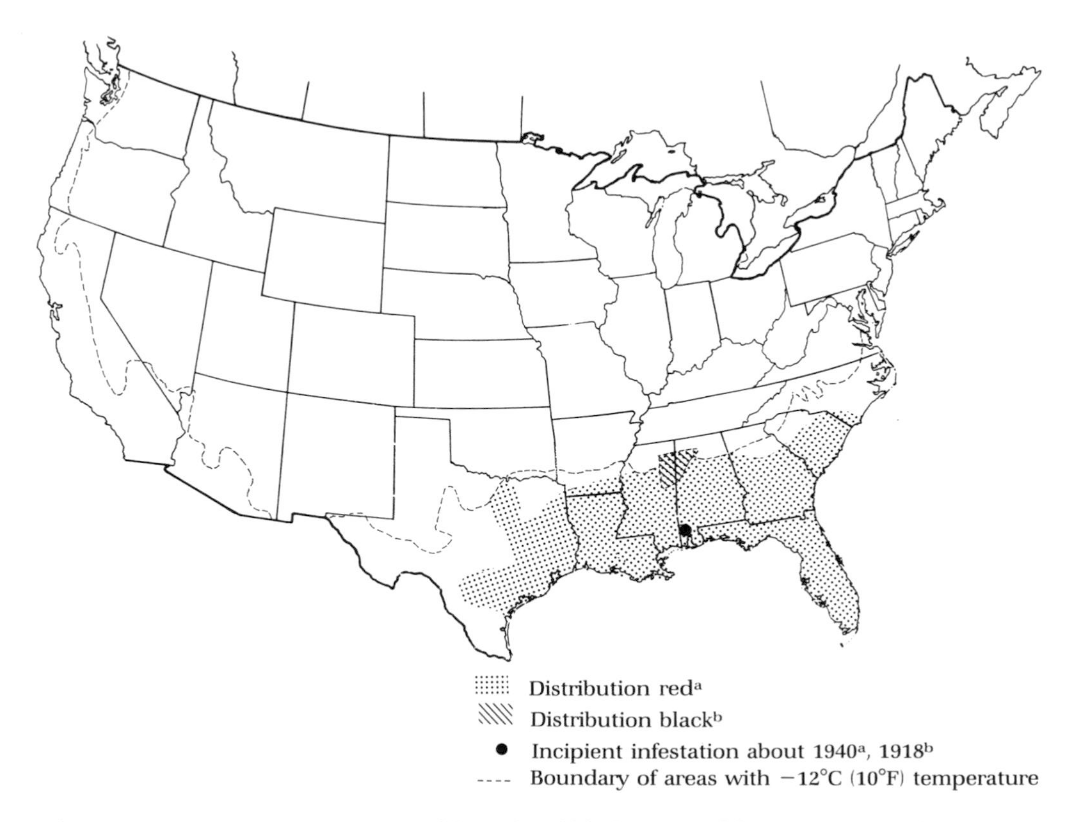

Figure 76. Approximate distribution of the red and black imported fire ant in 1980 and maximum possible spread of the former. (Adapted from Oliver 1982a, fig. 2, courtesy of *American Lawn Applicator*.)

coastal Virginia south into the infested states and from Texas into California and up the Pacific coast through Washington represents the area of −12°C (10°F) minimum temperature. This marks the area of maximum probable spread because of the RIFA's minimum temperature tolerance (Oliver 1982a).

Host Plants and Damage

All sunny turfgrass areas, regardless of grass species, can be invaded by ants in search of areas to start a new colony. Lawns, parks, playgrounds, golf courses, and other turfgrass areas in addition to cultivated farm and pasture land are likely sites as long as they are well drained (Plate 51). Ants that feed on excretions of root-infesting aphids help promote the aphid colonies, which are also detrimental to the health of grasses (Baker 1982, Schread 1964).

As the colonies grow and mature, mounds of earth are created by many species, while other species leave the surface relatively flat. On the surface the unsightly mounds smother the grass, while the galleries in the soil disturb the roots, dry the soil, and cause the grass to thin out (Plate 51). The mounds of various sizes make turf maintenance difficult, and mowing over even low mounds dulls and chips mower blades as the surface is scalped (Baker 1982, Schread 1964, Vance and App 1971).

Mounds vary in size, shape, and appearance, depending on the ant species involved. Mounds of the RIFA may be 35 cm (14 in.) or more in diameter and 20–25 cm (8–10 in.) high (Plate 51). On permanent sod where ample food for the RIFA is available, as many as 125 mounds per ha (50 per acre) can develop. Harvester ants not only construct large mounds but clear areas of grass around the mounds in an area 7 m (7.7 yd) or more in diameter and along trails radiating out from the mounds. Red ants have large colonies with several openings at the surface but do not pile soil in definite mounds. The surface soil feels spongy, and grass grows poorly (Baker 1982, Oliver 1982a, Schread 1964).

When the mounds are disturbed, worker ants attack by sinking their powerful jaws into the skin, afterward repeatedly thrusting their poisonous stinger into the flesh. This may be the most serious problem with ants in a turfgrass area. The RIFA has the most vicious of stings, one that produces a burning sensation, but it does not have the severity of a honeybee or yellow jacket sting. The area stung usually festers within a day, and the sting persists for several days. Harvester ants also inflict a painful sting. People who are allergic to bee and wasp stings may also be allergic to the RIFA sting and should take appropriate precautions (Baker 1982).

Description of Stages

Adults

Few people are unfamiliar with the characteristic shape and form of ants. Ants have a very narrow, constricted connection between the thorax and head and an even more conspicuous constriction between the thorax and the abdomen. The constriction is made by the greatly modified first two segments of the abdomen. Adults may be winged males, winged females, or wingless female workers (Plate 51). The body surface may be smooth or hairy, red, brown, black, or yellow. Harvester ants are 5–6 mm long and reddish brown to yellow or black. The red ant is nearly 13 mm long, with a red head and thorax and a black abdomen (Baker 1982, Schread 1964).

Eggs

The eggs of ants are white to pale yellow and vary in shape according to species. Eggs of harvester ants are <0.5 mm long and elliptical (Baker 1982).

Larvae

Larvae of ants are mostly white and legless, with a body that dilates gradually from the anterior to the posterior (Plate 51). Larvae have small, distinct heads. Harvester ant larvae are shaped like a crookneck squash or a gourd (Baker 1982).

Pupae

The ants' soft, colorless pupae resemble adults in size and shape. Some are enclosed in a papery cocoon; others are not (Plate 51). Harvester ant pupae occur in cocoons; the body is rigid and straight, with legs and wings distinct (Baker 1982).

Life History and Habits

Life Cycles

Male ants appear only in very large or old colonies. Winged males and females swarm from parent colonies, pair off, and mate, and the males die soon afterward. The fertile winged female, the queen, flies to a site suitable for a colony, loses her wings, finds a suitable nesting site, and begins laying eggs. She cares for and feeds the first brood of larvae that develop into workers, the wingless females. Thereafter her sole function is to lay eggs. The workers then care for the eggs and the queen, construct and repair nests, gather food, feed the immatures, care for the brood, and defend the colony (Baker 1982, Schread 1964).

Colony Formation of RIFA

Nuptial flights have been noted every month of the year, but peak activity occurs from late May through August. The young mated female upon alighting loses her wings and burrows vertically into the soil to a depth about 10 cm (4 in.) and constructs an egg chamber near the bottom. A new colony is begun as she deposits several eggs. She tends the eggs as they incubate and feeds the larvae as they develop. The transition from egg to worker requires 3–4 weeks. The first workers to develop are minims (minute, polymorphic workers). Within 2 or 3 months they are gradually replaced by the minor workers, and in about 5 months to a year the minor workers are replaced by the major workers. Colonies are generally considered mature after 3 years. The average colony size may vary from 100,000 to 500,000 workers and includes relatively few winged or reproductive forms. Where food is abundant, as many as 50 mounds per acre may develop. The preferred food of the RIFA is other arthropods—termites, springtails, crickets, caterpillars—and other soil-inhabiting species (Baker 1982, Lofgren et al. 1975, Oliver 1982a).

Natural Enemies of RIFA

Microorganisms

A species of *Pseudomonas* bacteria is reported to be highly toxic to the RIFA. Three common entomophagous fungi are also pathogenic to the RIFA. These are *Beauveria bassiana, Metarrhizium anisopliae,* and *Aspergillus flavus* Link. (Lofgren et al. 1975).

Insect Predators

Other ants of several species have been found in RIFA nests, but there is no evidence that any of the foraging predators exerts much influence on the RIFA (Lofgren et al. 1975).

Bees and Wasps

Many ground-nesting bees and wasps become troublesome pests of turfgrass in much the same way that ants are a pest. Their nesting habits create mounds of soil as they bring subsoil to the surface in making their galleries and nests. One of the more prominent is the cicada killer, a large wasp with unusual habits.

Cicada Killer

Taxonomy

The cicada killer, CK, *Specius speciosus* (Drury), is of the order Hymenoptera, family Sphecidae, subfamily Bembicinae. No doubt it produces the greatest soil surface disruption of the wasps because of its large size. This species is also called the sand hornet or golden digger wasp.

Importance

In size and forbidding appearance, the CK appears to be one of the most formidable of North American wasps. Its sting causes an initial sharp pain and lasts about a week. The CK has no nest-guarding instinct, however, and stings only when provoked. Apart from the possibility that it will sting people, this insect has a slight economic importance as a pest of turfgrass because it makes unsightly mounds of earth as it digs its burrows (Damback and Good 1943).

History and Distribution

The CK is present in all states east of the Rocky Mountains. Its nests are made in full sun where vegetation is sparse, in light, well-drained soils (Baker 1982).

Host Plants and Damage

The location of CK nests and the resulting mounds has no bearing on the kind of turfgrass present. Wherever the CK makes nests in turfgrass, the mounds smother the grass (Plate 52). A very gravelly or bare area is often preferred.

Description of Stages

Adult. The wasps are large, >30 mm in length, and black, with a yellow-banded abdomen. The head and thorax are rusty red, and the wings are russet yellow (Plate 52; Baker 1982, Matheson 1951).

Egg. The cylindrical, cigar-shaped, translucent greenish white eggs are 3–6 mm long and 1.0–1.5 mm wide. Each egg is placed under one femur on a middle leg of its victim, a cicada (Baker 1982).

Larva and Pupa. Larvae grow to a maximum length of 28–32 mm. When they are mature, they become somewhat shrunken and leathery, brown, and stiff. The pupa is undescribed; its case is about 32 mm long and about 11 mm wide (Baker 1982).

Life History and Habits

Life Cycle. The CK overwinters as a mature larva within a cocoon in its earthen cell. Pupation occurs in the spring and lasts 25–30 days. The wasps emerge in about the 1st week in June in Arkansas and rarely before the 1st week in July in Ohio, but in both locations the CK closely approximates the emergence of its prey. Once emergence has begun, it continues throughout the summer. In Ohio, adults live for 60–70 days, are present from July to mid-September, and disappear as the cicadas are on the wane. Eggs are deposited during late July through August. Eggs hatch within 24–48 hours, and the larvae complete their growth in 4–14 days. There is only one generation a year (Baker 1982, Damback and Good 1943).

Adult Activity. Immediately after their emergence, female wasps feed on exudates of trees and flowers. Male wasps station themselves on the tops of plants and when a female flies nearby capture her. They mate in flight. Mating and the digging of burrows take place several weeks before the insects prey on cicadas. Each female digs her own burrows 15–25 cm (6–10 in.) deep and 1.2 cm (0.5 in.) in diameter. Broadly oval cells are then dug perpendicular to the main tunnel. The female digs by dislodging soil with her mouth and kicking the loosened particles back as a dog would dig. The excess soil thrown out of the burrow forms a U-shaped mound at the entrance, producing the wasp's primary detrimental effect on the turf (Baker 1982, Damback and Good 1943, Davis 1920).

Once the burrows are ready, the female wasps seek adult cicadas (Plate 52) in trees, charge into the victim, and sting it between the abdominal segments to paralyze it. Straddling the cicada (Plate 52) and firmly grasping it, the female glides or flies to her burrow with her prey and places it in one of the earthen cells. She places an egg snugly under a femur of one middle leg of the cicada, then seals the cell. Occasionally two cicadas are placed in a single cell, and a larger wasp develops (Damback and Good 1943, Matheson 1951).

Danger of Stings. Wasps are capable of inflicting a painful sting, but it is usually difficult to provoke. Mating males are aggressive (but do not sting), and females near their burrows are easily disturbed and provoked. It is wise to keep a distance from burrows to avoid the possibility of a sting (Baker 1982).

Larval Activity. Eggs hatch in 24–36 hours, and the young larva immediately inserts its mouthparts into the body cavity of the cicada to feed on the liquid contents. Growth is very rapid, and the wasp larva completes its development in as short a time as 5–14 days. It then weaves a cylindrical case, the cocoon, around itself for overwintering (Damback and Good 1943).

Natural Enemies

A sarcophagid fly maggot is sometimes found feeding on the same cicada as the wasp, but whether it is predatory on the wasp is not known (Damback and Good 1943).

20

Secondary Insect and Mite Pests

Turfgrass insects and mites that are troublesome on a regular basis over a wide area are naturally considered major pests. A smaller number of pests confined to a limited geographic area are nevertheless regularly troublesome where they occur and are also considered major pests. Major pests are discussed in Chapters 3–19.

Another group of insects and mites become turfgrass pests only on an occasional basis, or, if they occur regularly, their activities produce only minor injury. These are designated as secondary turfgrass pests, and I discuss each briefly. They are listed in the same sequence of orders as the major pests.

Clover Mite

Importance, Host, and Damage

The clover mite, *Bryobia praetiosa* Koch (Acari: Tetranychidae), is often concentrated in the turf next to the foundation of a building (Plate 53). The common name indicates only one of its preferred host plants; it also feeds on a variety of herbaceous plants, including turfgrasses. The grass may be killed about 0.3–1.0 m (1–3 ft) out. The mites' feeding causes the turf to become silvery because of the extraction of plant sap and the drying of cells.

During the spring and fall when they are plentiful, mites will often invade homes and are first seen on windowsills. They do not bite, transmit any disease, or feed in the house, but when crushed, they leave a reddish stain. They are more of a nuisance than a pest.

Description

Adult mites are less than 1 mm in length, with a reddish brown to greenish body and four pairs of legs. The unusually long front pair, extending in front of the body, is their most prominent feature (Plate 53). Eggs are bright red and spherical and are laid on the walls of buildings, on the bark of trees, and on other plants (Plate 53). A 10× hand lens is sufficient for viewing these mites and their eggs.

Life Cycle and Habits

Clover mites prefer to be active during cool spring and fall periods. Like the winter grain mite, the clover mite usually oversummers in the egg stage. When day temperatures regularly rise above 21°C (70°F), this mite becomes less active, and the adult population is reduced. The summer eggs hatch when cool fall temperatures return, and the mites will be active until freezing temperatures cause the turf and mites to become dormant. Several generations can occur during the fall, winter, and spring seasons (Jeppson et al. 1975).

Banks Grass Mite

Importance, Host, and Damage

The banks grass mite, *Oligonychus "reckiella" pratensis* (Banks) (Acari: Tetranychidae), is reported as a pest of bluegrass and bermudagrass and has become a pest of St. Augustinegrass in Florida. It is also a pest of wheat, corn, sugarcane, sorghum, and dates. Feeding produces stippling and resembles mildew or St. Augustinegrass decline, a viral disease. Most serious damage normally occurs in water-stressed turf. It is not considered to damage irrigated turf. Bluegrass is attacked in the Pacific Northwest, and bermudagrass is damaged in Arizona, New Mexico, and Texas.

Description

Adult mites are deep green, with a very light salmon color over the palpi and front pair of legs. Their color makes them difficult to detect on St. Augustinegrass. The banks grass mite is smaller than most spider mites. It forms considerable webbing at the base of host grasses.

Life Cycle and Habits

The life cycle lasts 8–25 days, depending on the temperature. Adults live an average of 23 days (Cromroy and Short 1981, Jeppson et al. 1975, Malcolm 1955).

Short-Tailed Cricket

Importance, Host, and Damage

The short-tailed cricket, *Anurogryllus muticus* (De Geer), and other species (Orthoptera: Gryllidae), occurs along the Atlantic coast from New Jersey to Florida and west to Texas and eastern Oklahoma. It feeds on grasses, weeds, pinecones, and pine seedlings. Burrows constructed by nymphs and adults produce unsightly mounds of small soil pellets that are most abundant after a rain, when the soil is moist and the weather warm. Damage caused by night feeding on grass blades is negligible.

Description

Adults are brown, with fully developed brown to black wings, and resemble field crickets except for the short ovipositor (Plate 53). The body length varies from 14 mm to 17 mm. The off-white glabrous, oblong eggs are 1.8–2.8 mm long and 0.6–1.2 mm wide. All nymphal instars are light brown.

Life Cycle and Habits

This cricket overwinters in the penultimate instar. Individuals become adults in early spring, mate, and oviposit during late spring or early summer in a multichambered burrow constructed by the adult. Both eggs and young nymphs occupy the burrow, but the fourth- to sixth-instar nymphs leave to construct their own burrows. At first the burrows are 2–3 mm in diameter and 5–10 cm (about 2–4 in.) deep. As the cricket matures, burrows are enlarged and become 30–51 cm (12–20 in.) deep. Only one mature cricket occupies a burrow, but females with their new brood of young can be found in a single burrow (Baker 1982, Weaver and Sommers 1969).

Grasshoppers

Importance, Host, and Damage

Grasshoppers (Orthoptera: Acrididae) have a worldwide distribution and occur throughout the United States but are most troublesome in the semiarid regions from Montana-Minnesota south into New Mexico–Texas. Grasshoppers feed on a wide range of forage legumes and grasses but do not feed on turfgrasses unless other forage is scarce or the weather is particularly dry.

Description

Grasshoppers are about 19–38 mm long when full grown. Their color varies from reddish brown to dull yellow to greenish, with other markings (Plate 53). Egg pods are oval-elongate and curved, with eggs that are white, yellow green, tan, or various shades of brown and about the size of rice kernels. Nymphs resemble small wingless adults and are white until they assume the colors and markings of adults.

Life History and Habits

Many species overwinter as eggs and hatch from April to June. Nymphs molt five or six times over 35–40 days to become adults. Oviposition begins about 2 weeks after mating. Egg pods placed in the soil may contain 15–150 eggs, and each female may produce 300 eggs. Some species swarm to feed and for egg deposition on sunny, warm days. Generally one or two generations occur each year (Baker 1982).

Periodical Cicadas

Importance, Hosts, and Damage

Periodical cicadas, *Magicicada* spp. (Homoptera: Cicadidae), occurring in the eastern United States, are known best for their very long nymphal life under trees and because they emerge from the soil rather suddenly as full-grown nymphs. The numbers of exit holes may be quite striking, with as little as 5 cm (2 in.) separating each hole and marring the surface of the turf. Twigs and small branches are injured by the female as she slashes these tissues with her ovipositor in depositing eggs.

Description

The periodical cicada, *Magicicada septendecim* (L.), also called the 17-year cicada, is the best known of the periodical cicadas. It has reddish eyes and wing veins. Its body is 27–33 mm long but is about 40 mm long to the tips of the folded wings (Plate 53).

Life History and Habits

The presence of cicadas is obvious from the day-long shrill "singing" of males on summer days. Females lay eggs in the twigs of various trees and shrubs. Eggs hatch in about a month. Nymphs drop to the ground, burrow into the soil, and feed on roots of trees and shrubs. After spending 17 years as nymphs in the soil, the last nymphal instars emerge, climb some object, usually a tree trunk, and fasten themselves to the bark, from which point adults escape for a short existence of about a month (Borror et al. 1981, Shurtleff and Randell 1974).

Two-Lined Spittlebug

Importance, Host, and Damage

The two-lined spittlebug, *Prosapia bicincta* (Say) (Homoptera: Cercopidae), occurs from Maine to Florida and west to Iowa, Kansas, and Oklahoma, feeding on grasses and many crops. Its hosts include bermudagrass, St. Augustinegrass, centipedegrass, bahiagrass, and ryegrass. Adults and nymphs pierce plant tissues and feed on the sap, causing the plants to wither or to stop growing. The spittle masses are unpleasant under foot.

Description

The black adults with red eyes and legs have two red or orange lines across the wings (Plate 53). Adults are 6–10 mm long. The bright yellow-orange eggs are oblong, pointed at one end, and about 1 mm long. Wingless nymphs resembling small adults are yellow, orange, or white, with red eyes and a brown head. They are enveloped in a white frothy mass.

Life History and Habits

Eggs overwinter in hollow stems, behind leaf sheaths, or among plant debris. Nymphs emerge in the spring, seek sheltered, humid hiding places among plants, and start feeding. The nymph exudes a white frothy mass resembling spittle around itself that protects it from natural enemies and dessication (Plate 53). Nymphs develop through four instars to become adults in about 1 month. Adults are most active during early morning, spending the warmer part of the day hiding in the grass. They live about 23 days, with females depositing eggs for about 2 weeks. Two generations occur annually from northern Florida to North Carolina (Baker 1982, Byers 1965, Fagan and Kuitert 1969).

Leafhoppers

Importance, Hosts, and Damage

Many species of leafhoppers (Homoptera: Cicadellidae) are common throughout the eastern United States. All turfgrasses may be attacked. At least three of the species known to feed on turfgrasses are the painted leafhopper, *Endria inimica* (Say), on bluegrass, *Draeculacephala minerva* Ball, and *Deltacephalus sonorus* Ball. Both adults and nymphs suck plant sap from leaves and stems, causing a bleaching or drying out of the grass, which eventually turns brown. Leafhoppers rarely damage turf in the northern and eastern states but can cause serious damage in the prairie states and in some Gulf states. Newly seeded lawns can be killed outright by leafhoppers.

Description

Adults of various species average about 4–7 mm in length, have triangular heads, and have a general body color of yellow, green, or gray (Plate 54). Eggs are white, elongate, and 1 mm or less in length. Nymphs are of various colors as adults and are wingless at first but develop wing stubs as they molt and have the general shape of adults except that they are smaller. Nymphs have a habit of moving sideways or backward when disturbed.

Life History and Habits

Leafhoppers may overwinter as eggs or as adults. Some species overwinter only in the Deep South and migrate annually into the northern latitudes to establish an infestation. In the spring after feeding and mating, females insert from less than a hundred to several hundred eggs singly beneath the epidermis of leaves and stems. Eggs hatch in about 10 days and develop through five instars to adults in from about 12–30 days. From one to four generations may be produced, depending on the latitude and on the species (Baker 1982, Bowen 1980, Shetlar et al. 1983).

Leafbugs and Fleahoppers

Importance, Hosts, and Damage

One common species on California turf is the white-marked fleahopper, *Spanogonicus albofasciatus* (Reut.) (Heteroptera: Miridae). Adults and nymphs suck juices from leaves and stems of all grasses and dichondra. When infestations are heavy, growth is retarded, and grass may die in spots. An open palm run over the grass will make fleahoppers hop about if they are present.

Description

Adults of the above species are about 3 mm long and blackish or grayish, with white markings on the wings, which are folded flat over the back. They can be differentiated from the flea beetles by their long, thin antennae.

Although the tarnished plant bug, *Lygus lineolaris* (Palisot de Beauvois), is not a common turf-infesting species, it is included as a representative of this group (Plate 54).

Life History and Habits

Little is known about their life history. Several generations are thought to occur each season. Fleahoppers jump readily and fly short distances when they are disturbed (Bowen 1980).

Bermudagrass Scale

Importance, Host, and Damage

Bermudagrass scale, *Odonaspis ruthae* (Homoptera: Diaspididae), sometimes called Ruth's scale, is a pest wherever bermudagrass is grown. It seldom causes serious damage but is most injurious when the grass is under stress. Heavily infested turf has a dry appearance, with severely retarded new growth. The turf becomes weakened, thin, and susceptible to other stresses (Plate 54). The scale thrives in heavily thatched turf and does its greatest damage in heavy shade.

Description

Adults are about 1.6 mm in length and are covered with a hard, white secretion of a white circular or clam-shaped covering. They are found on stems, clustered around the nodes, and occasionally on the leaves (Plate 54).

Life History and Habits

Females lay eggs under their clam-shaped waxy coverings. Upon hatching, the crawlers move about the grass plant and soon settle down, insert their piercing mouthparts into the grass, and become sessile (Plate 54). In molting they lose their antennae and legs, exude their waxy covering, and remain for many months before laying eggs to repeat the life cycle (Converse 1982).

Turfgrass Scale

Importance, Host, and Damage

Turfgrass scale, *Lecanopsis formicarum* Newstead (Homoptera: Coccidae), was found during 1983 and 1984 on turfgrasses of home lawns, parks, and sod farms in the province of Ontario, Canada. It occurred on Kentucky bluegrass, red fescue, and creeping bentgrass but has not been seen on annual bluegrass. Sodded lawns 3–5 years old were the most commonly infested. In Poland, red fescue was the grass most commonly infested. Nymphs occur on the tillers of host grasses, usually at the base of leaves and stems and on rhizomes but not on roots (Plate 54).

Description

The oblong adult females are yellowish with broad brown lateral stripes on each side of the median area to a uniformly orange-brown body (Plate 54). They are about 1.5 mm wide by 2.5 mm long (Plate 54). Females produce a cottony mass of silk containing eggs (Plate 54). The minute elongate-oval crawlers are reddish. Several nymphal instars are present.

Life History and Habits

In Ontario, mature nymphs were present throughout the year and are the only stage that overwinters. These became adults in May and June and produced the cottony masses of silk containing eggs. Crawlers were present in late June and July.

At the peak of their activity, crawlers move to the tips of the grass blades, causing the turf to become reddish. This stage is dispersed by the wind. First-stage nymphs were present in July and August. Intermediate-stage nymphs were present from August into October (Boratynski et al. 1982, Sears and Maitland 1983, 1984).

Cottony Grass Scale

The cottony grass scale, *Eriopeltis festucae* (Fonscolombe) (Homoptera: Coccidae), was found infesting and killing home lawns in New Jersey (Plate 55). Cottony masses are produced on grass stems (Plate 55). Very little is known about this scale. It was found in large numbers in Nova Scotia in 1889 and thereafter also in New Brunswick. It is said to occur in the Dakotas, Indiana, and Illinois on timothy and redtop, *Agrostis alba* L. (King 1901, L. M. Vasvary, Rutgers University, personal communication, 1985).

Lucerne Moth

Importance, Host, and Damage

The larvae of the lucerne moth, *Nomophilia noctuella* D. and S. (Lepidoptera: Pyralidae), normally feed on clover and similar legumes but will feed on grass leaves and stems. If damage appears, it is usually late in the summer, and control measures are seldom necessary.

Description

Adults are mottled gray brown with two pairs of indistinct dark spots on the forewing (Plate 55). Wingspan is 25–30 mm. Larvae are slender, about 25 mm long, and spotted like those of sod webworms but larger. When disturbed, they become very active.

Life Cycle and Habits

Moths fly at night and lay eggs on clover, other legumes, dichondra, and turfgrasses. Larvae live in silken tubes near the base of the plant (Bowen 1980, Shurtleff and Randell 1974).

Burrowing Sod Webworms

Importance, Host, and Damage

Burrowing sod webworms (Lepidoptera: Acrolophidae) are primarily tropical insects but occur from Pennsylvania and New Jersey south through North Carolina into Florida and westward through Texas and Nebraska and into Arizona. They infest the roots of most lawn grasses and other plants, severing the grass near the thatch line and feeding on leaves within its burrows. These burrows are vertical and deep, unlike the burrows near the surface made by the temperate-region webworms. Some species periodically attack bluegrass in Utah.

Description

Adults of the burrowing sod webworm, *Acrolophus popeanellus* (Clemens), with a wingspan of 25–38 mm, have predominantly yellowish or grayish brown forewings with irregular dark brown to black spots (Plate 55). Hindwings are yellowish to bronzy brown. Males of this genus have large, hairy labial palpi recurved over the head. The cigarette-paper sod webworm, *A. plumifrontellus* Clemens, has been a species injurious to turf in Illinois (Plate 55). Larvae spin silken tubes about 0.5 cm (0.25 in.) diameter by 3.8–5.0 cm (1.5–2.0 in.) long in which to pupate. Robins pull these up to feed on the insects, and the empty tubes superficially resemble cigarette butts. *Acrolophus* spp. larvae are 20–30 mm long and brown to grayish or dirty white; they have no dark spots on the body (Plate 55).

Life History and Habits

Moths may emerge as early as May but normally do so in June or July. Larvae construct burrows the size of a pencil and 15–60 cm (6–24 in.) deep into the soil. The tubular webs mixed with frass and soil extend from the lower leaf blades into the burrow to the point where larvae retreat when they are disturbed (Baker 1982, Banerjee 1967c, D. L. Schuder, Purdue University, personal communication, 1985).

Granulate Cutworm

Importance, Host, and Damage

The granulate cutworm, *Feltia subterranea* (Fabricius) (Lepidoptera: Noctuidae), is a widely distributed pest throughout most of the United States and feeds on a large number of plants, including bermudagrass and dichondra. It cuts off plants near the soil surface, climbs plants, and feeds on foliage or bores into fruits or vegetables touching the ground.

Description

Moths are various shades of black, brown, and gray, appearing inconspicuous (Plate 56). They have a wingspan of 35–40 mm. Their eggs are about 0.63 mm in diameter. Larvae develop through six or seven instars, with head capsule widths of about 3.3 mm in mature larvae.

Life History and Habits

Both adults and larvae are nocturnal. Peak activities occur a few hours after sunset. The insects go into hiding before sunrise. Eggs are deposited singly or a few together on the upper surface of plants. Each female lays from 800 to more than 1,500 eggs. Pupation takes place in the soil after the larva burrows 5–15 cm (2–6 in.) deep and forms a cell. Light trap catches show the granulate cutworm to be active throughout the year in the Gulf States (Snow and Callahan 1968).

Striped Grassworm

Importance, Host, and Damage

The striped grassworm, *Mocis latipes* (Guenée) (Lepidoptera: Noctuidae), is an occasional pest of turfgrass but is an annual major pest of pasturegrasses in Florida. Turfgrasses infested include bahiagrass, bermudagrass, and St. Augustinegrass. Heavy, unchecked infestations of 3.0–7.6 per m^2 (32–82 per ft^2) eat the grass so closely that only bare stolons remain. Large populations that build up on pasturegrasses often migrate to adjacent parks and residential lawns.

Description

The moth is mottled gray to brown, with a dark brown stripe near and parallel to the apical margin of the forewing. It has a wingspan of 30–45 mm (Plate 56). Mature larvae are about 5 cm in length, striped with brown and yellow that blend with the vegetation. There are six or seven larval instars. Their cocoons are spindle shaped, with leaf clippings incorporated on the surface.

Life History and Habits

The striped webworm has several generations a year in south Florida and is present throughout the year, with peak adult populations in the fall. Females begin laying eggs on the third day after emergence and deposit more than 300 eggs each during the next several days.

Newly hatched larvae feed on the upper cell layers and start to eat the margins of leaves in the fourth instar. Larvae passing through six instars require a mean of 30.5 days, and those passing through seven require a mean of 32.4 days, to develop from eggs to adults at 27°C (80.6°F) and about 80% RH.

Pupation occurs after the larva has folded a grass blade about it. Pupae are enclosed in a silken cocoon covered with a whitish powder (Reinert 1975).

Wireworms

Importance, Hosts, and Damage

The larvae of the wireworm (Coleoptera: Elateridae) are strictly soil borne, feeding on roots of grasses and boring into fleshy underground stems. The gradual loss of turf roots causes patchy, withered spots and dead grass. The larval period may require 2–6 years for completion.

Description

Adults are click beetles and are generally recognized by their elongate body, usually parallel sided and rounded at both ends (Plate 56). The posterior corner of the pronotum comes to sharp points or spines. Most species are 12–30 mm in length and are light brown, dark brown, or black, depending on the species. The larvae are slender and are usually compressed dorso-ventrally, shiny, and heavily chitinized. They are generally rust to brown in color and 1.3–3.8 cm long (Plate 56).

Life History and Habits

When larvae are finally mature, they pupate during late summer, and adults overwinter. Females lay eggs in the soil during spring to early summer on roots of grasses and live for about 1 year.

Adults have a flexible union between the pro- and mesothorax. When they are upside down on a hard surface, they can arch at this union and, with a sudden jerk and a clicking sound, throw their bodies into the air to turn over (Borror et al. 1981, Shurtleff and Randell 1974, Shetlar et al. 1983, Vance and App 1971).

Polyphylla spp. Grubs

Importance, Hosts, and Damage

Some 23 species of the genus *Polyphylla* (Coleoptera: Scarabaeidae) occur in the United States. *Polyphylla decimlineata* (Say) is probably the most widely distributed, but *P. variolosa* Hentz and *P. comes* Csy are considered the more injurious species. Members of the genus rather closely resemble the May or June beetles, *Phyllophaga* spp., in life history, habits, and economic importance. Grubs feed on the roots of a wide variety of plants including grasses, coniferous nursery plants, strawberries, corn, and so forth. Adults of at least one species feed on the needles of conifers.

Description

Adults of *Polyphylla* are somewhat larger than *Phyllophaga,* mostly striped with brown and white, and easily separated from the latter by the presence of six to seven segments in the lamellae of their antennae (Plate 56). *Phyllophaga* spp. have only three. The larvae of *Polyphylla* have transverse anal slits. The raster has two short, longitudinal parallel palidia, with fewer than 16 pali per palidium. In the *Phyllophaga* the anal slit is Y-shaped, and the palidia are much longer, with generally more than 20 pali per palidium.

Life History and Habits

Like the *Phyllophaga, Polyphylla* spp. have life cycles of 3 or more years. *Polyphylla* larvae have a greater tendency to coil and remain coiled when disturbed. Their coiled bodies are more rigid than those of *Phyllophaga* grubs (Cazier 1940, Heit and Henry 1940, Ritcher 1966, Yeager 1949).

Vegetable Weevil

Importance, Hosts, and Damage

The vegetable weevil, *Listroderes costirostris obliquus* (Klug) (Coleoptera: Curculionidae), occurs in the Gulf Coast states and in California, where it damages dichondra lawns. It does not infest turfgrasses. It infests such vegetables as turnips and carrots. Both adults and larvae cut small holes in the leaves of dichondra, completely skeletonize leaves, or remove leaves completely, leaving only the base of stems. Damage is usually localized or spotty, because adults do not fly.

Description

There are no males. Females, grayish brown to dull black "snout" beetles about 9.5 mm long, have very rough or punctate wing covers with sparse, short setae (Plate 56). Each elytrum has a pointed protuberance on top near the rear and a gray mark that when paired with the mark on the adjacent elytrum, forms a V. Larvae are greenish, legless grubs about 9.5 mm long when fully grown (Plate 56).

Life Cycle and Habits

Since males do not exist, reproduction occurs without mating. Eggs are laid on plants or in soil during late summer and fall. Larvae develop during the winter. By early spring they pupate and become adults. There is one generation a year. Both adults and larvae are slow and sluggish in their movements. They feed normally at night and hide in the soil around plants during the day. Adults can occasionally be seen during the day hiding under plant foliage. They often crawl to nearby lights at night. In a close-cut dichondra lawn, larvae respond to a pyrethrin drench, which can be used for population survey (Bowen 1980).

21

Turfgrass-Associated Arthropods and Near Relatives

Another group of arthropods causes no injury to turfgrass but is present in the turfgrass ecosystem for a variety of reasons. Some are attracted to the usually damp soil surface with a protective cover that makes an ideal habitat for feeding and hiding. Others use turfgrass areas mainly as a resting place when they are not on the wing. Still others that are parasitic on warm-blooded animals are present in lawns because pets rest there. These turfgrass-associated arthropods and near relatives are listed in their normal phylogenetic sequence. I do not list in this section the beneficial parasitic and predatory insects and mites that are host specific on major turfgrass pests. These would not be present in any quantity except for the presence of their hosts and are mentioned in the appropriate individual chapters.

Turfgrass-Associated Invertebrates

Earthworms

Earthworms (phylum Annelida: class Oligochaeta) are found in almost any moist soil, feeding mainly at night on the organic matter in or on the soil. They do not feed on living plants.

The soil eaten with organic matter is deposited on the soil surface as "castings" that are detrimental on turfgrass greens. Castings on greens are seldom a problem now, probably because of the regular use of fungicides and the occasional use of insecticides. The castings of larger earthworms can make a smooth, level lawn very bumpy and irregular (Plate 57). This nuisance effect, balanced with their beneficial "tilling" activities, makes them a neutral soil turf organism.

Earthworms typically have many similar-looking segments and range in length from 2.5 cm to 15.0 cm (1 to 6 in.) or more. A glandular organ, the clitellum, used during reproduction, appears as a slightly enlarged section from segments 31 or 32 through 37 (Plate 57).

Earthworms are not only hermaphroditic (having both sexes in each individual) but homosexual as well. Each is capable of functioning as a male and a female, but only one sex is functional at any given time. Self-fertilization does not take place. In mating, two individuals come together and, acting as males, exchange sperm. Then the worms separate. The clitellum secretes a bandlike tube that is pushed forward. A worm's own eggs and sperm from the partner worm are discharged into this band. As the band is forced off the anterior end, it becomes a closed cocoon, or "egg," containing the fertilized egg, which develops within the hardened cocoon into a minute worm. Upon hatching, the young worm feeds on the contents of the cocoon before emerging as a tiny replica of its parents to start its independent life (Baker 1982, Hegner 1942).

Snails and Slugs

Snails and slugs (phylum Mollusca: class Gastropoda). These and earthworms are the only organisms in this section not in the phylum Arthropoda. Snails and slugs are most active under moist conditions, hiding under vegetation and coming out mainly at night to feed. They occasionally eat dichondra. They never bother well-maintained turfgrasses, but weedy lawns may have large numbers of these slimy pests.

Snails differ from slugs in having a protective shell in which they can enclose their entire body, while slugs are completely naked (Plate 57). They both move on a slime trail with wavelike contractions of the foot (Bowen 1980, Judge 1972).

Turfgrass-Associated Arthropods

Pillbugs and Sowbugs

Pillbugs and sowbugs (phylum Arthropoda: class Crustacea: order Isopoda) are common in damp places, feeding on decaying organic matter such as mulch, grass clippings, and manure. Some may attack roots and succulent stems. Most of the feeding occurs at night.

Adults are 5–10 mm in length, with a slightly fattened, segmented body. Pillbugs can roll into a ball when disturbed, but sowbugs remain flat. Both have seven pairs of legs, long antennae, and well-developed eyes. In color they are blackish, gray, or brownish (Plate 57).

Both require about a year to become full grown. One to three broods occur each year, with 25–200 young per brood. Eggs are carried by the female in a ventral brood pouch, where the young remain 1 month or 2 after hatching. Their life span is 2 or 3 years (Baker 1982, Paris 1963).

Millipedes

Millipedes (class Diplopoda) usually live in damp places and feed on decaying vegetation, but some attack growing plants. They do no significant damage to turf but become a nuisance when they leave this habitat and wander into buildings.

Several species may occur in turf. Some are dark brown to almost black and have a generally cylindrical body (Plate 57). Others are pinkish, orange, or cream colored with brown markings and have hard, flattened bodies. Most of these millipedes are 13–38 mm long and have a distinct head with short antennae. The anterior segments of many millipedes are modified for copulation and have only one pair of legs per segment. The rest of the segments have two pairs of legs each.

Millipedes overwinter as adults in protected areas and lay eggs during summer in nestlike cavities in the soil or just in a damp place. Eggs, usually white, hatch into a six-legged millipede but add more legs in subsequent molts. Many species give off an ill-smelling fluid that in some cases is reported to be hydrogen cyanide. They do not bite humans. They walk slowly, their legs moving in a wavelike motion. When disturbed, they will curl up tightly (Baker 1982, Borror et al. 1981, Matheson 1951).

Centipedes

Centipedes (class Chilopoda) are terrestrial and predacious, feeding on insects and spiders. They are generally found in protected places in the soil, under bark, or in rotting logs. Unlike millipedes, they are fast running. All possess poison jaws for paralyzing their prey. Those in the north are harmless to humans, but those in the south or tropics can inflict a painful bite. Generally, centipedes more than 4 cm long can pierce human skin.

Unlike the millipede, the centipede body is dorso-ventrally flattened, with a distinct head bearing a pair of long antennae, a pair of mandibles, and two pairs of maxillae. The first body segment behind the head has a pair of clawlike appendages functioning as poison jaws. All other segments except the last two bear a pair of legs (Plate 57).

Centipedes overwinter as adults in protected places and lay eggs during summer. The sticky eggs, laid singly, become covered with soil. In some species this covering protects them from being eaten by the males (Baker 1982, Borror et al. 1981, Matheson 1951).

Scorpions

Scorpions (class Arachnida: order Scorpiones) are well-known animals in the southern and western United States. They are found on the ground, whether barren or covered with low vegetation, and often in lawns. Scorpions will sting if disturbed. Their painful sting is followed by local swelling and discoloration but is generally not dangerous. Of some 40 species, only 1, *Centruroides sculpturatus* Ewing, is dangerously venomous. As far as is known, it occurs only in Arizona. In areas where scorpions occur, objects on the ground should be picked up with care. Scorpions on the body should be brushed off rather than swatted.

Scorpion bodies are divided into the prosoma, which bears the eyes, pedipalps, and four pairs of legs, and the broadly joined opisthosoma. The opisthosoma is differentiated into two sections, a broad, seven-segmented anterior mesosoma and a narrow, five-segmented posterior metasoma terminating in a sting (Plate 58). Most scorpions are small to medium, not exceeding about 25 mm in length. The largest are about 125 mm long.

Scorpions are largely nocturnal and carnivorous, feeding on insects and spiders that they catch with their pedipalps and sometimes sting. The young are born alive and for some time after birth are carried on the mother's back. When scorpions run, the pedipalps are held outstretched and forward and the metasoma is usually curved upward (Borror et al. 1981).

Spiders

The spiders (class Arachnida: order Araneae) are a large, distinct, and widespread group occurring in many habitats, including turfgrass areas. Nearly all have venom glands, but only a few are dangerous. Chief among these are the black widow spider, *Latrodectus mactans* (Fabricius), and the brown recluse spider, *Loxosceles reclusa* Gertsch and Mulaik.

The black widow spider, the most venomous, is common in the southern states. Females are coal black, with a red hour-glass figure on the ventral abdominal area. Appropriately named, the female often kills the male soon after mating. The brown recluse spider, second only to the black widow in its venomous nature, occurs east of the Rocky Mountains, mainly in Arkansas, Missouri, and Texas. Known also as the *fiddle back,* it is nonhairy and has a violin-shaped mark on the cephalothorax.

Spiders have two body regions, the cephalothorax, bearing the eyes, mouthparts, and four pairs of legs, and the abdomen, bearing the genital structures, spiracles, anus, and spinnerets for weaving (Plate 58).

Spiders lay eggs generally in a silken sac attached to leaves, bark, or crevices or carried by the female. They undergo very little metamorphosis and look like miniature adults when hatched. They molt 4–12 times to become adults. All spiders are predacious, feeding mainly on insects, killing their prey by injecting a poison into it by biting. Their habit of spinning silk into webs of various shapes and sizes is one of their most characteristic features (Borror et al. 1981).

Ticks

Ticks (class Arachnida: order Acari) have two families in the United States, the Ixodidae, or hard ticks, and the Argasidae, or soft ticks. Both are present on turfgrass and other areas covered by vegetation. The most common species are the American dog tick, abundant in the eastern two-thirds of the country, the brown dog tick, of worldwide distribution, and the lone star tick, which occurs from Texas and Oklahoma eastward to the Atlantic. Ticks attach themselves to their hosts to feed on blood, and some inject toxic saliva that produces paralysis. Tick-borne diseases transmitted in the feeding process are spotted fever, relapsing fever, tularemia, and Texas cattle fever.

Adult ticks are brown to reddish brown and 3–7 mm long, according to species, and have four pairs of legs (Plate 58). Eggs are about 0.5 mm long by 0.4 mm wide, shiny, oval, and yellowish to pale yellow. Larvae, 0.5–1.0 mm and brownish, have three pairs of legs. Nymphs are 1.5–2.5 mm long and brownish, with four pairs of legs.

Ticks lay their eggs in various places but not on their hosts, which are warm-blooded animals. After hatching they climb vegetation and attach themselves to their host as they pass by. Hard ticks take only one blood meal in each of their three instars and drop off their host to molt. Soft ticks feed several times before molting (Baker 1982, Borror et al. 1981).

Chiggers

Chiggers (class Arachnida: order Acari) are a worldwide pest. Larvae of the chigger mite, *Trombicula irritans* (Riley), are often abundant in grass and waste lands. Humans walking into infested areas can find larval chiggers clinging to the clothing or bare skin, where they insert their mouthparts, if possible in a hair follicle, and suck blood. This bite causes severe irritation, itching, and sometimes intense pain. When fully fed, larval mites drop to the ground to molt. The nymphs and adults are free-living predators and scavengers. No diseases are transmitted by chiggers in the United States.

Both adults and nymphs are usually bright red and about 1.25 mm long, with four pairs of legs. Eggs are minute and globular, changing from light to dark with age. Larvae are orange yellow to light red and hairy, with three pairs of legs.

Larvae, the parasitic stage, attach themselves to the vertebrate host until fully engorged, then drop to the ground to molt first to the nymphal stage, then to the adult stage. An entire life cycle requires about 50 days, and there is usually one generation a year in North Carolina (Baker 1982, Matheson 1951).

Turfgrass-Associated Insects

Earwigs

Earwigs (order Dermaptera) occur throughout North American and are found wherever there is moisture and cover, so that turfgrass areas are a natural habitat. The European earwig, *Forficula auricularia* L., is most common. Its diet is variable, with incidental feeding on foliage, but it is more of a nuisance because of its presence. Other species are beneficial, feeding on other insects and on decaying organic matter.

The common species are brown to black, elongate insects, 19–25 mm long, with a pair of large forceps at the end of the abdomen (Plate 58). Those of the female are straight, but those of the males are noticeably curved. Adults may be winged or wingless. Forewings, if present, are short, thick, and veinless, covering a membranous hindwing. Earwigs rarely fly. White, ovate eggs about 1 mm long are deposited in clusters. Grayish nymphs resemble adults but are smaller, and forceps of both sexes are straight.

Few earwigs can successfully overwinter outdoors in the northern states. In warmer areas all stages can overwinter. Upon hatching, nymphs feed and develop through five instars, taking about 45 days to do so in summer or >150 days in winter. There is only one generation a year (Baker 1982).

Ground Beetles

Ground beetles (order Coleoptera: family Carabidae) have a worldwide distribution. There are more than 2,500 species in North America alone. They are most commonly seen in the grass, in the soil, and under bark or debris. Some are often seen running over golf greens during the summer. Some species may feed on seed or pollen, but most prey on other insects and are considered beneficial.

Adults are variable in shape, color, and size, ranging from 3 mm to 25 mm in length. Most are black or dull brown in color. They are broad and have hard wing covers with many parallel longitudinal ridges (Plate 58). Their eggs are oval and cream colored. The larvae of ground beetles are elongate, 10–45 mm long, and slightly flattened and taper toward the rear. Their heads are large, with sickle-shaped mandibles directed forward.

Most ground beetles complete a life cycle within a year, overwintering as larvae or adults. Eggs are deposited singly in the soil. Larvae develop in or on the soil, generally hidden under stones, boards, or other debris. Adults are usually nocturnal. Both adults and larvae are predacious and are therefore beneficial (Baker 1982, Matheson 1951).

Fleas

Fleas (order Siphonoptera) have a worldwide distribution and are commonly found in bedding or near areas where host animals sleep. During the warmer months lawns can become heavily infested with fleas when infested pets rest and sleep on the turf. Most commonly encountered are the cat flea, *Ctenocephalides felis* (Bouche), and the dog flea, *C. canis* (Curtis). Adult fleas hop onto humans and suck blood, causing irritation, itching, and sometimes pustules.

Adults are from 1.0–2.5 mm long, brown, wingless, tough spiny insects compressed laterally, with legs modified for jumping (Plate 58). White oval eggs are about 0.5 mm long. Larvae are 4–6 mm long, slender, white, and legless. Pupae are dingy white, oval, and about 4 mm long in a silken cocoon.

Adult fleas require a blood meal to lay eggs, which are deposited on the host but fall off and hatch. Outdoors in lawns or in pet quarters, the larvae feed on the excrement of adult fleas, rodents, domestic pets, and general decaying organic matter. Pupation also occurs on the ground or bedding. A life cycle may be completed in as little as 2 or 3 weeks, several months, or as much as 2 years. Adults are capable of living for as long as 2 years without a blood meal (Baker 1982).

Mosquitoes and Biting Midges

Adult mosquitoes and midges (order Diptera) rest in turfgrass and adjoining shrubs during the day. They emerge in the evening to feed on blood and cause a nuisance with which everyone is familiar during the summer. Only the females bite.

Mosquitoes (family Culicidae), represented by more than 100 species in North America, have as their most common genera *Culex*, *Aedes*, and *Anopheles*. Adults are slender, long-legged flies 5–10 mm long, with an abundance of scales and hairs on the wings and appendages (Plate 58). The biting midges (family Ceratopogonidae), often called *no-see-ums* or *punkies*, are so small that they are often not seen even when they are biting. Adults are <2.0 mm long. Most of the bloodsuckers attacking humans belong to two genera, *Culicoides* and *Leptoconops*.

Both groups of insects require water for reproduction. Mosquitoes deposit eggs on the surface of the water or on soil that will be submerged later. Both larvae and pupae are strictly aquatic. Biting midges are also aquatic, with larvae occurring in mud and ooze, among algae in ponds, and in decaying vegetation. Adults that are not bloodsuckers are predacious on other insects.

Mosquitoes and biting midges transmit serious human diseases, making them much more important than their bites alone would indicate (Borror et al. 1981, Matheson 1951).

Wild Bees and Yellow Jackets

Wild bees (order Hymenoptera: family Andrenidae) and yellow jackets (family Vespidae), both cosmopolitan groups, do little, if any, damage to turf in their ground nesting activities. They select well-drained sites of moderate to sparse plant growth with little organic matter. They can inflict painful stings and are important because of this habit.

Wild bees are small to medium-sized insects 8–17 mm in length in a wide range of colors of metallic red, black, blue, green, or copper. The adults of the eastern yellow jacket, *Vespula maculifrons* (Buysson), are distinctly marked with black and yellow and range in length from 12 mm to 25 mm (Plate 58).

Yellow jackets overwinter as mature, fertilized queens. In the spring the female seeks out a suitable nesting site to build a single comb of several cells. She places an egg in each. The nest may be in the ground, under a board or rock, or in attics.

Developing larvae become the first workers, and the colony life begins. By fall, the future queens mate, the males die, and the females seek sites in which to hibernate.

Wild bees overwinter as adults in their burrows in the soil. They emerge in April and dig new burrows, provision them with pollen balls 3–5 mm in diameter, and deposit a single egg on each pollen ball. The larvae develop within the burrows, become adults, and remain there until the following spring (Baker, 1982).

22

Vertebrate Pests

Vertebrates become destructive to turfgrass in most cases simply because an attractive food supply exists in the turfgrass or soil below. Most often it is insects or earthworms. The vertebrate pests are either birds or mammals. Information in this chapter on the biology and distribution of these pests is primarily from Peterson (1980) on avian pests and from Burt and Grossenheider (1964) and Timm (1983) on mammalian pests.

To control turf-damaging birds it is necessary merely to eliminate the attractive food supply. The control of mammalian pests, apart from elimination of the attractive food supply of turfgrass insects, is discussed briefly, since the various methods of control are long standing and are not subject to frequent changes, as are insecticidal recommendations for arthropod pests in turfgrass. I have drawn recommendations for the control of mammalian pests primarily from Timm (1983).

Birds

Starlings

The European starling, *Sturnus vulgaris* (order Passeriformes: family Sturnidae), called simply as *starling*, is distributed throughout the United States and southern Canada. It is a gregarious, garrulous, short-tailed blackbird, 19–22 cm (7.5–8.5 in.) long from the tip of its bill to the tip of the tail (Plate 59). In winter it has a heavily speckled body and a dark bill. In spring it becomes iridescent, with a yellow bill. Starlings feed on insects, seeds, fruits, and berries. Webworm larvae in the thatch and scarabaeid grubs in the soil are favorite foods. Once food has been found, flocks of starlings descend to feed by inserting their beaks into the thatch or soil to obtain the insects. They leave peck holes the diameter of a pencil, which can be interpreted as minor bird damage. More often than not, such holes serve as a warning signal that destructive turf insects may be present and that an examination is in order.

Grackles

The common grackle, *Quiscalus quiscala* (order Passeriformes: family Icteridae), ranges over the southern half of Canada and the United States east of the Rocky Mountains. It is a large, iridescent blackbird, 28–34 cm (11.0–13.5 in.) long with a long wedge-shaped or keel-shaped tail and larger than a robin (Plate 59). The males have iridescent purple on their heads with deep bronze or dull purple on their backs. Females have little or no iridescence. Grackle food habits are similar to those of the starling, and the two are often seen together, feeding on grubs and other turf-inhabiting insects. Grackles are generally not as abundant as starlings but because of their larger size can disrupt a weakened lawn in scratching for and feeding on insects.

Crows

The American crow, *Corvus brachyrhynchos* (order Passeriformes: family Corvidae), ranges from the southern half of Canada south throughout the eastern two-thirds of the United States to the arid southwest and north from Baja California. Almost everyone is familiar with this large, chunky, ebony-colored bird with a strong bill and feet (Plate 59). In strong sunlight its plumage shows a purplish gloss. Its body length is 43–53 cm (17–21 in.). Crows are often gregarious. Their food consists of almost anything edible, including insects such as grasshoppers, scarabaeid beetles, ground beetles, and caterpillars. Annual bluegrass weevil–infested fairways and tees have been further damaged by crows tearing and uprooting the weakened turf as they scratch in search of the weevil larvae.

Mammals

Moles

Moles (order Insectivora: family Talpidae) live most of their lives underground. The presence of one species or another can be detected by the low, snaking ridges pushed up in the turf as they move just under the surface or by surface mounds of 2–8 l (0.5–2.0 gal) of soil, but moles leave no trace of the entrance to their burrow as pocket gophers do (Plate 60). Damage is most frequent and extensive during spring and fall during cool, moist periods, especially along ditches.

Moles consume from 70% to 100% of their own weight each day and need access to large amounts of food. The stomach contents show what is eaten. Of 100 moles examined, 67 had beetles, 64 white grubs, 40 earthworms, 44 other beetle larvae, and 43 seedpods and husks. Other food items in descending order included centipedes, spiders, ants, and crickets (Henderson 1983). As soon as insects and earthworms disappear from a turfgrass area, disruption of turf generally ceases, and the moles move on to another food supply.

Moles have broad front feet, with palms usually facing outward. Their eyes are pinhead size. They have no external ears, and their body is covered with soft, thick fur that has a sheen and varies from golden, to brown, to slate, and to nearly black.

At least four species in the United States invade turfgrass areas, consuming insects and worms but in the process damaging lawns, golf courses, and other turf as they burrow through soils of moist, sandy loam. These are the star-nosed mole, *Condylura cristata,* with an unusual fingerlike projection of 22 tentacles surrounding the end of the nose (Plate 60), the eastern mole, *Scalopus aquaticus* (Plate 60), the hairy-tailed mole, *Parascalops breweri,* and the California mole, *Scapanus latimanus.* Moles range in size from 11 cm to 16 cm (4.3–6.3 in.) in body length and from 2.5 cm to 9.0 cm (1.0–3.5 in.) in length of tail.

The first three species occur east of the Rocky Mountains, and one or more can be found in any part of this region. The California mole occupies much of California and southern Oregon. Certain differences exist in the habits of the three eastern species. Damage by the eastern mole is recognized by tunneling just 2.5–5.0 cm (1–2 in.) below the surface that creates a long, winding ridge. The star-nosed mole makes similar ridges during cool moist weather when the food supply is very near the surface, but ordinarily it tunnels at depths of 10–15 cm (4–6 in.), raising numerous and frequent mounds rather than ridges. The star-nosed mole prefers poorly drained soils, while the eastern mole prefers well-drained soils. Meadow mice and shrews frequently invade and use mole tunnels.

Control of moles can be accomplished by direct killing, trapping, or fumigation or by insecticidal applications that eliminate the food supply and drive the animals to another source of food. Moles burrow at night but continue until shortly after dawn. When the ridges are being pushed up, a sharp blow with the back of a shovel just behind the leading edge of the ridge can kill the eastern mole.

In trapping, baiting (with use of strychnine or zinc phosphide), or fumigation (with aluminum phosphide, calcium cyanide, or gas cartridges), it is essential to select active tunnels for placement. Flatten the ridges and mounds. Ridges pushed back up within 12–24 hours and mounds that are pushed back up in 24–48 hours indicate active habitats (Dudderar 1983). Trapping is the most effective and practical way of eliminating moles because it capitalizes on the mole's natural habits. Excavation of a mole tunnel is the first requisite for trapping (Figure 77). Live-capturing a mole is possible with the use of a large jar and a board to exclude light (Figure 77). Mechanical traps are sprung by the animal's natural instinct to reopen obstructed passageways. Of the three types of traps available, the Victor trap is the simplest to set for moles making the surface ridges, because no portion of the trap need be placed in the tunnel (Figure 78). If a small section of an active tunnel is flattened and the trigger pan is set in contact with the firmed turf, the trap will be activated when the mole pushes up the ridge again (Henderson 1983).

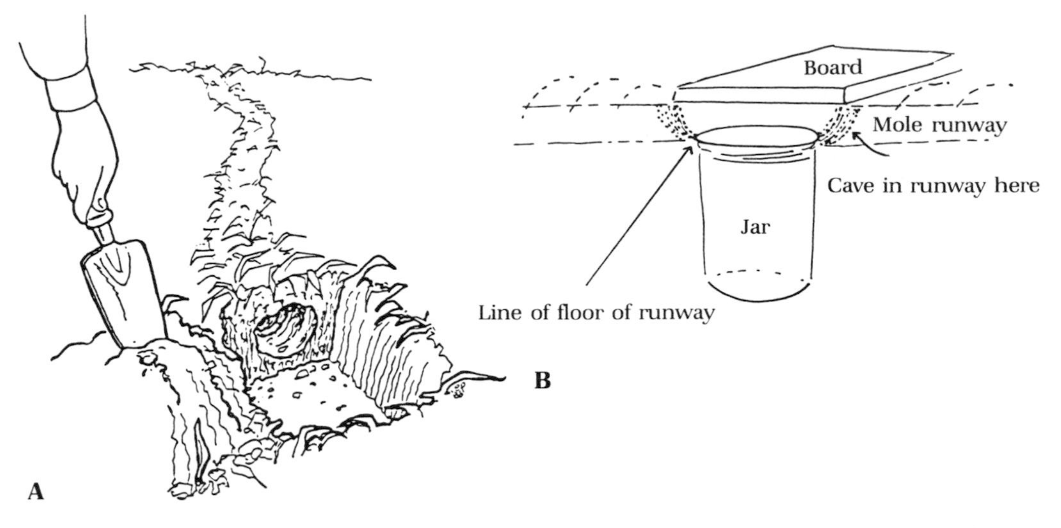

Figure 77. Tunneling characteristics of moles and adaptation to trapping. **A.** Excavation to locate tunnel, essential for trapping. **B.** Method of live-capturing mole with jar as pitfall trap and board to exclude light. (From Henderson 1983, figs. 5, 9, courtesy of F. R. Henderson; artwork by E. Peck, Kansas State University.)

Shrews

Shrews (order Insectivora: family Soricidae) are mouse size, with beadlike eyes not covered with skin. They can be further separated from mice and voles by the fact that they have no external ears. Although shrews are primarily insectivorous, they do eat roots, seeds, and other vegetable matter sparingly. They are not harmful to turfgrass. The short-tailed shrew, *Blarina brevicauda* (Plate 60), occupies the entire eastern half of the United States and southern Canada. Its saliva is poisonous. This shrew is often found occupying the tunnels of the star-nosed mole.

Voles

Throughout Canada and the United States, where there is good grass cover, one or more species of voles (order Rodentia: family Cricetidae) are likely to be found. Mice and voles belong to the same family and are similar in size and general appearance. Mice have large ears, large eyes, and long tails, however, while voles have small ears, small eyes, and short tails (Plate 61). Voles have mostly brownish gray fur. They range in body length from 10 cm to 15 cm (4–6 in.), with tails 3.0–7.5 cm (1.2–3.0 in.) long.

Voles are active both day and night. Since they are mainly vegetarians, insects need not be present before they will damage turf. They make narrow runways 2.5–5.0 cm (1–2 in.) wide through the grass. During winter snow, their round openings to the surface reveal their presence (Plate 61). After the snow is gone, their trails,

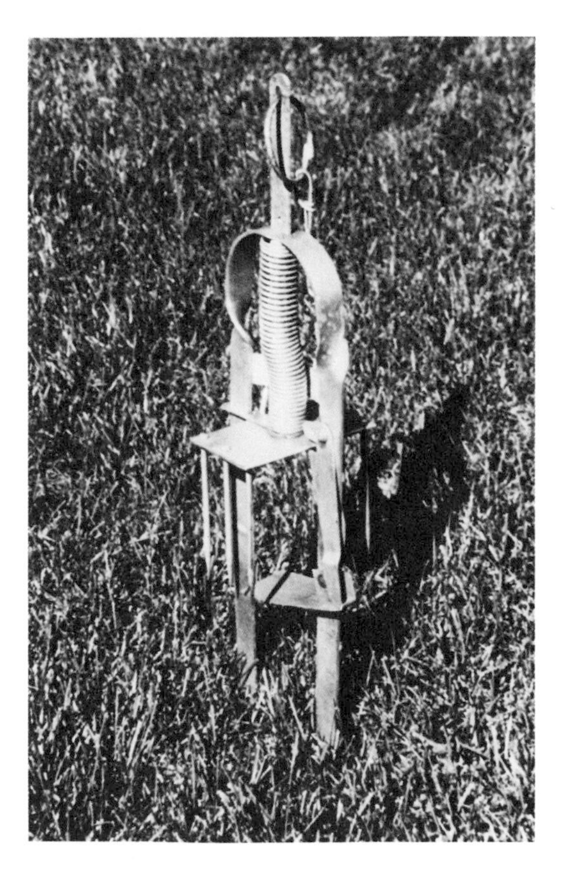

Figure 78. Victor mole trap, one of several types for moles. (Photo by B. Aldwinckle, NYSAES.)

which are completely devoid of grass, make an unsightly lawn that will not fill in completely for several months (Plate 61). In addition to the turfgrass damage, they can severely harm trees by removing bark from roots.

Toxicants have been the mainstay in vole control. Zinc phosphide in pelleted or grain bait formulations broadcast or placed in runways or burrow openings has been the most widely used, followed by strychnine grain baits placed in runways or burrows. Weeds, ground cover, and litter provide food and cover for moles, and so eliminating these around lawns and turfgrass areas reduces damage. Trapping, fumigation, and shooting are not effective methods of control (O'Brien 1983).

Chipmunks

Chipmunks spend most of their time on the ground but occasionally climb trees. They feed on plants, snails, and insects, often digging for scarabaeid grubs in the lawn. They disrupt soil much less than skunks but still leave excavations (Plate 61). The eastern chipmunk, *Tamias striatus* (order Rodentia: family Sciuridae), with a head and body 13–15 cm (5–6 in.) long and a tail 7.6–10.0 cm (3–4 in.) long, is squirrellike and runs with its bushy tail straight up (Plate 61). It ranges east of the 100th meridian throughout southern Canada and most of eastern United States except the southern half of the Gulf Coast states, Florida, and eastern North and South Carolina. It feeds on scarabaeid grubs in the Northeast (author's personal observations).

Since chipmunks damage turfgrass in the course of seeking grubs or other insects as a source of food, elimination of the food supply generally eliminates the chipmunk problem.

Pocket Gophers

Pocket gophers (order Rodentia: family Geomyidae) do not seek turfgrass insects as a source of food but burrow in moist, friable soil to depths of 10–46 cm (4–18 in.) below the surface, making large fan-shaped surface mounds on turfgrass areas. A plug of soil placed in the entrance hole is plainly visible (Plate 62). Many golf courses with soil of sandy loam have numerous unsightly mounds on fairways, roughs, and even on greens. Home lawns, parks, cemeteries, and other turfgrass areas are also affected.

Gophers are considered harmful wherever they occur. The surface mounds, in addition to being unsightly, interfere with normal turfgrass mowing operations. Gophers can destroy underground utility cables and plastic irrigation pipes in addition to girdling stems and feeding on roots, tubers, and surface vegetation. Gophers are capable of swimming.

There are no fewer than 10 species, scattered mostly throughout the West but with 1 species, the southeastern pocket gopher, *Geomys pinetis,* in Florida, Georgia, and Alabama. The two most widespread in the West are the valley pocket gopher, *Thomomys bottae,* occupying most of California eastward through Utah and south into western Texas and Mexico. The northern pocket gopher, *T. talpoides,* occupies an area east of the Sierra Nevada–Cascade Range eastward into North and South Dakota, into southern Canada, and southward into New Mexico. The plains pocket gopher, *Geomys bursarius,* occupies a large area from southern Texas into northern Minnesota and from Wyoming, Colorado, and New Mexico eastward into Indiana (Plate 62).

Members of this family are small to medium-sized rodents with a body length of 13–36 cm (5–14 in.), varying in color from nearly white and mostly brown to nearly black. Their two most distinguishing features are their yellowish incisor teeth, always exposed in front even when the mouth is closed, and their external cheek pouches, which are fur lined, reversible, and open on either side of the mouth (Plate 62). Characteristics on the front of upper incisors and forefeet distinguish certain genera (Figure 79; Case 1983).

To control pocket gophers it is first necessary to locate their tunnel system,

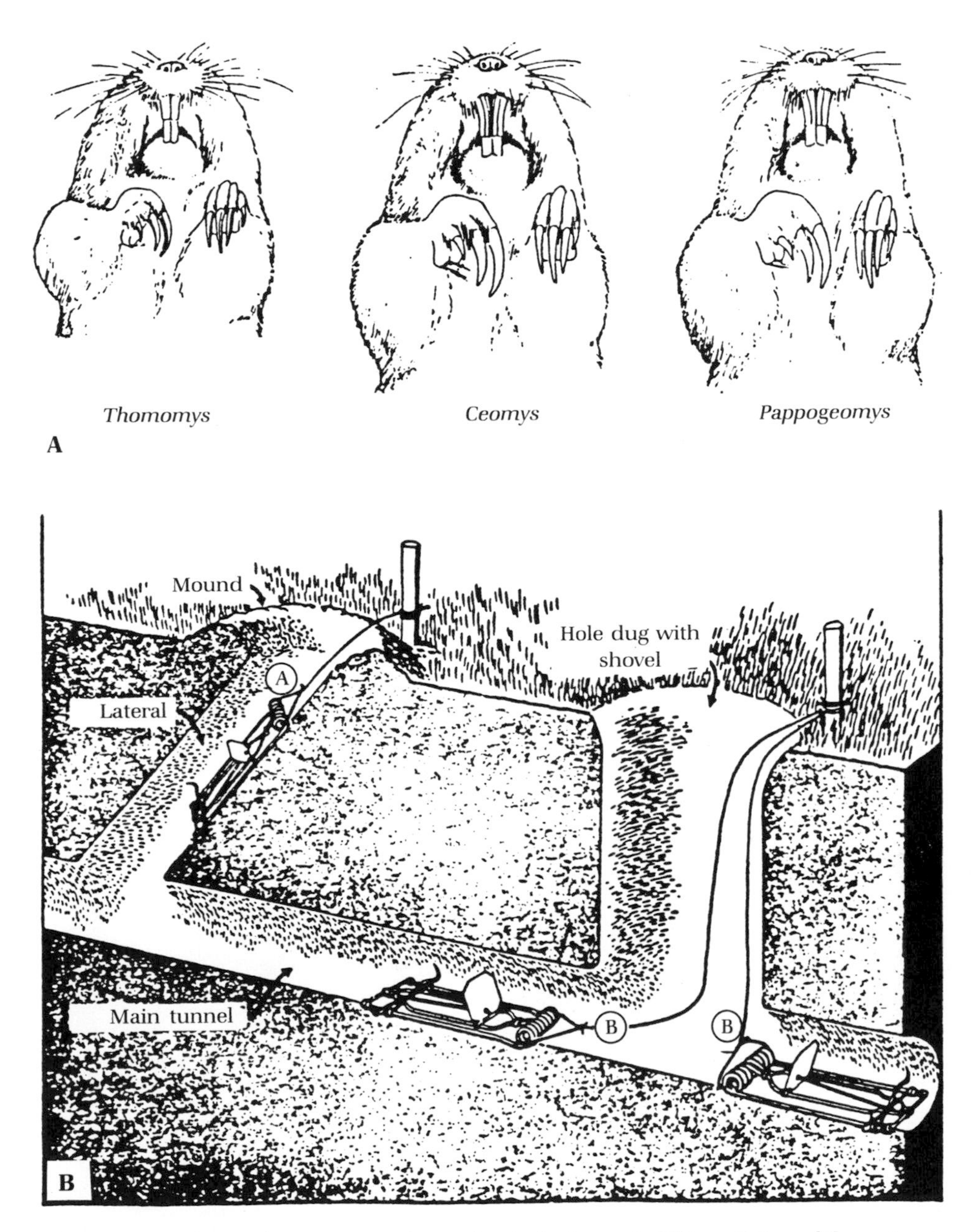

Figure 79. Pocket gopher genera and trapping techniques. **A.** Differentiation of three genra by upper incisor characteristics and relative size of forefeet. **B.** Tunneling system and trap locations. (Part A courtesy of the Colorado Agricultural Experiment Station, Fort Collins; part B from leaflet attached to trap at purchase, no manufacturer's identification.)

whether traps or toxicants are to be employed. The main burrow is found 30–46 cm (12–18 in.) from the plug on the fan-shaped mounds. Trapping is an effective means of controlling gophers in small areas and can eliminate the remaining animals following a poison control program over a large area. One trap is placed in a lateral tunnel, or two traps facing in opposite directions are placed in the main tunnel, to capture an animal coming from either direction (Figure 79). Of the several types and makes of traps, the Macabee trap is commonly used and highly effective (Figure 80).

Four federally registered rodenticides for pocket gopher control include strychnine alkaloid, zinc phosphide, arsenic trioxide, and an anticoagulant chlorophacinone (RoZol). All are used as baits placed in the gopher's tunnel system (Case 1983).

Skunks

Skunks (order Carnivora: family Mustelidae) are chiefly nocturnal and omnivorous, feeding on insects, mice, eggs, carrion, berries, roots, and other vegetation. In their search for insects, primarily scarabaeid grubs, skunks destroy turf, tearing up and uprooting the already weakened grass (Plate 63). Their grubbing activities are

Figure 80. Macabee pocket gopher trap set (left) and neutral (right). (Photo by H. Tashiro, NYSAES.)

most common during the fall and early spring, when larvae are feeding near the surface. They may leave golf ball–sized pits when digging out individual insects.

Four species of skunks occur in the United States. The striped skunk, *Mephitis mephitis* (Plate 63), ranges over the entire country, and the spotted skunk, *Spilogale putorius,* ranges over most of the country except the Northeast, the southern Great Lakes region, and Montana. The hooded skunk, *Mephitis macroura* (Plate 63), and the hog-nosed skunk, *Conepatus leuconotus,* range over the Southwest into Mexico. Skunks range in size from the spotted skunk, the smallest, 23–33 cm (9–13 in.) in body length and 11–23 cm (4.5–9.0 in.) in length of tail, to the hog-nosed skunk, the largest, with a 36–48 cm (14–19 in.) body length and tail of 18–30 cm (7–12 in.).

Eliminating or greatly reducing populations of grubs, one of the skunk's favorite foods, eliminates its turf-damaging activies. Skunks' beneficial habits in eating other insects, rodents, and other pestiferous small animals outweigh their occasional destruction in damaging beehives, killing fowl, and eating eggs (Knight 1983).

Raccoons

The raccoon, *Procyon lotor* (order Carnivora: family Procyonidae), also called *coon* and *ringtail,* is chiefly nocturnal and omnivorous, feeding on fruits and nuts, insects, frogs, bird eggs, and whatever else it can obtain. When feeding on scarabaeid grubs during the fall or spring, it tears up and turns over large chunks of sod, so that the grubbing activities of skunks appear relatively insignificant by comparison (Plate 63).

Raccoons range over the entire United States except the Rocky Mountains and parts of the Great Basin area. They are 46–71 cm (18–28 in.) long, with a tail 20–30 cm (8–12 in.) long. The pepper-and-salt body, the black mask over the eyes, and the yellowish rings alternating with black on the tail identify this animal (Plate 63).

In addition to grub-infested turf, raccoons may destroy poultry, bird nests, and sweet corn or other vegetables; they also raid garbage cans. By eliminating or reducing grub populations in turf, turf damage can be prevented (Boggess 1983).

Armadillos

The armadillo, *Dasypus novemcinctus* (order Xenarthra: family Dasypodidae), a peculiar-looking animal, is chiefly a tropical mammal about the size of a house cat (Plate 63). It ranges from most of Florida across the Gulf Coast and northward from Mississippi into eastern Kansas and southward through most of Texas. The armadillo's body and tail and the top of the head are covered with a horny material. It is the only mammal in the United States with such protective armor.

The armadillo is mainly nocturnal during the summer and may be diurnal during the winter. It digs burrows that are usually 18–20 cm (7–8 in.) in diameter and as much as 4.5 m (15 ft) in length. The burrows are located in rock piles and around stumps, brush piles, and dense woodlands. More than 90% of the armadillo's diet is insects (adults and larvae), but the animals also feed on lizards, frogs, snakes, eggs of birds, berries, fruits, and roots. An armadillo pursuing turfgrass insects will tear and uproot the turf much as skunks and raccoons do but at somewhat deeper depths of 2.5–7.6 cm (1–3 in.) (Plate 63; Hawthorne 1983).

Since the armadillo prefers to have burrows in areas with cover, the elimination of such habitats discourages it from becoming established. Removal of the insect food source from turfgrass areas largely eliminates its turf-destroying habits (Hawthorne 1983).

23

Insects and Mites: Turf Association

Habitats of Turfgrass Insects

Destructive turfgrass insects and mites can be grouped according to the habitat in which the destructive stage of the arthropod spends most of its life in the turfgrass ecosystem. In a great many cases this is the larval or nymphal stage, but in a few the adults are also destructive.

This method of grouping has merit, since the control tactics employed for each group, especially in insecticidal treatments, have a direct bearing on the ease or difficulty with which satisfactory results may be accomplished. Arthropods living and feeding exclusively on leaves and stems are most readily controlled, since insecticidal applications are easily made to the target zone, with timing of treatments being the most important criterion. Soil-inhabiting insects that feed exclusively on the roots are the most difficult to control, because insecticidal applications must be made to a nontarget zone of established turf and depend on gravity and water for downward movement to the soil surface and into the soil. Not only might the nature of the thatch impede rapid percolation, but the water solubility and residual properties of the insecticide become important considerations.

This system of grouping is not entirely exclusive, since a given insect may occupy more than one habitat in a given stage and may even occupy all habitats in different stages of its development. These habitats as grouped by Niemczyk (1981) for the most turfgrass-damaging stages are: (1) leaf and stem inhabitants, (2) thatch inhabitants, and (3) soil inhabitants.

In the sections that follow, I indicate for each species or group the chapter giving its detailed account.

Leaves and Stems

Major insects and mites inhabiting leaves and stems include:

Bermudagrass mite, *Eriophyes cynodoniensis* Sayed
Winter grain mite, *Penthaleus major* (Duges)

Greenbug, *Schizaphis graminum* (Rondani)
Rhodesgrass mealybug, *Antonina graminis* (Maskell)
Frit fly, *Oscinella frit* (L.)

Bermudagrass mites are destructive as adults and nymphs. They settle in the leaf sheaths of bermudagrass and remain in the internodal areas, producing rosetted internodes (Chapter 3).

Adults and nymphs of the winter grain mite feed on the leaves of grasses. They migrate on a daily basis, moving up to the leaves to feed at night and giving the grass the appearance of winter dessication. Mites descend to the base of the grasses during the day. The winter grain mite, most appropriately named, flourishes only at cold temperatures at <24°C (75°F) and aestivates as eggs from May to October (Chapter 3).

Greenbug adults, whether winged, wingless, or both, feed on the upper leaf surface of bluegrasses, fescues, and ryegrasses. They produce living young that mature to wingless females to repeat the cycle for many overlapping generations. More than 50 individuals can occur on a single grass blade (Chapter 6).

The Rhodesgrass mealybug, called a scale until recently, is destructive at the adult and nymphal stages primarily on bermudagrass and St. Augustinegrass. Crawlers (larvae) settle near the base of plants or lower nodes, inserting their mouthparts into the stem to feed. They remain sessile throughout the nymphal and adult life. Secretions by the insect produce clusters of cottony masses of about 0.3 cm in diameter on the stems, and this is the main diagnostic feature of the infestation (Chapter 6).

Only the larvae of the frit fly cause damage. Adults lay eggs on turfgrass leaves, primarily on bentgrass. Larvae migrate down to the upper portion of the crown to feed on the primordial leaves until pupation (Chapter 18).

Thatch

Insects that inhabit the thatch use it as their primary resting place for the destructive stage or stages. They include:

Hairy chinch bug, *Blissus leucopterus hirtus* Montandon
Southern chinch bug, *B. insularis* Barber
Temperate-region sod webworms, major species:
  Bluegrass webworm, *Parapediasia teterrella* (Zinck.)
  Corn root webworm, *Crambus caliginosellus* Clem.
  Cranberry girdler, *Chrysoteuchia topiaria* (Zell.)
  Larger sod webworm, *Pediasia trisecta* (Wlk.)
  Pretty crambus, *Microcrambus elegans* (Clem.)
  Silver barred lawn moth, *Crambus sperryellus* Klots

Striped sod webworm, *Fissicrambus mutabilis* (Clem.)
Tropical-region sod webworms:
Grass webworm, *Herpetogramma licarsisalis* (Walker)
Tropical sod webworm, *H. phaeopteralis* Guenee
Cutworms:
Black cutworm, *Agrotis ipsilon* (Hufnagel)
Variegated cutworm, *Peridroma saucia* (Hubner)
Bronzed cutworm, *Nephelodes minians* Guenee
Armyworms:
Armyworm, *Pseudaletia unipuncta* (Haworth)
Fall armyworm, *Spodoptera frugiperda* (J. E. Smith)
Lawn armyworm, *S. mauritia* (Boisduval)
Fiery skipper, *Hylephila phyleus* (Drury)
Annual bluegrass weevils:
Hyperodes near *anthracinus* (Dietz)
H. maculicollis (Kirby) (note taxonomy)

Chinch bugs are destructive to their host grasses as adults and as nymphs. They pierce stems, suck plant sap, and inject toxic salivary enzymes that disrupt the water-conducting system of plants. Injury is most severe when heavy infestations occur on plants under drought stress. The hairy chinch bug is most damaging to bluegrasses, fine fescues, and bentgrasses in the cool, humid zone, while the southern chinch bug is most damaging to St. Augustinegrass, bermudagrass, and zoysiagrass (in that order) in the warm, humid zone (Chapter 5).

All of the thatch-inhabiting larvae whose adults are moths have a number of important common features. Adults of all these species fly and deposit their eggs at night. Turfgrass damage is caused only by the larvae. All the larvae are nocturnal, feeding on grass blades and stems at night and remaining hidden in the thatch or soil surface during the day. All deposit copious quantities of green fecal pellets in the thatch (Chapters 7–8).

All of the webworms, whether of the temperate or tropical regions, derive their name from their habit of webbing together plant debris, soil particles, and fecal pellets to form a tunnel, in which they hide during the day. Young larvae feed on the surface of tender leaves, but as they grow, they feed on entire leaves and stems. Practically all turfgrasses are susceptible to their feeding. Extensive feeding during dry weather causes the most serious damage. At least one species, the cranberry girdler, feeds on the roots of their host grasses (Chapter 7).

Cutworm larvae are at least twice the size of webworms and produce green fecal pellets at least three times the size of webworm pellets. They do not prepare tunnels to hide in during the day but rest in the thatch or in other secluded situations in a curled position. Their habit of cutting off whole plants near the ground level without consuming any more of the remaining plant gives them their name. All turfgrasses are susceptible (Chapter 8).

Armyworms are very similar to cutworms in size, appearance, and destructive habits. When large numbers of armyworms decimate host plants in one area, they move en masse from the decimated area to a fresh food supply with a definite advancing front. This habit gives them their name. All turfgrasses are susceptible (Chapter 8).

The fiery skipper is a butterfly, not a moth, and differs in its habits from the moths. Eggs are deposited during the day. The turfgrass-destructive stage is the larva. Larvae hatching from eggs laid on the leaves feed on the blades while they are young. Older larvae feed on leaves and stems and spend most of their life in the thatch. Practically all turfgrasses serve as host plants (Chapter 9).

The annual bluegrass weevil is exclusively a golf course problem in the Northeast, damaging only short-cut annual bluegrass. Callow adults emerge from the soil and remain in the thatch to mature. They move to the surface at night to feed on leaves and mate, but they return to the thatch, where they spend most of their life. Adult feeding is relatively insignificant. Eggs are inserted in the lower stem. Young larvae feed in the stems. Older larvae feed externally on the crowns, causing the most serious damage. The grass first yellows but soon dies in large patches (Chapter 17).

Soil

Soil-inhabiting insects tend to occupy a single habitat more exclusively than do members of the previous two groups. The scarabaeid grubs dominate this group. Beetles deposit eggs in the soil, and the entire larval and pupal life is soil bound. Adults are predominantly soil inhabitants. They leave the ground only to feed and mate, or only to mate if the adults are nonfeeders, and return to the soil daily for diurnal or nocturnal oviposition, depending on the species. Mole crickets and ground pearls are equally soil bound. Those with multiple habitats that are still major soil inhabitants include the dichondra flea beetle, billbugs, and the European crane fly. The entire group includes the following:

Southern mole cricket, *Scapteriscus acletus* Rehn & Hebard
Tawny mole cricket, *S. vicinus* Scudder
Ground pearls, *Margarodes meridionalis* Morrison, and other species
Black turfgrass ataenius, *Ataenius spretulus* (Haldeman)
Turfgrass aphodius, *Aphodius granarius* (L.) and *A. pardalis* LeConte
Scarabaeids normally of 1-year life cycles:
  Green June beetle, *Cotinus nitida* (L.)
  Northern masked chafer, *Cyclocephala borealis* Arrow
  Southern masked chafer, *C. immaculata* (Oliver)
  Asiatic garden beetle, *Maladera castanea* (Arrow)

European chafer, *Rhizotrogus majalis* (Razoumowsky)
Japanese beetle, *Popillia japonica* Newman
Oriental beetle, *Anomola orientalis* Waterhouse
Scarabaeids normally of 2- to 3-year life cycles:
May or June beetles, *Phyllophaga* spp.
Dichondra flea beetle, *Chaetocnema repens* McCrea
Billbugs:
Bluegrass billbug, *Sphenophorus parvulus* Gyllenhal
Hunting billbug, *S. venatus vestitus* Chittenden
Phoenix billbug, *S. phoeniciensis* Chittenden
European crane fly, *Tipula paludosa* Meigen

Apart from adult mating flights in the spring, the southern and tawny mole crickets live in the soil their entire life. Both adults and nymphs feed on grass roots and burrow in loose soil, disrupting the turf and causing the soil to dry. Bahiagrass and bermudagrass are most susceptible, but all common warm-season turfgrasses are attacked. Mole cricket activity is especially damaging to newly planted turf (Chapter 4).

Ground pearls live in the soil their entire lives and feed on turfgrass roots. Bermudagrass, St. Augustinegrass, zoysiagrass, and centipedegrass are commonly infested, causing the turf to yellow, become unthrifty, and eventually die (Chapter 6).

Damage by the black turfgrass ataenius is restricted almost exclusively to golf courses, where fairways and greens are most commonly injured. The grasses most frequently damaged are annual bluegrass, bentgrass, and Kentucky bluegrass. The larvae feed on the roots at the soil-thatch interface or in the soil, causing the grass to wilt even under irrigation owing to the loss of roots. The insect has been the most troublesome where there are two generations each year. North of the latitude of New York State and Michigan there is one generation a year. The turfgrass Aphodius causes identical damage and is often associated with the black turfgrass ataenius (Chapter 11).

The group of scarabaeid beetles, normally of a 1-year life cycle, have much in common in their life cycles and in their injury to turf. Damage is caused by third-instar grubs during the fall and again during the spring. All species of turfgrasses are susceptible. Adults are present during early to midsummer. Eggs are deposited in moist soil in earthen cells. Upon hatching, they molt twice before becoming the large, destructive third instars (Chapters 12–15).

May or June beetles, genus *Phyllophaga,* have more than 200 species in North America, with different species being troublesome in different regions of the country. Most of these beetles and their larvae are much larger than most scarabaeids that complete a generation each year. Species with a 3-year life cycle produce minor turfgrass injury, feeding on roots as second-instar larvae during the 1st summer.

Major injury is produced by third-instar larvae during the 2d year. There is little spring feeding during the 3d year, since the insect spends this time primarily in completing its life cycle as pupae and adults. In many species the adults feed on tender, expanding leaves of many deciduous trees (Chapter 14).

In southern and central California, where the broad-leaved dichondra is used as a lawn, the dichondra flea beetle can be a devastating pest. Adults produce a characteristic crescent-shaped feeding injury to the upper surface of the leaves, but this injury is secondary to the root-feeding damage done by the larvae. Damage occurs throughout the growing season (Chapter 16).

The three species of billbugs have similar life cycles and appear nearly alike but have different primary host plants. Bluegrass billbugs prefer bluegrasses, the hunting billbug prefers zoysiagrass, and the phoenix billbug prefers bermudagrass. Adults feed by chewing holes in the lower stems, where they deposit their eggs. The young larvae develop within the stems, but the older larvae feed on the crowns and roots, causing the most serious damage. Most of the dichondra damage occurs during the summer (Chapter 17).

The European cranefly has a very limited distribution in the Pacific Northwest, but where it does occur, it is a very serious turfgrass pest on all lawn grasses. Larvae cause all the damage by feeding on leaves and stems during the night and on roots and crowns of turfgrass during the day (Chapter 18).

Seasonal Presence of Injurious Stages

Most turfgrass insects in the tropical regions of the United States do not have a true winter diapause stage but continue to develop through the winter months at a slower pace if it is cool and at a faster pace if it is abnormally warm.

In the temperate regions there is usually a true diapause stage, in which development ceases until spring. These insects show a distinct, predictable life stage during each of the seasons. Having this information, we can determine which insects are most likely to be causing damage at a given time on the basis of seasonal habits alone. Figure 81 lists temperate-region insects and the season when they normally cause their major damage in relation to their oviposition period. Mole crickets, while tropical and subtropical in distribution, have been added as an exception because they have only one generation a year, and the seasonal occurrence of each stage is as predictable as it is for temperate-region insects.

Winter

Only one species, the winter grain mite, causes turf damage during the winter (Chapter 3).

Season of damage and pests	Overwintering stage
Winter	
Winter grain mite	Egg[a]
Spring	
European crane fly	Larva[d]
Spring and fall	
Mole crickets	Adult and nymph
Scarabaeids: 1-year cycle	Larva[d]
Summer	
Hairy chinch bug	Adult
Greenbug	Egg
Temperate-region sod webworms	Larva
Cutworms, army-worms	Larva[b]
Black turfgrass ataenius	Adult
May or June beetles of 3-year cycle	
Year 1	Adult
Year 2	Larva[c]
Year 3	Larva[d]
May or June beetles of 2-year cycle	
Year 1	Larva[d]
Year 2	Larva[c]
Bluegrass billbug	Adult
Annual bluegrass weevil	Adult
Frit fly	Larva

Period of oviposition and turf damage: Jan | Feb | Mar | Apr | May | Jun | Jul | Aug | Sep | Oct | Nov | Dec

[a]Oversummering eggs.
[b]Infestations from migrating moths from the South.
[c]2d instar.
[d]3d instar.

.... Period of oviposition
______ Period of damage by immatures
\- - - - - Period of damage by adults

Figure 81. Periods of oviposition and major turfgrass damage by temperate-region pests or groups and by mole crickets, tropical-region pests. (Drawn by H. Tashiro, NYSAES.)

Spring

Only one species, the European crane fly, damages turf and only during the spring, as overwintering third instars become fourth instars in April and feed vigorously until mid-May (Chapter 18).

Spring and Fall

Two groups, the mole crickets in the warm, humid region and the scarabaeid species that normally have one generation a year in the cool, humid region, are destructive to turf during both the spring and the fall. Either the adult or the nymph of the southern and tawny mole crickets is present throughout the year. Both species deposit eggs from April into June. Since adults and nymphs are present throughout the year, turf damage can occur any time, but the period of greatest damage is late summer and early fall, when nymphs of both species are in their later instars and are actively foraging for food (Chapter 4).

The scarabaeid grubs of 1-year life cycles include at least seven species of importance as pests of turfgrass in the cool, humid region. All deposit eggs during a 4- to 6-week period during midsummer. The resulting larvae reach the third instar by late summer and early fall and become highly destructive, feeding on the grass roots near the soil surface. The cold weather of October and November forces them to migrate deeper into the soil for winter hibernation. They resume feeding again during early spring. The period of most vigorous feeding occurs as the grubs reach maturation 2 to 3 weeks before pupation. During May into early June, large patches of turf may be killed (Chapters 12–15).

Summer

All other major temperate-region turfgrass insects (Figure 81) damage turfgrass during the summer. The bluegrass billbug does so, developing through a single generation (Chapter 17). Most others do so, developing through multiple generations. The May or June beetles with a 3-year life cycle cause minor damage during late summer of the 1st year as second instars. Their major damage occurs from midspring throughout the summer of the 2d year as third instars. May or June beetles with a 2-year life cycle cause their major damage throughout the 2d summer (Chapter 14).

Ants, bees, and wasps (not listed on Figure 81) are destructive to turfgrass indirectly. They do not feed on turfgrass but nest in the drier, well-drained turfgrass soils and make unsightly mounds that interfere with mowing and with other maintenance practices. Some of the ants eat seeds, which is a relatively minor problem. Colony formation and mound-making activity can occur throughout the summer and into fall in the areas of the cool-season grasses but happen throughout the year in the Gulf Coast states (Chapter 19).

24

Detection and Diagnosis of Infestations and Damage

Early Detection and Diagnosis

Value of Early Detection

Insects and mites that damage turfgrass are generally so small that their invasion or the early development of potentially damaging populations is difficult to detect. Early symptoms of damage are frequently evident before the actual pest is observed. Once the earliest symptoms of damage appear, the progression of deteriorating turf can be very rapid, simply because of the sheer numbers of insects present. Any adverse appearance of the turf should therefore be investigated immediately to determine the cause.

Early Symptoms

The very first symptom of damage caused by leaf- and stem-feeding insects is yellowing of the leaves in small isolated patches. When strictly root-feeding insects are involved, the earliest symptoms may be a gradual thinning of the turf.

The best way to find the cause of turf deterioration is to go down on your hands and knees to examine the leaves and stems closely, separating the thatch to search for insects in this layer. To check for insects in the soil, lift a section of sod and soil to a depth of at least 10 cm (4 in.), breaking it apart for close examination. Insects are much more apparent in moist, friable soil than in dry soil, where dust particles adhere to the hairs of the body and make insects more difficult to see.

Relating Symptoms to Cause

The deterioration of a turfgrass stand, whether caused by insects, pathogenic fungi, improper soil pH, poor nutrition, or something else, may have symptoms that remain very much the same in appearance. Positive attribution of the symptoms to a pest therefore requires determination that the pest is present in a turf-injurious stage. Brown, dead turf may no longer be harboring the pest responsible for the damage, since the individuals may have moved on to a fresher food supply, may have entered the soil for pupation, or may have dispersed as adults, so that they are no longer present. Evidence of insect activity is often still visible, for example the dried sawdustlike frass of billbugs at the crown of plants or the dried fecal pellets of webworms, cutworms, or armyworms.

Seriously damaged turfgrass may in time become heavily infested with saprophytic insects attracted to this habitat by decaying vegetation. Such changes in the habitat and the progression of organisms attracted suggest the importance of positively identifying the turf-destructive pest.

The vast numbers of insect and mite species present in turf make proper identification difficult even for those familiar with turfgrass arthropods. A 10× to 20× hand lens is usually a necessity for field identification (Figure 82). Examination of the smaller details of an insect such as the arrangement of spines, the presence or absence of certain markings, and so forth is usually necessary. If a species cannot readily be identified, it is a simple matter to ask a specialist for help. It is of course necessary to submit good, representative specimens, properly preserved so that the body and all appendages are intact. Both adults and immature forms should be submitted if possible.

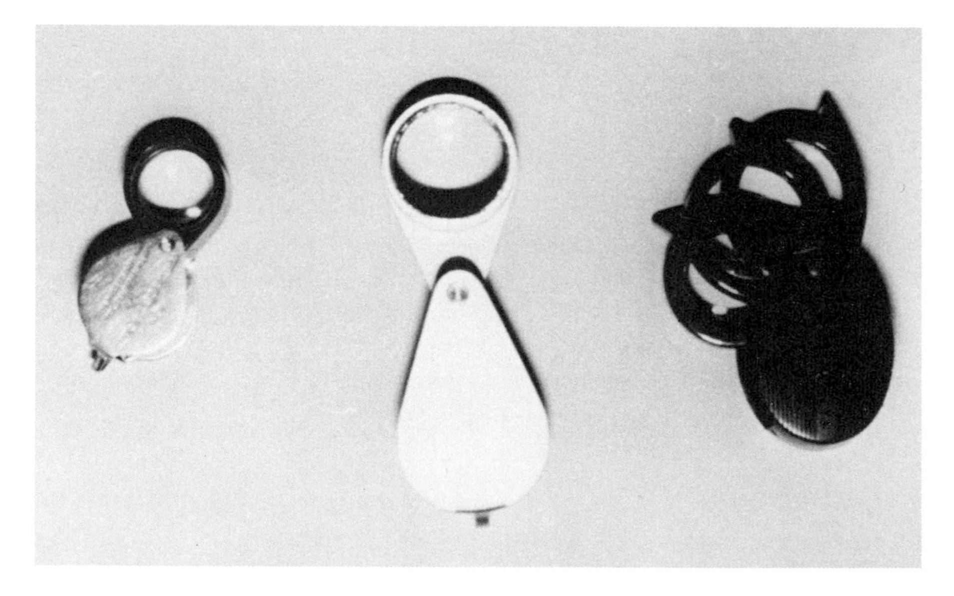

Figure 82. Some available hand lenses providing 10× to 20× magnifications. (Photo by B. Aldwinckle, NYSAES.)

Collecting and Preparing Insects for Identification

Collection Techniques

Small live insects that move rapidly about in or on the thatch or on foliage are often best captured with an aspirator (Figure 83). Quick suction on the plastic tube creates enough vacuum at the tip of the opposite tube to draw small objects into the clear plastic container. A fine screen over the end of the suction line prevents debris from entering the mouth. There is no way to keep from inhaling the odor given off by insects. Insects of the order Hemiptera, suborder Heteroptera, such as chinch bugs, are the insects most likely to have a disagreeable odor. The odor may be unpleasant, but it is not harmful.

An aerial insect net (Figure 83) is the most practical and simplest equipment to use in capturing large and small insects that fly over the turf. A sweeping net with a more sturdily built frame and much heavier cloth (Figure 83) is a useful tool for collecting insects on leaf blades, such as aphids and plant bugs.

Additional turfgrass insects are easily captured directly by hand, since they move relatively slowly in the thatch and soil.

Insects being collected for identification can be dropped directly into a killing solution in the case of most immature forms and in the case of insects that are relatively free of scales and long hairs, such as beetles. Moths and butterflies should be placed not in a liquid but in a dry killing jar.

Transporting Live Insects and Mites

I prefer to capture insects and hold them alive until I return to the laboratory, where proper killing and preserving techniques can more easily be used. Several techniques can be used for keeping insects alive. Insects collected on foliage can be placed in plastic cups, with a small amount of plant material to provide both food and a moist atmosphere for them in transit. Small, partially inflated plastic bags can also be used. It is helpful to place a piece of dry paper toweling in the bag to absorb free water.

Soil-borne insects can best be transported in small containers such as salve tins or plastic cups partially filled with soil. To avoid cannibalism, avoid overstocking containers. Scarabaeid grubs in overcrowded conditions invariably destroy each other. A good ratio would be no more than about 10 third-instar scarabaeid grubs per liter (quart) of soil. Containers should be kept in the shade and preferably in picnic coolers on warm sunny days.

Killing and Fixing

Most adult insects, except butterflies and moths, that are not designated for a permanent collection, can be placed directly into ethyl, rubbing, or isopropyl alco-

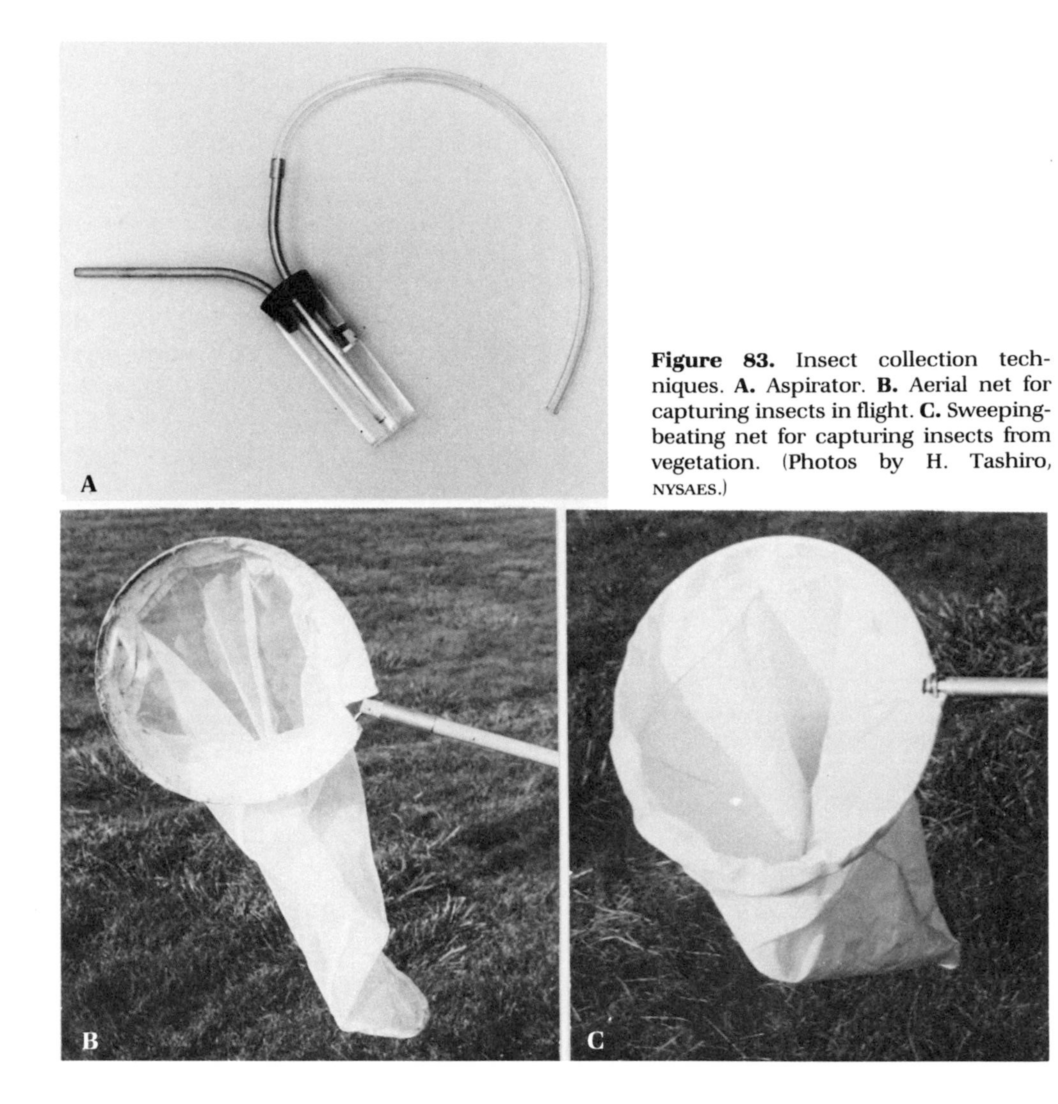

Figure 83. Insect collection techniques. **A.** Aspirator. **B.** Aerial net for capturing insects in flight. **C.** Sweeping-beating net for capturing insects from vegetation. (Photos by H. Tashiro, NYSAES.)

hol for killing and fixing (Table 26). Adults that are to be pinned and placed in a permanent insect collection should be killed in a dry atmosphere. Butterflies and moths should always be killed dry. This procedure is discussed below.

Immature stages of insects for a permanent collection can be preserved by killing and fixing according to any of three different methods, all of which are satisfactory. (1) Bring water to a boil and remove it from heat. As soon as the water stops bubbling, drop the insects in. After 1 to 2 min, depending on the proper distension, transfer to 70% ethyl alcohol (preferred) or rubbing alcohol for preserving. This is the preferred killing and preserving technique for obtaining the very best spec-

Table 26. Methods of preserving turf arthropods

Adults (wings present)		Adults or immatures	Immatures
Moths	Ants	Aphids	Billbug larvae
Armyworms	Beetles	Big-eyed bugs	Caterpillars
Cutworms	Billbugs	Centipedes	Armyworms
Webworms	Chafers	Chinch bugs	Cutworms
	Japanese beetles	Earwigs	Webworms
	May or June beetles	Fleas	White grubs
	Ataenius beetles	Grasshoppers	↓
	Big-eyed bugs	Leafhoppers	
	Chinch bugs	Millipedes	Boiling water,
	Flies	Mites	Kahle's, or
	Grasshoppers	Mole crickets	K.A.A.D.
	Mole crickets	Spiders	
	Wasps	Spittlebugs	
	↙ ↓	Wireworms	
↓		↓	↓
Dry	Alcohol	Alcohol[a]	Alcohol

[a]When in doubt, place specimen in alcohol.
Source: After Dunn and Kennedy 1983.

imens. (2) Drop live insects in K.A.A.D. solution or (3) into Kahle's solution. Hold very small larvae in either solution 1 hr or longer, depending on distention. Hold mature larvae in solutions 12–24 hr for proper fixation. Transfer to 75% ethyl alcohol (preferred) or rubbing alcohol.

The killing solutions are prepared in the following way. *K.A.A.D.*: Mix thoroughly 7–10 parts 95% ethyl alcohol (or commercially refined isopropyl alcohol), 2 parts glacial acetic acid, 1 part kerosene, and 1 part dioxane. If dioxane is unavailable, the remaining three ingredients will still make an adequate killing solution. *Kahle's solution:* Mix thoroughly 15 parts 95% ethyl alcohol, 10 parts distilled water, 6 parts Formalin (40% formaldehyde), and 1 part glacial acetic acid.

During the killing and fixing process, the containers should be wide enough to allow larvae to lie flat and to distend without obstruction. If a vial is used in field killing and fixing, lay it on its side. In preserving, the ratio of alcohol to specimens should be about 10:1 v/v (Dunn and Kennedy 1983, Peterson 1948, Rings and Musick 1976).

Most adult insects, and especially the moths and butterflies, are best killed in a dry killing jar using a fumigant. Entomologists usually make killing jars by placing several centimeters of potassium cyanide crystals in a wide-mouthed jar, covering with an equivalent depth of dry sawdust, and sealing the two ingredients in place with plaster of paris mixed with water to a thick, viscous consistency. The mixture is then allowed to dry under a fume hood. The jar should be capped and labeled *poison*. If the contents are kept dry, and the lid is removed only to insert or remove insects, the jar should be usable for months. Killing jars that can be charged with other types of fumigants are available from biological supply stores.

Insects killed dry can best be prepared for shipment by placing them in a small plastic bag without attempting to manipulate the position of appendages. Pack them very lightly in tissue paper, and place them in a small box. Careful handling ensures that the wing scale patterns of moths and butterflies, essential for proper identification, will be properly preserved. It is equally important to retain other appendages, since legs and antennae have diagnostic characters.

Proper Labeling

Specimens submitted for identification should bear the following information clearly printed or typed:

Host plant and description of damage
Locality (city, state)
Date of collection
Name, address, and telephone number of the collector

If specimens are being submitted in a vial, fill it full of the preserving fluid, and place a label written in pencil (not ink) inside. It is advisable that you send as many as 10 specimens of each stage because variations may be present among individuals. Information on the best sources for proper identification are listed in Appendix 1, "Sources of Information on the Identification of Insects and Recommendations for Their Control."

25

Population Survey Techniques

Various methods have evolved by which the populations of insects present in the turfgrass ecosystems may be determined in a rapid, efficient, and consistent manner. These methods have become standard techniques for reporting population densities.

Surveys for Injurious Stages

Submergence and Flotation

The technique described below has developed primarily for determining the presence or absence of chinch bugs in a turfgrass area. Select a large can with a diameter of at least 15 cm (6 in.) from which both ends can be removed. Where extensive studies on chinch bugs have been undertaken, cylinders made of heavy sheet metal or heavier material were specifically designed for this purpose. One end with a cutting edge is forced through the thatch into the surface soil. The cylinder is then filled with water, and more is added as the level recedes (Figure 84; Mailloux and Streu 1979, Niemczyk 1981).

Chinch bug adults and nymphs present in the thatch soon float to the surface, and within 5–10 min the entire population present within the cylinder will have surfaced. The insects are removed and counted as they surface. Both adults and nymphs are easily detected, since they will be constantly moving on the surface of the water (Figure 84).

A modification of this technique makes it possible to count eggs as well. Water pressure is used to loosen the thatch forcibly and to wash it, along with the dislodged insects, out of the cylinder into a 100-mesh sieve. All collected material in 40% ethyl alcohol is centrifuged to separate the insects, including eggs, from the thatch. Further centrifugation in a sugar solution results in recovery of 95%–100% of the chinch bug eggs in addition to the adults, nymphs, and other arthropods present in the cylinder (Chapter 5; Mailloux and Streu 1979).

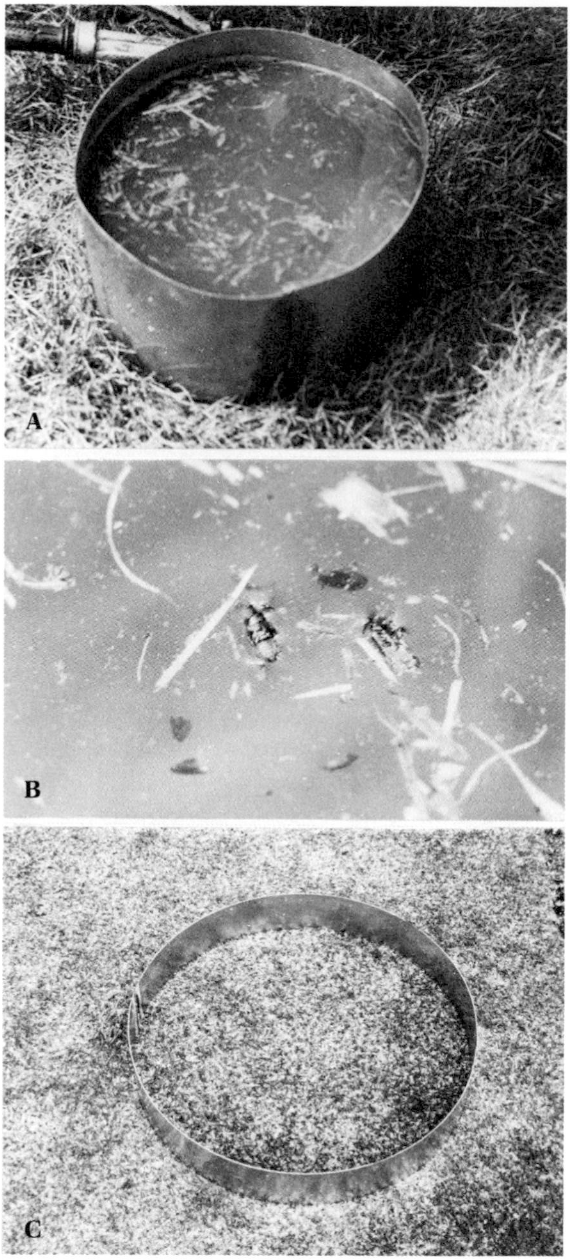

Figure 84. Methods of sampling for foliage- and thatch-inhabiting insects with water. **A.** Submergence and flotation, primarily of chinch bugs. **B.** Chinch bugs floating to the surface and easily seen. **C.** Enclosure of 1,860 cm^2 (2 ft^2) for application of pyrethrins or detergent solutions for forcing primarily lepidopterous larvae to surface. (Photos by H. Tashiro, NYSAES.)

Surfacing via Irritants

This technique involves sprinkling a gallon of water containing an irritant over a 0.84 m^2 (1 yd^2) of turf. Contact with the irritant forces thatch-inhabiting insects, primarily larvae of moths and butterflies, to move to the surface. Other insects that can also be forced to surface by changing the concentrations of the irritant or the volume of water are vegetable weevils, bluegrass billbug adults, mole crickets, and chinch bug adults and nymphs. The most commonly used irritants are a quarter cup of dry detergent, an ounce of liquid detergent, or a tablespoon of 1%–2% pyrethrin per gallon of water. The European cranefly larvae may be brought to the surface with another irritant, orthodichloro benzene (Chapters 4, 5, 17, 18, and 20).

A more thorough study of the role of the various components of this technique was made using the grass webworm infesting 'Sunturf' bermuda, *Cynodon magennisii* Hurcombe and 'Seashore' paspalum, *Paspalum vaginatum* Sw., as the test insect (Tashiro et al. 1983).

Five concentrations of liquid detergent (0.063%–1.000%) and five of pyrethrins (0.0004%–0.0060%) were tested. The 0.25% detergent (10 ml per 4 l) and 0.0015% pyrethrins (1 ml per 4 l) applied to 1,860 cm^2 of turf within a metal frame were the most efficient. These were used as the standard concentration for all other studies.

One liter of water containing either 10 ml of detergent or 1 ml of pyrethrins was applied to each 1,860 cm^2 of turf. This is equivalent to applying a gallon per square yard. This volume was insufficient to make the maximum number of larvae surface. Only 53% and 61% of the larvae surfaced for the detergent and pyrethrins, respectively, as compared with the 4× volume of water.

The detergent forced larvae to the surface more quickly than the pyrethrins. It was also determined that continuous observation was essential for accurate larval counts. Five minutes after application, 29% of the surfaced larvae had reentered the thatch, and at 10 min, 39% and 38% had reentered with the treatments with detergent and pyrethrins, respectively.

This study demonstrated the advisability of using an observation area of 1,860 cm^2 (2 ft^2) rather than the 0.84 m^2 (1 yd^2; Figure 84). An area a yard square is too large for a single person to observe critically for 10 min, since the observation of a mere movement often indicates the presence of the insect being surveyed. The application of four times the standard volume of water within a retaining frame provides for thorough saturation of the turf and thatch and gave the most efficient sampling technique for forcing thatch-inhabiting larvae to the surface (Tashiro et al. 1983).

Thatch and Foliar Sampling

The DeVac suction device (Figure 85; DeVac Company, Riverside, California) is an effective motorized apparatus for general sampling of insects on foliage. It has found little or no use in sampling insects from turfgrass but has the potential for large-

Figure 85. Thatch and foliage sampling techniques and equipment. **A.** The DeVac, for sampling via suction, is seldom used for sampling turfgrass-infesting insects but has much potential. **B.** Berlese funnels for driving live specimens out of foliage, thatch, and soil samples with heat into collection jar. (Photos by H. Tashiro, NYSAES.)

scale thatch and foliage sampling. The Berlese funnel is an effective apparatus for driving insects out of thatch and foliage with heat (Figure 85). Samples taken from the field are placed on a 0.6-cm (0.25-in.) screen near the top of the funnel. A 25-watt lamp mounted in a cover reflector placed on the funnel drives any live insect away from the heat and forces it to fall into a receptacle that usually contains alcohol and is placed below the funnel. Within 24–48 hr, depending on the moisture content of the sample, the jar of alcohol should contain insects that were alive in the turf sample. Since the insects are in alcohol, examination and identification can be undertaken at any later time. This technique is the most useful for sampling populations of billbug adults, chinch bugs, webworms, cutworms, and mites (Niemczyk 1981).

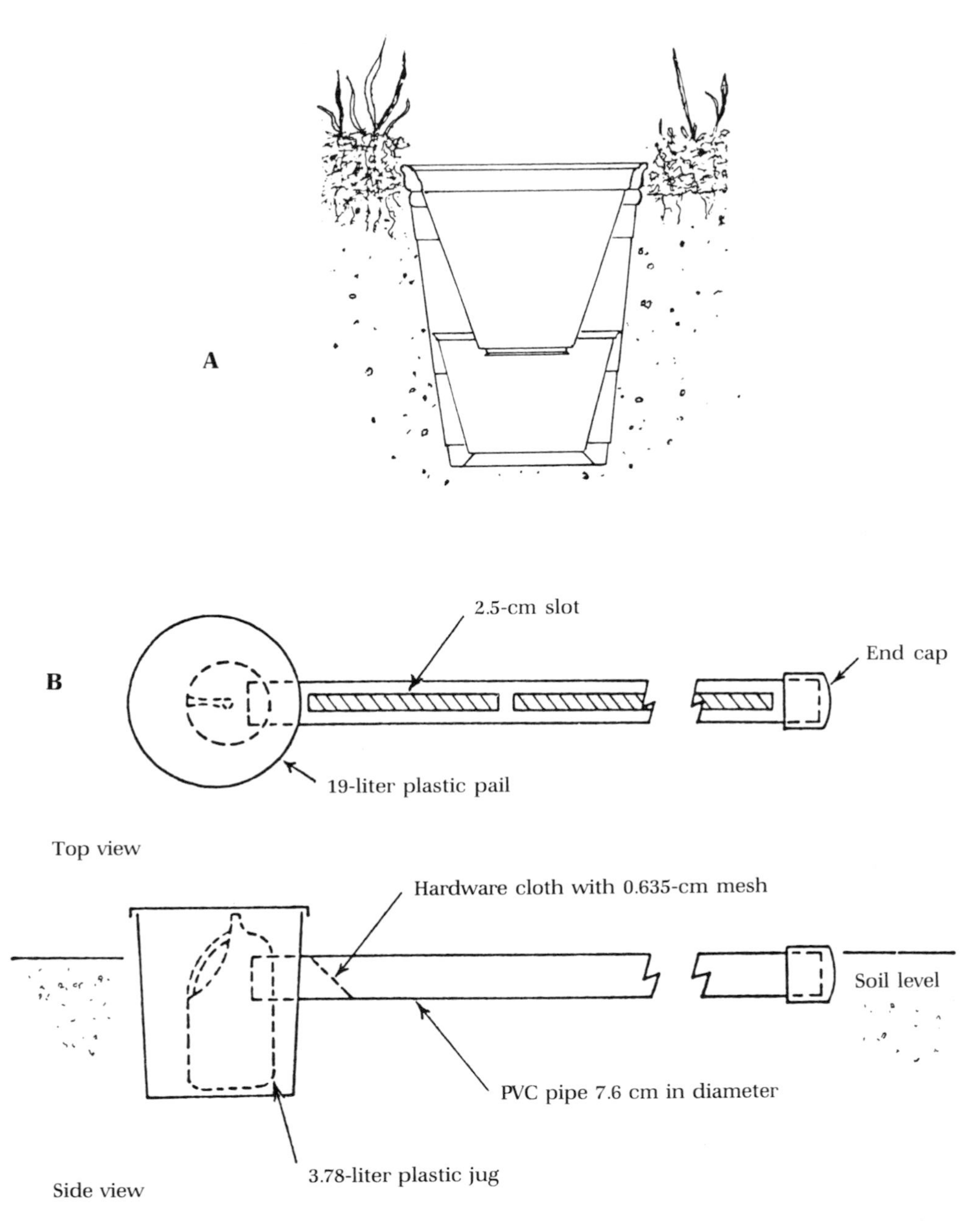

Figure 86. Pitfall traps. **A.** Conventional trap. **B.** Linear trap, designed mainly for mole crickets. (Part A from Niemczyk 1981, p. 43, courtesy of H. D. Niemczyk; part B from Lawrence 1982, fig. 1.)

Pitfall Traps

The conventional type of pitfall trap can be made of three plastic cups, with the upper interior cup modified to act as a funnel to direct captured specimens into the collection cup below (Figure 86). Alcohol or water is placed in the latter. The lip of the exterior cup is placed at the soil-thatch level. Arthropods crawling through the turf fall into the outer cup and are captured. These traps can be used to monitor and detect chinch bugs, adult billbugs, and many other insects (Niemczyk 1981).

A linear pitfall trap designed primarily for collecting mole cricket nymphs captures the insects at all stages plus many other arthropods (Figure 86, Plate 8). Major components of the trap are a PVC pipe 2.5 m (8.2 ft) long with a slot 2.5 cm (1 in.) wide cut to run nearly its entire length. A 19-liter (5-gal) plastic pail, a 3.8-liter (1-gal) plastic jug, a screen, and an end cap complete the assembly. The PVC pipe is placed in a trench with the open slot up, at the soil level or slightly below, and the pail is embedded into the ground. About 1 cm (0.4 in.) of soil placed inside the PVC pipe and jug keeps the mole crickets separated to prevent cannibalism. Insects that fall into the pipe eventually move to the open end and fall into the jug. These traps work equally well on sod or bare ground and catch large numbers of mole crickets at all stages as well as other turf-inhabiting insects (Lawrence 1982) (Chapter 4).

Figure 87. Jacobson sod cutter with a cut 30.5 cm (1 ft) wide, used for spring and fall sampling of sod at predetermined depths for scarabaeid grub surveys. (Photo by H. Tashiro, NYSAES.)

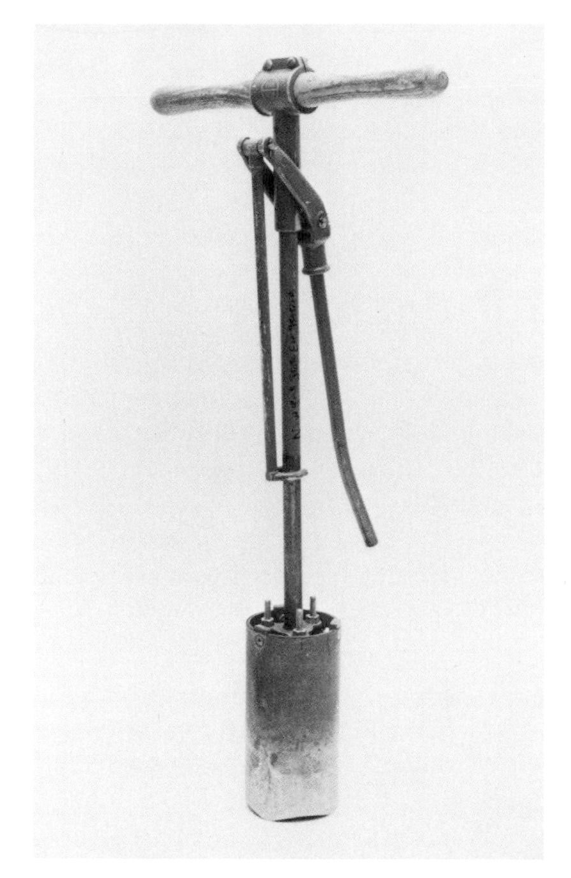

Figure 88. Standard golf cup cutter with handle for ejecting samples of turf and soil 10.8 cm in diameter. (Photo by J. Ogrodnick, NYSAES.)

Soil Sampling

The techniques required to sample the soil underlying turfgrass areas are not only the most arduous but also the most disruptive of the turf's appearance, especially when soil samples must be as large as 0.1 m^2 (1 ft^2) because the population is likely to be scattered. Because some turf-damaging insects remain strictly in the soil, disruption of the soil is necessary to obtain accurate counts. When 0.1 m^2 samples are taken with a spade, the depth of sampling is often variable because of the rooting habits, soil moisture, and texture. Most often sampling is much deeper than necessary in order to ensure that specimens of all the insects present have been obtained. The least disruptive method of examining a 0.1 m^2 of sod is to cut three sides and turn back the cut area as if it were a flap. This procedure allows many of the plants to keep their root system intact.

In New York State, the most convenient equipment for making scarabaeid larval counts has proven to be a motorized sod cutter that cuts a strip of sod 30 cm (1 ft) wide (Figure 87). During grub counts in a large number of turfgrass plots, one pass is made through a complete row of plots, but only a small portion of the cut strip is lifted in each plot for examination. On a trial basis the depth of cut can be adjusted to penetrate only deep enough to include all of the larvae present. During the most active feeding period of scarabaeid grubs in the spring and fall, almost the entire population is in the upper 5.0 cm (2 in.) of soil. A sod cutter has other beneficial features. The strip not lifted for larval counts reroots immediately, without showing any evidence of disturbance, and soil is not examined to an excessive depth, resulting in >50% saving of labor.

A standard golf cup cutter that takes a sample of sod and soil 10.8 cm (4.25 in.) in diameter is a very useful tool for sampling for the smaller soil-inhabiting pests such as the black turfgrass ataenius larvae and the annual bluegrass weevil (Figure 88). The depth of samples can be adjusted. The sample should be ejected gradually to prevent loss of specimens in the thatch or on the soil. To convert counts to populations per 0.1 m^2 (1 ft^2), multiply by a factor of 10.15 (Niemczyk 1981, author's personal observations).

Surveys for Noninjurious Adults

Black-Light Traps

Black-light traps generally fitted with a 15-watt black-light lamp with a peak emission of 3,650 Å are standard equipment for sampling most night-flying insects (Figure 89). Insects attracted to the radiation strike the metal baffles that are at right

Figure 89. Black-light trap with 15-watt lamp for capturing night-flying insects, including moths of webworms, cutworms, and armyworms and many scarabaeid beetles. (Photo by H. Tashiro, NYSAES.)

angles to the lamp and are then deflected into the funnel and into a receptacle. The latter can be charged with a volatile insecticide to kill the insects quickly and thus preserve their natural appearance. When an A-C circuit is employed, the location of the traps is greatly restricted, but the lamp is generally operated continuously, since it consumes very little power. If the unit is powered with a storage battery employing a photoelectric switch, it becomes much more expensive, but its placement is much more flexible. During sampling for certain insects it is a disadvantage to trap large numbers of extraneous insects, since much time must then be spent in sorting (Tashiro et al. 1967).

None of the turf-damaging stages are caught in a black-light trap, but the adults caught will represent a very large number of turf pests. All the moths, including webworms, cutworms, and armyworms, are attracted. The scarabaeid beetles attracted include the black turfgrass ataenius, the northern and southern masked chafers, the Asiatic garden beetle, the European chafer, and May or June beetles. The annual bluegrass weevil is also attracted.

Visual Observations

Food Plants

Turf-damaging populations of three scarabaeid species can be estimated by observing the density of adults on their favorite host plants. These are the green June beetle, the May or June beetles, and the Japanese beetle (Plates 29, 35, 38).

Once the early beetles of the season are seen, Japanese beetle populations build up rapidly and congregate on their favorite food plants during warm, sunny days to feed and mate. They feed on many plants, but one of the most favored is grape leaves. Other favorites in a landscaped area include ornamental roses, cherries, peaches, and plums. Some of the most favored trees include mountain ash, linden, and sassafras. When most of the leaves on these trees turn brown as a result of adult feeding, populations have been high enough for many eggs to have been deposited in nearby turf. Larval damage can be expected later that season (Chapter 15).

May or June beetles feed on the young tender leaves of oaks, as these trees are starting to leaf out in spring. Other favored trees include persimmon, hickory, walnut, elm, and birch. Defoliation of these trees by high adult populations indicates potential turf damage for three seasons, with the most severe during the second year (Chapter 14).

Green June beetle adults feed most heavily on ripening thin-skinned fruits, such as figs, peaches, and grapes. Heavy feeding damage can often mean sufficient larval populations to cause significant turf damage (Chapter 12).

Mating Flights

European chafer beetles leave the ground shortly after sunset on warm sunny days and congregate around silhouetted trees for mating. During nights of peak flight, so many beetles may congregate around a single tree that their beating wings sound like a swarm of honey bees. Peak flight activity occurs about 9:00 PM (daylight saving time; Plate 24). Flights of this intensity usually result in such massive oviposition in adjacent and nearby turf areas that damage usually occurs later in the fall and in the following spring (Chapter 14).

Preoviposition Adults

Twice during the year, billbug adults are often seen on sidewalks and driveways adjacent to home lawns and appear to be wandering aimlessly in no particular direction (Plate 45). The heavier populations of wandering adults are seen on warm, sunny days of late September and October as adults search for overwintering quarters. On warm sunny days of May and June, the insects are also seen wandering about, usually in smaller numbers than in the fall and this time apparently in search of grass for feeding and oviposition. When adults are so numerous that one can be seen every minute, turfgrass damage can usually be expected in adjacent lawns (Chapter 17).

Black turfgrass ataenius adults are often present on golf greens during early spring when overwintering adults begin to fly and again during midsummer as the first-generation larvae develop into adults. Other areas where adults may be seen are on the surface of swimming pools and around lights (Chapter 11).

Surveys of Hibernation Quarters

Annual bluegrass weevil adults start moving from golf course fairways into roughs during the fall to seek hibernation quarters. The most favored quarters are the needle litter under white pine trees (Plate 48). The density of weevil populations can be determined by collecting white pine litter and surface soil and completely submerging them in water. Weevils will start to surface in a few minutes and will have nearly finished doing so within an hour. Hibernating populations of more than 100 per 0.1 m^2 (per ft^2) usually mean extensive damage to adjacent annual bluegrass fairways the following spring and early summer (Chapter 17).

26

Insect Control: Principles and Strategies

In the management and control of turfgrass insect pests, consideration should be given not only to immediate needs but also to long-term biological factors affecting the turfgrass ecosystem.

Plant Resistance

Promise of Resistant Strains

The selection and breeding of plants for resistance to injurious insects and disease organisms has long been considered the ultimate goal in plant protection. The procedures are both time consuming and expensive, and many years of research may pass before a highly resistant cultivar possessing the desired characteristics becomes a reality. The search for turfgrass cultivars possessing resistance to certain insects has greatly accelerated in recent years. The successes to date provide encouraging incentives for an acceleration of this approach to turfgrass insect control.

Research Techniques

Both laboratory and greenhouse techniques have been employed with success. In most cases a choice feeding test is involved, where a known standard cultivar susceptible to a given insect and a candidate cultivar are isolated and are exposed to a given stage of a turfgrass insect for a predetermined period. The degree of feeding on the two host plants, the rate of development, or the death of the insect after a set period determines the presence or absence of resistance. In some studies the discovery of a tolerant or resistant cultivar in laboratory greenhouse tests has been pursued in field plot studies. Evidence to date indicates that the two types of studies complement each other.

Resistant Cultivars

Cool-Season Grasses

At least two cultivars of Kentucky bluegrass, 'Baron' and 'Newport', and three cultivars of perennial ryegrass, 'Score', 'Pennfine', and 'Manhattan', have shown some resistance to the hairy chinch bug—the bluegrasses through a greater tolerance to foliage injury during feeding and the ryegrasses through reduced degrees of infestation (Chapter 5).

At least three cultivars of Kentucky bluegrass, 'Park', 'Nebraska Common', and 'South Dakota Certified', showed moderate resistance to bluegrass billbug larvae through reduced infestation levels in Nebraska field plots. In New Jersey, at least three perennial ryegrasses showed resistance to the bluegrass billbug, manifesting a lower degree of damage and extent of infestation. These were 'HS-1352', 'Pennant', and 'Regal'.

Lolium endophytic fungi associated with perennial ryegrass have shown some very promising leads in resistance to sod webworms in New Jersey. The degree of resistance appears correlated with the amount of endophytic fungus associated with the ryegrass. Cultivars 'GT-11', 'Pennant', and 'All-Star' with high levels of fungi were highly resistant (Chapter 7).

Warm-Season Grasses

A cultivar and at least two accessions of St. Augustinegrass, 'Floratam' and FA-108 and TX-33, have shown resistance to the southern chinch bug through adult mortality and reduced egg production (Chapter 5).

A high degree of resistance to the bermudagrass mite has been observed in FB-119, 'Midiron', and 'Tifdwarf'. 'Tifgreen (328)' and 'Tifway (419)' have such high degrees of resistance that they border on immunity. The latter two owe their resistance to their *Cynodon transvaalensis* lineage (Chapter 3).

Natural Enemies

Practically every important turfgrass insect pest has one or more natural enemies. Some exert a noticeable influence, while others play a minor role. Beneficial organisms include microorganisms such as pathogenic viruses, bacteria, protozoa, or fungi, to name the major groups. Turfgrass insects have their share of insect parasites and predators. Even vertebrate predators play a significant role in helping to keep turfgrass insects in check but often at the expense of the turfgrass itself (Chapter 22).

The effect of these beneficial organisms, however, depends on weather conditions and on other factors that can greatly influence their effectiveness. Virus diseases generally need a very high host population density, causing population stress, before disease epizootics will occur. Epizootics destroy nearly the entire population present, producing what is known as a population crash, but not until after the pest insect has caused considerable damage. Even parasites and predators are density dependent, functioning most efficiently in a high, damaging population of pest insects. Manipulation of beneficial organisms has done little to make them more effective in managing pest populations. In spite of these limitations, the combined presence of natural enemies may be sufficiently effective to prevent major pest-population explosions. Such population explosions often occur when a new pest has been introduced without its natural enemies and native beneficial organisms have not yet adapted themselves to prey on the introduced pest.

When the environmental conditions are optimum for maximum manifestation of a beneficial organism, their actions are often spectacular and highly effective. Under hot, humid conditions and high moisture levels, the fungus *Beauveria bassiana* may be highly effective in infecting and destroying chinch bugs and several other turfgrass insects (Chapter 5).

Many of the older established turfgrass soils on the East Coast contain high levels of spores of the milky disease bacteria, *Bacillus popilliae* and *B. lentimorbus*. When Japanese beetle grubs feed actively at soil temperatures exceeding 25°C (77°F), a milky disease epizootic can decimate the population. Even oriental beetle grubs, which are not known to be highly susceptible to the milky disease bacteria, succumb to epizootics when high populations are present under optimum temperatures (Chapter 15). These are just two examples of beneficial organisms that require certain specific predisposing conditions to be of much value.

One way in which we can continue to derive the greatest benefits from beneficial organisms present is by choosing insecticides that have the least detrimental effects on the beneficial organisms and using them carefully. By avoiding the use of carbaryl for control of webworms during the summer, for example, it may be possible to reduce the effects of the winter grain mite the following winter (Chapter 3).

Cultural Practices

Turf Vigor

The vigor of a turfgrass sward directly influences its ability to withstand insect population pressures even though maintenance may be entirely independent of any consideration of turfgrass insects. Since injury by turfgrass insects is caused mainly by feeding, a vigorous, steadily growing stand can be one of the strongest deterrents to permanent turf damage because of the plant's active regenerative capacity.

A turf stand temporarily set back by a sudden invasion of an insect such as the sod webworm can quickly recover once the larvae stop feeding. Rapidly growing rhizomes and stolons quickly fill in small localized dead patches. Vigorous turf can withstand two or three times the normal threshold population of a root-feeding insect, such as a scarabaeid grub, that would destroy a weak, starved turf.

A localized or a generally thinned stand can be made to recover very rapidly with a supplemental application of a high-nitrogen fertilizer, with immediate watering for rapid absorption. Care must be exercised, however, not to overfertilize so that growth becomes so succulent as to make the turf susceptible to diseases. It is desirable to maintain turfgrass fertility at a level that will assure early-season green-up, resistance from summer drought through deep root penetration, and prolonged root activity into winter not only for the sake of a vigorous, steadily growing turf but also to promote rapid recovery from low to moderate insect damage as well.

Soil Moisture

Turfgrass in a high-maintenance area such as a golf course is usually irrigated to prevent drought. In the humid regions a large proportion of the turfgrass grown depends solely on natural precipitation. On many home lawns, the desire to have vigorous, beautiful green lawns throughout the season is often outweighed by the cost of irrigation. An extended period of drought does not in itself kill a well-established turf of perennial grasses, because such grasses have the ability to become dormant, but it does create a serious problem where the detection of insects is concerned.

Two groups of insects, the chinch bugs and sod webworms, do most of their damage during periods of high temperature and moisture stress, with the grass going into dormancy. The dormancy of the turfgrass makes it impossible or very difficult to detect early symptoms of damage. Only after heavy precipitation and greening of the healthy grass does the insect damage becomes apparent, often after irreversible damage has taken place. The only remaining option then is to renovate and reseed.

When the turf is growing steadily at adequate moisture levels, symptoms of early insect damage are much more readily detected in the form of yellow leaves and small patches of brown. Supplemental irrigation of a turfgrass stand with high value is therefore a sound investment not only to prevent drought but also to make possible the early detection of a serious insect problem. Remedial measures can easily be taken before irreversible damage occurs.

With proper maintenance of vigorous turf through an annual fertilizer program and supplemental irrigation to prevent drought stress–induced dormancy, much turfgrass insect damage can be prevented.

Thatch: Development and Influences

Thatch Defined

Thatch is an accumulation of organic matter, dead grass stems, and other organic debris between the soil surface and the turfgrass foliage that occurs when the rate of accumulation exceeds the rate of decomposition. A small accumulation of thatch of about 1.0–1.5 cm (0.4–0.6 in.) is beneficial in giving the turf resilience and in acting as a mulch that prevents accelerated drying of the soil surface. An excess accumulation of 1.8–3.8 cm (0.7–1.5 in.) of dense thatch or 4–5 cm (1.5–2.0 in.) of even loose thatch becomes detrimental to optimum turfgrass growth (Plate 64). Penetration of water into the soil may be impeded, and the crowns, rhizomes, and stolons may develop in the thatch rather than in or on the soil and may be subject to more rapid drying. A thick thatch also attracts chinch bugs and other destructive turfgrass insects and may prevent pesticide applications from reaching the soil.

Development of Thatch

Several cultural practices contribute to excessive thatch accumulation. The maintenance of an excessively vigorous turf growth with infrequent mowing adds stems that decompose slowly because of their higher *lignin* content. Lignin is a cellulose compound that forms a woody fiber. Pesticide applications that reduce pH or are otherwise detrimental to soil-borne organisms may also enhance thatch accumulation.

Prevention of Thatch

To prevent or greatly retard thatch accumulation, turfgrass should be mowed often enough so that not more than about 30% of the foliar growth is removed at any one time, which means that only leaves are returned. The collection and removal of leaves contributes little, if at all, to the prevention of thatch formation. Top dressing with soil, practical only on small areas, adds soil-borne organisms that digest organic matter. The mechanical removal of thatch with motorized verticutting machines and soil aeration by removal of plugs are the only practical methods of removing or retarding the accumulation of thatch.

Influence on Insecticidal Efficiency

Thatch, when it is thick or tight, greatly limits the effectiveness of insecticides in the control of soil-inhabiting turfgrass insects. Insecticides applied to the sod surface may be adsorbed by the thatch, preventing their movement even to the surface soil. When long residual chlorinated cyclodiene insecticides, such as chlordane, dieldrin, and heptachlor, were used for scarabaeid grub control, these materials eventually moved to and penetrated the soil through various actions of weather.

The organophosphate (O-P) and carbamate insecticides to be effective must move through the thatch into the surface soil rapidly because of their short residual activity. Some are degraded completely in less than a month. The most effective medium for percolation through the thatch and into the soil is water, either natural rainfall or irrigation. Chlorpyrifos, a commonly used organophosphate insecticide, is the most tightly bound by thatch and has one of the lowest water solubilities.

Table 27. Water solubility of insecticides and their binding characteristics on turfgrass thatch

Insecticide	Water solubility, ppm[a]	Units of thatch required to bind 50% of insecticides applied
Chlorpyrifos (Dursban)	<1	4
Diazinon	40	75
Isazophos (Triumph)	150	300
Trichlorfon (Dylox, Proxol)	120,000	500 +
Bendiocarb (Turcam)	40	640 +

Note: Trade names appear in parentheses.
[a]Obtained from manufacturer's technical data.
Source: Adapted from Niemczyk 1977b and Tashiro 1983.

Trichlorfon has one of the highest water solubilities and is one of the least tightly bound by thatch. These relationships of solubility and thatch adsorption hold true except for bendiocarb (Table 27).

Insecticidal Control

Insecticidal Dependency

Despite the promising results to date in selecting insect-resistant cultivars, the value of good cultural practices that provide greater recuperative processes, and the presence of beneficial organisms to check epidemic populations, the first line of defense for controlling turf-infesting insects continues to be insecticidal control. When there is a sudden and unexpected heavy infestation of a detrimental insect, there is no alternative but to depend on a recommended insecticide.

Timing of Applications

Effective insecticidal control of a given turfgrass pest depends on several factors. Proper timing directed at the most vulnerable stage is one of the more important. In many cases the most vulnerable stage is the early part of the destructive larval stage, as in webworms, or the nymphal stage, as in chinch bugs. In other cases it may be the adult stages just prior to egg laying, as in the annual bluegrass weevil. These variables suggest the need to have a good working knowledge of the ecology of the insect in question. Figure 90 indicates the period of seasonal occurrence of the destructive stage of the major groups of injurious insects and the most opportune period for effective insecticidal applications. Comparison of Figure 90 with Figure 81 shows that the former is a protection of the latter. Such comparison also adds to an understanding of proper timing of insecticides relative to the biology of the insect.

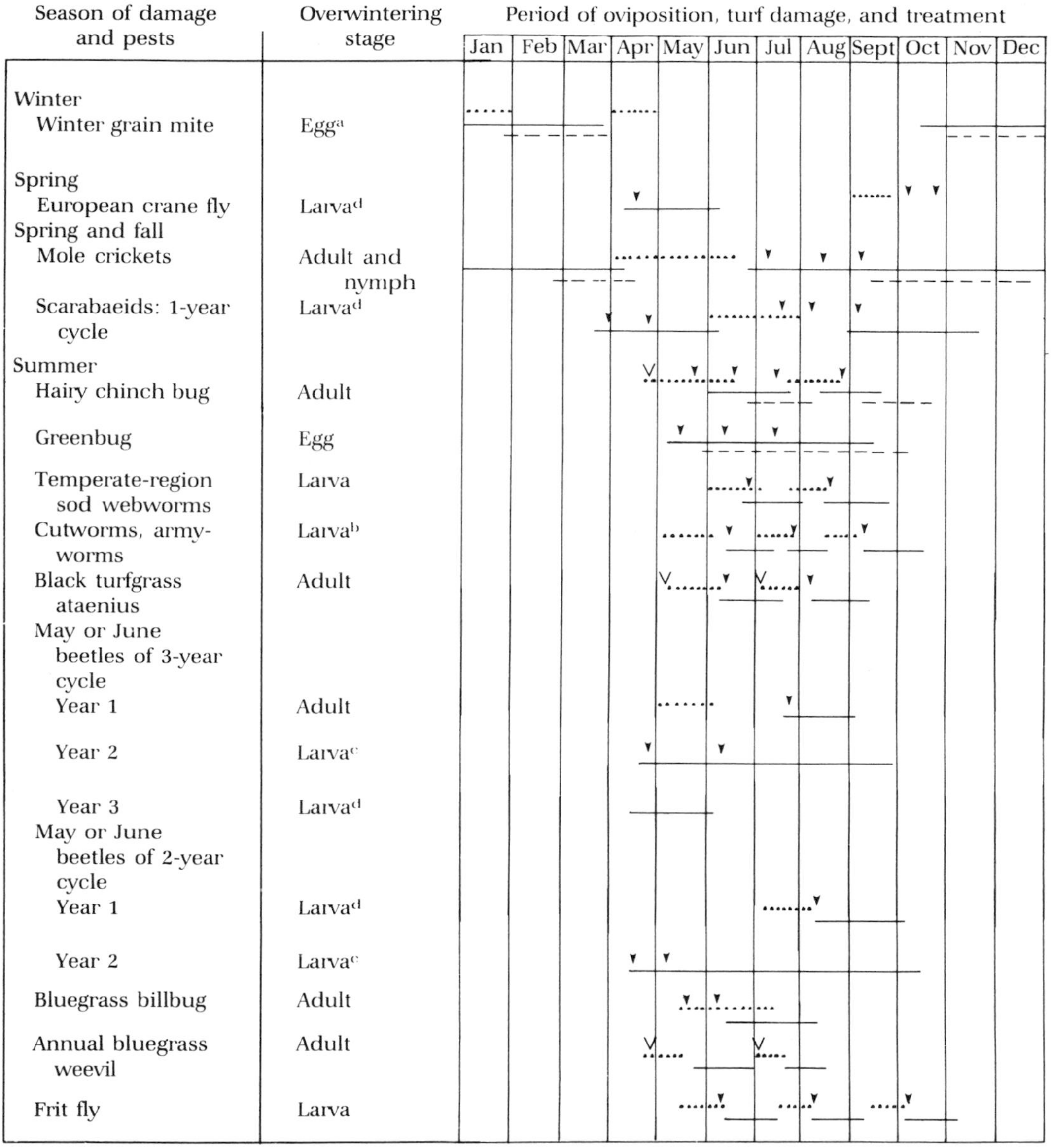

Season of damage and pests	Overwintering stage	Period of oviposition, turf damage, and treatment
		Jan Feb Mar Apr May Jun Jul Aug Sept Oct Nov Dec
Winter		
Winter grain mite	Egg[a]	
Spring		
European crane fly	Larva[d]	
Spring and fall		
Mole crickets	Adult and nymph	
Scarabaeids: 1-year cycle	Larva[d]	
Summer		
Hairy chinch bug	Adult	
Greenbug	Egg	
Temperate-region sod webworms	Larva	
Cutworms, armyworms	Larva[b]	
Black turfgrass ataenius	Adult	
May or June beetles of 3-year cycle		
Year 1	Adult	
Year 2	Larva[c]	
Year 3	Larva[d]	
May or June beetles of 2-year cycle		
Year 1	Larva[d]	
Year 2	Larva[c]	
Bluegrass billbug	Adult	
Annual bluegrass weevil	Adult	
Frit fly	Larva	

[a]Oversummering eggs.
[b]Infestations from migrating moths from the South.
[c]2d instar.
[d]3d instar.

- .... Period of oviposition
- ——— Period of damage by immatures
- – – – – – Period of damage by adults
- V Adult treatment period
- ▾ Treatment period for damaging stage.

Figure 90. Seasonal occurrence of destructive stages of turfgrass insects and timing of insecticide applications. (Drawn by H. Tashiro, NYSAES.)

Winter Damage

No control recommendations have been developed for the winter grain mite, the only major pest damaging turf during the winter.

Spring Damage Only

The sole pest damaging turfgrass only during the spring is the European crane fly. Major turf damage occurs during March and April as the overwintering third instars become fourth (ultimate) instars. Normally, applications of recommended insecticides should be made during April 1–15. Preventive fall applications made during October after most of the eggs have hatched and the larvae are small and vulnerable have also been effective (Chapter 18).

Spring and Fall Damage

The southern and tawny mole crickets complete their egg laying during June. During late summer the new generation becomes larger nymphs capable of heaviest damage. The most effective time for insecticidal control is during July through August (Chapter 4).

Scarabaeid species that normally have a 1-year life cycle are the only other group of insects that are most damaging during fall as a new generation of grubs become third instars. Midsummer (mid-July) to early fall (early September) is the most effective period for insecticidal treatment to kill first the second instars or young third instars before turf damage becomes apparent, provided that the soil is moist and the insecticide is watered in. If the grubs are not treated during the fall, they will move down in the soil to overwinter and will then migrate to the surface in early spring during late March and early April to resume feeding. Insecticidal applications should be made as larvae return to the surface. In spite of the generally cool weather, insecticidal treatments can be effective. The frequent spring rains promote rapid percolation of the material applied to turf surface into the surface soil zone where grubs are feeding (Chapters 12–15).

Summer Damage

All remaining turfgrass insects damage the turf generally during the summer. Treatments should be applied for the hairy chinch bug, greenbug, temperate-region webworms, cutworms, and armyworms as nymphal or larval populations become established but before turf damage becomes acute. Since these are all leaf and stem feeders and the easiest group to control, proper insecticidal application should give rapid response (Chapters 5–8).

Two approaches to controlling the black turfgrass ataenius are available. Phenological correlations indicate that overwintering adults start laying eggs when Vanhoutte spirea and horse chestnut are in full bloom and when black locust is in early bloom. This is generally the first half of May in southern Ohio and early June in New York. Insecticides applied to the turf during this period will kill adults before major oviposition occurs. The deposition of second-generation eggs coincides with the early bloom of rose of sharon. In southern Ohio this is normally during the second half of July. These phenological correlations also coincide with degree-day accumulations of 100–150 for first-generation egg laying and 650–710 degree-days for second-generation egg laying, with a threshold temperature of 13°C. The second approach directs insecticidal applications for killing larvae when their presence is known. In the latitude of Ohio, this is during the last two-thirds of June for the first-generation larvae and during August for the second-generation larvae (Chapter 11).

May or June beetles with a 3-year life cycle lay eggs in May of the 1st year. Young grubs as second instars do minor damage during July and August, and this is the most opportune treatment period to prevent serious damage the following year. Major damage occurs during the 2d year as overwintering second instars migrate to near the surface and third instars feed all summer into September. A spring or early summer treatment during the 2d year is next best. During the 3d year damage is minor, as third-instar grubs feed for a short period in May before moving down for pupation (Chapter 14).

May or June beetles of a 2-year life cycle deposit eggs during midsummer (normally during late July). Treatment during August to early September would be directed at young first- and second-instar grubs. Missing this opportunity, treatment the following spring during late April through May would be directed at second- or third-instar grubs that have migrated to near the soil surface to feed throughout the summer (Chapter 14).

Phyllophaga crinita in Texas has an annual generation in the South, and some may require 2 years for a generation in the North. Treatment periods vary from June to early July in the extreme south to late July to mid-August in the extreme north (Chapter 14).

Damage caused by the bluegrass billbug appears beginning in mid-June. Since the most effective way to prevent an infestation is to kill adults before oviposition, insecticidal applications made to the turf normally during early May should kill adults before major egg laying during May into June (Chapter 17).

An identical principle holds for the annual bluegrass weevil, which produces its major damage during late spring into early summer and spotty minor damage during mid to late summer. Treatment during the spring to kill overwintering adults before the major period of egg laying produces the most efficient control. Adults move out of hibernating quarters normally from mid-April to early May, timing that coincides with the blooming of forsythia and the full bract color of flowering dogwood. Treatments during this 2- to 3-week period are effective, with the greatest efficiency during the period when the dogwood bracts are in full color. Midsummer treatment periods are more variable, and problem areas are more localized but often involve the same areas from one summer to the next. Generally only localized areas need attention for the second generation (Chapter 17).

Frit flies have as many as three generations a season from the summer to early fall, with larval damage occurring shortly after each adult appearance. Seldom do all three generations create a threat of damage, but if damage appears imminent, treatments are recommended when the flies or larvae are present, with a second treatment 10–14 days later (Chapter 18).

Threshold Populations

When an insect infestation is known to occur, it is necessary to decide whether to treat with an insecticide or not. Several factors should be considered, the most important being populations present per unit area, vigor and condition of the turf, and time of year. Infestations found late in the season should cause little concern, since insect activity is greatly diminished with cooler weather. Practically all insects except the scarabaeid grubs become inactive and cause no turf damage at temperatures below 16°C (61°F).

Threshold populations that may produce sufficient damage to be of concern are so dependent on the vigor of the turf and on available moisture that the decision to undertake treatment must remain purely subjective. A few indications of damage threshold populations and recommendations for treatment, however, may be helpful.

In California if populations exceed 5 cutworms, 10 skipper larvae, 15 sod webworms, or 9 white grub or billbug larvae per 0.84 m^2 (1 yd^2) of established turf, control measures should be taken (Bowen 1980).

According to Potter (1982b), the general rule of thumb for treatment of a scarabaeid white grub infestation is 6–8 grubs per 0.1 m^2 (per ft^2). After his extensive study on the southern masked chafer, Potter concluded that treatment to prevent damage to Kentucky bluegrass should be made when 9–10 grubs per 0.1 m^2 (per ft^2) are present in a moisture-stressed turf or 15–20 grubs per 0.1 m^2 under optimum moisture. Turf can withstand more of the masked chafer grubs than other grubs because the former eat organic matter more than do other larvae.

Insecticides

In spite of recent promising developments in noninsecticidal control of turfgrass insects, such as resistant cultivars, we will depend on insecticides for a long time, if not forever. By the very nature of their use, all insecticides are poisons and must be considered hazardous chemicals in storage, handling, application, and immediate postapplication turf use. Compared with herbicides and fungicides, insecticides are generally more toxic, approximately as toxic as nematicides. The following information is condensed from Tashiro (1983).

Toxicity of Insecticides

One way of expressing the toxicity of economic poisons is by an abbreviation such as: AO or AD LD_{50} mg/kg. *AO* and *AD* refer to single oral dose or a single dermal application, respectively. The rest of the term refers to a single dose that will kill 50% of the test animals, usually laboratory white rats, in terms of milligrams of actual toxicant per kilogram of total body weight of the test animal. Table 28 lists common insecticides that are or have been recommended for turfgrass use and their toxicities. The larger the number, the less toxic the chemical. In actual use these are not the toxicities encountered, since formulations lower toxicities by reducing the percentage of the active ingredient. There is generally no correlation between high toxicity and insecticidal effectiveness, because many other factors, such as solubility, timing, weather, insect activities, and insecticidal degradation, are involved. All necessary precautions relating to toxicity and use are stated on the label. These insecticides are safe to use when they are applied according to label directions, regardless of the toxicity of the active ingredient. Ordinarily, the least toxic pesticide that gives acceptable control should be chosen. This minimizes danger when label directions are not strictly followed. The danger is real with casual users such as homeowners.

Names of Insecticides

The commonly used turfgrass insecticides in the 1980s belong to two classes, the organophosphate and carbamate compounds. It is considered wise to alternate between the two classes if possible in order to discourage insecticidal resistance from developing.

Table 28. Toxicity of technical grades of turfgrass insecticides

		LD_{50} mg/kg[a]	
Trade name	Common name	AO[b]	AD
Aspon	propyl thiopyrophosphate	890–1,700	3,830
Sevin[c]	carbaryl[c]	500–850	4,000
Dylox, Proxol	trichlorfon	560–630	2,000
Diazinon	diazinon	300–400	455–900
Dursban	chlorpyrifos	97–276	500–2,000
Triumph	isazophos	60–200	290–700
Turcam[c]	bendiocarb[c]	40–156	510–800
Mocap	ethoprop	61	26
Oftanol	isofenphos	28–38	188
Dasanit	fensulfothion	2–11	3–30

[a]LD_{50} mg/kg = lethal dose, kills 50% of a test animal group, usually white rats. Mg/kg = equivalent to 1 part chemical to a million parts test animal weight.

[b]AO = acute oral. AD = acute dermal. *Acute* refers to a single dose response, as opposed to successive sublethal doses fed or applied over a long period of time.

[c]Carbamate insecticides; all others are organophosphate insecticides.

Source: From Tashiro 1983.

All insecticides have a chemical name, a common name, and a trade name. Chemical names have little bearing on the applied uses of insecticides, but common names have meaning to the users. Regardless of crop use or manufacturer, a common name refers to only one compound. Trade names are often confusing, because a given compound may have two names, depending on its intended use. For instance, chlorpyrifos prepared for turf use is called Dursban, but the preparation for other agricultural crop uses is Lorsban. A single compound may have different trade names, depending on the company packaging and selling the product. Trichlorfon as it is sold by Mobay Chemical Corporation is called Dylox, while that sold by the Nor-Am Company is called Proxol. The use of insecticides' common names eliminates confusion.

Water Solubility

Insecticides vary widely in their water solubility (Table 29). Of the compounds now recommended for use on turfgrass, trichlorfon has the highest solubility—120,000 ppm—while chlorpyrifos has the lowest at <3 ppm. Solubility is generally an important consideration when treatments are made to thatchy turf. Chlorpyrifos is intercepted and bound by thatch, preventing penetration to the soil. Trichlorfon penetrates thatch more readily than other compounds.

Formulation

Often the choice of a formulation is of concern to a turf manager, as it may influence results. Many comparisons suggest, however, that generally there is essentially no difference in performance between the liquid and granular forms. The primary considerations are twofold, available equipment and the cost of the active ingredient. If cost is not the major consideration, the choice of formulation would depend on the equipment available for application. For a given insecticide, the most

Table 29. Water solubility of turfgrass insecticides

Trade name	Common name	Solubility (ppm)
Dylox, Proxol	trichlorfon	120,000
Dasanit	fensulfothion	1,600
Sevin	carbaryl	1,000
Mocap	ethoprop	750
Triumph	isazophos	150
Turcam	bendiocarb	40
Diazinon	diazinon	40
Oftanol	isofenphos	20
Dursban	chlorpyrifos	<3

Note: Information regarding water solubility was obtained from manufacturer's technical data sheets.

Source: From Tashiro 1983.

Table 30. Comparison of insecticide cost per unit of active ingredient (Ai) depending on formulations (1983 prices)

Insecticide formulation	Suggested retail (dollars per lb Ai)[a]	Cost ratio
Diazinon 5.0 G	20.60	2.15
Diazinon 4.0 E	9.60	
Dursban 2.32 G	90.30	3.46
Dursban 4.0 E	26.10	
Oftanol 1.5 G	122.00	2.90
Oftanol 5.0 G	42.00	

[a]Obtained from local wholesalers of agricultural chemicals.
Source: From Tashiro 1983.

significant factor is no doubt the unit price of the active ingredient (Table 30). A comparison made during 1983 showed that liquid formulations are two to three and a half times cheaper than granular formulations for active ingredient. Cost factors of such magnitude should be the overriding consideration in the choice of a formulation.

Insecticidal Resistance

History

Turfgrass insects have also developed resistance to the commonly used modern synthetic insecticides. The resistance of the scarabaeid larvae to the chlorinated cyclodiene insecticides, such as chlordane and dieldrin, that developed after 10–20 years use, terminated some of the most efficient insect control practices ever known. A single application provided nearly complete control for 5–10 years with chlordane and for as long as 20 years with dieldrin (Chapter 12).

Shortly thereafter, various turfgrass insects developed resistance to several of the organophosphate insecticides. Some of the most severe cases included the southern chinch bugs in southeastern coastal areas of Florida, which developed resistance to diazinon and chlorpyrifos after 11–20 years of use (Chapter 5).

Preventive Strategies

Continued use of one insecticide or one class of insecticide places a definite selection pressure on certain genes for resistance. Alternation between insecticides of different structure and class is recommended, although there is no empirical evidence that it prevents or retards the development of resistance.

Sources of Information on the Identification of Insects and Recommendations for Their Control

When turfgrass pests become a problem there is often an urgent need to identify the cause as quickly as possible and to determine the proper actions to take. The nearest governmental agency available for this task is usually the Cooperative Extension Service and farm and home centers, which are located in practically all county seats. These agencies are an arm of the land grant universities of each state and the U.S. Department of Agriculture. They are usually in the telephone directory under the city or county government or the U.S. Department of Agriculture. Their services are free, and they are usually the best sources of immediate information on routine problems.

The land grant universities and agricultural experiment stations located in every state are the original sources of recommendations for control of turfgrass pests as well as for other crops. The universities are also the best sources of specialists for identification of turfgrass pests. Land grant universities are the centers for each state's colleges of agriculture and the agricultural experiment stations. Below I list all land grant universities in the nation as given by the U.S. Department of Agriculture (1984). The telephone numbers shown (usually the Departments of Entomology) are likely to be the best way of reaching specialists at each institute who can identify unknown insects and can recommend ways of controlling pests. The superscripts on telephone numbers indicate:

[E]Departments of Entomology
[E+]Entomology combined with another discipline
[B]Entomologists in Departments of Biology

The absence of a superscript indicates that the entomologists are attached to various other disciplines or that there are no departments at the institute.

Alabama: Auburn University, Auburn, AL 36849 (205-826-4850)[E+]
Alaska, University of, Fairbanks, AK 99701 (907-474-7188)
Arizona, University of, Tucson, AZ 85721 (602-621-1151)[E]

Arkansas, University of, Fayetteville, AR 72701 (501-575-2451)[E]
California, Univeristy of, Berkeley, CA 94720 (415-642-3327)[E+]
California, University of, Davis, CA 95616 (916-752-0492)[E]
California, University of, Riverside, CA 92521 (714-787-5830)[E]
Colorado State University, Fort Collins, CO 80523 (303-491-7011)[E]
Connecticut, University of, Storrs, CT 06268 (203-486-2000)
Connecticut Agricultural Experiment Station, New Haven, CT 06504 (203-789-7241)[E]
Connecticut Valley Laboratory, Windsor, CT 06095 (203-688-3647)
Delaware, University of, Newark, DE 19711 (302-738-2000)[E+]
District of Columbia, University of, Washington, DC 20008 (202-282-7300)
Florida, University of, Gainesville, FL 32611 (904-392-2015)[E+]
Florida Agricultural Research and Education Center, Ft. Lauderdale, FL 33314 (305-475-8990)
Georgia, University of, Athens, GA 30602 (404-542-1765)[E]
Georgia Coastal Plain Experiment Station, Tifton, GA 31793 (912-386-3374)[E]
Georgia Agricultural Experiment Station, Experiment, GA 30212 (404-228-7288)[E]
Hawaii, University of, Honolulu, HI 96822 (808-948-6737)[E]
Idaho, University of, Moscow, ID 83843 (208-885-6276)[E+]
Illinois, University of, Urbana, IL 61801 (217-333-6656)[E]
Indiana: Purdue University, West Lafayette, IN 47907 (317-494-4553)[E]
Iowa State University of Science and Technology, Ames, IA 50011 (515-294-7400)[E]
Kansas State University, Manhattan, KS 66506 (913-532-6154)[E]
Kentucky, University of, Lexington, KY 40506 (606-257-7450)[E]
Louisiana State University, Baton Rouge, LA 70893 (504-388-1634)[E]
Maine, University of, Orono, ME 04469 (207-581-2957)[E]
Maryland, University of, College Park, MD 20742 (301-454-3843)[E]
Massachusetts, University of, Amherst, MA 01003 (413-545-2285)[E]
Massachusetts Suburban Experiment Station, Waltham, MA 02154 (617-891-0650)
Michigan State University, East Lansing, MI 48824 (517-355-4665)[E]
Minnesota, University of, St. Paul, MN 55108 (612-273-1701)[E]
Mississippi State University, Mississippi State, MS 39762 (601-325-2085)[E]
Missouri, University of, Columbia, MO 65211 (314-882-4445)[E]
Montana State University, Bozeman, MT 59717 (406-994-4548)[B]
Nebraska, University of, Lincoln, NE 68583 (402-472-2123)[E]
Nevada, University of, Reno, NV 89557 (702-784-6911)
New Hampshire, University of, Durham, NH 03824 (603-862-1707)[E]
New Jersey, Rutgers—The State University of, New Brunswick, NJ 08903 (201-932-9774)[E+]
New Mexico State University, Las Cruces, NM 88003 (505-646-3225)[E+]
New York: Cornell University, Ithaca, NY 14853 (607-256-3253)[E]
New York State Agricultural Experiment Station, Geneva, NY 14456 (315-787-2321)[E]
North Carolina State University, Raleigh, NC 27650 (919-737-2746)[E]
North Dakota State University, Fargo, ND 58105 (701-237-7582)[E]
Ohio State University, Columbus, OH 43210 (614-422-8209)[E]
Ohio Agricultural Research and Development Center, Wooster, OH 44691 (216-422-8209)[E]
Oklahoma State University, Stillwater, OK 74078 (405-624-5527)[E]
Oregon State University, Corvallis, OR 97331 (503-754-4733)[E]
Pennsylvania State University, University Park, PA 16802 (814-865-1895)[E]

Puerto Rico, University of, Mayaguez, PR 00708 (809-832-4040)
Rhode Island, University of, Kingston, RI 02881 (401-792-2481)[E+]
South Carolina: Clemson University, Clemson, SC 29631 (803-656-3112)[E+]
South Dakota State University, Brookings, SD 57007 (605-688-6141)[B]
Tennessee, University of, Knoxville, TN 37901 (615-974-7135)[E+]
Texas A & M University, College Station, TX 77843 (409-845-2516)[E]
Texas A & M University Research and Extension Center, Dallas, TX 75252 (214-231-5362)
Utah State University of Agriculture and Applied Science, Logan, UT 84322 (801-750-2485)[B]
Vermont, University of, Burlington, VT 05405 (802-656-2630)
Virginia Polytechnic Institute and State University, Blacksburg, VA 24061 (703-961-6341)[E]
Washington State University, Pullman, WA 99164 (509-335-5504)[E]
Western Washington Research and Extension Center, Puyallup, WA 98371 (206-593-8506)
West Virginia University, Morgantown, WV 26506 (304-293-4817)
Wisconsin, University of, Madison, WI 53706 (608-262-3227)[E]
Wyoming, University of, Laramie, WY 82071 (307-766-3103)

English and SI Units of Measure and Conversion Formulas

The International System of Units, SI, which was adopted and endorsed in 1960 by the International Bureau of Weights and Measures, forms the basis for the modernized metric system. In the United States, as in the rest of the world, the scientific community has been using the metric system for a long time. The U.S. Congress enacted the Metric Conversion Act of 1975, but this has done little to promote general public acceptance of the metric system. Today, however, with few exceptions, the entire world is using the metric system or is shifting to it, although the process has admittedly been slow in this country.

Measurements in my 26 chapters are given in SI (metric) units, followed by English units in parentheses except in the case of insects and mites. Their measurements are given only in SI units, since their small size would, in many cases, create unfamiliar fractions of an inch. The following information is adapted primarily from Beard (1982) and U.S. Department of Commerce (1981).

Units of Measure

	English Units		SI Units
Length			
1 yd	= 3 ft = 36 in.	1 mm	= 1,000 µm
1 rod	= 5.5 yd = 16.5 ft	1 m	= 100 cm = 1,000 mm
1 mi	= 1,760 yd = 5,280 ft	1 km	= 1,000 m
Area			
1 ft^2	= 144 $in.^2$	1 a	= 100 m^2
1 acre	= 4,840 yd^2 = 43,560 ft^2	1 ha	= 100 a = 10,000 m^2
1 mi^2	= 1 section = 640 acres	1 km^2	= 100 m^2
Volume			
1 cup	= 16 tbsp = 48 tsp	1 ml	= 1,000 µl
1 pt	= 2 cups = 16 fl oz	1 l	= 1,000 ml = 0.001 m^3
1 qt	= 2 pt = 4 cups		

1 gal	= 4 qt = 8 pt		
1 gal	= 231 in.^{3} = 128 fl oz		
1 ft^3	= 1,728 in.^{3} = 7.48 gal		
Mass or Weight			
1 lb	= 16 oz	1 g	= 1,000 mg
1 ton	= 2,000 lb	1 kg	= 1,000 g
		1 mt	= 1,000 kg

Conversion Chart

English to SI		SI to English	
Length			
1 in.	= 25.4 mm = 2.54 cm	1 mm	= 0.039 in.
1 ft	= 30.48 cm = 0.305 m	1 cm	= 0.394 in. = 0.033 ft
1 yd	= 914 cm = 0.914 m	1 m	= 39.37 in. = 3.281 ft = 1.094 yd
1 mi	= 1,609.3 m = 1.609 km	1 km	= 1,093.6 ft = 0.621 mi
Area			
1 in.^{2}	= 6.452 cm^2	1 cm^2	= 0.155 in.^{2}
1 ft^2	= 929 cm^2 = 0.093 m^2		
1 yd^2	= 0.836 m^2	1 m^2	= 10.764 ft^2 = 1.196 yd^2
1,000 ft^2	= 0.929 a	1 a	= 1,076 ft^2 = 119.6 yd^2
1 acre	= 0.405 ha	1 ha	= 2.471 acre
1 mi^2	= 2.59 km^2	1 km^2	= 0.4 mi^2
Volume			
		1 ml	= 0.034 fl oz
1 in.^{3}	= 16.387 cm^3	1 cm^3	= 0.061 in^3
1 ft^3	= 0.028 m^3		
1 yd^3	= 0.765 m^3	1 m^3	= 35.315 ft^3 = 1.308 yd^3
1 fl oz	= 29.573 ml = 0.03 l		
1 pt	= 0.473 l		
1 qt	= 0.946 l	1 l	= 33.814 fl oz = 2.12 pt = 1.957 qt = 0.264 gal
1 gal	= 3.785 l		
Mass or Weight			
1 oz	= 28.349 g	1 g	= 0.035 oz
1 lb	= 453.592 g = 0.453 kg	1 kg	= 2.205 lb
1 ton	= 0.907 mt	1 mt	= 1.102 ton
Rate			
1 oz/gal	= 7.8 ml/l	1 g/cm^2	= 2.048 lb/ft^2
1 lb/1,000 ft^2	= 0.488 kg/ha	1 g/cm^3	= 62.4 lb/ft^3
1 lb/acre	= 1.12 kg/ha	1 kg/a	= 2.05 lb/1,000 ft^2
		1 kg/ha	= 0.89 lb/acre
1 mph	= 1.61 km/hr	1 km/hr	= 0.621 mph

Temperature

°F	°C
−10	−23.3
0	−17.8
10	−12.2
20	−6.7
32	0
60	15.6
80	26.7
100	37.8
212	100.0

°C	°F
−30	−22
−20	−4
−10	14
0	32
10	50
20	68
30	86
40	104
100	212

Conversion Formula

$(^\circ F - 32) \div 1.8 = {}^\circ C.$

$(^\circ C \times 1.8) + 32 = {}^\circ F.$

Abbreviations

Units of Measure

English

Abbreviation	Unit
°F	Fahrenheit
fl oz	fluid ounce
ft	foot
ft-c	footcandle
ft^2	square foot
gal	gallon
in.	inch
lb	pound
mi	mile
mph	miles/hour
pt	pint
qt	quart
tbsp	tablespoon
tsp	teaspoon
yd	yard

SI

Abbreviation	Unit
Å	angstrom
°C	Celsius
cm	centimeter
a	are
dm	decimeter
g	gram
ha	hectare
kg	kilogram
km	kilometer
l	liter
m	meter
m^3	cubic meter
mg	milligram
ml	milliliter
mm	millimeter
mt	metric ton
µm	micrometer

Other Abbreviations and Symbols

>, < – greater than, less than
AC/DC – alternating current/direct current
AD – acute dermal; used in measuring toxicity of toxic materials
AO – acute oral; used in measuring toxicity of toxic compounds

BL – black light; electromagnetic radiation peaking at 3650 Å and highly attractive to night-flying insects; generally a fluorescent lamp
BLB – black-light fluorescent lamp with a blue-violet tube that filters out most of the visible radiation
LC_{50} – lethal concentration of a toxic compound killing 50% of the text animals
LD_{50} – lethal dose of a toxic compound killing 50% of the test animals
pH – symbol used to indicate acidity or alkalinity with pH 7 as neutral, >7 as alkaline, and <7 as acidic; logarithm of the reciprocal of the hydrogen ion concentration in gram atoms per liter of solution
ppm – parts per million
RH – relative humidity
SEM – scanning electron microscopy
TU – thermal unit

Glossary

accession – an addition
adsorb – collect in condensed form on the surface
adventitious – occurring accidentally, out of the ordinary course
aedeagal – pertaining to the aedeagus, or penis, of male insects
aeration – exposure to the air
aestivate – spend the summer in a dormant condition; opposed to *hibernate*
alga – a one-celled to many-celled plant containing chlorophyll, having no true root, stem, or leaf, and found in damp places
antenna – one of a pair of segmented sensory organs borne one on each side of the head
anterior – toward the front or head, as opposed to *posterior*
antibiosis – in biology, an association between two organisms in which one is adversely affected
apical – at or near the apex of a structure
apron – in golf, the fairway area in closest proximity to and in front of the putting green that adjoins the putting green collar and is mowed at the fairway height; sometimes called the approach
arcuate – arched, bowlike
arthropod – a member of a phylum of invertebrate animals having jointed legs and segmented bodies
articulation – a joint or the state of being jointed
aspirator – any apparatus for moving air, fluids, and so forth by suction
auricle – an earlike part or organ
axil – the upper angle between a leaf and the stem from which it grows

bacterium – a one-celled microorganism that has no chlorophyll and multiplies by simple division
Berlese funnel – a combination of a heat source, usually an incandescent lamp, and an acute angle funnel with a screen near the upper edge to hold foliage, thatch, or soil. The apparatus forces living organisms into a collecting jar via heat
binomial – scientific name of a plant or animal, consisting of generic and species names
bioassay – determine the toxicity of a compound or infectivity of a parasitic organism by the use of a living organism

bionomics – branch of biology dealing with the adaptation of living things to their environment; ecology
biotype – a group of plants or animals with similar hereditary characteristics
black light – electromagnetic illumination peaking at 3,650 Å and attractive to night-flying insects; also causing fluorescence of certain compounds
blade – the flat portion of the leaf blade above the sheath
brachyptery – short-winged
braconid – a hymenopterous insect belonging to the family Braconidae, the members of which are parasitic on other insects
bract – in botany, a modified leaf, small and scalelike or large and brightly colored, growing at the base of a flower, for example in flowering dogwood

caecal – of the caecum, a blind pouch or tubelike structure
callow – teneral, or undeveloped and immature
caproic acid – a colorless, liquid fatty acid ($C_6H_{12}O_2$) found in butter and other animal fats
carbamate – any of a group of synthetic organic insecticides that are esters of *N*-methylcarbamic acid (examples: carbaryl, bendiocarb)
carina – an elevated ridge or keel
caudal – of or pertaining to the cauda, or the anal end of the insect body
caudomesal – directed toward the rear and toward the medial region, or middle of the body
cellulose – the chief substance composing the cell walls or woody part of plants
cephalothorax – the united head and thorax of arachnids or spiders and crustacea
chelicera – one of the pinching, pincerlike first pair of appendages of adult Chelicerata, such as ticks and mites
chitin – a colorless, nitrogenous polysaccharide secreted by the epidermis and applied to the hardened parts of an insect body
chlorinated cyclodiene – synthetic insecticides composed of highly chlorinated cyclic hydrocarbons (examples: chlordane, dieldrin)
chlorophyll – the green, light-sensitive pigment of plants that in sunlight is capable of combining carbon dioxide and water to make carbohydrates
chlorotic – characterized by the fading of green color in plant leaves to light green or yellow
chorion – the outer covering of the insect egg
ciliate – fringed with a row of parallel hairs
claviform spot – spots clublike in form
claw (of leg) – a hollow, sharp, multicellular organ, generally paired, at the end of the insect leg
clitellum – in earthworms, the thickened, glandular, saddlelike portion of the body wall on (roughly) segments 31 to 37 and having a reproductive function
clone – an asexually produced progeny
clypeus – the part of the head of an insect which commonly bears the labrum on its anterior margin
coelomic cavity – the space between the viscera and the body wall
coition – sexual intercourse
collar – in botany, a narrow band marking the junction of the blade and leaf sheath; in golf, a narrow area adjoining the putting surface that is mowed at a height intermediate between that used for the fairway and the putting surface

compound eye – an aggregation of separate visual elements known as ommatidia
conifer – a cone-bearing tree or shrub, usually an evergreen
contact toxicity – toxicity of a compound when it is applied directly to the body rather than ingested
contiguous – touching
cornicle – in aphids, either of two honey tubes, the dorsal erect or semierect tubules in pairs that secrete a waxy liquid as a defense against enemies
costa – thickened anterior margin of any wing of insects
coxa – basal segment of the leg on insects or other arthropods
crown – compact series of nodes from which culms and roots arise
culm – stem of grass plant
cultivar – cultivated variety
cuticle – or cuticula, the outer covering of an insect formed by a noncellular layer of chitin
cyst – a sac or vesicle

dactyl – a finger or toe; a tarsal joint after the first one
degree-day accumulation – an estimate of solar energy using the average daily temperature minus the threshold temperature for a given organism
dermapteran – an insect belonging to the order Dermaptera, or earwigs
deutonymph – the second nymphal stage of Acari, which assumes the general nonsexual characteristics of the adult
diapause – a condition of suspended animation
diplopod – member of the class Diplopoda, known as millipedes
distal – near or toward the free end of any appendage
diurnal – active during daylight
diverticulum – an offshoot from a vessel or from the alimentary canal, usually blind or saclike
dormant – living in a state of reduced physiological activity
dorsum – the upper surface, or back

eclosion – emergence of the adult insect from the pupa; act of hatching from the egg
ecosystem – an ecological system; a functional system that includes the organisms of a natural community together with their environment
ectoparasite – parasitic organism living outside its host
elliptical – oblong-oval, the ends equally rounded
elytra – the anterior leathery or chitinous wings of beetles that serve as coverings to the hindwings
endoparasite – parasitic organism living inside its host
endophyte – any plant growing within another plant
entomogenous – growing in or on an insect, as do fungi, for example
entomophagus – feeding upon insects
entomopthorous – parasitic on insects; specifically, a fungus of the order Entomophthorales
enzyme – a complex organic substance produced by living cells and causing chemical changes in organic substances by catalytic action; indicated by the suffix *-ase*
epicranial suture – a Y-shaped suture on the dorsal surface of the head of a generalized insect
epidermis – the cellular layer of the skin underlying and secreting the cuticula

epimeron – the posterior of a thoracic pleuron or lateral region of a segment, usually small, narrow, or triangular
epipharynx – an organ, probably of taste, that is attached to the inner surface of the labrum
epizootic – a disease temporarily prevalent among many animals
etiology – science or theory of the causes or origins of diseases
evagination – a turning inside out, with protrusion of an inner surface
excretion – act of eliminating a waste product
exoskeleton – entire body wall to the inner side, to which muscles are attached; the outside skeleton of insects
extravaginal growth – young vegetative stems that grow outside the basal leaf sheath by penetrating through the sheath
exuviae – the cast-off skin of larvae or nymphs at metamorphosis

fairway – in golf, area between the tee and putting green of variable lengths from near 100 to 400+ yd and mowed at heights of 1.3–3.0 cm (0.5–1.2 in.)
fallow ground – land plowed but not seeded for one or more growing seasons
fauna – animals of any given region or time
feather claw – distal portion of legs of eriophyid mites ending in rays and resembling a feather
fecundity – the quality or power of being prolific or fertile
femur – the thigh; in insects, usually the stoutest segment of the leg articulated proximally to the trochanter and distally to the tibia
filiform – threadlike; slender and of equal diameter
flaccid – soft and limp; flabby
flagellum – that part of the antenna beyond the pedicel; a whip or whiplike process
foot-candle – illumination equal to 1 lumen per square foot
forage – a food or fodder plant
fossorial – formed for or with the habit of digging or burrowing
frass – solid larval insect excrement
fungus – nucleated, usually filamentous spore-bearing organisms devoid of chlorophyll
fusiform – spindle-shaped; broad at the middle and narrowing toward the ends

ganglion – a nerve center composed of a cell mass and fibers and serving as a center from which impulses are transmitted
garrulous – talking much, noisy
gena – the cheek; the part of the head on each side below the eyes
genital sheath – portion of empty pupal case surrounding the terminal portion where the genital organs are located
genus – taxonomic category above species and below family
glabrous – smooth, hairless, and without punctures or structures
glutinous – gluey, slimy, viscid
gnathosoma – portion of Acari that resembles the head of generalized arthropods only in that the mouthparts are appended to it
green – in golf, the putting green, the area of finest turf mowed at 4–8 mm (0.16–0.37 in.) that contains the cup and flag
gregarious – living in societies or communities but not social

grub – an insect larva; a term loosely applied but with specific reference to larvae of Coleoptera and Hymenoptera
gustatory – of or relating to the sense of taste

haltere – one of a pair of knobbed, movable filaments in Diptera that are situated on each side of the thorax and that represent the hindwings
hamate – furnished with hooks
head capsule – the combined sclerites of the head, which form a hard compact case
hemelytron – the anterior wing in the Heteroptera, which has a thickened basal half and a membranous apical half
hermaphrodite – an individual having characters of both sexes
hibernaculum – a larva's tent or a sheath, in which it hibernates
hibernate – to pass the winter in a dormant state
holometabolous – having a complete transformation, with egg, larval, pupal, and adult stages distinctly separated
hymenopterous – of or relating to an insect in the order Hymenoptera, which includes ants, bees, and wasps, many of the latter being parasitic on other insects
hypha – one of the filaments that make up the mycelium of a fungus
hysterosoma – section of Acari body that occupies the third and fourth pair of legs and the posterior part of the abdomen

ichneumonid – a hymenopterous insect belonging to the family Ichneumonidae, the members of which are parasitic on other insects
idiosoma – the portion of Acari that assumes functions parallel to those of insects' abdomen, thorax, and portions of the head
incandescent (lamp) – a lamp in which the light is produced by a filament contained in a vacuum and heated to incandescence by an electric current
incisor – a cutting tooth
in copula – in copulation
indigenous – native to
insectivorous – feeding on insects
inseminate – fertilize with semen
instar – stage between molts
integument – outer covering, or cuticle, of the insect body
internode – in botany, the section of a plant stem between two successive nodes or joints
intravaginal growth – vegetative stems that grow upward within the enveloping basal leaf sheath
isotherm – a line on a map connecting two points on the earth's surface having the same mean temperature or the same temperature at a given time

labial palpus – a one- to four-jointed sensory appendage of the insect labium
labium – the lower lip; a compound structure that forms the floor of the mouth in mandibulate insects
labrum – the upper lip, which covers the base of the mandibles and forms the roof of the mouth in insects

lacinia – inner lobe of the first maxilla, articulated to the stipe and bearing brushes of hair or spines

lamellate – sheet or leaflike

lamellicorn – characteristic of a group of beetles that have their antennae terminating in lamellae, or thin platelike or leaflike processes

larva – a young insect that has hatched and is an immature form and is called a caterpillar, slug, maggot, or grub, depending on the kind of insect

larvaevorid – producing living young instead of eggs, as in some Diptera

leaf sheath – basal portion of a grass leaf that surrounds the stem

life cycle – the period between egg deposition and the attainment of sexual maturity as shown by egg laying

lignin – an organic substance forming the essential part of wood fiber

ligule – a thin outgrowth membrane attached to a leaf of grass at the point where the blade meets the leaf sheath

lodge – of crops, to fall or lie down; refers especially to hay or grain crops

lumen – unit of luminous flux equal to the light emitted in a unit solid angle by a uniform point source of one candle

lux – unit of illumination equal to 1 lumen per square meter

macroptery – long-winged

maggot – footless larva of Diptera

Malpighian tubules – the insect urinary system, composed of long, slender blind excretory tubes that vary in number and open into the hind intestine at its junction with the midintestine

mandibles – first pair of jaws in insects, stout and toothlike in chewing insects, needlelike in sucking insects, mouth hooks in muscid flies

maritime – on, near, or living near the sea

maxilla – the second pair of jaws in a mandibulate insect

maxillary palp – the pair of palps on the maxilla carried by the stipes on its outer end and sensory in function

meatus – a channel or duct

meconium – substance excreted by certain insects soon after their emergence from the chrysalis, or pupa

meridian – any lines of longitude running north and south through the poles on the globe or map

meristem (activity) – undifferentiated plant tissue consisting of cells actively growing and dividing, as at the tip of roots or stems

mesad – toward or in the direction of the median plane of a body

mesasoma – in scorpions, the broad seven-segmented section of the body immediately posterior to the last pair of legs

mesophyll – the leaf substance lying between the upper and lower epidermis; the parenchyma

metamorphosis – a series of changes through which an animal passes in its growth from the egg to the adult

metasoma – in scorpions, the narrow five-segmented posterior section of the body terminating in a sting

metasternum – underside of the metathorax
micropyle – one of the minute openings in the insect egg through which spermatozoa enter in fertilization
microsporidium – any of a group of protozoa some of which are pathogens to insects and other animals
milorganite – activated sewage sludge that is steam-sterilized and marketed by the City of Milwaukee, Wisconsin, as a turfgrass fertilizer
minim – something very small; in ants, the smallest and the first worker in a new colony
molt – to cast off or shed the outer skin and so forth at certain intervals prior to replacement of the cast-off parts by a new growth
multivoltine – having more than one generation in a year or season
muscid – a fly belonging to the superfamily Muscoidea

nema – short for *nematode*
nematode – any of a class or phylum of elongated cylindrical worms that are parasitic in animals or plants or are free living in soil or water
nocturnal – active at night
node – point along a stem at which a leaf is attached
nymph – young stage of insects with incomplete metamorphosis, for example, Heteroptera

obligate (parasites) – a parasite living on one host exclusively
obtuse – not pointed; at an angle greater than a right angle; opposite of *acute*
ocellus – simple eye of insects, occurring singly or in small groups
olivaceous (dusting) – olive green; color of green olives
ommatidium – one of the visual elements that make up the compound eye of an arthropod
omnivorous – feeding generally on both animal and vegetable food
opisthosoma – in scorpions, the entire section of the body behind the legs and made of a broad, seven-segmented anterior portion and a narrow, five-segmented posterior section
orbicular – round and flat
organophosphate – synthetic insecticides that are esters of phosphoric or other phosphorus-derived acids (examples: diazinon, chlorpyrifos)
oscillate – to move or travel back and forth between two points
ovarian yolk – concentration of yolk cells in the egg at the point where the cells of the developing embryo undergo cleavage
overwinter – to survive the winter
oviparous – reproducing by eggs laid by the female
oviposition – the act of depositing eggs
ovipositor – a tubular or valved structure by means of which eggs are deposited
ovoid – egg shaped
ovoviviparous – in insects, producing living young by the hatching of the ovum while it is still within the mother

palidium – in scarabaeid larva, a group of pali arranged in a single row or more and usually placed on the median of the venter of the lower anal lip
palpus – a process on a mouthpart of an arthropod that has a tactile or gustatory function
palus – a straight-pointed spine, a component of the palidium

parasite – any animal that lives in or on, or at the expense of, another
parasporal body – sporelike body apart from the spore proper in a bacterial spore
parenchyma – tissue consisting of large, thin-walled cells
parthenogenesis – reproduction by direct growth from egg cells without fertilization by sperm
pectinate – comblike, said especially of antennae with even processes like the teeth of a comb
pedicel (antenna) – a narrow basal part by which a larger part of an organ of an animal is attached; second joint of an antenna
pedipalp – one of the second pair of appendages in the cephalothorax, used in crushing prey
penultimate – next to the last
peripheral – relating to the outer margin
petiole – a stem or stalk
phenology– a branch of science dealing with the relations between climate and periodic biological phenomena
pheromone – any substance secreted by an animal that influences the behavior of other individuals of the same species
phloem – a complex, food-conducting vascular tissue in higher plants
photosynthesis – a process by which green leaves manufacture carbohydrates using carbon dioxide, water, chlorophyll, and light
phylogenetic – of or related to phylogeny or based on natural evolutionary relationships
phylum – primary division of the animal kingdom; used in classification to indicate a series of related organisms
phytophagous – feeding upon plants
piceous – pitchy black; black with a reddish tinge
piperonyl butoxide – a synthetic hydrocarbon compound used as a synergist in a mixture with insecticides to increase the effectiveness of the latter
pipet, pipette – a narrow glass tube into which liquid is drawn by suction and is retained by closing the upper end
pollination – the transfer of pollen from a stamen to a pistil; fertilization in flowering plants
polyembryony – the production of several embryos from a single egg
polyhedral – having many faces or sides
polyhedrosis – any of several viral diseases of insect larvae characterized by the breakdown of tissues and the presence of polyhedral granules
polymorph – in biology, one of several adult forms
polyphagous – eating many kinds of foods
polystand – a turfgrass community composed of plants of two or more cultivars and/or species
posterior – toward the rear, as opposed to *anterior*
postocular lobe – exoskeletal lobe behind the eyes
predator – any animal that preys on another
prepupa – a quiescent transitional period between the end of the larval period and the pupal period
primordial – of leaves, the earliest formed tissues
process – a prolongation of the surface or of an appendage, or any prominent body part not otherwise definable

proctodaeum – posterior portion of the alimentary tract as far forward as the Malpighian tubes
pronotum – the upper or dorsal surface of the prothorax
propodosoma – the anterior portion of the idiosoma, or body proper, of a mite, to which the legs are attached
prosoma – in scorpions, the anterior portion of the body bearing the eyes, pedipalps, and legs
protonymph – the first nymphal stage of Acari to assume the general nonsexual characteristics of the adult
protozoa – a phylum of the animal kingdom containing the one-celled animals; one-celled animals in general
proximal – part of an appendage nearest the body
pubescent – downy; clothed with soft, short, fine, closely set hairs
punctate – set with impressed points or punctures
pupa – the resting, inactive stage between the larva and the adult in all insects that undergo a complete metamorphosis
puparium – in Diptera, the thickened, hardened, barrellike larval skin within which the pupa is formed
pygidium – the tergum (upper surface) of the last segment of the abdomen
pyrethrin – one of six separate compounds or a mixture of compounds extracted from certain *Chrysanthemum* spp. flowers that have very rapid insecticidal action

raster – in scarabaeid larvae, a complex of definitely arranged bare places, hairs, and spines on the ventral surface of the last abdominal segment, in front of the anus
rectal sac – the enlarged anterior part of the rectum
recumbent – lying down; reclining
relative humidity – the ratio (expressed as a percentage) of the amount of water vapor actually present in the air to the greatest amount possible at the same temperature
reniform – kidney shaped
resistance – a phenomenon of differing susceptibilities between populations resulting from artificial selection to a given chemical, organism, and so forth
reticulation – formation or pattern; network
rhizome – a jointed underground stem that can produce roots and shoots at each node
rickettsia – rod-shaped nonfilterable microorganisms that cause various diseases
rosette – a cluster of leaves in crowded circles or spirals arising basally from a crown or apically from an axis, with greatly shortened internodes
rufous – pale red
runners – elongate growths produced by a plant; stolons

sagittal – longitudinal
saprophyte – any plant or animal living on dead or decaying vegetable matter
sarcophagid – flesh eating
scape – long basal joint in an antenna
scarabaeid – pertaining to members of the beetle family Scarabaeidae
sclerite – any part of the insect body wall bounded by sutures

sclerotized – of an insect, hardened in definite areas by deposition or formation of substances other than chitin

scrobe – groove formed for the reception or concealment of an appendage, especially, in weevils, grooves at the sides of the rostrum to receive the scape of the antennae

septicemia – invasion of the blood by virulent microorganisms from a focus of infection

septula – in scarabaeid grubs, the median narrow, bare region of the raster between the palidia

serosal cuticle – outer membrane

sessile – closely seated, without a stalk

setae – slender hairlike appendages

sheath (leaf) – basal tubular portion of a leaf surrounding the stem

sinuate – cut into sinuses; wavy; said especially of edges or margins

snout – the prolongation of the head of weevils at the end of which the mouthparts are located

sod – plugs, blocks, squares, or strips of turfgrass plus the adhering soil that are used for planting

spatulate – round and broad at the top, slender at the base

spermatheca – the sac or reservoir in the female that receives the sperm during coition

sphecid – a hymenopterous insect belonging to the family Sphecidae, the members of which are parasitic on other insects

spinneret – an organ in the larva or adult that is used in spinning silk

spiracle – a breathing pore through which air enters the tracheae; in insects, located laterally on body segments

spittle – a frothy fluid secreted by insects; saliva

sporangium – a case within which the asexual spores are produced

sporulate – to undergo sporulation

sporulation – the formation of spores

stadium – the interval between the molts of larvae; instar

sternum – the entire ventral division of any segment

stipe – the foot stalk of the maxilla, bearing the movable parts

stolon – a jointed, above-ground, creeping stem that can produce roots and shoots at each node and may originate extravaginally from the main stem

stomodaeum – anterior section of the arthropod alimentary tract; the foreintestine of an insect

striate – marked with parallel, fine, impressed lines

stridulate – in insects, to make a shrill, creaking, grating, or hissing sound or noise by rubbing two ridges or roughened surfaces against each other

stubble – the basal part of plants left after harvest

stylet – a small style; a median dorsal element in the shaft of the ovipositor

subterranean – existing under the surface of the earth

sward – a carpet of grass or other ground cover; turf

tachinid – a dipteran insect belonging to the family Tachinidae, the members of which are parasitic on other insects

tarsus – the foot; the distal part of the insect leg that consists of one to five segments

tee – in golf, an area that is specially prepared from which the first stroke on each hole is made

teges – in scarabaeoid larvae a continuous dense or sparse patch of hooked or straight spines occupying almost the whole of the venter of the 10th abdominal segment when the palidium is absent

tegillum – in scarabaeid larvae, a paired patch of hooked or straight spines on each side of the palidia on the tenth abdominal segment

tegmen – the hardened leathery or horny forewing in Orthoptera

teneral – condition of the adult shortly after emergence, when it is not entirely hardened or fully of mature color

tergum – the upper or dorsal surface of any body segment of an insect

terminal ampullae – in scarabaeid grubs, a pair of ovoidal structures, the posterior portion of the reproductive organs, found in the eighth abdominal sternum of females and ninth abdominal sternum of males

thatch – layer of plant litter from long-term accumulation of dead plant roots, crowns, rhizomes, and stolons between the zone of green vegetation and the soil surface

thermal unit – units relating to heat

thigmotactic – contact loving; an involuntary reaction to simple contact with some outside object or body

thorax – the second or intermediate region of the insect body, bearing the true legs and wings and composed of three rings, the pro-, meso-, and metathorax

threshold – the beginning point of something; in physiology, the point at which a stimulus is just strong enough to produce a response

tibia – in insects, the fourth division of the leg articulated at the proximal end to the femur and at the distal end to the tarsi

tiller – shoot, culm, or stalk arising from a crown bud

toxin – poisonous substance secreted by plants and animals

trachea – a spirally ringed internal elastic air tube in insects; an element of the respiratory system

translucent – partially transparent, like frosted glass

trochanter – the second segment of an insect leg, between the coxa and femur

ultimate – last; terminal; final

univoltine – having one generation in a year or season

vector – any organism that is the carrier of a disease-producing organism

venter – the undersurface of the abdomen as a whole

ventriculus – true stomach of an insect; the midintestine or midgut

vernation – in botany, the arrangement of leaves in a leaf bud

vertebrate – having a backbone or spinal column

vestigial – small or degenerate; the remains of a previously functional organ

violaceous – violet in color

virulence – quality of being poisonous; the relative infectiousness of a microorganism causing disease

virus – any of a large group of infectious agents 10–250 nm in diameter, composed of a protein sheath surrounding a nucleic acid core and capable of infecting animals, plants, and bacteria and totally dependent on living cells for reproduction

viviparous – bearing or bringing forth living young instead of laying eggs

white grub – whitish, C-shaped larvae of insects belonging to the family Scarabaeidae

whorl – in botany, an arrangement of leaves, petals, and so forth around a point on a stem

References

Adams, J. A. 1946. Rate of development of milky disease in Japanese beetle populations. J. Econ. Entomol. 39: 248–254.

Adams, J. A. 1949a. The Oriental beetle as a turf pest associated with the Japanese beetle in New York. J. Econ. Entomol. 42: 366–371.

Adams, J. A. 1949b. *Cyclocephala borealis* as a turf pest associated with the Japanese beetle in New York. J. Econ. Entomol. 42: 626–628.

Agriculture Canada. 1983. Japanese beetle/Scarabée japonais: Survey in Canada/Enquête au Canada. Biol. Programs Sect., Plant Health Div., Ottawa. 4 pp.

Ahmad, S., and C. R. Funk. 1983. Bluegrass billbug (Coleoptera: Curculionidae): Tolerance of ryegrass cultivars and selections. J. Econ. Entomol. 76: 414–416.

Ahmad, S., and Y.-S. Ng. 1981. Further evidence for chlorpyrifos tolerance and partial resistance by the Japanese beetle (Coleoptera: Scarabaeidae). New York Entomol. Soc. 89: 34–39.

Ahmad, S., H. T. Streu, and L. M. Vasvary. 1983. The Japanese beetle: A major pest of turfgrass. Amer. Lawn Appl., Mar./Apr.: 4–11.

Ainslie, G. G. 1922. Webworms injurious to cereal and forage crops and their control. U.S. Dept. Agr. Farmers' Bull. No. 1258. 16 pp.

Ainslie, G. G. 1923a. Silver-striped webworm, *Crambus praefectellus,* Zincken. J. Agr. Res. 24: 415–426.

Ainslie, G. G. 1923b. Striped sod webworm, *Crambus mutabalis* Clemens. J. Agr. Res. 24: 399–414.

Ainslie, G. G. 1927. The larger sod webworm. U.S. Dept. Agr. Tech. Bull. No. 31. 17 pp.

Ainslie, G. G. 1930. The bluegrass webworm. U.S. Dept. Agr. Tech. Bull. No. 173. 25 pp.

Aldrich, J. M. 1920. European frit fly in North America. J. Agr. Res. 18: 451–473.

Alexander, C. P. 1919. The crane flies of New York. Pt. 1. Distribution and taxonomy of adult flies. Cornell Univ. Agr. Exp. Sta. Memoir 25: 766–993.

Allen, W. A., and R. L. Pienkowski. 1974. The biology and seasonal abundance of the frit fly, *Oscinella frit,* in reed canary grass in Virginia. Ann. Entomol. Soc. Amer. 67: 539–544.

Allen, W. A., and R. L. Pienkowski. 1975. Life tables for frit fly, *Oscinella frit,* in reed canary grass in Virginia. Ann. Entomol. Soc. Amer. 68: 1001–1007.

Allen, W. H. 1944. The Asiatic beetles in New Jersey. New Jersey Dept. Agric. Circ. No. 348. 18 pp.

Anonymous. 1983. Larra's theme: The undoing of mole crickets. Golf Course Manage., April: 46–47.

Antonelli, A. L., and R. L. Campbell. 1984. The European crane fly: A lawn and pasture pest. Washington State Univ. Coop. Ext. Bull. No. 0856. 3 pp.

Archer, T. L., and G. J. Musick. 1976. Response of black cutworm larvae to light at several intensities. Ann. Entomol. Soc. Amer. 69: 476–478.

Baker, J. R. 1982. Insects and other pests associated with turf: Some important, common, and potential pests in the southeastern United States. North Carolina Agr. Ext. Service. AG-268. 108 pp.

Baker, P. B., R. H. Ratcliffe, and A. L. Steinhauer. 1981. Laboratory rearing of the hairy chinch bug. Environ. Entomol. 10: 226–229.

Banerjee, A. C. 1967a. Flight activity of the sexes of crambid moths as indicated by light-trap catches. J. Econ. Entomol. 60: 383–390.

Banerjee, A. C. 1967b. Sod webworm parasites in Illinois. J. Econ. Entomol. 60: 1173–1174.

Banerjee, A. C. 1967c. Injury to grasses in lawns by *Acrolophus* sp. J. Econ. Entomol. 60: 1174.

Banerjee, A. C. 1968. Microsporidia diseases of sod webworms in bluegrass lawns. Ann. Entomol. Soc. Amer. 61: 544–545.

Banerjee, A. C. 1969a. Development of *Crambus trisectus* at controlled constant temperatures in the laboratory. J. Econ. Entomol. 62: 703–705.

Banerjee, A. C. 1969b. Sex attractants in sod webworms. J. Econ. Entomol. 62: 705–708.

Banerjee, A. C., and G. C. Decker. 1966. Studies on sod webworms, I. Emergence rhythm, mating, and oviposition behavior under natural conditions. J. Econ. Entomol. 59: 1237–1244.

Banks, N. 1904. A treatise on the acarina, or mites. Proc. U.S. Nat. Mus. Smithsonian Inst. 28: 1–114.

Barnes, H. F. 1937. Methods of investigating the bionomics of the common crane fly, *Tipula paludosa* Meigen, together with some results. Ann. Appl. Biol. 24: 356–368.

Beard, J. B. 1973. Turfgrass science and culture. Prentice-Hall, Englewood Cliffs, N.J. 658 pp.

Beard, J. B. 1975. How to have a beautiful lawn: Easy steps in turfgrass establishment and care for aesthetic and recreational purposes. Intertee Publishing, Kansas City, Mo. 113 pp.

Beard, J. B. 1982. Turf management for golf courses. Pub. U.S. Golf Ass. Burgess Publishing, Minneapolis, Minn. 642 pp.

Beard, J. B. 1984. Grasses for the transition zone. Ground Maintenance, Jan.: 60–62.

Beckhan, C. M., and M. Dupree. 1952. Attractants for the green June beetle, with notes on seasonal occurrence. J. Econ. Entomol. 45: 736–737.

Bianchi, F. A. 1935. Investigations on *Anomala orientalis* Waterhouse at Oahu Sugar Company, Ltd. Hawaii Plant Rec. 39: 234–255.

Bianchi, F. A. 1957. Notes and exhibitions. Proc. Hawaiian Entomol. Soc. 16: 184.

Blatchley, W. S. 1910. An illustrated descriptive catalog of the Coleoptera, or beetles (exclusive of the Rhyncophora), known to occur in Indiana. Nature Publishing, Indianapolis. 1,386 pp.

Blatchley, W. S. 1920. Orthoptera of northeastern America, with special reference to the faunas of Indiana and Florida. Nature Publishing, Indianapolis. 784 pp.

Blatchley, W. S. 1926. Heteroptera, or true bugs of eastern North America, with special reference to the faunas of Indiana and Florida. Nature Publishing, Indianapolis. 1,116 pp.

Blatchley, W. S., and C. W. Leng. 1916. Rhyncophora, or weevils, of North America. Nature Publishing, Indianapolis. 682 pp.

Boggess, E. K. 1983. Carnivores (meat-eating mammals), raccoons, pp. C-73–C-79. *In* R. M. Timm (ed.), Prevention and control of wildlife damage. Great Plains Agr. Council and Univ. Nebraska Coop. Ext. Service, Inst. Agr. and Natur. Resources, Lincoln. Looseleaf, 632 pp.

Bohart, R. M. 1940. Studies on the biology and control of sod webworms in California. J. Econ. Entomol. 33: 886–890.

Bohart, R. M. 1947. Sod webworms and other lawn pests in California. Hilgardia 17: 267–307.

Boratynski, K., E. Pancer-Koteja, and J. Koteja. 1982. The life history of *Lecanopsis formicarum* Newstead (Homoptera: Coccinea) [in English; Polish and Russian summaries]. Ann. Zool. 36: 517–536.

Borror, D. J., D. M. DeLong, and C. A. Triplehorn. 1981. An introduction to the study of insects. 5th ed. Saunders College Publishing, Dryden Press, Philadelphia. 827 pp.

Boving, A. G. 1939. Descriptions of the three larval instars of the Japanese beetle, *Popillia japonica* Newm. Proc. Washington Entomol. Soc. 41: 183–191.

Boving, A. G. 1942. A classification of larvae and adults of the genus *Phyllophaga* (Coleoptera: Scarabaeidae). Mem. Entomol. Soc. Washington No. 2. 96 pp.

Bowen, W. R., F. S. Morishita, and R. O. Oetting. 1980. Insects and related pests of turf, pp. 31–39. *In* W. R. Bowen, Comp., Turfgrass pests. Agr. Sci. Univ. California Pub. No. 4053. 53 pp.

Brindle, A. 1960. The larvae and pupae of British Tipulinae (Diptera: Tipulidae) Trans. Soc. Brit. Entomol. 14: 63–114.

Britton, W. E. 1925. A new pest of lawns. Connecticut Agr. Exp. Sta. Bull. Immed. Inform. 52: 25–28.

Britton, W. E. 1932. Weevil grubs injure lawns. Connecticut Agr. Exp. Sta. Bull. 338: 593.

Bruner, L. 1890. Report on Nebraska insects, pp. 95–105. *In* Reports of observations and experiments in practical work of the division. U.S. Dept. Agr. Entomol. Bull. No. 22. 110 pp.

Brussell, G. E., and R. L. Clark. 1968. An evaluation of Baygon for control of the hunting billbug in a Zoysia grass lawn. J. Econ. Entomol. 61: 100.

Burrage, R. H., and G. G. Gyrisco. 1954. Distribution of third instar larvae of the European chafer and the efficiency of various sampling units for estimating their population. J. Econ. Entomol. 47: 1009–1014.

Burt, W. H., and R. P. Grossenheider. 1964. A field guide to the mammals. Houghton Mifflin, Boston. 284 pp.

Busching, M. K., and F. T. Turpin. 1976. Oviposition preferences of black cutworm moths among various crop plants, weeds, and plant debris. J. Econ. Entomol. 69: 587–590.

Butler, G. D., Jr. 1963. The biology of the bermudagrass eriophyid mite. Arizona Agr. Exp. Sta. Rep. 219: 8–13.

Butt, B. A., and E. Cantu. 1962. Sex determination of lepidopterous pupae. U.S. Dept. Agri. ARS-33-75. 7 pp.

Butt, F. H. 1944. External morphology of *Amphimallon majalis* (Razoumowski) (Coleoptera, the European chafer). Cornell Univ. Agr. Exp. Sta. Memoir No. 266. 18 pp.; 13 plates.

Byers, G. W. 1961. The crane fly genus *Dolichopeza* in North America. Univ. Kansas Sci. Bull. 42: 665–924.

Byers, R. A. 1965. Biology and control of a spittlebug. *Prosapia bicincta* (Say), on coastal bermudagrass. Georgia Agr. Exp. Sta. Tech. Bull. n.s. 42. 26 pp.

Cameron, R. S. 1970. Control of a species of *Hyperodes*. New York Turfgrass Assoc. Bull. 86: 333–336.

Cameron, R. S., and N. E. Johnson. 1971a. Biology and control of turfgrass weevil, a species of *Hyperodes*. New York State Coll. Agr., Cornell Univ. Ext. Bull. No. 1226. 8 pp.

Cameron, R. S., and N. E. Johnson. 1971b. Biology of a species of *Hyperodes* (Coleoptera: Curculionidae), a pest of turfgrass. Search Agr. No. 1. 31 pp.

Cameron, R. S., and N. E. Johnson. 1971c. Chemical control of the "annual bluegrass weevil," *Hyperodes* sp. nr. *anthracinus*. J. Econ. Entomol. 64: 689–693.

Cameron, R. S., H. J. Kastl, and J. F. Cornman. 1968. *Hyperodes* weevil damages annual bluegrass. New York Turfgrass Assoc. Bull. No. 79. 2 pp.

Cartwright, O. L. 1974. *Ataenius, Aphotaenius,* and *Pseudataenius* of the United States and Canada (Coleoptera: Scarabaeidae: Aphodiinae). Smithsonian Contrib. Zool. 154: 1–106.

Case, R. M. 1983. Rodents (gnawing animals), pocket gophers, pp. B-13–B-26. *In* R. M. Timm (ed.), Prevention and control of wildlife damage. Great Plains Agr. Council and Univ. Nebraska Coop. Ext. Service, Inst. Agr. and Natur. Resources, Lincoln. Looseleaf, 632 pp.

Cazier, M. A. 1940. The species of *Polyphylla* in America, north of Mexico (Coleoptera: Scarabaeidae). Entomol. News 51: 134–139.

Chada, H. L. 1956. Biology of the winter grain mite and its control in small grains. J. Econ. Entomol. 49: 515–520.

Chada, H. L., and E. A. Wood. 1960. Biology and control of the rhodesgrass scale. U.S. Dept. Agr. Tech. Bull. No. 1221. 21 pp.

Chamberlin, T. R., C. L. Fluke, L. Seaton, and J. A. Callenbach. 1938. Population and host preference of June beetles. U.S. Dept. Agr. Bur. Entomol. and Plant Quarantine, 18 Suppl. to No. 4: 225–240.

Chambliss, C. E. 1895. The chinch bug, *Blissus leucopterus* (Say). Tennessee Agr. Exp. Sta. Bull. 8(4): 41–55.

Chapman, D., and R. Kohut. 1985. Air pollution, acid rain, and New York agriculture, pp. 185–195. *In* New York Agriculture 2000. Governor's Office, Albany. 254 pp.

Chapman, P. J., and S. E. Lienk. 1981. Flight periods of adults of cutworms, armyworms, loopers, and others (family Noctuidae) injurious to vegetable and field crops. Search Agric. No. 14. 43 pp.

Chittenden, F. H., and D. E. Fink. 1922. The green June beetle. U.S. Dept. Agr. Bull. No. 891. 52 pp.

Choban, R. G., and A. P. Gupta. 1972. Meiosis and early embryology of *Blissus leucopterus hirtus* Montandon (Heteroptera: Lygaeidae). Int. J. Insect Morph. and Embryol. 1: 301–314.

Cobb, P. 1982. Mole crickets. Amer. Lawn Appl., Sept./Oct.: 4–8.

Cockfield, S. D., and D. A. Potter. 1984. Predation on sod webworm (Lepidoptera: Pyralidae) eggs as affected by chlorpyrifos application to Kentucky bluegrass turf. J. Econ. Entomol. 77: 1542–1544.

Comstock, J. A. 1927. Butterflies of California. McBride Printing, Los Angeles. 334 pp.

Converse, J. 1982. Scott's guide to the identification of the turfgrass diseases and insects. O. M. Scott & Sons, Marysville, Ohio. 105 pp.

Coulson, J. C. 1962. The biology of *Tipula subnodicornis* Zetterstedt, with comparative observations on *Tipula paludosa* Meign. J. Animal Ecol. 31: 1–21.

Crawford, C. S. 1968. Oviposition rhythm in *Crambus teterrellus:* Temperature depression effects and apparent circadian periodicity. Ann. Entomol. Soc. Amer. 61: 1481–1486.

Crawford, C. S., and R. F. Harwood. 1964. Bionomics and control of insects affecting Washington grass seed fields. Washington Agr. Exp. Sta. Tech. Bull. No. 44. 25 pp.

Creighton, W. S. 1950. The ants of North America. Cosmos Press, Cambridge, Mass. 585 pp.; 57 plates.

Crocker, R. L., and J. B. Beard. 1982. Southern mole cricket moves further into Texas, pp. 58–61. *In* Texas Turfgrass Research, 1982. Texas Agr. Exp. Sta. Consolidated PR-4032-4055. 79 pp.

Crocker, R. L., and C. L. Simpson. 1981. Pesticide screening tests for the southern chinch bug. J. Econ. Entomol. 74: 730–731.

Crocker, R. L., C. L. Simpson, H. Painter, T. W. Fuchs, and R. E. Woodruff. 1982. White grub of southern masked chafer, *Cyclocephala immaculata,* found in Texas turfgrass, pp. 39–40. *In* Texas Turfgrass Research, 1982. Texas Agr. Exp. Sta. Consolidated PR-4032-4055. 79 pp.

Cromroy, H. L., and D. E. Short. 1981. Pests of three types of turfgrass in Florida. Amer. Lawn Appl., Aug.: 32–37.

Crumb, S. E. 1929. Tobacco cutworms. U.S. Dept. Agr. Tech. Bull. No. 88. 179 pp.

Crumb, S. E. 1956. The larvae of the Phalaenidae. U.S. Dept. Agr. Tech. Bull. No. 1135. 356 pp.

Dahlsson, S. O. 1974. Frit fly damage to turfgrass, pp. 418–420. *In* E. C. Roberts (ed.), Proc. 2d Int. Turfgrass Res. Conf. Amer. Soc. Agron., Madison, Wis. 602 pp.

Damback, C. A., and E. Good. 1943. Life history and habits of the cicada killer in Ohio. Ohio J. Sci. 43: 32–41.

Davis. C. J. 1969. Notes on the grass webworm, *Herpetogramma licarsisalis* (Walker) (Lepidoptera: Pyraustidae), a new pest of turfgrass in Hawaii, and its enemies. Proc. Hawaiian Entomol. Soc. 20: 311–316.

Davis, J. J. 1919. Contributions to a knowledge of the natural enemies of *Phyllophaga.* Ill. Nat. Hist. Survey Bull. 13: 53–133.

Davis, J. J., and P. Luginbill. 1921. The green June beetle, or fig eater. North Carolina Agr. Exp. Sta. Bull. No. 242. 35 pp.

Davis, W. T. 1920. Mating habits of *Sphecius speciosus,* the cicada killing wasp. Bull. Brooklyn Entomol. Soc. 15: 128–129.

Dean, H. A., and M. F. Schuster. 1958. Biological control of rhodesgrass scale in Texas. J. Econ. Entomol. 51: 363–366.

Dominick, C. B. 1964. Notes on the ecology and biology of the corn root webworm. J. Econ. Entomol. 57: 41–42.

Dudderar, C. R. 1983. Mole control—A problem for lawn applicators. Amer. Lawn Appl., Mar./Apr.: 18–21.

Duff, D. T. 1984. Value of turf to the economy of Rhode Island. Univ. of Rhode Island Turfgrass Res. Review. Rhode Island Agr. Exp. Sta. No. 16. 4 pp.

Duff, M. J. 1982. The mole crickets of Jekyll Island. Golf Course Manage., May: 57–59.

Dunbar, D. M. 1971. Big-eyed bugs in Connecticut lawns. Connecticut Agr. Exp. Sta. Circ. No. 244. 6 pp.

Dunbar, D. M., and R. L. Beard. 1975. Present status of milky disease of Japanese and Oriental beetles in Connecticut. J. Econ. Entomol. 68: 453–457.

Dunn, G. A., and M. K. Kennedy, 1983. How to prepare turf insects for identification. Amer. Lawn Appl., Jan./Feb.: 18–21.

Dutky, S. R. 1940. Two new spore-forming bacteria causing milky diseases of Japanese beetle larvae. J. Agric. Res. 61: 57–68.

Fagan, E. B., and L. C. Kuitert. 1969. Biology of the two-lined spittlebug, *Prosapia bicincta,* on Florida pastures (Homoptera: Cercopidae). Florida Entomol. 52: 199–206.

Fairfax Biological Laboratory, Inc. 1979. The winsome fly. Clinton Corners, N.Y. 2 pp.

Felt, E. P. 1894. On certain grass-eating insects: A synopsis of the species of *Crambus* of the Ithaca fauna. Cornell Univ. Agr. Exp. Sta. Bull. 64: 47–102.

Fleming, W. E. 1958. Biological control of the Japanese beetle, especially with entomogenous diseases. Proc. 10th International Cong., 1956 (3): 115–125.

Fleming, W. E. 1962. The Japanese beetle in the United States. U.S. Dept. Agr. Handbook No. 236. 30 pp.

Fleming, W. E. 1968. Biological control of the Japanese beetle. U.S. Dept. Agr. Tech. Bull. No. 1383. 78 pp.

Fleming, W. E. 1972. Biology of the Japanese beetle. U.S. Dept. Agr. Tech. Bull. No. 1449. 129 pp.

Fleming, W. E. 1976. Integrating control of the Japanese beetle—A historical review. U.S. Dept. Agr. Tech. Bull. No. 1545. 65 pp.

Fletcher, D. S. 1956. *Spodoptera mauritia* (Boisduval) and *S. triturata* (Walker), two distant species. Bull. Entomol. Res. 47: 215–217.

Forbes, W. T. M. 1954. Lepidoptera of New York and neighboring states. Noctuidae. Pt. 3. Cornell Univ. Agr. Exp. Sta. Mem. No. 329. 433 pp.

Fox, D. J. J. 1957. Note on occurrence in Cape Breton Island of *Tipula paludosa* Meig. (Diptera: Tipulidae). Can. Entomol. 89: 288.

Frankie, G. W., H. A. Turney, and P. J. Hamman. 1973. White grubs in Texas turfgrass. Texas Agr. Ext. Service L-1131. 3 pp.

French, J. C. 1964. Chinch bugs. Univ. Georgia Coll. Agr. Leaflet No. 20. 2 pp.

Friend, R. B. 1929. The Asiatic beetle in Connecticut. Connecticut Agr. Exp. Sta. Bull. 304: 585–664.

Funk, C. R., and S. Ahmad. 1983. The bluegrass billbug: Susceptibility of bluegrass to damage. New York State Turfgrass Ass. Bull. 117: 842–843.

Funk, C. R., and R. H. Hurley. 1984. Seed facts update on perennial ryegrass. Lawn Care Ind., Jan.: 38–38A.

Fushtey, S. G., and M. K. Sears. 1981. Turfgrass diseases and insect pests (descriptions, illustrations, and controls). Min. Agric. and Food Pub. No. 162. University of Guelph, Ontario. 32 pp.

Gambrell, F. L. 1943. Observations on the economic importance and control of the European chafer. New York State Agr. Exp. Sta. Bull. 703: 8–13.

Gambrell, F. L., S. C. Mendel, and E. H. Smith. 1942. A destructive European insect new to the United States. J. Econ. Entomol. 35: 289.

Gammon, E. T. 1961. The Japanese beetle in Sacramento. California Dept. Agr. Bull. 50: 221–235.

Garman, H. 1926. Two important enemies of bluegrass pastures. Kentucky Agr. Exp. Sta. Bull. 265: 29–47.

Gaylor, M. J., and G. W. Frankie. 1979. The relationship of rainfall to adult flight activity, and of soil moisture to oviposition behavior and egg and first instar survival in *Phyllophaga crinita*. Environ. Entomol. 8: 591–594.

Gibeault, V. A., K. Mueller, and J. Davidson. 1977. Dichondra. Div. Agr. Sci. Univ. California Leaflet No. 2983. 11 pp.

Gruttadaurio, J., E. E. Hardy, and A. S. Lieberman. 1978. Final report on an investigation of turfgrass land use acreages and selected maintenance expenditures across New York state. Dept. Floriculture and Ornamental Hort., New York State Coll. Agr. and Life Sci., Cornell Univ., Ithaca. 36 pp.

Guthrie, F. E., and G. C. Decker. 1954. The effect of humidity and other factors on the upper thermal death point of the chinch bug. J. Econ. Entomol. 47: 882–887.

Gyrisco, G. G., W. H. Whitcomb, R. H. Burrage, C. Logothetis, and H. H. Schwardt. 1954. Biology of European chafer, *Amphimallon majalis* Razoumowsky (Scarabaeidae). Cornell Univ. Agr. Exp. Sta. Mem. No. 328. 35 pp.

Hadley, C. H., and I. M. Hawley. 1934. General information about the Japanese beetle in the United States. U.S. Dept. Agr. Circ. No. 332. 22 pp.

Hallock, H. C. 1929. Known distribution and abundance of *Anomala orientalis* Waterhouse, *Aserica castanea* Arrow, and *Serica similis* Lewis in New York. J. Econ. Entomol. 22: 293–299.

Hallock, H. C. 1930. The Asiatic beetle, a serious pest in lawns. U.S. Dept. Agr. Circ. No. 117. 7 pp.

Hallock, H. C. 1933. Present status of two Asiatic beetles (*Anomola orientalis* and *Autoserica castanea*) in the United States. J. Econ. Entomol. 26: 80–85.

Hallock, H. C. 1935. Movements of larvae of the Oriental beetle through soil. J. New York Entomol. Soc. 43: 413–425.

Hallock, H. C. 1936. Notes on biology and control of the Asiatic garden beetle. J. Econ. Entomol. 29: 348–356.

Hallock, H. C., and I. M. Hawley. 1936. Life history and control of the Asiatic garden beetle. Rev. ed. U.S. Dept. Agr. Circ. No. 246. 20 pp. [First edition, 1932, written by Hallock]

Hamman, P. J. 1969. Control of southern chinch bug, *Blissus insularis,* in Brazos County, Texas, pp. 15–17. *In* H. T. Streu and R. T. Bangs (eds.), Proc. Scott's Turfgrass Res. Conf. 1. 89 pp.

Hammond, G. H. 1940. White grubs and their control in eastern Canada. Dom. Can. Dept. Agr. Pub. No. 668. 18 pp.

Hanson, A. A., F. V. Juska, and G. W. Burton. 1969. Species and varieties, pp. 370–409. *In* A. A. Hanson and F. V. Juska (eds.), Turfgrass science. Monogr. No. 14. Amer. Soc. Agr., Madison, Wisc. 715 pp.

Harris, C. R., J. H. Mazurek, and G. V. White. 1962. The life history of the black cutworm, *Agrotis ipsilon* (Hufnagel), under controlled conditions. Can. Entomol. 94: 1183–1187.

Hawley, I. M. 1944. Notes on the biology of the Japanese beetle. U.S. Dept. Agr., Bur. Entomol. and Plant Quarantine E 615. 19 pp.

Hawthorne, D. W. 1983. Other mammals: Armadillos, pp. D-5–D-7. *In* R. M. Timm (ed.), Prevention and control of wildlife damage. Great Plains Agr. Council and Univ. Nebraska Coop. Ext. Service, Inst. Agr. and Natur. Resources, Lincoln. Looseleaf, 632 pp.

Hayslip, N. C. 1943. Notes on biological studies of mole crickets at Plant City, Florida. Florida Entomol. 26: 33–46.

Hegner, R. W. 1942. College zoology. 5th ed. Macmillan, New York. 817 pp.

Heinrichs, E. A. 1973. Bionomics and control of sod webworms. Bull. Entomol. Soc. Amer. 19: 89–97.

Heinrichs, E. H., and C. J. Southwards. 1970. Susceptibility of the sod webworm *Pediasia trisecta* to biological control agents. Tennessee Farm and Home Sci., Jan., Feb., Mar.: 30–32.

Heit, C. E., and H. K. Henry. 1940. Notes on the species of white grubs present in the Saratoga Forest Tree Nursery. J. Forest. 38: 944–948.

Henderson, F. R. 1983. Other mammals: Moles, pp. D-53–D-61. *In* R. M. Timm (ed.), Prevention and control of wildlife damage. Great Plains Agr. Council and Univ. Nebraska Coop. Ext. Service, Inst. Agr. and Natur. Resources, Lincoln. Looseleaf, 632 pp.

Henry, H. K., and C. E. Heit. 1940. Flight records of *Phyllophaga* (Coleoptera: Scarabaeidae). Entomol. News 40: 279–282.

Hodges, Ronald, W., et al. 1983. Check list of the Lepidoptera of America north of Mexico. E. W. Classey and Wedge Entomol. Research Found., London. 284 pp.

Hoffman, C. H. 1935. Biological notes on *Ataenius cognatus* (Lec.), a new pest of golf greens in Minnesota (Scarabaeidae—Coleoptera). J. Econ. Entomol. 28: 666–667.

Hunter, S. J. 1909. The greenbug and its enemies. Kansas Univ. Bull. No. 9. 163 pp.

Hurpin, B. 1953. Reconnaissance des sexes chez les larves de Coléoptères, Scarabaeidae. Bull. de la Soc. Entomol. de France 58: 104–107.

Jackson, D. M., and R. L. Campbell. 1975. Biology of the European crane fly, *Tipula paludosa*

Meigen, in western Washington (Tipulidae: Diptera). Washington State Univ. Tech. Bull. No. 81. 23 pp.
Jackson, D. W., K. J. Vessels, and D. A. Potter. 1981. Resistance of selected cool and warm season turfgrasses to the greenbug (*Schizaphis graminum*). HortScience 16: 558–559.
Janisch, E. 1941. Das temperaturoptimum der Wiesenschnake *Tipula paludosa*. Mitt. Biol. Reichsanst 65: 38.
Jaynes, H. A., and T. R. Gardner. 1924. Selective parasitism by *Tiphia* sp. J. Econ. Entomol. 17: 366–369.
Jefferson, R. N., and C. O. Eades. 1952. Control of sod webworm in southern California. J. Econ. Entomol. 45: 114–118.
Jefferson, R. N., I. M. Hall, and F. S. Morishita. 1964. Control of lawn moths in southern California. J. Econ. Entomol. 57: 150–152.
Jeppson, L. R., H. H. Keifer, and E. W. Baker. 1975. Mites injurious to economic plants. Univ. of California Press, Berkeley. 614 pp.; 74 plates.
Jepson, W. F., and A. J. Heard. 1959. The frit fly and allied stem boring Diptera in winter wheat and host grasses. Ann. Appl. Biol. 47: 114–130.
Jerath, M. L. 1960. Notes on larvae of nine genera of Aphodiinae in the United States (Coleoptera: Scarabaeidae). Proc. U.S. Nat. Mus. Smithsonian Inst. 111: 43–94.
Johnson, F. A. 1975. The bermudagrass mite *Eriophyes cydoniensis* (Sayed) (Acari: Eriophyidae) in Florida, with reference to its injury, symptomology, ecology, and integrated control. Ph.D. Thesis. Univ. of Florida, Gainesville. 182 pp.
Johnson, J. P. 1941. *Cyclocephala* (*Ochrosidia*) *borealis* in Connecticut. J. Agric. Res. 62: 79–86.
Johnson, N. E., and R. S. Cameron. 1969. Phytophagous ground beetles. Ann. Entomol. Soc. Amer. 62: 909–914.
Judge, F. D. 1972. Aspects of the biology of the gray garden slug (*Deroceras reticulatum* Muller). Search Agr. 2: 1–18.
Kamm, J. A. 1969. Biology of the billbug *Sphenophorus venatus confluens*, a new pest of orchardgrass. J. Econ. Entomol. 62: 808–812.
Kamm, J. A. 1970. Effects of photoperiod and temperature on *Crambus trisectus* and *C. leachellus cypridalis* (Lepidoptera: Crambidae). Ann. Entomol. Soc. Amer. 63: 412–416.
Kamm, J. A. 1971. Environmental biology of a sod webworm *Crambus tutillus* (Lepidoptera, Crambinae). Entomol Exp. and Appl. 14: 30–38.
Kamm, J. A. 1973. Biotic factors that affect sod webworms in grass fields in Oregon. Environ. Entomol. 2: 94–96.
Kamm, J. A., and J. Capizzi. 1977. Control of grass seed insect pests, pp. 127–129. *In* Oregon insect control handbook. Oregon State Univ., Corvallis.
Kamm, J. A., and L. M. McDonough. 1979. Field tests with the sex pheromone of the cranberry girdler. Environ. Entomol. 8: 773–775.
Kamm, J. A., and L. M. McDonough. 1980. Synergism of the sex pheromone of the cranberry girdler. Environ. Entomol. 9: 795–797.
Kamm, J. A., P. D. Morgan, D. L. Overhulser, L. M. McDonough, M. Triebwasser, and L. N. Kline. 1983. Management practices for cranberry girdler (Lepidoptera: Pyralidae) in Douglas-fir nursery stock. J. Econ. Entomol. 76: 923–926.
Kawanishi, C. Y., C. M. Splittstoesser, H. Tashiro, and K. H. Steinkraus. 1974. *Ataenius spretulus*, a potentially important turf pest, and its associated milky disease bacterium. Environ. Entomol. 3: 177–180.
Keifer, H. H., E. W. Baker, T. Kono, M. Delfinado, and W. E. Styer. 1982. An illustrated guide to

plant abnormalities caused by eriophyid mites in North America. U.S. Dept. Agr. Handbook No. 573. 178 pp.

Kelsheimer, E. G. 1956. The hunting billbug, a serious pest of zoysia. Proc. Florida Hort. Soc. 69: 415–418.

Kelsheimer, E. G., and S. H. Kerr. 1957. Insects and other pests of lawns and turf. Univ. Florida Agr. Exp. Sta. Circ. S-96. 22 pp.

Kennedy, M. K. 1980. New webworm pests in Michigan lawns. Amer. Lawn Appl., Nov./Dec.: 11–14.

Kennedy, M. K. 1981. Chinch bugs: Biology and control. Amer. Lawn Appl., July/Aug.: 12–15.

Kerr, S. H. 1955. Life history of the tropical sod webworm *Pachyzancla phaeopteralis* Guenée. Florida Entomol. 38: 3–11.

Kerr, S. H. 1966. Biology of the lawn chinch bug, *Blissus insularis.* Florida Entomol. 49: 9–18.

Kerr, T. W. 1941. Control of white grubs in strawberries. Cornell Univ. Agr. Exp. Sta. Bull. No. 770. 40 pp.

Kindler, S. D., and E. J. Kinbacher. 1975. Differential reaction of Kentucky bluegrass cultivars to the bluegrass billbug, *Sphenophorus parvulus* Gyllenhal. Crop Sci. 15: 873–874.

Kindler, S. D., and E. J. Kinbacher. 1982. Sampling for eggs, larvae, and adults of the bluegrass billbug. Crop Sci. 22: 677–678.

Kindler, S. D., R. Staples, S. M. Spomer, and O. Adeniji. 1983. Resistance of bluegrass cultivars to biotypes C and E greenbug (Homoptera: Aphididae). J. Econ. Entomol. 76: 1103–1105.

King, G. B. 1901. The Coccidae of British North America. Can. Entomol. 33: 193–200.

Kissinger, D. G. 1964. Curculionidae of America north of Mexico: A key to the genera. Taxonomic Publishing, South Lancaster. 143 pp.

Klein, M. G. 1981. Mass trapping for suppression of Japanese beetles, pp. 183–190. *In* E. R. Mitchell (ed.), Management of insect pests with semiochemicals. Plenum Publishing, New York. 514 pp.

Klots, A. B. 1951. A field guide to the butterflies of North America east of the Great Plains. Riverside Press, Cambridge, Mass. 349 pp.

Knight, J. E. 1983. Carnivores (meat-eating mammals): Skunks, pp. C-81–C-86. *In* R. M. Timm (ed.), Prevention and control of wildlife damage. Great Plains Agr. Council and Univ. Nebraska Coop. Ext. Service, Inst. Agr. and Natur. Resources, Lincoln. Looseleaf, 632 pp.

Koehler, P. G., and D. E. Short. 1976. Control of mole crickets in pasture grass. J. Econ. Entomol. 69: 229–232.

Kouskolekas, C. A., and R. L. Self. 1974. Biology and control of the ground pearl in relation to turfgrass infestation, pp. 421–423. *In* E. C. Roberts (ed.), Proc. 2d Int. Turfgrass Res. Conf. Amer. Soc. Agron., Madison, Wisc. 602 pp.

Krantz, G. W. 1957. Winter grain mite (*Penthaleus major*). Plant Pest Control Div., U.S. Dept. Agr., Coop. Econ. Insect Rep. 7: 302.

Krantz, G. W. 1978. A manual of acarology. 2d ed. Oregon State Univ. Book Stores, Corvallis. 509 pp.

Ladd, T. L., Jr. 1970. Sex attraction in the Japanese beetle. J. Econ. Entomol. 63: 905–908.

Ladd, T. L., M. G. Klein, and J. H. Tumlinson. 1981. Phenethyl proprionate + eugenol + geraniol (3:7:3) and Japonilure: A highly effective joint lure for Japanese beetles. J. Econ. Entomol. 74: 665–667.

Laigo, F. M., and M. Tamashiro. 1966. Virus and insect parasite interaction in the lawn armyworm, *Spodoptera mauritia acronyctoides* (Guenée). Proc. Hawaiian Entomol. Soc. 19: 233–237.

LaPlante, A. A., Jr. 1966a. How to control the lawn armyworm. Univ. Hawaii Coop. Ext. Entomol. Notes No. 1. 2 pp.

LaPlante, A. A., Jr. 1966b. How to control the hunting billbug. Univ. Hawaii Coop. Ext. Service Notes No. 2. 2 pp.

Laughlin, R. 1958. Desiccation of eggs of the crane fly *Tipula oleracea.* Nature 182: 613.

Lawrence, K. O. 1982. A linear pitfall trap for mole cricket and other soil arthropods. Florida Entomol. Sci. Notes 65: 376–377.

Leonard, D. E. 1966. Biosystemics of the leucopterus complex of the genus *Blissus* (Heteroptera: Lygaeidae). Connecticut Agr. Exp. Sta. Bull. 677: 1–47.

Leonard, D. E. 1968. A revision of the genus *Blissus* (Heteroptera: Lygaeidae) in eastern North America. Ann. Entomol. Soc. Amer. 61: 239–250.

Liu, H. J., and F. L. McEwen. 1979. The use of temperature accumulations and sequential sampling in predicting damaging populations of *Blissus leucopterus hirtus.* Environ. Entomol. 8: 512–515.

Lofgren, C. S., W. A. Banks, and B. M. Glancey. 1975. Biology and control of imported fire ants. Ann. Rev. Entomol. 20: 1–30.

Ludwig, D. 1932. The effect of temperature on the growth curves of the Japanese beetle (*Popillia japonica* Newman). Physiol. Zool. 5: 431–447.

Luginbill, P. 1922. Bionomics of the chinch bug. U.S. Dept. Agr. Bull. No. 1016. 14 pp.

Luginbill, P. 1928. The fall armyworm. U.S. Dept. Agr. Tech. Bull. No. 34. 92 pp.

Luginbill, P. 1938. Control of common white grubs in cereal and forage crops. U.S. Dept. Agr. Farmers' Bull. No. 1798. 19 pp.

Luginbill, P., and H. R. Painter. 1953. May beetles of the United States and Canada. U.S. Dept. Agr. Tech. Bull. No. 1060. 102 pp.; 78 plates.

McCrea, R. J. 1972. The dichondra flea beetle (Genus *Chaetocnema*) in southern California (Coleoptera: Chrysomelidae). M.S. Thesis. California State Coll., Long Beach. 83 pp.

McDonough, L. M., and J. A. Kamm. 1979. Sex pheromone of the cranberry girdler, *Chrysoteuchia topiaria* (Zellar) (Lepidoptera: Pyralidae). J. Chem. Ecol. 5: 211–219.

McDonough, L. M., J. A. Kamm, D. A. George, C. L. Smithhisler, and S. Voerman. 1982. Sex attractant for the western lawn moth *Tehama bonifatella* Hulst. Environ. Entomol. 11: 711–714.

McGregor, R. A. 1976. Florida turfgrass survey, 1974. Florida Crop and Livestock Rep. Service. 33 pp.

Maddock, D. R., and C. F. Fehn. 1958. Human ear invasion by adult scarabaeid beetles. J. Econ. Entomol. 51: 546–547.

Maercks, H. 1939. Die Wiesenschnaken und ihre Bekämpfung. Kranke Pflanze 16: 107–110.

Mahr, D. L., and R. Kachadoorian. 1983. Turfgrass disorder: Bluegrass billbug. Univ. Wisconsin Urban Phytonarian Ser. A 3234. 2 pp.

Mahr, D. L., and R. Kachadoorian. 1984. Turfgrass disorder: Sod webworms. Univ. Wisconsin Urban Phytonarian Ser. A 3271. 2 pp.

Mailloux, G., and H. T. Streu. 1979. A sampling technique for estimating hairy chinch bug (*Blissus leucopterus hirtus* Montandon, Hemiptera: Lygaeidae) populations and other arthropods from turfgrass. Ann. Entomol. Soc. Quebec 24: 139–143.

Mailloux, G., and H. T. Streu. 1981. Population biology of the hairy chinch bug (*Blissus leucopterus hirtus,* Montandon: Hemiptera: Lygaeidae). Ann. Entomol. Soc. Quebec 26: 51–90.

Mailloux, G., and H. T. Streu. 1982. Bionomics of the larger sod webworm, *Pediasia trisecta* (Walker) (Lepidoptera: Pyralidae: Crambinae). Ann. Entomol. Soc. Quebec 27: 68–74.

Malcolm, D. R. 1955. Biology and control of the timothy mite, *Paratetranychus pratensis* (Banks). Washington Agr. Exp. Sta. Tech. Bull. No. 17. 35 pp.

Matheny, E. L., Jr. 1971. Seasonal abundance, distribution, and egg studies of sod webworm moths (Lepidoptera: Pyralidae: Crambinae) in Tennessee. Ph.D. Thesis. Univ. of Tennessee, Knoxville. 102 pp.

Matheny, E. L., Jr., and E. A. Heinrichs. 1972. Chorion characteristics of sod webworm eggs. Ann. Entomol. Soc. Amer. 65: 238–246.

Matheny, E. L., Jr., and R. L. Kepner. 1980. Maxillae of the mole crickets, *Scapteriscus acletus* Rehn and Hebard, and *S. vicinus* Scudder (Orthoptera: Gryllotalpidae): A new means of identification. Florida Entomol. 63: 512–514.

Matheson, R. 1951. Entomology for introductory courses. 2d ed. Comstock Publishing Co., Cornell Univ. Press, Ithaca, N.Y. 630 pp.

Maxwell, K. E., and G. F. McLeod. 1936. Experimental studies of the hairy chinch bug. J. Econ. Entomol. 29: 339–343.

Metcalf, C. L., W. P. Flint, and R. L. Metcalf. 1962. Destructive and useful insects: Their habits and control. 4th ed. McGraw-Hill, New York. 1,087 pp.

Mitchell, W. C., and C. L. Murdoch. 1974. Insecticides and their application frequency for control of turf insects in Hawaii. Down to Earth 30: 17–23.

Morishita, F. S., W. Humphrey, L. C. Johnston, and R. F. Jefferson. 1971. Control of billbugs on turf. California Turfgrass Culture 21: 13–14.

Morrison, W. P., B. C. Pass, and C. S. Crawford. 1972. Effect of humidity on eggs of two populations of the bluegrass webworm. Environ. Entomol. 1: 218–221.

Muma, M. H. 1944. The attraction of *Cotinus nitida* by caproic acid. J. Econ. Entomol. 37: 855–856.

Murdoch, C. L., and H. Tashiro. 1976. Host preference of the grass webworm, *Herpetogramma licarsisalis,* to warm season turfgrasses. Environ. Entomol. 5: 1068–1070.

National Oceanic and Atmospheric Admin. 1980. Climatological data, New York, No. 92. 21 pp.

Neiswander, C. R. 1938. The annual white grub, *Ochrosidia villosa* Burm., in Ohio lawns. J. Econ. Entomol. 31: 340–344.

Neiswander, C. R. 1963. The distribution and abundance of May beetles in Ohio. Ohio Agr. Exp. Sta. Res. Bull. No. 951. 35 pp.

Ng, Y.-S., and S. Ahmad. 1979. Resistance to dieldrin and tolerance to chlorpyrifos and bendiocarb in a northern New Jersey population of the Japanese beetle. J. Econ. Entomol. 72: 698–700.

Nickle, D. A., and J. L. Castner. 1984. Introduced species of mole crickets in the United States, Puerto Rico, and the Virgin Islands (Orthoptera: Gryllotalpidae). Ann. Entomol. Soc. Amer. 77: 450–465.

Niemczyk, H. D. 1976. A new grub problem in golf course turf. Golf Course Superintendent, Mar.: 26–29.

Niemczyk, H. D. 1977a. *Ataenius spretulus*—New concern over old turf insect pest. Ohio Rep. 62: 3–5.

Niemczyk, H. D. 1977b. Thatch: A barrier to control of soil-inhabiting insects pests of turf. Weeds, Trees, and Turf, Feb.: 16–19.

Niemczyk, H. D. 1978. The winter grain mite: Winter pest of turf. Weeds, Trees, and Turf, Feb.: 22–23.

Niemczyk, H. D. 1980a. Insects and their control. Lawn Care Ind., Mar.: 34–46.

Niemczyk, H. D. 1980b. New evidence indicates greenbug overwinters in North. Weeds, Trees, and Turf, June: 64–65.

Niemczyk, H. D. 1980c. Proper pesticide application for effective greenbug control. Lawn Care Ind., Oct.: 16–19.

Niemczyk, H. D. 1981. Destructive turf insects. HDN Books, Wooster, Ohio. 48 pp.

Niemczyk, H. D. 1982. Chinch bug and bluegrass billbug control with spring application of chlorpyrifos, pp. 85–89. *In* H. D. Niemczyk and B. J. Joyner (eds.), Advances in turfgrass entomology. Hammer Graphics, Piqua, Ohio. 150 pp.

Niemczyk, H. D. 1983. The bluegrass billbug: A frequently misdiagnosed pest of turfgrass. Amer. Lawn Appl., May–June: 4–7.

Niemczyk, H. D., and D. M. Dunbar. 1976. Field observations, chemical control, and contact toxicity experiments on *Ataenius spretulus,* a grub pest of turf grass. J. Econ. Entomol. 69: 345–348.

Niemczyk, H. D., and C. Frost. 1978. Insecticide resistance found in Ohio bluegrass billbugs. Ohio Rep. 63: 22–23.

Niemczyk, H. D., and J. R. Moser. 1982. Greenbug occurrence and control on turfgrasses in Ohio, pp. 105–111. *In* H. D. Niemczyk and B. G. Joyner (eds.), Advances in turfgrass entomology. Hammer Graphics, Piqua, Ohio. 150 pp.

Niemczyk, H. D., and K. T. Power. 1978. Another grub pest, similar to the black turfgrass ataenius, found damaging fairways, p. 31. *In* Proc. Ohio Turfgrass Conf.

Niemczyk, H. D., and K. T. Power. 1982. Greenbug buildup in Ohio is linked to overwintering. Weeds, Trees, and Turf, June: 36.

Niemczyk, H. D., and G. S. Wegner. 1979. Life history and control of the black turfgrass ataenius. Ohio Rep. 64: 85–88.

Niemczyk, H. D., and G. S. Wegner. 1982. Life history and control of the black turfgrass ataenius (Coleoptera-Scarabaeidae), pp. 113–117. *In* Niemczyk, H. D., and B. G. Joyner (eds.), Advances in turfgrass entomology. Hammer Graphics, Piqua, Ohio. 150 pp.

O'Brien, J. M. 1983. Rodents (gnawing mammals): Voles, pp. B-147–B-152. *In* R. M. Timm (ed.), Prevention and control of wildlife damage. Great Plains Agr. Council and Univ. Nebraska Coop. Ext. Service, Inst. Agr. and Natur. Resources, Lincoln. Looseleaf, 632 pp.

Okamura, G. T. 1959. Illustrated key to the lepidopterous larvae attacking lawns in California. California Dept. Agr. Bull. 48: 15–21.

Oliver, A. D. 1982a. The red imported fire ant as a lawn insect problem. Amer. Lawn Appl., May/June: 4–8.

Oliver, A. D. 1982b. The fall armyworm as an annual pest. Amer. Lawn Appl., Nov./Dec.: 18–22.

Oliver, A. D. 1984. The hunting billbug—One among the complex of turfgrass insect and pathogen problems. Amer. Lawn Appl., Mar./April: 24–27.

Oliver, A. D., and J. B. Chapin. 1981. Biology and illustrated key for the identification of twenty species of economically important noctuid pests. Louisiana Agr. Exp. Sta. Bull. No. 733. 26 pp.

Oliver, A. D., and K. N. Komblas. 1981. Southern chinch bug in Louisiana. Amer. Lawn Appl., July/Aug.: 26–31.

O. M. Scott and Sons. 1979. Scotts information manual for lawns. O. M. Scott and Sons, Marysville, Ohio. 98 pp.

Opler, P. A., and G. O. Krizek. 1984. Butterflies east of the Great Plains. John Hopkins Univ. Press, Baltimore. 274 pp.

Oschmann, M. 1979. Violet trays for recording the population dynamics of the frit fly (*Oscinella frit* L.). Archiv für Phytopathologie und Pflanzenschutz 15: 197–203.

Paris, O. H. 1963. The ecology of *Armadillidum vulgare* (Isopoda: Oniscoidea) in California grassland: Food, enemies and weather. Ecol. Monogr. 33: 1–22.

Parker, H. L. 1959. Studies of some Scarabaeidae and their parasites. Boll. del Lab. di Entomol. Agr. (Filippo Silvestri) di Portico 17: 29–50.

Pemberton, C. E. 1955. Notes and exhibitions. Proc. Hawaiian Entomol. Soc. 15: 373.

Peterson, A. 1948. Larvae of insects–An introduction to Nearctic species. Pt. 1. Lepidoptera and plant-infesting Hymenoptera. Edwards Brothers, Ann Arbor, Mich. 315 pp.

Peterson, R. T. 1980. A field guide to the birds. 4th ed. Houghton Mifflin, Boston. 384 pp.

Polivka, J. B. 1960a. Effect of lime applications to soil on Japanese beetle larval population. J. Econ. Entomol. 53: 476–477.

Polivka, J. B. 1960b. Grub population in turf varies with pH levels in Ohio soils. J. Econ. Entomol. 53: 860–863.

Polivka, J. B. 1963. Control of hairy chinch bug, *Blissus leucopterus hirtus* Mont., in Ohio. Ohio Agr. Exp. Sta. Res. Circ. No. 122. 8 pp.

Porter, K. B., G. L. Peterson, O. Vise. 1982. A new greenbug biotype. Crop Sci. 22: 847–850.

Potter, D. A. 1980. Flight activity and sex attraction of northern and southern masked chafers in Kentucky turfgrass. Ann. Entomol. Soc. Amer. 73: 414–417.

Potter, D. A. 1981a. Biology and management of masked chafer bugs. Amer. Lawn Appl., July/Aug.: 2–6.

Potter, D. A. 1981b. Seasonal emergence and flight of northern and southern masked chafers in relation to air and soil temperature and rainfall patterns. Environ. Entomol. 10: 793–797.

Potter, D. A. 1982a. Greenbugs on turfgrass: An informative update. Amer. Lawn Appl., Mar./Apr.: 20–25.

Potter, D. A. 1982b. Influence of feeding by grubs of the southern masked chafer on quality and yield of Kentucky bluegrass. J. Econ. Entomol. 75: 21–24.

Potter, D. A. 1983. Effect of soil moisture on oviposition, water absorption, and survival of southern masked chafer (Coleoptera: Scarabaeidae) eggs. Environ. Entomol. 12: 1223–1227.

Potter, D. A., and F. C. Gordon. 1984. Susceptibility of *Cyclocephala immaculata* (Coleoptera: Scarabaeidae) eggs and immatures to heat and drought in turfgrass. Environ. Entomol. 13: 794–799.

Ratcliffe, R. H. 1982. Evaluation of cool-season turfgrasses for resistance to the hairy chinch bug, pp. 13–18. *In* H. D. Niemczyk and B. J. Joyner (eds.), Advances in turfgrass entomology. Hammer Graphics, Piqua, Ohio. 150 pp.

Ratcliffe, R. H., and J. J. Murray. 1983. Selection for greenbug (Homoptera: Aphididae) resistance in Kentucky bluegrass cultivars. J. Econ. Entomol. 76: 1221–1224.

Reese, J. C., L. M. English, T. R. Yonke, and M. L. Fairchild. 1972. A method for rearing black cutworms. J. Econ. Entomol. 65: 1047–1050.

Regnier, R. 1939. Contribution à l'étude des hannetons: Un grand ennemi des gazons: *Amphimallon majalis* Razoumowsky. Ann. des Epiphyt. et de Phytogenet. 5: 257–265.

Reid, W. 1983. European chafer/Hanneton européen: Situation in Canada/Situation au Canada. Biol. Programs Sect., Plant Health Div., Ottawa. 5 pp.

Reinert, J. A. 1972. New distribution and host record for the parasitoid *Eumicrosoma benefica*. Florida Entomol. 55: 143–144.

Reinert, J. A. 1973. Sod webworm control in Florida turfgrass. Florida Entomol. 56: 333–337.

Reinert, J. A. 1974. Tropical sod webworm and southern chinch bug control in Florida. Florida Entomol. 57: 275–280.

Reinert, J. A. 1975. Life history of the striped grassworm, *Mocis latipes*. Ann. Entomol. Soc. Amer. 68: 201–204.

Reinert, J. A. 1978. Natural enemy complex of the southern chinch bug in Florida. Ann. Entomol. Soc. Amer. 71: 728–731.

Reinert, J. A. 1982a. The bermudagrass stunt mite. U.S. Golf Ass. Green Sec. Rec. 20: 9–12.

Reinert, J. A. 1982b. Insecticide resistance in epigeal insect pests of turfgrass: 1, A review, pp. 71–75. *In* H. D. Niemczyk and B. J. Joyner (eds.), Advances in turfgrass entomology. Hammer Graphics, Piqua, Ohio. 150 pp.

Reinert, J. A. 1982c. Southern chinch bug resistance to insecticides: A method for quick diagnosis of chlorpyrifos (OP) resistance and alternate controls. Florida Turfgrass Proc. 30: 64–78.

Reinert, J. A. 1983a. Field experiments for insecticidal control of sod webworms (Lepidoptera: Pyralidae) in Florida turfgrass. J. Econ. Entomol. 76: 150–153.

Reinert, J. A. 1983b. Foraging sites of the southern mole cricket, *Scapteriscus acletus* (Orthoptera: Gryllotalpidae). Proc. Fla. State Hort. Soc. 96: 149–151.

Reinert, J. A., and P. Busey. 1983. Resistance of bermudagrass selections to the tropical sod webworm (Lepidoptera: Pyralidae). Environ. Entomol. 12: 1844–1845.

Reinert, J. A., and H. L. Cromroy. 1981. Bermudagrass stunt mite and its control in Florida. Proc. Florida State Hort. Soc. 94: 124–126.

Reinert, J. A., and S. H. Kerr. 1973. Bionomics and control of lawn chinch bugs. Bull. Entomol. Soc. Amer. 19: 91–92.

Reinert, J. A., and K. M. Portier. 1983. Distribution and characterization of organophosphate-resistant southern chinch bugs (Heteroptera: Lygaeidae) in Florida. J. Econ. Entomol. 76: 1187–1190.

Reinert, J. A., and D. E. Short. 1981. Turf devastation in the Southeast. Golf Course Manage., Apr.: 22–28.

Reinert, J. A., B. D. Bruton, and R. W. Toler. 1980. Resistance of St. Augustinegrass to southern chinch bug and St. Augustine decline strain of Panicum mosaic virus. J. Econ. Entomol. 73: 602–604.

Reinert, J. A., A. E. Dudeck, and G. H. Snyder. 1978. Resistance in bermudagrass to the bermudagrass mite. Environ. Entomol. 7: 885–888.

Reinhard, H. J. 1940. The life history of *Phyllophaga lanceolata* (Say) and *Phyllophaga crinita* Burmeister. J. Econ. Entomol. 33: 572–578.

Rennie, J. 1917. On the biology and economic significance of *Tipula paludosa* Meign. Pt. 2. Hatching, growth, and habits of larva. Ann. Appl. Biol. 3: 116–137.

Rex, E. G. 1931. Facts pertaining to the Japanese beetle. New Jersey Dept. Agr. Circ. No. 180. 31 pp.

Rings, R. W. 1977. An illustrated field key to common cutworms, armyworms, and looper moths in north central states. Ohio Agr. Res. and Dev. Center Res. Circ. No. 227. 60 pp.

Rings, R. W., and G. J. Musick. 1976. A pictorial field key to the armyworms and cutworms attacking corn in the North Central States. Ohio Agr. Res. and Dev. Center Res. Circ. No. 221. 36 pp.

Rings, R. W., F. J. Arnold, A. J. Keaster, and G. J. Musick. 1974. A worldwide annotated bibliography of the black cutworm *Agrotis ipsilon* (Hufnagel). Ohio Agr. Res. and Dev. Center Res. Circ. No. 198. 106 pp.

Rings, R. W., B. A. Baughman, and F. J. Arnold. 1974. An annotated bibliography of the bronzed cutworm. Ohio Agr. Res. and Dev. Center Res. Circ. No. 200. 36 pp.

Rings, R. W., B. A. Johnson, and F. J. Arnold. 1976. A worldwide, annotated bibliography of the variegated cutworm, *Peridroma saucia* Hubner. Ohio Agr. Res. and Dev. Center Res. Circ. No. 219. 126 pp.

Ritcher, P. O. 1940. Kentucky white grubs. Kentucky Agr. Exp. Sta. Bull. 401: 71–157.

Ritcher, P. O. 1949. May beetles and their control in the inner bluegrass region of Kentucky. Kentucky Agr. Exp. Sta. Bull. No. 542. 12 pp.

Ritcher, P. O. 1966. White grubs and their allies: A study of North American scarabaeid larvae. Oregon State Univ. Press, Corvallis. 219 pp.

Rivers, R. L., K. S. Pike, and Z. B. Mayo. 1977. Influence of insecticides and corn tillage systems on larval control of *Phyllophaga anxia*. J. Econ. Entomol. 70: 794–796.

Roberts, E. C., ed. 1983a. Research synthesis: How green is green? The Lawn Institute Harvests, Pleasant Hill, Tenn., 30(1): 18–23.

Roberts, E. C., ed. 1983b. How green is green in New York State and Tennessee? The Lawn Institute Harvests, Pleasant Hill, Tenn., 30(2): 8–10.

Roberts, E. C., ed. 1984. Changing turfgrass perspective. The Lawn Institute Harvests, Pleasant Hill, Tenn., 30(4): 18–19.

Robinson, W. H., and M. P. Tolley. 1982. Sod webworms associated with turfgrass in Virginia. Amer. Lawn Appl., July/Aug.: 22–25.

Roselle, R. E. 1975. Bluegrass billbug. Univ. Nebraska Coop. Ext. Nebguide G75-236. 2 pp.

Sargent, S. 1982. The sod webworm, turfgrass pest. Amer. Lawn Appl., Mar./Apr.: 4–6.

Satterthwait, A. F. 1931. *Anaphoidea calendrae* Gahan, a mymarid parasite of eggs of weevils of the genus *Calendra*. J. New York Entomol. Soc. 39: 171–190.

Satterthwait, A. F. 1932. How to control billbugs destructive to cereals and forage crops. U.S. Dept. Agr. Farmers' Bull. No. 1003. 22 pp.

Satterthwait, A. F. 1933. Larval instars and feeding of the black cutworm, *Agrotis ipsilon* Rott. J. Agr. Res. 46: 517–530.

Saxena, P. N., and H. L. Chada. 1971. The greenbug *Schizaphis graminum:* Mouth parts and feeding habits. Ann. Entomol. Soc. Amer. 64: 897–904.

Schread, J. C. 1964. Insect pests of Connecticut lawns. Connecticut Agr. Exp. Sta. Circ. No. 212. 10 pp.

Schread, J. C. 1970a. Chinch bug control. Connecticut Agr. Exp. Sta. Circ. No. 233. 6 pp.

Schread, J. C. 1970b. The annual bluegrass weevil. Connecticut Agr. Exp. Sta. Circ. No. 234. 6 pp.

Schuder, D. L. 1964. The control of sod webworms (*Crambus spp.*) in Indiana. Proc. Indiana Acad. Sci. 72: 164–166.

Schurr, K. M., and R. W. Rings. 1964. Uniform terminology for generations of multivoltine insects. Bull. Entomol. Soc. Amer. 10: 89–91.

Sears, M. K. 1978. Hairy chinch bugs in lawns. Ontario (Canada) Ministry of Agric. and Food Factsheet AGDEX 626. 2 pp.

Sears, M. K. 1979. Damage to golf course fairways by *Aphodius granarius* (L.) (Coleoptera: Scarabaeidae). Proc. Entomol. Soc. Ontario. 109: 48.

Sears, M. K., and K. Maitland. 1983. Biology of the turfgrass scale insect, *Lecanopsis formicarum*. 1983 Turfgrass Res. Ann. Rep. Univ. of Guelph, Ontario. 43 pp.

Sears, M. K., and K. Maitland. 1984. Biology of the turfgrass scale. 1984 Turfgrass Research Annual Report. Univ. of Guelph, Ontario. 38 pp.

Sellke, K. 1937. Beobachtungen über die Bekämpfung von Wiesenschnakenlarven (*Tipula paludosa* Meig und *Tipula czizeki* de J.). Z. Angew. Entomol. 24: 277–284.

Shapiro, I. 1975. Courtship and mating behavior of the fiery skipper, *Hylephelia phyleus* (Hesperiidae). J. Res. Lepid. 14: 125–141.

Shaw, M. W., P. Blasdale, and R. M. Allan. 1974. A comparison between the ODCB technique and heat extraction of soil cores for estimating leatherjacket populations. Plant Pathol. 23: 60–66.

Shetlar, D. J., P. R. Heller, and P. D. Irish. 1983. Turfgrass insect and mite manual with an index of registered materials. Pennsylvania Turfgrass Council, Bellefonte. 63 pp.

Shorey, H. H., R. H. Burrage, and G. G. Gyrisco. 1960. The relationship between environmental factors and the density of European chafer (*Amphimallon majalis*) larvae in permanent pasture sod. Ecology 41: 253–258.

Short, D. 1973. Field evaluation of insecticides for controlling mole crickets in turf. Down to Earth 29: 3–5.

Short, D. E., and P. G. Koehler. 1979. A sampling technique for mole crickets and other pests in turfgrass and pasture. Florida Entomol. 62: 282–283.

Short, D. E., and J. A. Reinert. 1982. Biology and control of mole crickets in Florida, pp. 119–124. *In* H. D. Niemczyk and B. G. Joyner (eds.), Advances in turfgrass entomology. Hammer Graphics, Piqua, Ohio. 150 pp.

Shurtleff, M. C., and R. Randell. 1974. How to control lawn diseases and pests. Intertec Publishing, Kansas City, Mo. 97 pp.

Smiley, R. W. 1983. Compendium of turfgrass diseases. Amer. Phytopathol. Soc. St. Paul, Minn. 102 pp.

Smith, L. B., and C. H. Hadley. 1926. The Japanese beetle. U.S. Dept. Agr. Circ. No. 363. 67 pp.

Snodgrass, R. E. 1935. Principles of insect morphology. McGraw-Hill, New York. 667 pp.

Snow, J. W., and P. S. Callahan. 1968. Biological and morphological studies of the granulate cutworm, *Feltia subterranea* (F.), in Georgia and Louisiana. Univ. Georgia Coll. Agr. Exp. Sta. Res. Bull. No. 42. 23 pp.

Sorensen, K. A., and H. E. Thompson. 1971. Mating behavior of the buffalograss webworm, *Surattha identella*. J. Kansas Entomol. Soc. 44: 329–331.

Southwood, T. R. E., W. F. Jepson, and H. F. Van Emden. 1961. Studies on the behavior of *Oscinella frit* L. (Diptera) adults of the panicle generation. Entomol. Exp. Appl. 4: 196–210.

Splittstoesser, C., and H. Tashiro. 1977. Three milky disease bacilli from a scarabaeid, *Ataenius spretulus*. J. Invertebrate Pathol. 30: 436–438.

Splittstoesser, C. M., C. Y. Kawanishi, and H. Tashiro. 1978. Infection of the European chafer, *Amphimallon majalis*, by *Bacillus popilliae:* Light and electron microscope observations. J. Invertebrate Pathol. 31: 84–90.

Splittstoesser, C. M., H. Tashiro, S. L. Lin, K. H. Steinkraus, and B. J. Fiori. 1973. Histopathology of the European chafer, *Amphimallon majalis*, infected with *Bacillus popilliae*. J. Invertebrate Pathol. 22: 161–167.

Stockton, W. D. 1956. A review of the Nearctic species of the genus *Hyperodes* Jekel (Coleoptera: Curculionidae). Ph.D. Thesis. Cornell Univ., Ithaca, N.Y. 122 pp.

Stockton, W. D. 1963. New species of *Hyperodes* Jekel and a key to the Nearctic species of the genus (Coleoptera: Curculionidae). Bull. So. California Acad. Sci. 62: 140–149.

Street, J. R., R. Randell, and G. Clayton. 1978. Greenbug damage found on Kentucky bluegrass. Weeds, Trees, and Turf, Oct.: 26.

Streu, H. T. 1981. Winter grain mite. Amer. Lawn Appl., Mar./Apr.: 28–31.

Streu, H. T., and J. B. Gingrich. 1972. Seasonal activity of the winter grain mite in turfgrass in New Jersey. J. Econ. Entomol. 65: 427–430.

Streu, H. T., and H. D. Niemczyk. 1982. Pest status and control of winter grain mite, pp. 101–104. *In* H. D. Niemczyk and B. J. Joyner (eds.), Advances in turfgrass entomology. Hammer Graphics, Piqua, Ohio. 150 pp.

Streu, H. T., and L. M. Vasvary. 1966. 1966 report of turfgrass research, pp. 78–82. Rutgers University, New Brunswick.

Stringfellow, T. L. 1969. Turfgrass insect research in Florida, pp. 19–33. *In* H. T. Streu and R. T. Bangs (eds.), Proceedings of Scott's turfgrass research conf. O. M. Scott & Sons, Marysville, Ohio. 89 pp.

Strobel, J. 1971. Turfgrass Proc. Florida Turfgrass Manage. Conf. 19: 19–28.

Sutherland, D. W. S. [Chairman]. 1978. Common names of insects and related organisms (1978 revision): An update. Bull. Entomol. Soc. Amer. 24: 408.

Swezey, O. H. 1946. Insects of Guam, II. Bernice P. Bishop Mus. Bull. 189: 184.

Tanada, Y. 1955. Notes and exhibitions. Proc. Hawaiian Entomol. Soc. 15: 384.

Tanada, Y., and J. W. Beardsley. 1957. Probable origin and dissemination of a polyhedrosis virus of an armyworm in Hawaii. J. Econ. Entomol. 50: 118–120.

Tanada, Y., and J. W. Beardsley. 1958. A biological study of the lawn armyworm, *Spodoptera mauritia* (Boisduval), in Hawaii (Lepidoptera:Phalaenidae). Proc. Hawaiian Entomol. Soc. 16: 411–436.

Tashiro, H. 1976a. Biology of the grass webworm, *Herpetogramma licarsisalis* (Lepidoptera: Pyraustidae), in Hawaii. Ann. Entomol. Soc. Amer. 69: 797–803.

Tashiro, H. 1976b. Hyperodes weevil, a serious menace to *P. annua* in Northeast. Golf Superintendent, Mar.: 34–37.

Tashiro, H. 1977. Colonization of the grass webworm, *Herpetogramma licarsisalis,* and its adaptability for laboratory tests. Proc. Hawaiian Entomol. Soc. 22: 533–539.

Tashiro, H. 1983. Insecticides: Properties and insects. Proc. 37th Ann. New York State Turfgrass Conf. 7: 77–81.

Tashiro, H., and W. E. Fleming. 1954. A trap for European chafer surveys. J. Econ. Entomol. 47: 618–623.

Tashiro, H., and F. L. Gambrell. 1963. Correlation of European chafer development with the flowering period of common plants. Ann. Entomol. Soc. Amer. 56: 239–243.

Tashiro, H., and W. C. Mitchell. 1985. Biology of the fiery skipper, *Hylephila phyleus* (Lepidoptera: Hesperiidae), a turfgrass pest in Hawaii. Proc. Hawaiian Entomol. Soc. 25: 131–138.

Tashiro, H., and W. Neuhauser. 1973. Chlordane-resistant Japanese beetle in New York. Search Agr. No. 3. 6 pp.

Tashiro, H., and K. E. Personius. 1970. Current status of the bluegrass billbug and its control in western New York home lawns. J. Econ. Entomol. 63: 23–29.

Tashiro, H., and R. W. Straub. 1973. Progress in the control of turfgrass weevil, a species of *Hyperodes.* Down to Earth 29: 8–10.

Tashiro, H., S. I. Gertler, M. Beroza, and N. Green. 1964. Butyl sorbate as an attractant for the European chafer. J. Econ. Entomol. 57: 230–233.

Tashiro, H., G. G. Gyrisco, F. L. Gambrell, B. J. Fiori, and H. Breitfeld. 1969. Biology of the European chafer *Amphimallon majalis* (Coleoptera Scarabaeidae) in northeastern United States. New York State Agr. Exp. Sta. Bull. No. 828. 71 pp.

Tashiro, H., J. G. Hartsock, and G. G. Rohwer. 1967. Development of blacklight traps for European chafer surveys. U.S. Dept. Agr. Tech. Bull. No. 1366. 52 pp.

Tashiro, H., C. L. Murdoch, and W. C. Mitchell. 1983. Development of a survey technique for larvae of the grass webworm and other lepidopterous species in turfgrass. Environ. Entomol. 12: 1428–1432.

Tashiro, H., C. L. Murdock, R. W. Straub, and P. J. Vittum. 1977. Evaluation of insecticides on *Hyperodes* sp., a pest of annual bluegrass turf. J. Econ. Entomol. 70: 729–733.

Timm. R. M., ed. 1983. Prevention and control of wildlife damage. Great Plains Agr. Council and Univ. Nebraska Coop. Ext. Service, Inst. Agr. and Natur. Resources, Lincoln. Looseleaf, 632 pp.

Tolley, M. P. 1983. Resting site preferences for sod webworm moths. Amer. Lawn Appl., Nov./Dec.: 27–28.

Troester, S. J., W. G. Ruesink, and R. W. Rings. 1982. A model of black cutworm (*Agrotis ipsilon*) development:description, uses, and implications. Illinois Agr. Exp. Sta. Bull. No. 774. 33 pp.

Tumlinson, J. H., M. G. Klein, R. E. Doolittle, T. L. Ladd, and A. T. Proveaux. 1977. Identification of the female Japanese beetle sex pheromone: Inhibition of male response by an enantiomer. Science 197: 789–792.

Tuttle, D. M., and G. D. Butler, Jr. 1961. A new eriophyid mite infesting bermudagrass. J. Econ. Entomol. 54: 836–838.

Ulagaraj, S. M. 1975. Mole crickets: Ecology, behavior, and dispersal flight (Orthoptera: Gryllotalpidae: *Scapteriscus*). Environ. Entomol. 4: 265–273.

Ulagaraj, S. M. 1976. Sound production in mole crickets (Orthoptera: Gryllotalpidae: *Scapteriscus*). Ann. Ent. Soc. Amer. 69: 299–306.

Ulagaraj, S. M., and T. J. Walker. 1973. Phonotaxis of crickets in flight: Attraction of male and female crickets to male calling sounds. Science 182: 1278–1279.

U.S. Dept. Agr. 1951–1980. Cooperative economic insect report (1951–1975) and Cooperative Plant Pest Report (1976–1980).

U.S. Dept. Agr. 1972. Japanese beetle quarantines. U.S. Dept. Agr., Animal and Plant Health Serv., Plant Protection and Quarantine Programs, and Can. Dept. Agr., cooperating with affected states. Cooperative Economic Insect Report 22: between 216–217.

U.S. Dept. Agr. 1984. 1984–85 directory of professional workers in state agricultural experiment stations and other cooperating state institutions. Agr. Handbook No. 305. 237 pp.

U.S. Dept. Com. 1981. Brief history of measurement systems with a chart of the modernized metric system. National Bur. Standards Spec. Pub. No. 304A. 4 pp.

Vance, A. M., and B. A. App. 1971. Lawn insects: How to control them. U.S. Dept. Agr. Home and Garden Bull. No. 53. 23 pp.

Van Zwaluwenburg, R. H. 1918. The changa, or West Indian mole cricket. Porto Rico Agr. Exp. Sta. Bull. No. 23. 28 pp.

Vaurie, P. 1951. Revision of the genus *Calendra* (formerly *Schenophorus*) in the United States and Mexico (Coleoptera: Curculionidae). Bull. Amer. Mus. Natur. Hist. 98: 33–186.

Vickery, R. A. 1929. Studies on the fall armyworm in the Gulf Coast district of Texas. U.S. Dept. Agr. Tech. Bull. No. 138. 64 pp.

Vittum, P. J. 1979. A taxonomic study of two species of *Hyperodes* (Coleoptera: Curculionidae) damaging turfgrass in New York State. M.S. Thesis. Cornell Univ., Ithaca, N.Y. 50 pp.

Vittum, P. J. 1980. The biology and ecology of the annual bluegrass weevil, *Hyperodes* sp. near *anthracinus* (Dietz) (Coleoptera: Curculionidae). Ph.D. Thesis. Cornell Univ., Ithaca, N.Y. 117 pp.

Vittum, P. J. 1984. Effect of lime applications on Japanese beetle (Coleoptera: Scarabaeidae) grub populations in Massachusetts soils. J. Econ. Entomol. 77: 687–690.

Vittum, P. J., and H. Tashiro. 1980. Effect of soil pH on survival of Japanese beetle and European chafer larvae. J. Econ. Entomol. 73: 577–579.

Wadley, F. M. 1931. Ecology of *Toxoptera graminum*, especially as to factors affecting importance in the northern United States. Ann. Entomol. Soc. Amer. 24: 325–395.

Walkden, H. H. 1950. Cutworms, armyworms, and related species attacking cereal and forage crops in the Central Great Plains. U.S. Dept. Agr. Circ. No. 849. 52 pp.

Walker, T. J. 1982. Sound traps for sampling mole cricket flights (Orthoptera: Gryllotalpidae: *Scapteriscus*). Florida Entomol. 65: 105–110.

Walker, T. J., and D. A. Nickle. 1981. Introduction and spread of pest mole crickets: *Scapteriscus vicinus* and *S. acletus* reexamined. Ann. Entomol. Soc. Amer. 74: 158–163.

Walton, W. R. 1921. The green-bug, or spring grain aphis: How to prevent its periodical outbreaks. U.S. Dept. Agr. Farmers' Bull. No. 1217. 11 pp.

Walton, W. R. 1929. Cutworms on golf green. U.S. Golf Assoc. Green Sec. Bull. 9: 156–157.

Weaver, J. E., and J. D. Hacker. 1978. Bionomical observations and control of *Ataenius spretulus* in West Virginia. West Virginia Univ. Agric. Forest. Exp. Sta. Current Rep. No. 72. 16 pp.

Weaver, J. E., and R. A. Sommers. 1969. Life history and habits of the short-tailed cricket, *Anurogryllus muticus,* in central Louisiana. Ann. Entomol. Soc. Amer. 62: 337–342.

Webster, F. M., and W. J. Phillips. 1912. The spring grain aphid, or "greenbug." U.S. Bur. Entomol. Bull. No. 110. 153 pp.

Wegner, G. S., and H. D. Niemczyk. 1981. Bionomics and phenology of *Ataenius spretulus.* Ann. Entomol. Soc. Amer. 74: 374–384.

Werner, F. G. [Chairman]. 1982. Common names of insects and related organisms. Entomol. Soc. Amer. 132 pp.

Wessel, R. D., and J. B. Polivka. 1952. Soil pH in relation to Japanese beetle populations. J. Econ. Entomol. 45: 733–735.

Wheeler, E. H. 1946. The pathogenicity of *Bacillus lentimorbus* Dutky and strains of *Bacillus popilliae* Dutky to larvae of *Amphimallon majalis* (Razoumowski), Scarabaeidae. Ph.D. Thesis. Cornell Univ., Ithaca, N.Y. 83 pp.

White, R. T. 1941. Development of milky disease on Japanese beetle larvae under field conditions. J. Econ. Entomol. 34: 213–215.

White, R. T. 1947. Milky disease infecting *Cyclocephala* larvae in the field. J. Econ. Entomol. 40: 912–914.

Wildermuth, V. L. 1916. California green lacewing fly. J. Agr. Res. 6: 517.

Wilkinson, A. T. S. 1969. Leatherjackets—A new pest in British Columbia. Canada Agr. Summer. 2 pp.

Wilkinson, A. T. S., and H. S. Gerber. 1972. Description, life history, and control of leatherjackets. Brit. Columbia Dept. Agr. Pub. No. 72-5. 2 pp.

Wilkinson, A. T. S., and H. R. MacCarthy. 1967. The marsh crane fly, *Tipula paludosa* Mg., a new pest in British Columbia (Diptera: Tipulidae). J. Entomol Soc. Brit. Columbia 64: 29–34.

Wills, W., L. Lengel, Jr., and G. Whitmyre. 1969. False parasitism by the Asiatic garden beetle *Maladera castanea* (Arrow) (Coleoptera: Melolonthidae). Entomol. Soc. Pennsylvania, Melsheimer Entomol. Ser. No. 3. 3 pp.

Wilson, J. W. 1932. Coleoptera and Diptera collected from a New Jersey sheep pasture. J. New York Entomol. Soc. 40: 77–93.

Wolcott, G. N. 1941. The establishment in Puerto Rico of *Larra americana* Saussure. J. Econ. Entomol. 34: 53–56.

Woodruff, R. E. 1966. The hunting billbug, *Sphenophorus venatus vestitus* Chittenden, in Florida (Coleoptera: Curculionidae). Florida Dept. Agr. Entomol. Circ. No. 45. 2 pp.

Woodruff, R. E. 1973. Arthropods of Florida and neighboring land areas, vol. 8. The scarab beetles of Florida. Florida Dept. Agric. Consumers Service Bur. Entomol. Contrib. No. 260. 220 pp.

Wylie, W. D. 1944. *Crambus haytiellus* (Zincken) as a pest of carpet grass. Florida Entomol. 27: 5–9.

Yeager, L. E. 1949. Forest leaf chafers and white grubs lamellicornia, pp. 161–186. *In* F. C. Craighead, Insect enemies of eastern forests. U.S. Dept. Agr. Misc. Pub. No. 657. 679 pp.

Index

Page numbers in *italics* refer to figures and tables. "Pl." indicates the color plates, found between pages 192 and 193.